Microorganisms for Sustainability

Volume 56

Series Editor
Naveen Kumar Arora, Environmental Microbiology
School for Environmental Science, Babasaheb Bhimrao Ambedkar University
Lucknow, India

Microorganisms have been in existence since the origin of life on earth and can survive the most extreme habitats or conditions on earth. Microorganisms are involved in regulating biogeochemical cycles, maintaining plant and animal health, and sustaining the global food chain. Moreover, they play crucial roles in addressing the challenges of climate change and achieving the targets of Sustainable Development Goals (SDGs).

This multidisciplinary book series captures the role of microbes towards building a sustainable world, while encompassing cutting-edge technologies and current needs across various fields such as agriculture sustainability, bioremediation, restoration of degraded habitats and wastelands, and food security. Additionally, this series explores microbial applications in industries, and building their utilization in clean and green energy solutions. Furthermore, themes like microbial secondary metabolites, extremophilic microbes and modern omics, including next generation sequencing and metagenomics, are also covered in this series.

With contributions from researchers across the globe, this series addresses the important call of 'One Planet-One Health-One Future'. It comprises a collection of diverse volumes that provides insights for scientists, young researchers, educators and decision-makers in the government, private sector, and non-governmental organizations, empowering their efforts to achieve the global goals.

The series invites, evaluates, and accepts book proposals to ensure a diverse, inclusive, and evolving program. The final decision regarding acceptance rests with the series editor.

Peer-review

This book series follows a stratified review process. Proposals for individual volumes are reviewed by the series editor and then the editorial board members. On a case-to-case basis, external reviewers are also invited for further evaluation of the book proposal. Review of the chapters is the responsibility of the volume editor(s). A manuscript submission platform has been recently made available to the authors, volume editors and the series editor.

Piyush Pandey • Shrivardhan Dheeman
Dinesh K. Maheshwari
Editors

Microorganisms Resilience to Climate Change

Springer

Editors
Piyush Pandey
Department of Microbiology
Assam University
Silchar, Assam, India

Shrivardhan Dheeman
Department of Biotechnology and Microbiology
MVN University
Palwal (NCR), Haryana, India

Dinesh K. Maheshwari
Department of Botany and Microbiology
Gurukula Kangri (Deemed to be University)
Haridwar, Uttarakhand, India

ISSN 2512-1901 ISSN 2512-1898 (electronic)
Microorganisms for Sustainability
ISBN 978-981-96-3747-8 ISBN 978-981-96-3748-5 (eBook)
https://doi.org/10.1007/978-981-96-3748-5

This Springer imprint is published by the registered company Springer Nature Singapore Pte Ltd.
The registered company address is: 152 Beach Road, #21-01/04 Gateway East, Singapore 189721, Singapore

Preface

In recent years, the urgency to address climate change has reached a critical juncture, highlighting the intricate interplay between our environment and the myriads of microbial communities that inhabit it. This book, *Microorganisms Resilience to Climate Change*, serves as a comprehensive exploration of the foundational principles of microbial community dynamics in the context of an evolving climate and the pivotal role they play in fostering climate resilience.

The concerns over climate change are strongly reflected in the sustainable development goals (SDG) of the United Nations. The Climate action aims to combat climate change and its effects by taking immediate action. Human activities have been the main driver of climate change, primarily due to the burning of fossil fuels like coal, oil, and gas. In agriculture practice, burning crop-stubble is also a source of greenhouse effect and leads to climate change. The consequences of climate change invite serious concerns, such as extreme heat in some regions, causing intense droughts, water scarcity, and biodiversity loss. Parallelly it can affect our health and the ability of a farmer to grow crops in fields in high temperatures. The temperate effect of climate change on crops and food production is important concern of climate action, and also relates to SDG 2: Zero hunger. Imagine, a humankind living in such regions where climate change caused water loss, intense drought, and affect health; in such situations, cultivation of crops will be limited due to water, and efforts of man. The sustainable agriculture is key to achieve sustainable development. Life below land has resilience to combat climate change. The microorganisms that confer plant health and growth promotion have been studied well, already. Their multifaceted features of drought tolerance and supporting the plant life are tremendous. This book is an exclusive compilation of worldwide research addressing the resilience of microorganisms toward climate change and contribution to sustainable agriculture reviewed by eminent scientists.

The first section lays the groundwork by examining how microbial communities adapt to the challenges presented by climate change while simultaneously exploring the opportunities that arise from such transformations. Contributions from leading experts delve into the significance of microbial interactions in carbon dioxide fixation, shifts in plant-microbe dynamics, and the symbiotic relationships that flourish under diverse climatic conditions. This foundational knowledge is crucial for understanding the resilience of these communities in the face of global climate challenges. Transitioning into the second section, we turn our attention to agroecosystems and the vital role that microbial communities play in enhancing climate resilience in agricultural practices. The authors have presented compelling evidence of how plant growth-promoting rhizobacteria can alleviate drought stress and bolster the recruitment of beneficial rhizomicrobiomes. These insights are essential for developing sustainable agricultural strategies that can withstand the rigors of a changing climate. In the third section, we explore innovative microbial solutions for climate resilience, including beneficial plant-microbe interactions, biofortification, and the untapped potential of extremophiles. These contributions offer a glimpse into the future of sustainable practices that can harness microbial capabilities to combat climate change effectively. The fourth section shifts focus on the microbial responses to climate change, examining crucial topics such as carbon sequestration, cryosphere ecology, and the emerging threat of antimicrobial resistance. The diverse perspectives presented here underscore the complexity of microbial interactions with their environment and the critical role they play in maintaining ecological balance amidst changing climatic conditions. Finally, the book concludes with a synthesis of key findings, emphasizing the resilience of microorganisms as both a challenge and an opportunity in the fight against climate change.

The present book is an endeavor of collaborative efforts by editors and contributors across the globe and most from Indian sub-continent. The editors extend their warm gratitude to the authors, who contributed their scholarly article on time, our colleagues who voluntarily reviewed the manuscripts for quality improvisation, students for type setting, and other associates for their invaluable contributions to the book. The Editorial Team at Springer also deserves credit for their excellent coordination with authors and their assistance in editorial procedures of the manuscripts and hence is greatly acknowledged. It is our belief that researchers involved in, or embarking on, projects related to the role of microorganisms in climate change will find this book as an essential guide.

Silchar, Assam, India — Piyush Pandey
Haridwar, Uttarakhand, India — Dinesh K. Maheshwari
Palwal (NCR), Haryana, India — Shrivardhan Dheeman

Contents

Editors and Contributors

About the Editors

Piyush Pandey is working at the Department of Microbiology, Assam University, India. He has 23 years of academic and research experience. His research group works on plant-associated bacteria, with their potential applications in agriculture and environment. He has published more than 160 papers in journals of high impact, 3 books, and undertaken 16 research grants. He has been awarded with "DBT-Centre for Excellence award" for working on plant-rhizosphere-mediated bioremediation of polyaromatic hydrocarbons. Also, he has been "Overseas Associate" to Agroscope, Switzerland, SICI-SMP fellow to York University, Canada, and had visited Shizuoka University, Japan. Prof. Pandey is a Fellow of International Society of Environmental Botanists, member of several scientific societies, and has editorial role in reputed Journals. He has been extensively working on soil and plant microbiomes, rhizoremediation of environmental pollutants, and environmental genomics. He has been awarded from Springer, Society for Environmental Sustainability, and Asian PGPR Society for his research contributions. He has successfully supervised 16 PhDs while 8 are ongoing. He has two Indian patents granted.

Shrivardhan Dheeman is working as an Associate Professor and Head at the Department of Biotechnology and Microbiology, MVN University, India. Dr. Dheeman is involved in teaching and research of microbiology and biotechnology to different disciplines in undergraduate and postgraduate classes. His research interests lie in sustainable agriculture. He is an active member of several scientific bodies of international repute including Indian Science Congress Association (ISCA), Asian PGPR Society for Sustainable Agriculture, Microbiologists Society of India (MBSI), and Indian Botanical Society. He has published over 50 articles, including research papers, reviews, and book chapters in prestigious journals indexed in Web of Science and Scopus. He has edited five books, published by Springer Nature.

Dinesh K. Maheshwari former Vice-Chancellor, Gurukul Kangri (Deemed to be University), is a renowned microbiologist, holding the MHRD-UGC-BSR Eminent Faculty Fellowship. Prof. Maheshwari worked at the Biology Research Centre (Szeged), Hungary, in the year 1983–1984. He has visited prestigious universities in

Japan, Germany, Belgium, Spain, and South Korea as a professor. He has received the prestigious awards "Prof. Y.S. Murty Medal" and "Birbal Sahni Award" from the Indian Botanical Society in 1992 and 2018, respectively. Prof. Maheshwari was bestowed with the "Platinum Jubilee Lecture Award" at the Indian Science Congress in 2012. For the 103rd Indian Science Congress, he served as President of Plant Science. In 2020, he received the Lifetime Achievement Award at BBA University, Lucknow, and the "Vigyan Vibhuti Award" from the Government of Uttarakhand. Prof. Maheshwari has more than 260 publications, has guided 51 PhD students, and published two dozen books, including 18 edited volumes and 1 monograph published by Springer Nature.

Contributors

P. Anjali Sustainable Environmental Processes-Environmental Bioprocesses, CSIR-National Environmental Engineering Research Institute (NEERI), Nehru Marg, Nagpur, India

Academy of Scientific and Innovative Research (AcSIR), Ghaziabad, India

Mohd. Mogees Ansari CSIR-National Botanical Research Institute (CSIR-NBRI), Rana Pratap Marg, Lucknow, India

Academy of Scientific and Innovative Research (AcSIR), Ghaziabad, India

A. S. Arthananair Department of Microbiology, University of Delhi, New Delhi, India

A. Arunima Department of Microbiology, School of Life Science, Central University of Rajasthan, Ajmer, India

Afeefa Ateeq Swami Vivekanand Subharti University, Meerut, India

Puneet Singh Chauhan CSIR-National Botanical Research Institute (CSIR-NBRI), Rana Pratap Marg, Lucknow, India

Academy of Scientific and Innovative Research (AcSIR), Ghaziabad, India

Rajarshi Chowdhury Department of Microbiology, Soil and Environmental Microbiology Laboratory, Assam University, Silchar, India

Nandita Das Department of Microbiology, Soil and Environmental Microbiology Laboratory, Assam University, Silchar, India

Sayak Das Department of Life Science and Bioinformatics, Har Gobind Khurana School of Life Sciences, Assam University, Silchar, Assam, India

Manasi Debnath Department of Geography, Nagaland University, Lumamai, Nagaland, India

Fusun Eyidogan Institue of Food, Agriculture and Livestock Development, Başkent University, Ankara, Turkey

Department of Molecular Biology and Genetics, Faculty of Science and Letters, Başkent University, Ankara, Turkey

Amar P. Garg Swami Vivekanand Subharti University, Meerut, India

F. Seyma Gokdemir Institue of Food, Agriculture and Livestock Development, Başkent University, Ankara, Turkey

Dweipayan Goswami Department of Microbiology and Biotechnology, University School of Sciences, Gujarat University, Ahmedabad, Gujarat, India

Palash Hazra Institute of Bioresources and Sustainable Development, Regional Centre, Gangtok, Sikkim, India

Alvin Huang Case Western Reserve University, Cleveland, OH, USA

Chaitanya Kumar Jha Microbiology Department, Gujarat Arts and Science College, Ahmedabad, Gujarat, India

Atya Kapley Environmental Bioprocesses Division, CSIR-National Environmental Engineering Research Institute, Nehru Marg, Nagpur, India

Academy of Scientific and Innovative Research (AcSIR), Ghaziabad, Uttar Pradesh, India

Ram Karan Department of Microbiology, University of Delhi, New Delhi, India

Santosh Kumar Department of Microbiology, School of Life Sciences, Sikkim University, Gairigaon, Sikkim, India

Xiang Li College of Resources and Environment, Southwest University, Chongqing, China

Dinesh K. Maheshwari Department of Botany and Microbiology, Gurukula Kangri (Deemed to be University), Haridwar, Uttarakhand, India

Ying Ma College of Resources and Environment, Southwest University, Chongqing, China

Samina Mehnaz Kauser Abdulla Malik School of Life Sciences, Forman Christian College (A Chartered University), Lahore, Pakistan

Shilpa Mishra Department of Microbiology and Biotechnology, University School of Sciences, Gujarat University, Ahmedabad, Gujarat, India

Piyush Pandey Department of Microbiology, Soil and Environmental Microbiology Laboratory, Assam University, Silchar, India

Shubham Pandey Department of Microbiology, University of Delhi, New Delhi, India

Bhavna Parmar Department of Microbiology, University of Delhi, New Delhi, India

Gudhal Patanjali Amity Institute of Horticulture Studies and Research, Amity University, Noida, Noida, Uttar Pradesh,, India

Asifa Qureshi Sustainable Environmental Processes-Environmental Bioprocesses, CSIR-National Environmental Engineering Research Institute (NEERI), Nehru Marg, Nagpur, India

Academy of Scientific and Innovative Research (AcSIR), Ghaziabad, Uttar Pradesh, India

Samikcha Rai Academy of Scientific and Innovative Research (AcSIR), Ghaziabad, India

Ganesh Chandra Sahoo Department of Virology, ICMR-Rajendra Memorial Research Institute of Medical Sciences, Patna, Bihar, India

Meenu Saraf Department of Microbiology and Biotechnology, University School of Sciences, Gujarat University, Ahmedabad, Gujarat, India

Izzah Shahid Institute of Multidisciplinary Research in Applied Biology, Public University of Navarra, Pamplona, Spain

Abhishek Sharma Amity Food and Agriculture Foundation, Amity University, Noida, Noida, Uttar Pradesh, India

Aditya Sharma Enzyme Technology and Protein Bioinformatics Laboratory, School of Biotechnology, Institute of Science, Banaras Hindu University, Varanasi, Uttar Pradesh, India

Swati Sharma Department of Virology, ICMR-Rajendra Memorial Research Institute of Medical Sciences, Patna, Bihar, India

Mingma Thundu Sherpa Department of Microbiology, Nar Bahadur Bhandari Government College, Tadong, Sikkim, India

Subir Kumar Shill Department of Life Science and Bioinformatics, Har Gobind Khurana School of Life Sciences, Assam University, Silchar, Assam, India

Pratyoosh Shukla Enzyme Technology and Protein Bioinformatics Laboratory, School of Biotechnology, Institute of Science, Banaras Hindu University, Varanasi, Uttar Pradesh, India

K. Malabika Singha School of Life Science, Central University of Rajasthan, Ajmer, India

L. Paikhomba Singha Department of Microbiology, School of Life Science, Central University of Rajasthan, Ajmer, India

Kamal Singh Regional Viral Research and Diagnostic Laboratory, All India Institute of Medical Sciences Patna, Bihar, India

Niraj Singh Department of Microbiology, The Assam Royal Global University, Guwahati, Assam, India

Tanya Singh CSIR-National Botanical Research Institute (CSIR-NBRI), Rana Pratap Marg, Lucknow, India

Academy of Scientific and Innovative Research (AcSIR), Ghaziabad, India

Simran Yadav Department of Microbiology, University of Delhi, New Delhi, India

Abbreviations

ABA	Abscisic Acid
ABC	ATP-Binding Cassette
ABE	Acetone–Butanol–Ethanol
ACC	1-aminocyclopropane-1-carboxylate
ACO	1-aminocyclopropane-1-carboxylate oxidase
ACS	Aminocyclopropane-1-carboxylate synthase
AFPs	Antifreeze Proteins
AMD	Acid Mine Drainage
AMF	Arbuscular Mycorrhizal Fungi
AMO	Ammonia Monooxygenase
AMR	Antimicrobial Resistance
AO	Anaerobic and Aerobic
AOB	Ammonia Oxidizing Bacteria
AOS	Antioxidative System
APX	Ascorbate Peroxidase
ARB	Arsenic-Resistant Bacteria
ASR	Acquired Systemic Resistance
BCA	Bio Control Agents
BNF	Biological Nitrogen Fixing
CA	Carbonic Anhydrase
CAT	Catalase
CCM	Carbon Concentrating Mechanism
CfeSP	Corrinoid Iron-Sulfur Protein
CS	Citrate Synthase
CSPs	Cold shock Proteins
DOM	Dissolved Organic Matter
DON	Deoxynivalenol
ECO_2	Environmental Carbon Dioxide
EMC	Ectomycorrhiza Communities
EPS	Extracellular Polymeric Substances
ETI	Effector-Triggered Immunity
FCA	Farm Credit Administration
FLS	Formolase
FRO	Ferric-Chelate Reductase

GAPDH	Glyceraldehyde-3-Phosphate Dehydrogenase
GHG	Greenhouse Gases
GVNPs	Gas Vesicle Nanoparticles
HCN	Hydrogen Cyanide
HSPs	Heat Shock Proteins
IAA	Indole-3-Acetic Acid
ISR	Induced Systemic Resistance
ISR	Integrated Stress Response
IST	Induce Systemic Tolerance
JA	Jasmonic Acid
KSB	Potassium Solubilizing Bacteria
MAMPs	Microbe-Associated Molecular Patterns
MAOM	Mineral-Association Organic Matter
MAPK	Mitogen-Activated Protein Kinase
MCP	Methyl-accepting Chemotaxis Proteins
MDA	Malondialdehyde
MFC	Microbial Fuel Cell
MFS	Major Facilitator Superfamily
MICP	Microbial-Induced Calcite Precipitation
MOI	Multi-Omics Integration
NCEI	National Centres for Environmental Information
NH3	Ammonia
OMRP	Outer Membrane Receptor Protein
P5CS	Pyrroline-5-Carboxylate Synthase
PAHs	Polycyclic Aromatic Hydrocarbons
PBP	Periplasmic Binding Protein
PC	Phytochelatin
PCA	Protocatechuic Acid
PCBs	Polychlorinated Biphenyls
PGPM	Plant Growth Promoting Protein Modulator
PGPR	Plant Growth Promoting Rhizobacteria
PHAs	Polyhydroxyalkanoates
PLFAs	Phospholipid Fatty Acids
POD	Peroxide Dismutase
PRK	Phosphoribulokinase
PRRs	Protein Recognition Receptors
PSB	Phosphate Solubilizing Bacteria
PTI	Pattern-Triggered Immunity
RDA	Recommended Dietary Allowances
RHP	Reductive Hexulose-Phosphate
RNS	Reactive Nitrogen Species
ROS	Reactive Oxygen Species
RubisCO	Ribulose-1,5-bisphosphate carboxylase/oxygenase
RWC	Relative Water Content
SAM	S-Adenosyl Methionine

SAR	Systematic Acquired Resistance
SCO	Single Cell-Oil
SDG	Sustainable Development Goals
SOC	Soil Organic Carbon
SOD	Superoxide Dismutase
SPA	Stress-Protecting Agent
TAG	Triacylglycerols
VOCs	Volatile Organic Compounds
VOCs	Volatile Organic Compounds
WEOM	Water-Extractable Organic Matter
WUE	Water Use Efficiency

Part I

Fundamentals of Climate Change and Microbial Community Dynamics

1 Microbial Symbiosis in Diverse and Changing Climatic Conditions

K. Malabika Singha, A. Arunima, L. Paikhomba Singha, and Piyush Pandey

Abstract

Climate change has the potential to disrupt the biodiversity of microorganisms, which are critical to ecosystems and nutrient cycling. External factors, especially climate change, have a significant impact on microbial symbiosis. Mycorrhizal fungi and other soil microbes contribute significantly to soil aggregation. Fungal hyphae bind soil particles, improving soil structure, aeration, and water retention, thereby improving soil health and stability, which is required for plant growth and ecosystem function. Microbial symbioses are critical to biogeochemical cycles such as the carbon, nitrogen and phosphorus cycles. In terms of environmental adaptation, microbes can assist hosts in surviving environmental stresses; for example, mycorrhizal fungi improve plant drought tolerance by increasing water absorption. Symbiotic relationships can help organisms adapt to changing climates. Lichens exhibit a symbiosis between fungi and algae/cyanobacteria, which plays an important ecological role in a variety of environments.

K. M. Singha
School of Life Science, Central University of Rajasthan, Ajmer, India

A. Arunima · L. P. Singha (✉)
Department of Microbiology, School of Life Science, Central University of Rajasthan, Ajmer, India

P. Pandey
Department of Microbiology, Soil and Environmental Microbiology Laboratory, Assam University, Silchar, India

P. Pandey et al. (eds.), *Microorganisms Resilience to Climate Change*, Microorganisms for Sustainability 56, https://doi.org/10.1007/978-981-96-3748-5_1

1.1 Introduction

Climate change is probably the main concern of the twenty-first century, not only an issue to our health but also to agriculture, forestry, biodiversity, and the ecosystem. Climate change occurs mainly due to the emission of greenhouse gases like nitrous oxide (N_2O), methane (CH_4), and carbon dioxide (CO_2), and the drastic changes due to these gases are predicted to change the level and various parameters of life in the changing environment. While plant and animal impacts are well-documented, the effects on microorganisms are often neglected. Microorganisms, essential for ecosystems and nutrient cycles, may experience biodiversity shifts due to climate change. They play a dual role by influencing and mitigating climate change through greenhouse gas interactions (Cavicchioli et al. 2019). Climate change affects the function and composition of terrestrial microbial communities. The increasing temperature successively alters the structure of the microbial community and meanwhile accelerates several processes like methanogenesis, respiration, decomposition, and mineralization (Chen et al. 2020a, b).

Microbial symbiosis is integral to ecology and nutrient cycling because it facilitates nutrient acquisition and recycling, contributes to soil health and structure, supports biodiversity and ecosystem stability, and enhances the ability of organisms to adapt to environmental conditions (Cavicchioli et al. 2019). These interactions are essential for maintaining the balance and functionality of ecosystems. The mechanisms underlying microbial symbiosis involve various forms of cooperation, resource exchange, and communication, which are crucial for maintaining the health and stability of ecosystems and hosts. In mutualistic symbiosis or mutualism, both partners benefit from the relationship through nutrient exchange and metabolic cooperation. For example, in mycorrhizal associations, fungi provide plants with nutrients (like phosphorus) while receiving carbohydrates in return. In legume-rhizobium interactions, rhizobium bacteria provide fixed nitrogen to plants, while plants supply carbohydrates and other nutrients to the bacteria (Yang et al. 2022). In commensalism (non harmful interaction), one partner benefits while the other is unaffected. For example, certain bacteria on human skin benefit from a stable environment without impacting the host's health (Chen et al. 2020a, b). In parasitism (harmful interaction), one partner benefits at the expense of the other. For instance, some microbes can cause diseases in their hosts, leading to negative health outcomes (Drew et al. 2021).

External factors, particularly climate change, can significantly impact microbial symbiosis in various ways. These impacts can alter the stability, function, and dynamics of symbiotic relationships between microbes and their hosts. Coral bleaching is due to elevated sea temperatures, which causes the loss of zooxanthellae and impacts coral health and reef ecosystems (Blanckaert et al. 2020). Increased temperatures and changing precipitation can stress lichens, affecting their growth and reproduction, and disrupting their role in soil formation and nutrient cycling (Grube and Werdin 2016). Drought and temperature extremes can reduce mycorrhizal colonization and function, affecting plant nutrient uptake and ecosystem productivity. Changes in temperature and moisture can impact the efficiency of nitrogen

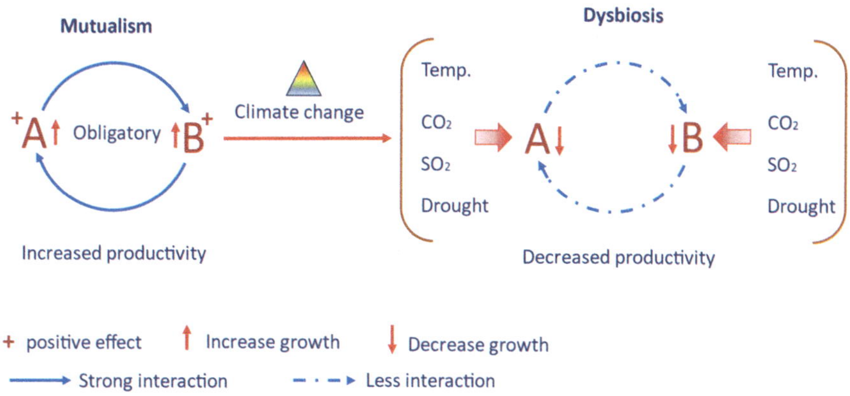

Fig. 1.1 Schematic outline of the obligatory mutualism affected by climate change. (A species—Fungal partner or Bacterial partner, B species—Plant or Cyanobacteria or Bacteria)

fixation, affecting plant growth and agricultural productivity. Overall, climate change introduces various stresses that can disrupt microbial symbioses (Fig. 1.1), impacting their ecological functions and the health of ecosystems. Understanding and mitigating these impacts are crucial for maintaining ecosystem stability and resilience in the face of climate change (Loo et al. 2022).

1.2 Fundamental Roles of Microbial Symbiosis

Microbial symbiosis encompasses a range of interactions between microorganisms and their host organisms, where both partners often derive benefits. These interactions are fundamental to various ecological processes and functions. These interactions and symbiosis formation start with the microbial communication and signaling system. Quorum Sensing is a well-known described mechanism for many microbial interactions which involve chemical signaling. This process allows microbes to coordinate behavior based on population density (Abisado et al. 2018). For example, bacteria in biofilms use quorum sensing to regulate gene expression and enhance their collective function. Symbiotic microbes and hosts often use signaling molecules to recognize each other and establish a mutualistic relationship. For example, mycorrhizal fungi and plant roots use chemical signals to initiate and maintain their association (Zhao et al. 2020).

This interaction leads to physical interaction such as adhesion and colonization. Microbes often adhere to host surfaces or tissues to establish a symbiotic relationship. For instance, mycorrhizal fungi form physical networks of hyphae that colonize plant roots. Some microbial symbionts form biofilms, which are communities of microorganisms embedded in a matrix of extracellular substances. Biofilms can enhance the stability and functionality of microbial communities in various environments. The interaction also plays a crucial role in horizontal gene transfer. Microbes can exchange genetic material through horizontal gene transfer, which can enhance their symbiotic abilities. For example, genetic exchanges can enable microbes to acquire new traits that improve their mutualistic interactions with hosts (Li et al. 2024). Mutualistic interactions have evolved to support growth and adaptation in varying environmental conditions. The benefits resulting from mutualistic symbiosis are briefly described in the following paragraphs.

1.2.1 Nutrient Exchange and Enhancement

Many microbial symbioses are formed to enhance nutrient acquisition. For example, mycorrhizal fungi form symbiotic relationships with plant roots, extending their hyphae into the soil to access nutrients like phosphorus and nitrogen that are otherwise unavailable to plants. In return, the fungi receive carbohydrates produced by the plants through photosynthesis (Wang et al. 2017; Bell et al. 2022). In nutrient fixation, certain microbes, such as nitrogen-fixing bacteria in legume root nodules or cyanobacteria in lichens, convert atmospheric nitrogen into usable forms by plants. This process enriches the soil with nitrogen, a critical nutrient for plant growth (Mishra et al. 2021).

1.2.2 Decomposition and Organic Matter Breakdown

In decomposition, symbiotic relationships, such as those between fungi and bacteria, are integral to the decomposition of organic matter. Fungi often break down complex organic materials like lignin and cellulose, while bacteria can further decompose these materials, releasing nutrients back into the soil. This process is essential for nutrient recycling and maintaining soil fertility (Janusz et al. 2017). In nutrient cycling, by breaking down organic matter, microbial symbioses contribute to the cycling of essential nutrients, such as carbon, nitrogen, and phosphorus (Nicolas et al. 2019). This recycling process ensures that these nutrients are continuously available to support plant growth and ecosystem productivity.

1.2.3 Soil Structure and Health

In soil aggregation, mycorrhizal fungi, and other soil-dwelling microbes help in the formation of soil aggregates. Fungal hyphae bind soil particles together, improving soil structure, aeration, and water-holding capacity. This enhances soil health and stability, which is vital for plant growth and ecosystem functioning (Al-Maliki et al. 2020). In soil fertility, the interactions between fungi, bacteria, and plant roots contribute to soil fertility by increasing the availability of nutrients. For instance, mycorrhizal fungi can mobilize nutrients from the soil and make them accessible to plants, while decomposing fungi and bacteria release nutrients from organic matter

(Khaliq et al. 2022). In protection and defense, microbial symbionts can protect hosts from pathogens. For example, beneficial bacteria in the gut microbiota outcompete harmful pathogens and produce antimicrobial compounds. Symbiotic relationships can enhance host immunity. For example, certain gut microbiota can modulate immune responses to improve overall health (Garcia-Gutierrez et al. 2019; Iacob et al. 2019).

1.2.4 Ecological Interactions and Community Dynamics

Microbial symbioses contribute to biodiversity by creating complex interactions among different organisms. For example, lichens, which are symbiotic associations between fungi and algae or cyanobacteria, can host diverse microbial communities, including bacteria. These interactions can influence the structure and function of microbial communities in various habitats (Rolshausen et al. 2023). In ecosystem stability, by participating in nutrient cycling and decomposition, microbial symbioses help maintain ecosystem stability. They contribute to processes like nutrient uptake, soil formation, and organic matter breakdown, which are essential for ecosystem resilience and productivity (Grube et al. 2012). In ecosystem structure and function- microbes can contribute to habitat formation. For example, coral reefs, formed by the calcium carbonate secretions of corals and their symbiotic zooxanthellae, provide critical habitats for diverse marine life. Microbial symbioses are integral to biogeochemical cycles, such as the carbon and nitrogen cycles. They drive processes like carbon sequestration and nitrogen fixation (Bourne et al. 2016).

1.2.5 Adaptation and Survival

In stress tolerance, symbiotic relationships can enhance the ability of organisms to tolerate environmental stress. For instance, mycorrhizal fungi can help plants survive in nutrient-poor soils, while lichens can endure extreme environmental conditions due to their combined fungal and algal capabilities. In resource utilization, microbial symbioses often arise from the need to efficiently utilize available resources (Elkhateeb et al. 2021). For example, in nutrient-limited environments, symbiotic relationships can optimize resource acquisition and utilization, benefiting all partners involved. In environmental adaptation, microbes can help their hosts tolerate environmental stresses. For instance, mycorrhizal fungi can enhance plant drought tolerance by improving water uptake. Symbiotic relationships can enable organisms to adapt to changing climates. For example, lichens can withstand extreme environmental conditions due to their symbiotic nature (Grube et al. 2015).

1.3 Microbial Symbiosis in Different Climatic Conditions

Microbial symbiosis adapts to various climatic conditions through a range of mechanisms that enable these interactions to persist and function effectively despite environmental stressors. Here's how microbial symbiosis manifests and adjusts in different climatic conditions.

1.3.1 Tropical and Subtropical Climates

In high temperatures and humidity, corals and their symbiotic zooxanthellae (photosynthetic algae) thrive in warm, nutrient-poor waters. Zooxanthellae provide corals with organic compounds produced through photosynthesis, which supports coral growth and reef-building. Corals and zooxanthellae are adapted to high temperatures and light levels, but rising temperatures due to climate change can lead to coral bleaching, where zooxanthellae are expelled, stressing coral reefs (Fig. 1.2). In high biodiversity, mycorrhizal fungi form extensive networks with plant roots in tropical forests, aiding in nutrient uptake, especially phosphorus. These fungi and their plant partners are adapted to the high biodiversity and complex soil environments of tropical forests, but deforestation and climate change can disrupt these symbiotic relationships by altering soil conditions and plant communities (Kahng et al. 2019).

1.3.2 Temperate Climates

In moderate temperatures and seasonal variation, lichens with green algae or cyanobacteria can thrive on a variety of substrates, including trees and rocks. They are adapted to seasonal changes in temperature and moisture. Lichens can tolerate extreme temperature fluctuations and varying moisture levels. However, climate change impacts such as increased frequency of extreme weather events can stress lichens, altering their distribution and function. Trees and their mycorrhizal fungi in temperate forests are adapted to seasonal variations in temperature and soil moisture. These fungi and their plant hosts synchronize their growth and nutrient uptake cycles with seasonal changes, but shifts in temperature and precipitation patterns can disrupt these cycles and affect ecosystem health (Spribille et al. 2022).

1.3.3 Arid and Semi-arid Climates

In low precipitation and high evaporation of arid environments, cyanobacterial crusts form symbiotic relationships with plants, enhancing soil fertility by fixing atmospheric nitrogen (Ramond et al. 2022). Cyanobacteria and their plant partners are adapted to extreme dryness and high temperatures. They can remain dormant during droughts and resume activity when moisture is available. However, prolonged droughts and increased temperatures can reduce their viability and impact

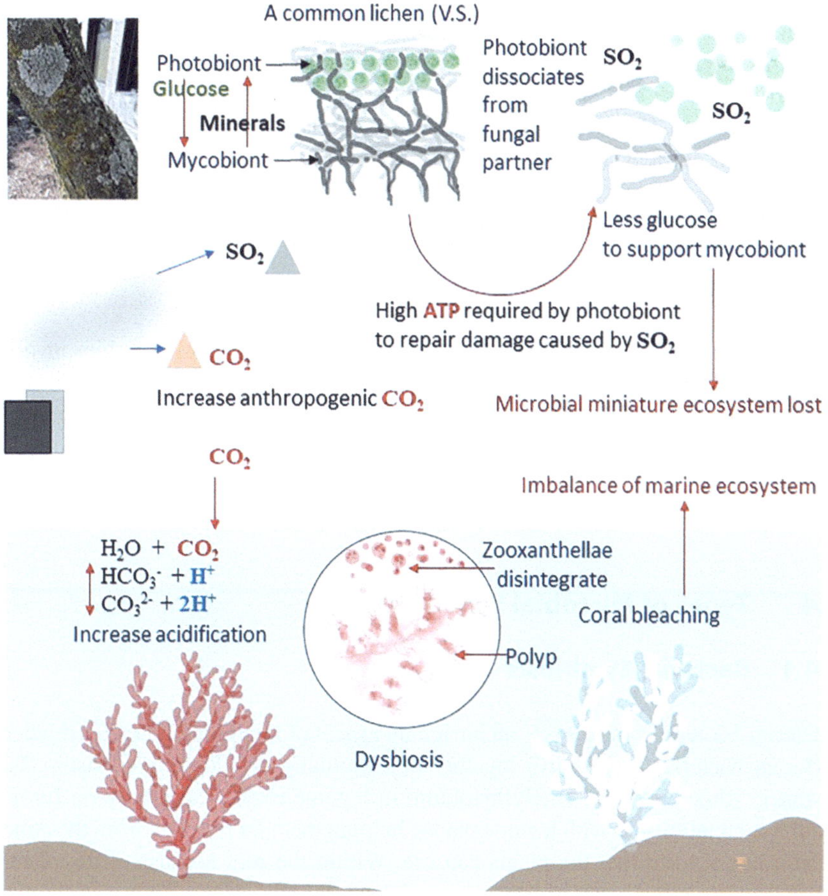

Fig. 1.2 Effect of climate change and air pollution on microbial symbiosis. Increased CO_2 causes ocean acidification leads to coral bleaching and disturbance of marine ecosystems. Elevated level of SO_2 pollution causing dysbiosis in lichen

soil health. Mycorrhizal fungi help desert plants access water and nutrients in sparse, nutrient-poor soils. These fungi are adapted to low water availability and can enhance the plant's drought resistance. Climate change-induced increases in aridity can strain these relationships, affecting plant health and ecosystem dynamics (Ramond et al. 2022).

1.3.4 Polar and High-Mountain Climates

In polar regions and high mountains, lichens and mosses form symbiotic relationships with cyanobacteria (Grimm et al. 2021). They are adapted to cold temperatures and short growing seasons. These organisms can tolerate freezing temperatures and desiccation. Climate change, including temperature increases and altered precipitation patterns, can impact their growth cycles and distribution. In alpine environments, mycorrhizal fungi help plants acquire nutrients from nutrient-poor soils. Mycorrhizal fungi and alpine plants are adapted to low temperatures and limited growing seasons. However, shifts in temperature and precipitation can affect their growth and nutrient uptake, potentially impacting alpine ecosystems (Patova et al. 2023).

1.3.5 Aquatic Environments

In variability in temperature, pH, and oxygen levels, symbiotic relationships involve interactions between microbes and aquatic plants or animals. Marine and freshwater organisms, including coral reefs and mangroves, have adapted to specific temperature and salinity ranges. Climate change impacts such as ocean acidification and altered water temperatures can disrupt these symbioses, affecting ecosystem health and functionality (Stock et al. 2021).

1.4 Types of Microbial Symbiosis

1.4.1 Bacterial Symbiosis

Bacterial-bacterial symbiosis is an intriguing aspect of microbial interactions where different bacterial species live together in a mutually beneficial relationship. For instance, Rhizobium and Bradyrhizobium in legume roots. These bacteria form a symbiotic relationship with legume plants, helping them fix nitrogen from the atmosphere into a form that the plants can use. Within the root nodules of the plants, these bacteria often work in concert to enhance the efficiency of nitrogen fixation (Gano-Cohen et al. 2016). Vibrio and Photobacterium in marine environments in some marine environments, Vibrio and Photobacterium species can work together. For example, in certain bioluminescent relationships, Vibrio bacteria can produce light in symbiosis with Photobacterium species, which helps in communication or attracting prey in the dark oceanic depths (Soto et al. 2022). In bacterial-archaeal symbiosis, sulfate-reducing bacteria and methanogens in anaerobic environments like sewage treatment plants or the guts of ruminants, sulfate-reducing bacteria and methanogens often work together. Sulfate-reducing bacteria break down organic materials, producing hydrogen sulfide, which is then utilized by methanogens to produce methane (Shi et al. 2020). This cooperative process is crucial for efficient organic matter decomposition and energy recovery. These examples illustrate the complexity and diversity of bacterial interactions in various ecosystems, highlighting how bacterial symbioses can play crucial roles in ecological balance and nutrient cycling.

1.4.2 Plant-Cyanobacterial Symbiosis

In aquatic habitats, *Azolla* spp. (water ferns) and *Anabaena* spp. (cyanobacteria). *Anabaena* fixes atmospheric nitrogen, enriching the water and soil with nitrogen compounds that are vital for plant growth. This is especially important in rice paddies, where *Azolla* and *Anabaena* contribute to soil fertility and reduce the need for synthetic fertilizers (de Vries et al. 2022). By enhancing nutrient availability, this symbiosis supports aquatic plant growth and contributes to the health of wetland ecosystems. In terrestrial plant-cyanobacterial symbiosis, cyanobacteria fix nitrogen, which is essential for the growth of bryophytes in nutrient-poor environments. This relationship enhances soil fertility and supports plant communities in arid or nutrient-limited habitats. Bryophytes and their associated cyanobacteria stabilize soil and reduce erosion, contributing to the maintenance of terrestrial ecosystems (Carrell et al. 2022).

1.4.3 Plant-Fungal Interaction

Arbuscular mycorrhizal (AM) fungi form symbiotic relationships with roughly two-thirds of all land plants and are found in nearly all terrestrial ecosystems (Smith and Smith 2012). These fungi expand their hyphal networks to absorb soil nutrients, which they then transfer to their host plants in exchange for carbon. As a result, AM fungi are essential for nutrient cycling (Smith and Smith 2012). Furthermore, AM fungi can improve soil structure and plant water relations (Pauwels et al. 2023), resulting in greater nutrient interception and availability for plants. This, in turn, directly and indirectly influences the amount of nutrients lost from the soil. Recent studies suggest that AM fungi help reduce phosphorus (P) losses through leaching (Asghari and Cavagnaro 2011; Bender et al. 2015). Furthermore, AM fungi are capable of assimilating nitrogen (N) in various forms (Veresoglou et al. 2012) and can reduce the losses of ammonium (NH_4^+) and nitrate (NO_3^-) from the ecosystem (Cavagnaro et al. 2015).

In agricultural soils, certain fungi like *Fusarium* can be plant pathogens causing diseases. Beneficial bacteria such as *Pseudomonas* spp. can suppress these pathogens by producing antimicrobial compounds or competing for resources. This interaction helps manage plant disease and supports plant health, enhancing agricultural productivity and ecosystem stability (Chandra et al. 2020). Mycoheterotrophic orchids, such as *Gastrodia*, rely entirely on their fungal partners for nutrients. The fungi involved are often mycorrhizal fungi that interact with soil bacteria to support the orchid's growth. This interaction illustrates the complexity of plant-fungal-bacterial networks and highlights the role of fungi and bacteria in supporting specialized plant life forms (Li et al. 2022).

1.4.4 Fungal-Bacterial Interactions

These symbioses are significant in various ecosystems, influencing nutrient cycling, ecosystem health, and the stability of microbial communities (Jiao et al. 2021). These examples demonstrate how fungal-bacterial interactions contribute to nutrient cycling, decomposition, soil health, and plant productivity, ultimately supporting ecosystem functioning and resilience. Here are some key examples illustrating the ecological importance of these interactions. Mycorrhizal fungi form symbiotic relationships with plant roots, extending their hyphae into the soil to access nutrients. These fungi interact with soil bacteria, such as those involved in nitrogen fixation (e.g., *Azotobacter* spp.) and decomposition (e.g., *Actinobacteria*). The bacteria break down organic matter, releasing nutrients that are then absorbed by the mycorrhizal fungi and provided to the plants (Ding et al. 2024). This interaction enhances soil fertility and plant growth, supports nutrient cycling, and improves soil structure and health. In decomposing wood and plant material, *Basidiomycete* fungi such as *Phanerochaete chrysosporium* break down complex organic compounds like lignin (Fig. 1.3). These fungi work in concert with *Actinobacteria*, which further decompose cellulose and other remaining organic materials. This cooperation accelerates the decomposition process, releases essential nutrients back into the soil, and contributes to the carbon cycle (Sun et al. 2022).

1.4.5 Symbiotic Relationships in Lichens

Fungal-algal/cyanobacterial symbiosis, particularly in the form of lichens, has significant ecological importance across various environments. Lichens are symbiotic associations between a fungal partner and a photosynthetic partner, which is often a green alga or cyanobacterium. The fungal partner provides a protective environment, while the alga or cyanobacterium performs photosynthesis. In some lichens, bacteria may also play a role in producing antimicrobial compounds that benefit the lichen. Lichens contribute to primary production in harsh environments (e.g., deserts, high mountains), aid in soil formation by breaking down rock surfaces, and support a diverse range of microorganisms (Almendras et al. 2018). *Cladonia* spp. (a type of lichen) with green algae or cyanobacteria. Lichens grow on bare rock and other harsh substrates where few other organisms can thrive. They secrete acids that chemically weather rocks, breaking them down into finer particles that contribute to soil formation. By facilitating soil formation, lichens help create habitats that can support other plant life. Additionally, their ability to stabilize soil helps prevent erosion, protecting soil resources and supporting ecosystem stability (Çobanoğlu Özyiğitoğlu 2020). *Xanthoria* spp. (a type of lichen) with green algae or cyanobacteria. These lichens can thrive in extreme environments such as deserts, high mountains, and Arctic tundras. They perform photosynthesis, producing organic matter even in nutrient-poor conditions. In these environments, lichens contribute significantly to primary productivity, providing a crucial source of energy for other organisms. They form the base of food webs and support ecosystem functioning in

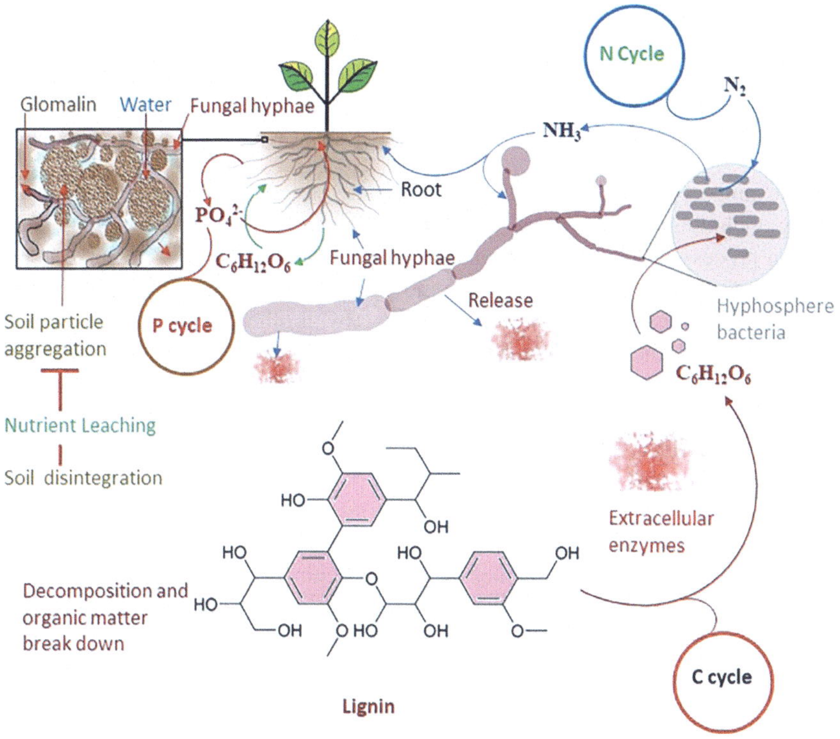

Fig. 1.3 Microbial symbiosis showing nutrient exchange and decomposition of organic matter linked to the Phosphorus (P), Carbon (C), and Nitrogen (N) cycle

otherwise inhospitable areas (Lorenz et al. 2023). Lichen associations involving *Cyanobacteria* like *Nostoc* spp. Some lichens harbour cyanobacteria that fix atmospheric nitrogen, converting it into forms that can be used by other organisms. This is particularly important in nitrogen-poor environments. By fixing nitrogen, lichens contribute to nutrient cycling and improve soil fertility, which can benefit surrounding plant communities and support broader ecosystem health (Hodkinson et al. 2014). Lichens with diverse algal and fungal partners in temperate forests. Lichens often host a variety of algal and fungal species, creating complex microhabitats. They provide shelter and food for small invertebrates, including mites, insects, and microorganisms. By forming diverse microhabitats, lichens enhance biodiversity. They offer niches for various small organisms and contribute to the complexity and stability of forest ecosystems. Lichens *Usnea* spp. with green algae are sensitive to air pollution, particularly sulfur dioxide and heavy metals. They can be used as bioindicators to monitor air quality. By serving as indicators of air quality, lichens help assess environmental health and the impact of pollution on ecosystems. These examples highlight the diverse and crucial roles that fungal-algal symbioses, particularly in lichens, play in various ecosystems. They contribute to soil formation, primary production, nutrient cycling, habitat creation, and environmental monitoring, supporting overall ecosystem health and stability (Martín-Pinto et al. 2023).

1.4.6 Inter-fungal Symbiosis

These examples illustrate how fungal-fungal interactions impact ecosystem functioning, including nutrient cycling, community dynamics, competition, and decomposition processes. These interactions are essential for maintaining the health and stability of various ecosystems. In some grass species, endophytic fungi like *Epichloë* (a type of mutualistic fungus) interact with other fungal species such as *Claviceps* (which includes ergot fungi). These interactions can involve competition for space or resources within the plant tissue (Laihonen et al. 2022). Endophytic fungi can enhance plant resistance to herbivores and pathogens, contribute to stress tolerance, and affect plant growth. Their interactions with other fungi can influence the overall health and productivity of the host plants. In forest ecosystems, wood-decaying fungi such as *Ganoderma* and *Trametes* can interact by either competing for decomposing wood or cooperating in the breakdown of complex substrates. They may produce different enzymes to degrade lignin and cellulose. These interactions contribute to the decomposition of dead wood, recycling of nutrients, and carbon cycling. By breaking down complex organic materials, these fungi help release nutrients back into the soil, supporting forest health and productivity (Fukasawa 2021).

1.5 Climate Change Affecting Microbial Symbiosis

1.5.1 Temperature Changes Induced Stress and Survival

Extreme temperatures can stress both microbial partners in symbioses. For instance, elevated temperatures can lead to the expulsion of symbiotic algae (zooxanthellae) from coral tissues, resulting in coral bleaching. Increased temperatures can accelerate the metabolic rates of microbes, potentially disrupting the balance of nutrient exchange in symbiotic relationships. For example, higher temperatures may alter the efficiency of nitrogen fixation in plant-microbe symbioses (Olivares et al. 2013). Temperature changes can influence microbial community composition, potentially favoring heat-tolerant species while disadvantaging others. This can shift the dynamics of symbiotic interactions (Fig. 1.3).

1.5.2 Precipitation and Moisture

Climate change models predict shifts in rainfall patterns, including increased annual precipitation in some regions and more intense rainfall events (IPCC 2007). Higher precipitation is often associated with increased nutrient leaching (Austin and

Vitousek 1998), which exacerbates the risk of related human and environmental health issues, as highlighted by the Millennium Ecosystem Assessment (MEA 2005). Therefore, it is essential to reduce nutrient losses through leaching to protect the environment and human health. Recent studies suggest that soil biota, including arbuscular mycorrhizal (AM) fungi, enhance nutrient cycling in agroecosystems and reduce leaching losses (Bender and van der Heijden 2015; Cavagnaro et al. 2015). However, it remains unclear whether the ability of AM fungi to decrease nutrient losses is consistent across different precipitation scenarios (Martínez-García et al. 2017).

Changes in precipitation patterns can lead to drought conditions, affecting microbial hydration and metabolic activity. For example, arid conditions can reduce the effectiveness of mycorrhizal fungi in nutrient uptake (Bouskout et al. 2022). Increased rainfall and flooding can lead to soil saturation, which can create anaerobic conditions that affect the activity of soil microbes and disrupt symbiotic relationships such as those between mycorrhizal fungi and plant roots. The amount of nutrients lost via leaching increased significantly with higher intensity of rainfall. However, the existence of AM fungi reduced phosphorus losses by 50% under both rainfall conditions and nitrogen losses by 40% under high rainfall intensity. This indicates that AM fungi enhance the soil's ability to intercept nutrients and reduce the risk of nutrient leaching as rainfall intensity increases (Martínez-García et al. 2017).

1.5.3 CO_2 Levels and Acidification

Rising CO_2 levels can lead to ocean acidification, which affects marine microbial symbioses (Fig. 1.2). For example, acidification can impair the calcium carbonate deposition in corals, weakening their ability to maintain a stable relationship with zooxanthellae (Lin et al. 2022). Increased atmospheric CO_2 can influence the growth rates and metabolic processes of symbiotic microbes (Fig. 1.3). This can impact their interactions with hosts and the efficiency of nutrient exchange (Ryalls et al. 2013).

1.5.4 Nutrient Availability

Climate change can alter the availability and cycling of essential nutrients such as nitrogen, phosphorus, and carbon. Changes in nutrient availability can affect microbial growth and their symbiotic relationships. For instance, nutrient imbalances can disrupt nitrogen-fixing bacteria in legume nodules (Etesami 2022). Increased runoff from agricultural activities, exacerbated by climate change, can lead to nutrient enrichment in aquatic systems, potentially causing harmful algal blooms and disrupting symbiotic relationships in aquatic ecosystems (Guo et al. 2024).

1.5.5 Extreme Weather Events

Extreme weather events, such as heatwaves, hurricanes, and heavy rains, can cause physical damage to ecosystems and disrupt microbial symbioses. For example, storms can lead to coral reef damage and the loss of symbiotic algae. Extreme weather events can lead to habitat destruction, affecting the environments in which microbial symbioses occur. This can result in the loss of suitable habitats for symbiotic organisms (Ban et al. 2014).

1.5.6 Shifts in Ecosystem Dynamics

Climate change can alter species distributions, leading to changes in the composition and interactions of microbial communities. This can impact the effectiveness of symbiotic relationships as new species invade or existing ones migrate. Changes in climate can affect the interactions between microbial symbionts and other organisms, including plants and animals. For instance, shifts in plant communities due to climate change can influence mycorrhizal associations (Kivlin et al. 2013).

1.5.7 Biotic Stress

Climate change can also alter the prevalence and distribution of pathogens and competitors. For example, warmer temperatures may favor pathogenic fungi or bacteria that disrupt symbiotic relationships, such as those between plants and mycorrhizal fungi. Changes in environmental conditions can disrupt the finely tuned-interactions between symbiotic partners. For instance, altered precipitation patterns may impact the ability of lichens to maintain their fungal-algal symbiosis (Singh et al. 2023).

1.6 Conclusion

Climate change can lead to shifts in microbial symbioses that impact ecosystem health and stability. Disruptions in key symbiotic relationships can affect nutrient cycles, biodiversity, and ecosystem services. While some microbial symbioses may adapt to changing conditions, others may struggle to cope with rapid environmental changes, leading to declines or extinctions. The ability to adapt and maintain functionality varies among different symbiotic relationships. Understanding and mitigating the impacts of climate change on microbial symbiosis is essential for conserving ecosystems and maintaining their functions. Monitoring changes, supporting adaptive strategies, and implementing conservation measures can help protect these vital relationships. Climate change can have diverse and significant impacts on microbial symbiosis across different habitats. The outcomes can range from disruptions and declines to potential adaptations and resilience, depending on the specific symbiotic relationships and environmental changes involved. Addressing these challenges requires a comprehensive understanding of microbial ecology and targeted conservation efforts.

References

Abisado RG, Benomar S, Klaus JR, Dandekar AA, Chandler JR (2018) Bacterial quorum sensing and microbial community interactions. MBio 9(3):10–1128

Al-Malik Khaliq S, Ebreesum H (2020) Changes in soil carbon mineralization, soil microbes, roots density and soil structure following the application of the arbuscular mycorrhizal fungi and green algae in the arid saline soil. Rhizosphere 14:100203

Almendras K, García J, Carú M, Orlando J (2018) Nitrogen-fixing bacteria associated with Peltigera cyanolichens and Cladonia chlorolichens. Molecules 23(12):3077

Asghari HR, Cavagnaro TR (2011) Arbuscular mycorrhizas enhance plant interception of leached nutrients. Funct Plant Biol 38:219–226

Austin AT, Vitousek PM (1998) Nutrient dynamics on a precipitation gradient in Hawai'i. Oecologia 113:519–529

Ban SS, Graham NA, Connolly SR (2014) Evidence for multiple stressor interactions and effects on coral reefs. Glob Chang Biol 20(3):681–697

Bell CA, Magkourilou E, Urwin PE, Field KJ (2022) Disruption of carbon for nutrient exchange between potato and arbuscular mycorrhizal fungi enhanced cyst nematode fitness and host pest tolerance. New Phytol 234(1):269–279

Bender SF, van der Heijden MG (2015) Soil biota enhance agricultural sustainability by improving crop yield, nutrient uptake and reducing nitrogen leaching losses. J Appl Ecol 52(1):228–239

Blanckaert AC, Reef R, Pandolfi JM, Lovelock CE (2020) Variation in the elemental stoichiometry of the coral–zooxanthellae symbiosis. Coral Reefs 39(4):1071–1079

Bourne DG, Morrow KM, Webster NS (2016) Insights into the coral microbiome: underpinning the health and resilience of reef ecosystems. Ann Rev Microbiol 70(1):317–340

Bouskou Martínez Martín M, Bourhia M, Al Feddy MN, Dounas H, Salamatullah AM, Soufan W et al (2022) Mycorrhizal fungi inoculation improves Capparis spinosa's yield, nutrient uptake and photosynthetic efficiency under water deficit. Agronomy 12(1):149

Carrell AA, Veličković D, Lawrence TJ, Bowen BP, Louie KB, Carper DL et al (2022) Novel metabolic interactions and environmental conditions mediate the boreal peatmoss-cyanobacteria mutualism. ISME J 16(4):1074–1085

Cavagnaro TR, Bender SF, Asghari HR, van der Heijden MG (2015) The role of arbuscular mycorrhizas in reducing soil nutrient loss. Trends Plant Sci 20(5):283–290

Cavicchioli R, Ripple WJ, Timmis KN, Azam F, Bakken LR, Baylis M et al (2019) Scientists' warning to humanity: microorganisms and climate change. Nat Rev Microbiol 17(9):569–586

Chandra H, Kumari P, Bisht R, Prasad R, Yadav S (2020) Plant growth promoting Pseudomonas aeruginosa from Valeriana wallichii displays antagonistic potential against three phytopathogenic fungi. Mol Biol Rep 47(8):6015–6026

Chen L, Li J, Zhu W, Kuang Y, Liu T, Zhang W et al (2020a) Skin and gut microbiome in psoriasis: gaining insight into the pathophysiology of it and finding novel therapeutic strategies. Front Microbiol 11:589726

Chen H, Zhu T, Li B, Fang C, Nie M (2020b) The thermal response of soil microbial methanogenesis decreases in magnitude with changing temperature. Nat Commun 11(1):5733

Çobanoğlu Özyiğitoğlu G (2020) Environmental significance of Lichens and Biodeterioration. Environmental Concerns and Sustainable Development: Volume 2: Biodiversity, Soil and Waste Management, pp 247–277

de Vries S, de Vries J (2022) Evolutionary genomic insights into cyanobacterial symbioses in plants. Quantitative Plant Biology 3:e16

Ding Y, Gao X, Shu D, Siddique KH, Song X, Wu P et al (2024) Enhancing soil health and nutrient cycling through soil amendments: Improving the synergy of bacteria and fungi. Sci Total Environ 923:171332
Drew GC, Stevens EJ, King KC (2021) Microbial evolution and transitions along the parasite–mutualist continuum. Nat Rev Microbiol 19(10):623–638
Elkhateeb WA, Somasekhar T, Thomas PW, Wen TC, Daba GM (2021) Mycorrhiza and lichens as two models of fungal symbiosis. J Microbiol Biotechnol Food Sci 11(3)
Etesami H (2022) Root nodules of legumes: a suitable ecological niche for isolating non-rhizobial bacteria with biotechnological potential in agriculture. Curr Res Biotechnol 4:78–86
Fukasawa Y (2021) Ecological impacts of fungal wood decay types: a review of current knowledge and future research directions. Ecol Res 36(6):910–931
Gano-Cohen KA, Stokes PJ, Blanton MA, Wendlandt CE, Hollowell AC, Regus JU et al (2016) Nonnodulating Bradyrhizobium spp. modulate the benefits of legume-rhizobium mutualism. Appl Environ Microbiol 82(17):5259–5268
Garcia-Gutierrez E, Mayer MJ, Cotter PD, Narbad A (2019) Gut microbiota as a source of novel antimicrobials. Gut Microbes 10(1):1–21
Grimm M, Grube M, Schiefelbein U, Zühlke D, Bernhardt J, Riedel K (2021) The lichens' microbiota, still a mystery? Front Microbiol 12:623839
Grube M, Wedin M (2016) Lichenized fungi and the evolution of symbiotic organization. Microbiol Spectrum 4(6):10–1128
Grube M, Köberl M, Lackner S, Berg C, Berg G (2012) Host–parasite interaction and microbiome response: effects of fungal infections on the bacterial community of the Alpine lichen Solorina crocea. FEMS Microbiol Ecol 82(2):472–481
Grube M, Cernava T, Soh J, Fuchs S, Aschenbrenner I, Lassek C et al (2015) Exploring functional contexts of symbiotic sustain within lichen-associated bacteria by comparative omics. ISME J 9(2):412–424
Guo D, Wang Y, Zhang Y, Duan J, Guan F, Hou B (2024) Effects of marine eutrophication environment on microbial corrosion: A review. Mar Pollut Bull 205:116637
Hodkinson BP, Allen JL, Forrest LL, Goffinet B, Sérusiaux E, Andrésson ÓS et al (2014) Lichen-symbiotic cyanobacteria associated with Peltigera have an alternative vanadium-dependent nitrogen fixation system. Eur J Phycol 49(1):11–19
IPCC [Intergovernmental Panel on Climate Change], 2007. Climate Change (2007) The Physical Science Basis. Contribution of Working Group I to the Fourth Assessment Report of the Intergovernmental Panel on Climate Change. Cambridge University Press, Cambridge, United Kingdom and New York, NY, USA, 996 pp
Iacob S, Iacob DG, Luminos LM (2019) Intestinal microbiota as a host defense mechanism to infectious threats. Front Microbiol 9:3328
Janusz G, Pawlik A, Sulej J, Świderska-Burek U, Jarosz-Wilkołazka A, Paszczyński A (2017) Lignin degradation: microorganisms, enzymes involved, genomes analysis and evolution. FEMS Microbiol Rev 41(6):941–962
Jiao S, Peng Z, Qi J, Gao J, Wei G (2021) Linking bacterial-fungal relationships to microbial diversity and soil nutrient cycling. Msystems 6(2):10–1128
Kahng, S. E., Akkaynak, D., Shlesinger, T., Hochberg, E. J., Wiedenmann, J., Tamir, R., & Tchernov, D. (2019). Light, temperature, photosynthesis, heterotrophy, and the lower depth limits of mesophotic coral ecosystems. Mesophotic coral ecosystems, 801–828, Springer International Publishing
Khaliq A, Perveen S, Alamer KH, Zia Ul Haq M, Rafique Z, Alsudays IM et al (2022) Arbuscular mycorrhizal fungi symbiosis to enhance plant–soil interaction. Sustain For 14(13):7840
Kivlin SN, Emery SM, Rudgers JA (2013) Fungal symbionts alter plant responses to global change. Am J Bot 100(7):1445–1457
Laihonen M, Saikkonen K, Helander M, Vázquez de Aldana BR, Zabalgogeazcoa I, Fuchs B (2022) Epichloë endophyte-promoted seed pathogen increases host grass resistance against insect herbivory. Front Microbiol 12:786619

Li YY, Boeraeve M, Cho YH, Jacquemyn H, Lee YI (2022) Mycorrhizal switching and the role of fungal abundance in seed germination in a fully mycoheterotrophic orchid, Gastrodia confusoides. Front Plant Sci 12:775290

Li Y, Chen Y, Fu Y, Shao J, Liu Y, Xuan W et al (2024) Signal communication during microbial modulation of root system architecture. J Exp Bot 75(2):526–537

Lin Z, Wang L, Chen M, Zheng X, Chen J (2022) Proteome and microbiota analyses characterizing dynamic coral-algae-microbe tripartite interactions under simulated rapid ocean acidification. Sci Total Environ 810:152266

Loo WT, Chua KO, Mazumdar P, Cheng A, Osman N, Harikrishna JA (2022) Arbuscular mycorrhizal symbiosis: a strategy for mitigating the impacts of climate change on tropical legume crops. Plan Theory 11(21):2875

Lorenz C, Bianchi E, Poggiali G, Alemanno G, Benesperi R, Brucato JR et al (2023) Survivability of the lichen Xanthoria parietina in simulated Martian environmental conditions. Sci Rep 13(1):4893

Martínez-García LB, De Deyn GB, Pugnaire FI, Kothamasi D, van der Heijden MG (2017) Symbiotic soil fungi enhance ecosystem resilience to climate change. Glob Chang Biol 23(12):5228–5236

Martín-Pinto P, Fernández C, Santos M, Fontúrbel T, Oria-de-Rueda JA, Vázquez-Veloso A et al (2023) Unaltered fungal community after fire prevention treatments over widespread Mediterranean rockroses (Halimium lasianthum). Sci Rep 13(1):608

MEA (2005) Ecosystems and human well-being: sythesis. Washington (DC): Island Press

Mishra A, Rajput S, Gupta PS, Goyal V, Singh S, Sharma S et al (2021) Role of cyanobacteria in Rhizospheric nitrogen fixation. In: Soil nitrogen ecology. Springer International Publishing, Cham, pp 497–519

Nicolás C, Martin-Bertelsen T, Floudas D, Bentzer J, Smits M, Johansson T et al (2019) The soil organic matter decomposition mechanisms in ectomycorrhizal fungi are tuned for liberating soil organic nitrogen. ISME J 13(4):977–988

Olivares J, Bedmar EJ, Sanjuán J (2013) Biological nitrogen fixation in the context of global change. Mol Plant-Microbe Interact 26(5):486–494

Patova E, Novakovskaya I, Gusev E, Martynenko N (2023) Diversity of cyanobacteria and algae in biological soil crusts of the Northern Ural mountain region assessed through morphological and metabarcoding approaches. Diversity 15(10):1080

Pauwels R, Graefe J, Bitterlich M (2023) An arbuscular mycorrhizal fungus alters soil water retention and hydraulic conductivity in a soil texture specific way. Mycorrhiza 33(3):165–179

Ramond JB, Jordaan K, Díez B, Heinzelmann SM, Cowan DA (2022) Microbial biogeochemical cycling of nitrogen in arid ecosystems. Microbiol Mol Biol Rev 86(2):e00109–e00121

Rolshausen G, Dal Grande F, Otte J, Schmitt I (2023) Lichen holobionts show compositional structure along elevation. Mol Ecol 32(23):6619–6630

Ryalls JM, Riegler M, Moore BD, Lopaticki G, Johnson SN (2013) Effects of elevated temperature and CO2 on aboveground-belowground systems: a case study with plants, their mutualistic bacteria and root/shoot herbivores. Front Plant Sci 4:445

Shi X, Gao G, Tian J, Wang XC, Jin X, Jin P (2020) Symbiosis of sulfate-reducing bacteria and methanogenic archaea in sewer systems. Environ Int 143:105923

Singh BK, Delgado-Baquerizo M, Egidi E, Guirado E, Leach JE, Liu H, Trivedi P (2023) Climate change impacts on plant pathogens, food security and paths forward. Nat Rev Microbiol 21(10):640–656

Smith SE, Smith FA (2012) Fresh perspectives on the roles of arbuscular mycorrhizal fungi in plant nutrition and growth. Mycologia 104(1):1–13

Soto W (2022) Emerging research topics in the vibrionaceae and the squid–vibrio symbiosis. Microorganisms 10(10):1946

Spribille T, Resl P, Stanton DE, Tagirdzhanova G (2022) Evolutionary biology of lichen symbioses. New Phytol 234(5):1566–1582

Sun Y, Chen L, Zhang S, Miao Y, Zhang Y, Li Z et al (2022) Plant interaction patterns shape the soil microbial community and nutrient cycling in different intercropping scenarios of aromatic plant species. Front Microbiol 13:888789

Stock W, Callens M, Houwenhuyse S, Schols R, Goel N, Coone M, Decaestecker E (2021) Human impact on symbioses between aquatic organisms and microbes. Aquatic Microbial Ecology 87:113–138

Veresoglou SD, Chen B, Rillig MC (2012) Arbuscular mycorrhiza and soil nitrogen cycling. Soil Biol Biochem 46:53–62

Wang W, Shi J, Xie Q, Jiang Y, Yu N, Wang E (2017) Nutrient exchange and regulation in arbuscular mycorrhizal symbiosis. Mol Plant 10(9):1147–1158

Yang J, Lan L, Jin Y, Yu N, Wang D, Wang E (2022) Mechanisms underlying legume–rhizobium symbioses. J Integr Plant Biol 64(2):244–267

Zhao X, Yu Z, Ding T (2020) Quorum-sensing regulation of antimicrobial resistance in bacteria. Microorganisms 8(3):425

2 Exploring the Potential of Natural and Synthetic Microbial Carbon Dioxide Fixation Pathways for Catalysing Climate Action

P. Anjali, Atya Kapley, and Asifa Qureshi

Abstract

The remarkable resilience and adaptability of microorganisms to extreme environments offer a wealth of potential for addressing climate change through carbon dioxide (CO_2) fixation. This review explores bacterial carbon fixation pathways, encompassing both natural and engineered synthetic mechanisms, to elucidate their role in catalysing climate change mitigation efforts. Natural pathways, such as the Calvin-Benson-Bassham cycle, Reductive tricarboxylic acid cycle, Reductive acetyl-CoA cycle, 3-Hydroxypropionate cycle, Hydroxypropionate/4-hydroxybutyrate cycle, Dicarboxylate/4-hydroxybutyrate cycle, RHP pathway, Reverse oxidative tricarboxylic acid cycle, and Natural reductive glycine pathway, are discussed alongside engineered synthetic pathways like the Crotonyl-CoA/ethylmalonyl-CoA/hydroxybutyryl-CoA cycle, Synthetic formolase pathway, and Synthetic acetyl-CoA pathway. As balancing ecosystem dynamics and nutrient cycling play vital roles in maintaining the balance of carbon dioxide levels in the atmosphere, understanding the microbial process of carbon fixation and storage in diverse ecosystems is crucial. Improving the knowledge of biological carbon fixation will help in the effective application of these microorganisms in different niches. The chapter highlights the significance of microbial carbon fixation in climate change regulation and addresses the challenges and future perspectives associated with utilizing microbial potential for effective climate change mitigation strategies.

P. Anjali · A. Kapley (✉) · A. Qureshi
Sustainable Environmental Processes-Environmental Bioprocesses, CSIR-National Environmental Engineering Research Institute (NEERI), Nehru Marg, Nagpur, India

Academy of Scientific and Innovative Research (AcSIR), Ghaziabad, India
e-mail: a_kapley@neeri.res.in

P. Pandey et al. (eds.), *Microorganisms Resilience to Climate Change*,
Microorganisms for Sustainability 56, https://doi.org/10.1007/978-981-96-3748-5_2

Keywords

Bacterial carbon sequestration · Natural carbon fixation pathways · Synthetic carbon fixation pathways · Microbial carbon capture

2.1 Introduction

Carbon dioxide plays very significant role in keeping the earth habitable by trapping heat in the atmosphere. During the last two decades, human activity has increased atmospheric carbon dioxide by 50%, pushing current CO_2 concentrations to 423 ppm (NASA, NOAA 2023). The continual increase in carbon dioxide levels threatens to throw off the delicate balance required to keep the planet warm without causing harm. Rise in carbon dioxide concentration and climate change need to be addressed swiftly as it poses serious threat to all life forms and ecosystems of this planet.

Carbon fixation is a biological and chemical process of great importance as atmospheric carbon dioxide get assimilated by living organisms and converted into organic substances. This process plays crucial role in balancing the carbon dioxide level as well the products generated through it serve as energy sources which form the foundation of life in this wonderful blue planet (Fig. 2.1). Not only plants but also wide range of microorganisms play significant role in carbon fixation through their diverse metabolic pathways.

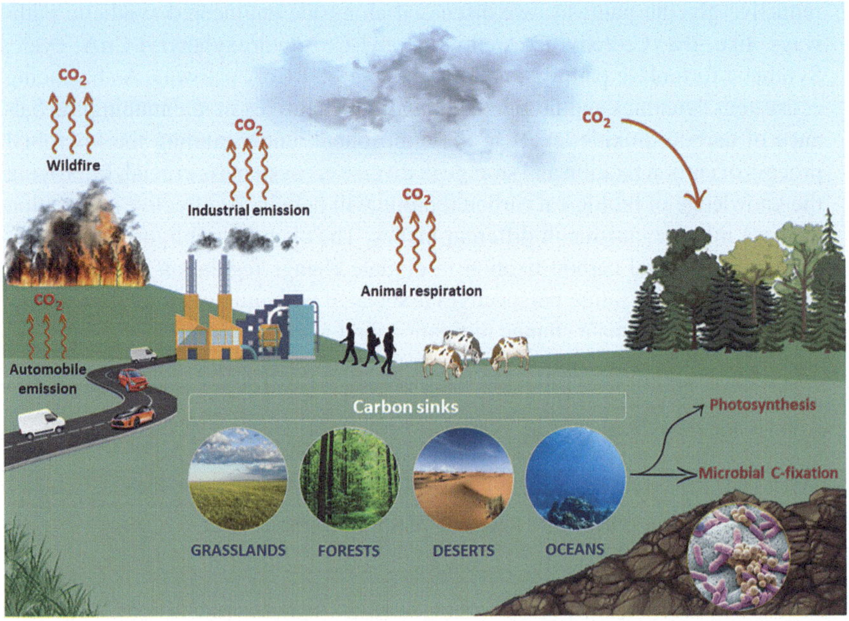

Fig. 2.1 Sources of carbon emission and carbon fixation routes in different natural carbon sinks

Among these microbial candidates, bacteria stand out because of their diverse carbon fixation strategies. Even under extreme environmental conditions, bacteria employ different biochemical pathways to assimilate carbon dioxide for cellular growth and metabolism. Utilizing bacteria capable of fixing atmospheric carbon dioxide to sequestrate CO_2 and biotransform it into value-added products such as biofuels, bioplastics, biosurfactants, and other industrially important compounds is one of the best ways to reduce carbon dioxide level sustainably. The adaptability of bacteria to diverse environmental conditions, high energy utilization efficiency, and unique metabolic pathways make them ideal candidates for CO_2 biofixation. Microbial CO_2 sequestration also contribute in balancing ecosystem dynamics by incorporating carbon into microbial biomass and soil organic carbon (SOC), which enhance soil fertility and promote crop growth.

The importance of bacterial carbon fixation extends beyond ecological balances to vital contributions in environmental sustainability, biotechnological applications, and astrobiology. Understanding the mechanisms and significance of bacterial carbon fixation in depth will help to unfold the ecosystem dynamics, mitigate climate change, and exploit microbial potential for various scientific and industrial purposes. This book chapter aims to explore different pathways of bacterial carbon fixation including natural and engineered synthetic pathways which can contribute greatly in future research.

2.2 Natural Carbon Fixation Pathways in Microbes

2.2.1 Calvin–Benson–Bassham Cycle (CBB Cycle)

Calvin cycle is a photosynthetic carbon-fixing pathway which uses light as major energy source and involves enzymatic reduction of pentose phosphate (Bassham et al. 1954). This pathway has been reported in a wide range of organisms such as cyanobacteria, algae, Proteobacteria, various photoautotrophic and chemoautotrophic bacteria, and plants (Santos Correa et al. 2023). Ribulose-1,5-bisphosphate carboxylase/oxygenase (RubisCO) and phosphoribulokinase (PRK) are considered as the most important enzymes in this cycle. Enzyme ribulose-1,5-bisphosphate carboxylase/oxygenase catalyses the electrophilic addition of carbon dioxide to C5 sugar ribulose-1,5-bisphosphate, resulting in the generation of an unstable C6 intermediate, which spontaneously hydrolysed into two 3-phosphoglycerate. It is then followed by reduction reaction with NADPH to form glyceraldehyde-3-phosphate. Two gluconeogenic processes catalyse this reduction and are driven by the hydrolysis of ATP. One of the most important phases in the Calvin cycle is regeneration of ribulose-5-phosphate. This step is catalysed by phosphoribulokinase (PRK) through interconversion of triose phosphates into a variety of sugar phosphates. Total of six NADPHs and nine ATP equivalents are needed for the generation of single glyceraldehyde-3-phosphate molecule through CBB cycle (Berg 2011).

RubisCO has been identified in four forms. *Rhodobacter*, *Rhodospirillum*, and *Rhodopseudomonas* are examples of purple non-sulphur bacteria that grow photoheterotrophically in anoxic environments and carry RubisCO form II (Badger and Bek 2008; Kusian and Bowien 1997). Furthermore, a lot of them contain form I enzyme, which serves when there is a carbon deficit (Joshi et al. 2009). *Magnetospirillum* strains only contain the form II enzyme, which allows them to grow autotrophically in microoxic environments where the form II's oxygenase side reaction plays a minor role (Geelhoed et al. 2010). Three sets of RubisCO are present in the autotrophic H_2-oxidizing *Hydrogenovibrio marinus*. Specifically, one form II RubisCO and two form I which are cytoplasmic and carboxysomal enzymes (Yoshizawa et al. 2004). Banda et al. used metagenomic analysis of environmental community samples containing a significant amount of nonculturable bacteria to identify an unknown RubisCO variant, RubisCO I′. RubisCO I′ differs from RubisCO form I in that it comprises eight large subunits exclusively, devoid of any small subunits. 24 rbcL genes, which exhibit significant sequence similarity to RubisCO type I, were documented. However, upon further analysis of the metagenome, researchers found the absence of rbcS genes (Banda et al. 2020). Consequently, they proposed that RubisCO I′ stand for a RubisCO variant that separated from form I prior to the emergence of life itself. Frolov et al.'s study showcased the presence of an active RubisCO form III in *Thermithiobacillus tepidarius* and *Thiobacillus thioparus* (Frolov et al. 2019). RubisCO form IV is categorized into 7 phylogenetic subgroups: IV-Photo present in phototrophic bacteria, IV-Nonphoto in nonphototrophic bacteria, IV-GOS -identified in ocean sampling sequences, IV-AMC in microbial consortia inhabiting acid mine lakes, IV-YkrW in *Bacillus*, IV-DeepYkr with a branch close to the YkrW clade and IV-Aful in *Archaeoglobus fulgidus* (Tabita et al. 2007). Additionally, a newly recognized variant functions as an oxygenase, converting 3-keto-D-ribitol-1,5-bisphosphate into phosphoglycolate and 3-D-phosphoglycerate (Kim et al. 2018).

RubisCO functions as both carboxylase and oxygenase (Cleland et al. 1998). However, this dual function of enzyme leads to a competition between carbon dioxide and oxygen for the active site. Carboxylation of RubisCO result in the generation of two molecules of a 3-carbon organic acid called 3-phosphoglycerate (3-PGA), on the other hand oxygenation yield a molecule of 3-PGA and 2-phosphoglycolate, a process known as photorespiration (Ducat and Silver 2012). While 3-PGA can be converted into carbohydrates, 2-PG generated from oxygenation is utilized in respiration, consuming oxygen and releasing previously fixed carbon dioxide (Caldwell et al. 2007). To mitigate the effects of photorespiration and enhance carbon fixation efficiency, cyanobacteria and certain chemoautotrophs have evolved an intracellular microcompartment known as a carboxysome (Yeates et al. 2008). Recent studies utilizing fluorescent imaging techniques have provided insights into the dynamics of carboxysomes within living cells. These microcompartments contain RubisCO and carbonic anhydrases and act as partially permeable barriers which selectively permit the entry of bicarbonate ions and RubisCO, at the same time exclude oxygen. Carbonic anhydrases present inside the carboxysome catalyse the transformation of bicarbonate ions to carbon dioxide, which is then utilized by RubisCO to produce 3-phosphoglycerate (Savage et al. 2010). This mechanism holds great importance in optimizing carbon fixation efficiency as it reduces the competitive binding of oxygen to the enzyme.

The regulation of the CBB cycle and the regeneration of RubisCO depend on glyceraldehyde-3-phosphate dehydrogenase (GAPDH) and phosphoribulokinase (PRK). CP12 is a regulatory protein made of eighty amino acids and act as a light/dark regulator by binding with GAPDH and PRK to form an inactive protein-complex. During dark, the CP12 protein complex remains inactive, but in presence of light it dissociates which release active glyceraldehyde-3-phosphate dehydrogenase and phosphoribulokinase (Gurrieri et al. 2021). In most of the oxygenic-photosynthetic organisms Calvin cycle is regulated by CP12 regulatory proteins (Santos Correa et al. 2023).

Grassland ecosystems cover nearly 40% of the land area on earth and shows relatively high carbon sequestration rates which make it potential sink for atmospheric carbon dioxide (Bai and Cotrufo 2022). Studies conducted on grassland microbial communities have shown that the community structure is mainly distributed in Proteobacteria, Actinobacteria, Chloroflexi, and Cyanobacteria (Yang 2017). These bacterial communities play a crucial role in incorporating carbon into soil organic matter through biochemical activities and interactions with grass roots. The presence of RubisCO in both grasses and associated bacteria can enhance the CO_2 assimilation potential of grasslands, and the proper management of grassland systems will contribute greatly to climate change mitigation. As anaerobic treatment systems and composting facilities also release large amounts of CO_2, using bacteria capable of the CBB cycle in anaerobic digesters as well as anoxic environments such as large-scale composting facilities, will have carbon sequestration potential, as the oxygen sensitivity of RubisCO can be avoided (Vasiliadou et al. 2018; Li et al. 2020).

2.2.2 Reductive Tricarboxylic Acid Cycle (rTCA) or Arnon–Buchanan Cycle

The Arnon–Buchanan cycle (Fig. 2.2) was initially observed in the green sulphur bacterium *Chlorobium limicola* in 1966 by Evans et al. This metabolic route represents a reversal of the citric acid cycle and facilitates the conversion of 2 carbon dioxide molecules into acetyl-CoA. In this pathway ferredoxin functions as an electron donor for C-fixation and three irreversible reactions of the conventional citric acid cycle are replaced by another enzymes. To reverse the reactions, succinate dehydrogenase, NAD^+-dependent 2-oxoglutarate dehydrogenase and citrate synthase are replaced by fumarate reductase, ferredoxin-dependent 2-oxoglutarate synthase and ATP-citrate lyase, respectively (Berg 2011). Principle product of the reductive TCA cycle is acetyl-CoA and it can be further transformed into key compounds in carbon metabolism. For example, carboxylation of acetyl-CoA to pyruvate is performed by ferredoxin-dependent pyruvate synthase, and it can be then transformed to phosphoenolpyruvate. Similarly, oxaloacetate is produced through the actions of either pyruvate or PEP carboxylase (Evans et al. 1966). As this cycle

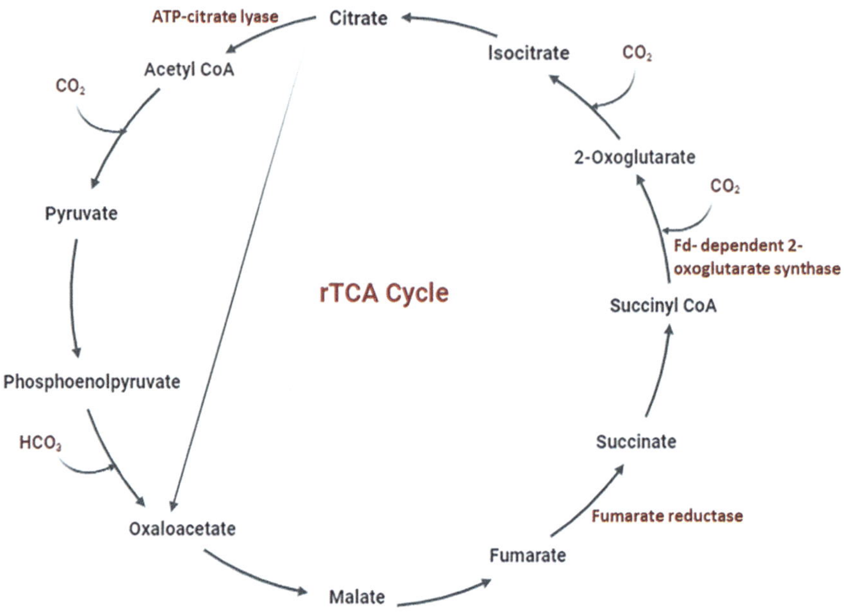

Fig. 2.2 Reductive tricarboxylic acid cycle. Energy source—Light & sulphur, Microorganisms—Green sulphur bacteria, *Proteobacteria* etc.

involves oxygen-sensitive enzymes, anaerobes or facultative anaerobes that can withstand low oxygen are known to exhibit this cycle (Thauer 2007).

Certain species show slight modifications from the original pathway. Instead of relying on ATP-citrate lyase, enzymes such as citryl-CoA lyase and citryl-CoA synthetase which are capable of catalysing citrate cleavage are used (Hügler et al. 2007). Recent investigation on *Pandoravirus massiliensis* found genes bearing low resemblance to all enzymes within the tricarboxylic acid cycle, along with a functional isocitrate dehydrogenase. The successful cloning and expression of the virus's isocitrate dehydrogenase gene in *Escherichia coli* confirmed its functional capacity. This discovery suggests the possible existence of a TCA cycle within the virus or an alternative, unidentified pathway associated with amoeba carbon metabolism, where the virus replicates (Aherfi et al. 2022). In *Hydrogenobacter* instead of ferredoxin, NADH acts as electron donor for the reaction (Miura et al. 2008). Green sulphur bacteria, which feature a FeS photosynthetic reaction center type I that can directly reduce ferredoxin and are acclimated to low light levels, are well suited to the rTCA cycle, which uses less energy than the CBB cycle (Beatty et al. 2005).

Green sulphur bacteria, Deltaproteobacteria, Alphaproteobacteria, and Nitrospirae are some of the microbial taxa that perform the rTCA cycle (Salehizadeh et al. 2020). However, the Carbon dioxide-fixing microbial community varies with abiotic conditions in the ecosystem. Therefore, microbial species and dominant carbon-fixing pathways differ in different ecosystems. For example, the abundance of microorganisms that fix carbon dioxide through the rTCA cycle increased from semi-arid grasslands to semi-humid grasslands (Huang et al. 2022). This study suggests that a lower energy requirement may be the reason for the preference of the rTCA cycle in semi-arid grasslands. The high efficiency of the rTCA cycle under anaerobic and micro-oxygen conditions also increased the selectivity of this pathway in semi-humid grasslands. Construction of artificial grasslands and bioaugmentation based on the specific abiotic and biotic conditions of the system may help mitigate climate change.

2.2.3 Reductive Acetyl-CoA Pathway (Wood-Ljungdahl Pathway)

The noncyclic process, which found in anaerobic bacteria and reduces CO_2 into acetyl-CoA, was identified by Ljungdahl et al. in 1986. These microbes were likely among the first autotrophs to utilize inorganic compounds like CO and H_2 for energy and electron acceptors (Brock 1997). The eastern/methyl, and the western/carbonyl, branches of this cycle each function in parallel to convert two molecules CO_2 into acetyl-CoA (Fig. 2.3). One CO_2 molecule in the eastern branch underwent 6 electron reduction to produce a methyl group, which is attached to a tetrahydropterin coenzyme. In the western branch another molecule of carbon dioxide is reduced to CO, which is attached to the nickel ion reaction center of carbon monoxide-dehydrogenase, which also functions as an acetyl-CoA synthase (Roberts et al. 1992; Adam et al. 2019; Youssef et al. 2019). Acetyl-CoA is used in cellular

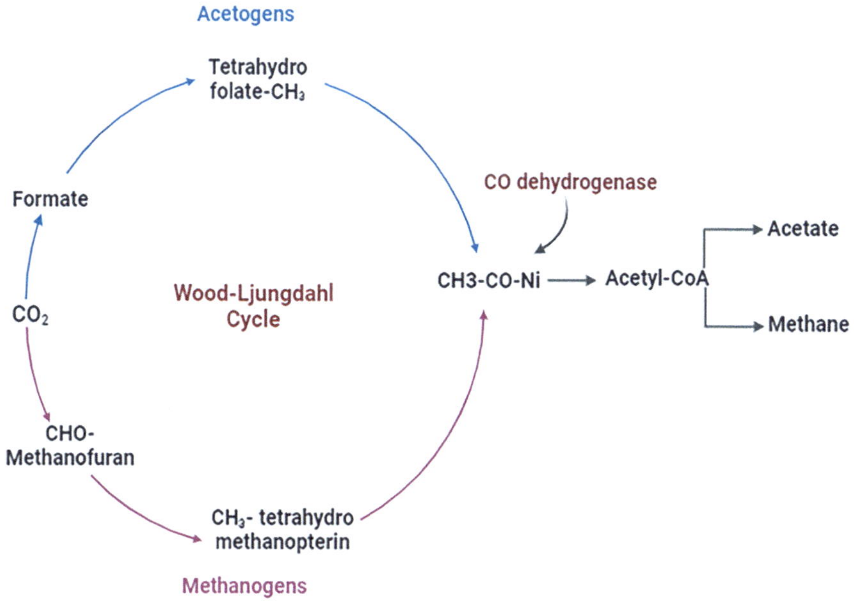

Fig. 2.3 Reductive acetyl-CoA pathway (Wood-Ljungdahl pathway). Energy source—hydrogen, microorganisms—acetogens, methanogens, planctomycetes etc.

metabolism or can be converted into acetyl-phosphate, eventually leading to ATP generation and acetate production, the primary growth products of acetogenic bacteria (Drake et al. 1997).

Methanogens and acetogens utilize the Wood-Ljungdahl pathway not only for carbon dioxide assimilation but also for conserving energy through establishing an electrochemical gradient (Biegel and Müller 2010). During this process, acetogens produce acetate, while methanogens remove methyl-tetrahydromethanopterin from the system and convert it to methane by means of methyl coenzyme M. This pathway is also involved in the growth of autotrophic Archaeoglobales that use anaerobic respiration, sulphate-reducing bacteria such as *Desulfobacterium sp.* and Deltaproteobacteria, and anaerobic ammonia-oxidizing planctomycetes (Vorholt et al. 1997; Strous et al. 2006). A novel metabolic pathway switching has been discovered in the Archaea *Methanosarcina acetivorans*, which allows them to switch from methanogens to acetogens. *Methanosarcina acetivorans* relies on acetyl-CoA reduction to store energy as acetate. The elimination of gene encoding for the methyl branch redirected all carbon flux through acetyl-CoA. This transformation demonstrates the possibility of converting *M. acetivorans* into an acetogenic organism, indicating a potential evolutionary connection between methanogenesis and the acetyl-CoA pathway (Schöne et al. 2022).

Since CO dehydrogenase and acetyl-CoA synthase are extremely sensitive to oxygen, the process must operate under strict anoxic conditions (Wood 1991). The species that carry the route have had to create mechanisms to cope with oxidative stress. The production of oxidative response enzymes, shifting to alternative electron acceptors, and forming symbiotic associations with O_2-consuming are a few of these. Despite the pathway's high energy efficiency, it is limited to a small number of biological niches due to the requirement for metals, cofactors, and anaerobiosis (Drake et al. 2008).

Acetogens are interesting candidates for microbial production platform as they can reduce carbon dioxide or carbon monoxide into acetyl-CoA. Some acetogens are used in industries for syngas fermentation as they are capable of producing acetic acid or ethanol (Bachleitner et al. 2023). Even if they can perform anoxic gas fermentation, range of products that can be synthesized is limited. More research on metabolic engineering of acetogens may increase the possibilities for C1-substrate based industrial applications.

2.2.4 3-Hydroxypropionate bicycle

The 3-hydroxypropionate cycle is one of the vital CO_2 sequestration pathways in grasslands. Studies point out the presence of an increased microbial community performing the 3-HP pathway in grasslands where mean annual precipitation is high (Huang et al. 2022). Metagenomic studies reported that 3-HP bicycle is an essential pathway in many agroecosystems, but the application of chemical fertilizers has a negative impact on related-functional gene abundance (Garritano et al. 2022). The 3-hydroxypropionate cycle is a promising metabolic pathway in industrial CO_2 utilization, as microbes can produce 3-hydroxypropionic acid by using CO_2 as the sole carbon source (Qin et al. 2024).

Helge Holo and Georg Fuchs discovered the 3-hydroxypropionate bicycle (Fig. 2.4), also known as the Fuchs–Holo route, in the green sulphur thermophilic bacterium *Chloroflexus aurantiacus* in 1989 (Herter et al. 2002). Glyoxylate was first suggested as a prime CO_2 fixation product in *Chloroflexus aurantiacus*. Even so, glyoxylate is not a central precursor molecule and necessitates another pathway to convert into the building blocks of cells. So, the cycle turned out to be a bicycle, in which glyoxylate synthesis and glyoxylate assimilation takes place (Herter et al. 2002). The first cycle begins with the carboxylation of acetyl-CoA, and subsequent enzyme reactions result in regeneration of acetyl-CoA and synthesis glyoxylate. Second cycle for the assimilation of glyoxylate starts with addition of propionyl-CoA to form methylmalyl-CoA, and subsequent reactions result in generation of pyruvate and acetyl-CoA. This pathway consists of 13 enzymes and 19 stages and none of the steps are oxygen-sensitive (Shih et al. 2017). Green sulphur bacterium uses combination of 3-HP cycle and glyoxylate cycle to channels organic substrates like lactate, butyrate, succinate, acetate, glycolate, and 3-hydroxypropionate for their carbon metabolism regardless of whether it's growing autotrophically or heterotrophically (Zarzycki and Fuchs 2011). Recently, scientists identified all genes needed for the 3HP bicycle in phylum Gemmatimonadota, whose members are typically found rhizospheres of plants.

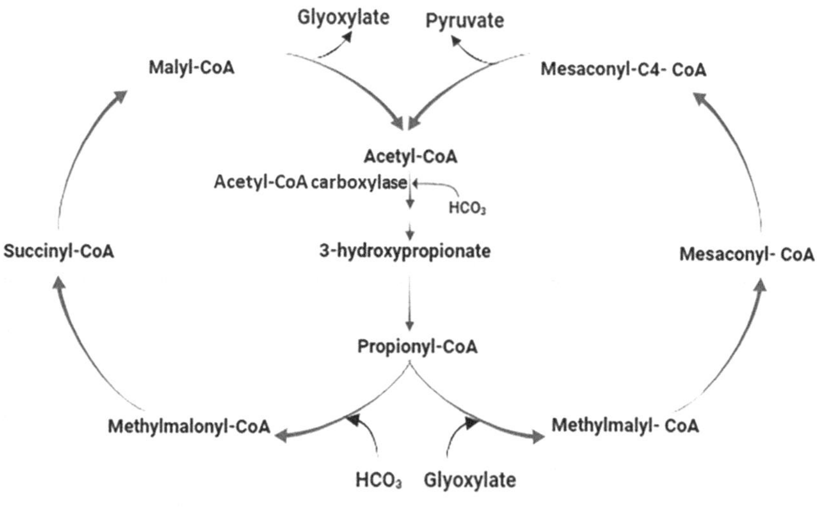

Fig. 2.4 3-Hydroxypropionate bicycle. Energy source—light, microorganism-*Chloroflexaceae*

2.2.5 Hydroxypropionate/4-Hydroxybutyrate Cycle & Dicarboxylate/4-Hydroxybutyrate Cycle

The HP/HB and DC/HB cycles can be considered as two variants of the 3-hydroxypropionate bicycle. *Metallosphaera sedula*, a thermoacidophilic member of the Sulfolobales was the first to describe the 3-hydroxypropionate/4-hydroxybutyrate cycle [HP/HB] and *Ignicoccus hospitalis*, a hyperthermophilic member of the Desulfurococcales was reported to have the dicarboxylate/4-hydroxybutyrate [DC/HB] (Berg et al. 2007; Huber et al. 2008). Both cycles involve the condensation of two CO_2 molecules with acetyl-CoA to form succinyl-CoA, which is then converted to acetoacetyl-CoA and split into 2 molecules of acetyl-CoA (Fig. 2.5) (Hawkins et al. 2013). The main difference between the cycles is that in HP/HB cycle propionyl CoA carboxylase catalyses the carboxylation of acetyl-CoA but pyruvate synthase and PEP carboxylase catalyse the same reaction in the DC/HB cycle. Another difference is the O_2-sensitivity and cofactor specificity of the cycles: the HP/HB cycle uses NAD(P)H as the electron donor and is O_2-tolerant, while the DC/HB cycle uses ferredoxin/NAD(P)H as the electron donor and is O_2-sensitive (Auernik and Kelly 2010). These cycles also differ in how they connect to the central carbon metabolism: the Dicarboxylate/4-hydroxybutyrate cycle directly produces pyruvate from acetyl-CoA, while the 5. Hydroxypropionate/4-hydroxybutyrate cycle needs an additional half turn of the cycle to generate succinyl-CoA, which is then oxidized to pyruvate (Estelmann et al. 2011). Interestingly, some Archaea have adapted their

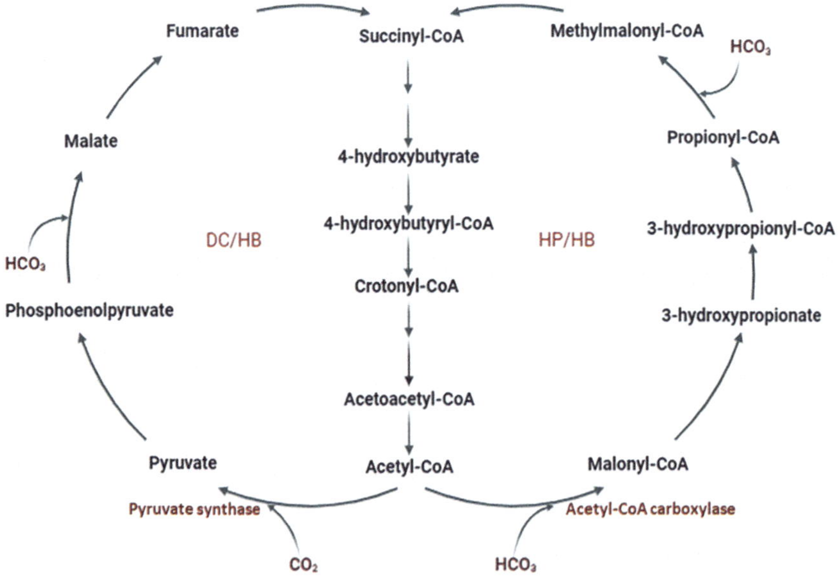

Fig. 2.5 Hydroxypropionate/4-hydroxybutyrate cycle & Dicarboxylate/4-hydroxybutyrate cycle. Energy source—hydrogen & sulphur, HP/HB microorganisms—aerobic *Sulfolobales*. DC/HB microorganisms—*Desulfurococcales*, Anaerobic *Thermoproteale*

CO_2 fixation pathways to different environmental conditions: *Stygiolobus azoricus*, an obligate anaerobe of the Sulfolobales, uses the aerobic HP/HB cycle, while *Pyrolobus fumarii*, a facultative aerobe of the Desulfurococcales, uses the anaerobic DC/HB cycle (Berg et al. 2010).

Both the DC/HB and the HP/HB cycles consume a lot of ATP to produce one pyruvate molecule. The DC/HB cycle uses five ATP and creates one pyrophosphate, while the HP/HB cycle uses nine ATP and creates three pyrophosphates. Pyrophosphates can be used as an energy source if they are hydrolysed by a membrane-bound pyrophosphatase, which would make the HP/HB cycle more energy efficient. However, these cycles have some advantages over other autotrophic pathways, such as lower metal and coenzyme requirements and higher oxygen robustness (Berg 2011).

2.2.6 RHP Pathway

The reductive hexulose-phosphate (RHP) pathway is a carbon metabolic pathway that involves RubisCO and PRK, two enzymes that are also found in photosynthetic organisms. Kono et al. showed that the methanogenic archaeon *Methanospirillum hungatei* has both RubisCO and a PRK that has a similar crystal structure to the bacterial PRK. The RHP pathway is similar to the Calvin–Benson cycle, except for a few steps. Both pathways use RubisCO to convert ribulose-5-phosphate to fructose-6-phosphate. However, in the RHP pathway, fructose-6-phosphate is further converted to D-arabino-3-hexulose-6-phosphate (Hu6P) and formaldehyde, which is used to regenerate RubisCO. The RHP pathway requires less energy than the Calvin–Benson cycle (Kono et al. 2017). *Bifidobacterium longum* JBL05, a gram-positive bacterium capable of fixing CO_2 via the RHP pathway, has reported increased yields of extrapolymeric substances with a rise in CO_2 concentration (Kajla et al. 2022).

2.2.7 Natural Reductive Glycine (nrGly) Pathway

A different carbon-fixation route, natural glycine-reducing pathway was proposed by Figueroa et al. using 16S rRNA gene community analysis and genome-resolved metagenomics. They demonstrated this in an uncultivated bacterium *Candidatus Phosphitivorax anaerolimi* strain Phox-21 which oxidizes phosphite to reduce carbon dioxide, and utilizes the resulting energy to assimilize formate via nrGly pathway. The rGly pathway is efficient and versatile because it combines ATP-independent carbon dioxide reduction and carboxylation, and supports growth across diverse environmental circumstances. Phox-21 couples phosphite oxidation to CO_2 reduction due to the lack of alternative electron acceptors and they also possess CO_2 reductase and formate dehydrogenase enzymes. The bacteria employ formate dehydrogenase to reduce CO_2 to formate, which is then converted into pyruvate (Figueroa et al. 2018).

As all essential precursors like acetyl-CoA, succinyl-CoA, oxaloacetate, and 2-oxoglutarate can be derived through pyruvate route in the partial reductive TCA cycle, studies indicates that some microorganisms can use phosphite oxidation as an energy source and achieve autotrophic growth via the rGly pathway using carbon dioxide as the sole electron acceptor (Cotton et al. 2018). *Desulfovibrio desulfuricans* is proven to fix carbon through the reductive glycine pathway by Sánchez-Andrea et al. In this pathway carbon dioxide is reduced to formate which condense with another CO_2 molecule to produce glycine. Glycine is subsequently reduced to acetyl-phosphate, which is further converted to acetyl-CoA. Acetyl-CoA can then react with another CO_2 molecule to produce pyruvate (Sánchez-Andrea et al. 2020).

2.2.8 Reverse Oxidative Tricarboxylic Acid Cycle (roTCA)

Desulfurella acetivorans is a thermophilic and sulphur-reducing bacterium that belongs to the Deltaproteobacteria class. It can use acetate as a carbon source and electron donor in heterotrophic growth, or H_2 in autotrophic growth. The existence of ATP–dependent citrate lyase (ACL) is usually regarded as a key biomarker for the reductive TCA cycle, which is a common pathway for autotrophic CO_2 fixation. However, *D. acetivorans* lacks the genes encoding ACL. Instead, recent studies have revealed that it uses a novel variant of the rTCA cycle that relies on citrate synthase (CS) to cleave citrate into acetyl-CoA and oxaloacetate. This variant is called the reverse oTCA cycle (roTCA) (Mall et al. 2018). Malate dehydrogenase then reduces oxaloacetate to malate. The roTCA cycle has also been identified in *Geobacter sulfurreducens*, another member of the Deltaproteobacteria class that was considered before as incapable of carbon fixation. This finding suggests that this bacterium may have the potential for chemolithoautotrophic growth (Zhang et al. 2020).

2.3 Engineered Pathways for Carbon Fixation

2.3.1 Crotonyl-(CoA)/Ethylmalonyl-CoA/Hydroxybutyryl-CoA Cycle

The CETCH cycle, proposed by Schwander et al., is a novel CO_2 fixation pathway that outperforms natural ones in terms of efficiency. It consists of seventeen different enzymes that can transform carbon dioxide into organic compounds at a rate of 5 nmol carbon dioxide/min, using the same energy carriers as aerobic photoautotrophs. This cycle was designed using a combinatorial and reductionist strategy that involved enzyme modifications and metabolic analysis (Schwander et al. 2016). The key steps of this pathway are the carboxylation reactions catalysed by enoyl-CoA carboxylases/reductases (ECRs) and CoA-dependent carboxylases, which were selected based on their superior kinetic and biochemical properties compared to other carboxylases. ECRs have a wide substrate range, are oxygen-tolerant, do not use oxygen as a substrate, and have a high catalytic efficiency for CO_2 fixation (Erb et al. 2009, Peter et al. 2015). ECRs are present in Streptomyces and Alphaproteobacteria, exhibiting higher levels of carboxylation activity compared to propionyl-CoA carboxylase, which is involved in the RubisCO carboxylation pathway, HP-HB cycle, and 3-HP bicycle (Erb et al. 2007). This cycle involves conversion of 2 molecules of CO_2 from propionyl-CoA to glyoxylate via thirteen main reactions involving seventeen enzymes from 9 different organisms. The CETCH cycle and the HP-HB cycle share 4 reactions and 5 intermediary compounds (Gong and Li 2016).

2.3.2 Synthetic Formolase Pathway & Synthetic Acetyl-CoA Pathway

A new synthetic pathway for converting one-carbon compounds into three-carbon compounds has been computationally designed and experimentally validated (Siegel et al. 2015). The pathway consists of four enzymatic steps: (i) formation of formyl-CoA from formate and CoA by Acetyl-CoA synthase of *E. coli*, (ii) conversion of formyl-CoA to formaldehyde by aldehyde dehydrogenase from *Listeria monocytogenes*, (iii) condensation of three formaldehyde molecules into dihydroxyacetone by formolase, a rationally designed enzyme based on benzaldehyde lyase from *Pseudomonas fluorescens* biovar I, and (iv) the integration of dihydroxyacetone into central metabolism. This synthetic pathway, named formolase (FLS) pathway, is predicted to be more efficient and thermodynamically favourable than any natural pathway for single-carbon utilization starting from formate under aerobic conditions. However, the sub-optimal activity of formolase limits the practical application of this pathway. Therefore, the researchers explored the possibility of producing two-carbon compounds directly from one-carbon compounds using FLS intermediates. Recently, a novel synthetic pathway for aerobic acetyl-CoA synthesis from formaldehyde was proposed and designed, which consists of three enzymatic steps -SACA pathway. The first step is catalysed by glycolaldehyde synthase in which 2 molecules of formaldehyde undergo condensation to form glycolaldehyde. In the second step acetyl-phosphate synthase phosphorylates glycolaldehyde to acetyl-phosphate. The third step is the transfer of CoA from coenzyme A to acetyl-phosphate, catalysed by phosphate acetyltransferase, resulting in acetyl-CoA formation. This pathway has been shown to work both in vitro and in vivo, and it has the advantages as it is carbon-conserved and ATP-independent. But formaldehyde toxicity to the host and the low activity of some enzymes limits the applicability of this pathway. Considering strategies such as host cell engineering to increase formaldehyde tolerance and enzyme modifications for optimum performance will improve the pathway efficiency (Liang et al. 2020).

2.4 Challenges of Biological Carbon Fixation and Future Perspectives

Conversion of inorganic carbon into useful organic compounds via biological carbon fixation holds great potential for mitigating climate change. However, several challenges restrict its widespread application in industrial scale. Firstly, natural carbon fixation pathways exhibit low efficiency due to the limited activity of key enzymes. While efforts to design novel enzymes show promise, more research is required on enzyme engineering so that efficient synthetic pathways can be developed. Secondly, the low solar energy efficiency of carbon fixation processes limits economic viability. Converting light energy into electricity or hydrogen can be used to improve the efficiency (Gong et al. 2018). Additionally, integrating carbon-fixation pathways into heterotrophic host cells faces challenges due to the difference in cellular metabolism. Despite these limitations, future research in omics tools and microbiome is expected facilitate the development of more efficient carbon fixation systems. As remarks, the combination of synthetic biology, enzyme optimization, and alternative energy sources holds solution for overcoming the challenges of biological carbon fixation and realizing its potential for climate change mitigation and sustainable resource production.

2.5 Conclusion

In conclusion, exploring CO_2 fixation pathways in bacteria reveals how remarkably resourceful the microbial world is and how much they contribute to global carbon cycling. Understanding the mechanisms of bacterial carbon fixation will help to recognize its potential for applications in biotechnology and environmental remediation. Microorganisms have shown remarkable adaptability and resilience in extreme environments, such as acidic hot springs, deep-sea vents, volcanic lakes, and polar ice caps. Their unique biochemical pathways enable them to thrive in these harsh conditions, making them valuable resources for addressing environmental challenges. By harnessing the capabilities of these microorganisms, we can enhance natural carbon sinks like forests, grasslands, and oceans, facilitating the sequestration of atmospheric carbon dioxide. Looking ahead, biotechnological advancements, including synthetic pathway engineering and enzyme engineering, are crucial for optimizing microbial CO_2 fixation. The microbial potential for CO_2 capture and conversion can be enhanced by designing and constructing more efficient metabolic pathways, as well as engineering enzymes to improve their activity, substrate specificity, and adaptability to harsh environmental conditions.

Acknowledgement The authors are thankful to the Director, CSIR-National Environmental Engineering Research Institute (CSIR-NEERI), Nagpur and Academy of Scientific and Innovative Research (AcSIR) for encouragement and facilities provided. The manuscript has been internally checked for plagiarism through iThenticate software (KRC No: CSIR-NEERI/KRC/2024/EB/1).

Conflict of Interest All the authors have mutually agreed to submit this manuscript to this journal and declare that they have no conflict of interest that could have appeared to influence the work reported in this paper.

References

Adam PS, Borrel G, Gribaldo S (2019) An archaeal origin of the Wood-Ljungdahl H_4MPT branch and the emergence of bacterial methylotrophy. Nat Microbiol 4(12):2155–2163. https://doi.org/10.1038/s41564-019-0534-2

Aherfi S, Brahim Belhaouari D, Pinault L, Baudoin JP, Decloquement P, Abrahao J, Colson P, Levasseur A, Lamb DC, Chabriere E, Raoult D, La Scola B (2022) Incomplete tricarboxylic acid cycle and proton gradient in *Pandoravirus massiliensis*: is it still a virus? ISME J 16(3):695–704. https://doi.org/10.1038/s41396-021-01117-3

Auernik KS, Kelly RM (2010) Physiological versatility of the extremely thermoacidophilic archaeon Metallosphaera sedula supported by transcriptomic analysis of heterotrophic, autotrophic, and mixotrophic growth. Appl Environ Microbiol 76(3):931–935. https://doi.org/10.1128/AEM.01336-09

Bachleitner S, Ata Ö, Mattanovich D (2023) The potential of CO_2-based production cycles in biotechnology to fight the climate crisis. Nat Commun 14(1):6978. https://doi.org/10.1038/s41467-023-42790-6

Badger MR, Bek EJ (2008) Multiple Rubisco forms in proteobacteria: their functional significance in relation to CO_2 acquisition by the CBB cycle. J Exp Bot 59(7):1525–1541. https://doi.org/10.1093/jxb/erm297

Bai Y, Cotrufo MF (2022) Grassland soil carbon sequestration: current understanding, challenges, and solutions. Science (N.Y.) 377(6606):603–608. https://doi.org/10.1126/science.abo2380

Banda DM, Pereira JH, Liu AK, Orr DJ, Hammel M, He C, Parry MAJ, Carmo-Silva E, Adams PD, Banfield JF, Shih PM (2020) Novel bacterial clade reveals origin of form I Rubisco. Nat Plants 6(9):1158–1166. https://doi.org/10.1038/s41477-020-00762-4

Bassham JA, Benson AA, Kay LD, Harris AZ, Wilson AT, Calvin M (1954) The path of carbon in photosynthesis. XXI. The cyclic regeneration of carbon dioxide acceptor1. J Am Chem Soc 76(7):1760–1770. https://doi.org/10.1021/ja01636a012

Beatty JT, Overmann J, Lince MT, Manske AK, Lang AS, Blankenship RE, Van Dover CL, Martinson TA, Plumley FG (2005) An obligately photosynthetic bacterial anaerobe from a deep-sea hydrothermal vent. Proc Natl Acad Sci USA 102(26):9306–9310. https://doi.org/10.1073/pnas.0503674102

Berg IA (2011) Ecological aspects of the distribution of different autotrophic CO_2 fixation pathways. Appl Environ Microbiol 77(6):1925–1936. https://doi.org/10.1128/AEM.02473-10

Berg IA, Kockelkorn D, Buckel W, Fuchs G (2007) A 3-hydroxypropionate/4-hydroxybutyrate autotrophic carbon dioxide assimilation pathway in Archaea. Science (N.Y.) 318(5857):1782–1786. https://doi.org/10.1126/science.1149976

Berg IA, Ramos-Vera WH, Petri A, Huber H, Fuchs G (2010) Study of the distribution of autotrophic CO_2 fixation cycles in Crenarchaeota. Microbiology 156(Pt 1):256–269. https://doi.org/10.1099/mic.0.034298-0

Biegel E, Müller V (2010) Bacterial Na+-translocating ferredoxin: NAD^+ oxidoreductase. Proc Natl Acad Sci USA 107(42):18138–18142. https://doi.org/10.1073/pnas.1010318107

Brock TD (1997) The value of basic research: discovery of Thermus aquaticus and other extreme thermophiles. Genetics 146(4):1207–1210. https://doi.org/10.1093/genetics/146.4.1207

Caldwell PE, MacLean MR, Norris PR (2007) Ribulose bisphosphate carboxylase activity and a Calvin cycle gene cluster in Sulfobacillus species. Microbiology 153(Pt 7):2231–2240. https://doi.org/10.1099/mic.0.2007/006262-0

Cleland WW, Andrews TJ, Gutteridge S, Hartman FC, Lorimer GH (1998) Mechanism of Rubisco: the carbamate as general base. Chem Rev 98(2):549–562. https://doi.org/10.1021/cr970010r

Cotton CA, Edlich-Muth C, Bar-Even A (2018) Reinforcing carbon fixation: CO_2 reduction replacing and supporting carboxylation. Curr Opin Biotechnol 49:49–56. https://doi.org/10.1016/j.copbio.2017.07.014

Drake HL, Daniel SL, Küsel K, Matthies C, Kuhner C, Braus-Stromeyer S (1997) Acetogenic bacteria: what are the in situ consequences of their diverse metabolic versatilities? BioFactors (Oxford) 6(1):13–24. https://doi.org/10.1002/biof.5520060103

Drake HL, Gössner AS, Daniel SL (2008) Old acetogens, new light. Ann N Y Acad Sci 1125:100–128. https://doi.org/10.1196/annals.1419.016

Ducat DC, Silver PA (2012) Improving carbon fixation pathways. Curr Opin Chem Biol 16(3–4):337–344. https://doi.org/10.1016/j.cbpa.2012.05.002

Erb TJ, Berg IA, Brecht V, Müller M, Fuchs G, Alber BE (2007) Synthesis of C5-dicarboxylic acids from C2-units involving crotonyl-CoA carboxylase/reductase: the ethylmalonyl-CoA pathway. Proc Natl Acad Sci USA 104(25):10631–10636. https://doi.org/10.1073/pnas.0702791104

Erb TJ, Brecht V, Fuchs G, Müller M, Alber BE (2009) Carboxylation mechanism and stereochemistry of crotonyl-CoA carboxylase/reductase, a carboxylating enoyl-thioester reductase. Proc Natl Acad Sci 106(22):8871–8876. https://doi.org/10.1073/pnas.0903939106

Estelmann S, Hügler M, Eisenreich W, Werner K, Berg IA, Ramos-Vera WH, Say RF, Kockelkorn D, Gad'on N, Fuchs G (2011) Labelling and enzyme studies of the central carbon metabolism in Metallosphaera sedula. J Bacteriol 193(5):1191–1200. https://doi.org/10.1128/JB.01155-10

Evans MC, Buchanan BB, Arnon DI (1966) A new ferredoxin-dependent carbon reduction cycle in a photosynthetic bacterium. Proc Natl Acad Sci USA 55(4):928–934. https://doi.org/10.1073/pnas.55.4.928

Figueroa IA, Barnum TP, Somasekhar PY, Carlström CI, Engelbrektson AL, Coates JD (2018) Metagenomics-guided analysis of microbial chemolithoautotrophic phosphite oxidation yields evidence of a seventh natural CO_2 fixation pathway. Proc Natl Acad Sci USA 115(1):E92–E101. https://doi.org/10.1073/pnas.1715549114

Frolov EN, Kublanov IV, Toshchakov SV, Lunev EA, Pimenov NV, Bonch-Osmolovskaya EA, Lebedinsky AV, Chernyh NA (2019) Form III RubisCO-mediated transaldolase variant of the Calvin cycle in a chemolithoautotrophic bacterium. Proc Natl Acad Sci USA 116(37):18638–18646. https://doi.org/10.1073/pnas.1904225116

Garritano AN, Song W, Thomas T (2022) Carbon fixation pathways across the bacterial and archaeal tree of life. PNAS nexus 1(5):pgac226. https://doi.org/10.1093/pnasnexus/pgac226

Geelhoed JS, Kleerebezem R, Sorokin DY, Stams AJ, van Loosdrecht MC (2010) Reduced inorganic sulphur oxidation supports autotrophic and mixotrophic growth of Magnetospirillum strain J10 and *Magnetospirillum gryphiswaldense*. Environ Microbiol 12(4):1031–1040. https://doi.org/10.1111/j.1462-2920.2009.02148.x

Gong F, Li Y (2016) Fixing carbon, unnaturally. Science (N.Y.) 354(6314):830–831. https://doi.org/10.1126/science.aal1559

Gong F, Zhu H, Zhang Y, Li Y (2018) Biological carbon fixation: from natural to synthetic. J CO2 Util 28:221–227. https://doi.org/10.1016/j.jcou.2018.09.014

Gurrieri L, Fermani S, Zaffagnini M, Sparla F, Trost P (2021) Calvin-Benson cycle regulation is getting complex. Trends Plant Sci 26(9):898–912. https://doi.org/10.1016/j.tplants.2021.03.008

Hawkins AS, Han Y, Bennett RK, Adams MW, Kelly RM (2013) Role of 4-hydroxybutyrate-CoA synthetase in the CO_2fixation cycle in thermoacidophilic archaea. J Biol Chem 288(6):4012–4022. https://doi.org/10.1074/jbc.M112.413195

Herter S, Fuchs G, Bacher A, Eisenreich W (2002) A bicyclic autotrophic CO_2 fixation pathway in Chloroflexus aurantiacus. J Biol Chem 277(23):20277–20283. https://doi.org/10.1074/jbc.M201030200

Huang Q, Huang Y, Wang B, Dippold MA, Li H, Li N, Jia P, Zhang H, An S, Kuzyakov Y (2022) Metabolic pathways of CO_2 fixing microorganisms determined C-fixation rates in grassland soils along the precipitation gradient. Soil Biol Biochem 172:108764. https://doi.org/10.1016/j.soilbio.2022.108764

Huber H, Gallenberger M, Jahn U, Eylert E, Berg IA, Kockelkorn D, Eisenreich W, Fuchs G (2008) A dicarboxylate/4-hydroxybutyrate autotrophic carbon assimilation cycle in the hyperthermophilic Archaeum *Ignicoccus hospitalis*. Proc Natl Acad Sci USA 105(22):7851–7856. https://doi.org/10.1073/pnas.0801043105

Hügler M, Huber H, Molyneaux SJ, Vetriani C, Sievert SM (2007) Autotrophic CO_2 fixation via the reductive tricarboxylic acid cycle in different lineages within the phylum Aquificae: evidence for two ways of citrate cleavage. Environ Microbiol 9(1):81–92. https://doi.org/10.1111/j.1462-2920.2006.01118.x

Joshi GS, Romagnoli S, Verberkmoes NC, Hettich RL, Pelletier D, Tabita FR (2009) Differential accumulation of form I RubisCO in Rhodopseudomonas palustris CGA010 under photoheterotrophic growth conditions with reduced carbon sources. J Bacteriol 191(13):4243–4250. https://doi.org/10.1128/JB.01795-08

Kajla S, Kumari R, Nagi GK (2022) Microbial CO_2 fixation and biotechnology in reducing industrial CO_2 emissions. Arch Microbiol 204(2):149. https://doi.org/10.1007/s00203-021-02677-w

Kim SM, Lim HS, Lee SB (2018) Discovery of a RuBisCO-like protein that functions as an oxygenase in the novel d-hamamelose pathway. Biotechnol Bioprocess Eng 23:490–499. https://doi.org/10.1007/s12257-018-0305-6

Kono T, Mehrotra S, Endo C, Kizu N, Matusda M, Kimura H, Mizohata E, Inoue T, Hasunuma T, Yokota A, Matsumura H, Ashida H (2017) A RuBisCO-mediated carbon metabolic pathway in methanogenic archaea. Nat Commun 8:14007. https://doi.org/10.1038/ncomms14007

Kusian B, Bowien B (1997) Organization and regulation of cbb CO_2 assimilation genes in autotrophic bacteria. FEMS Microbiol Rev 21(2):135–155. https://doi.org/10.1111/j.1574-6976.1997.tb00348.x

Li M, Xu J, Jiang Z, Li Q (2020) Molecular understanding of autotrophic CO_2-fixing bacterial communities in composting based on RuBisCO genes analysis. J Biotechnol 320:36–43. https://doi.org/10.1016/j.jbiotec.2020.06.010

Liang B, Zhao Y, Yang J (2020) Recent advances in developing artificial autotrophic microorganism for reinforcing CO_2 fixation. Front Microbiol 11:592631. https://doi.org/10.3389/fmicb.2020.592631

Mall A, Sobotta J, Huber C, Tschirner C, Kowarschik S, Bačnik K, Mergelsberg M, Boll M, Hügler M, Eisenreich W, Berg IA (2018) Reversibility of citrate synthase allows autotrophic growth of a thermophilic bacterium. Science (N.Y.) 359(6375):563–567. https://doi.org/10.1126/science.aao2410

Miura A, Kameya M, Arai H, Ishii M, Igarashi Y (2008) A soluble NADH-dependent fumarate reductase in the reductive tricarboxylic acid cycle of *Hydrogenobacter thermophilus* TK-6. J Bacteriol 190(21):7170–7177. https://doi.org/10.1128/JB.00747-08

NASA, NOAA (2023) Carbon Dioxide. https://climate.nasa.gov/vital-signs/carbon-dioxide. Accession date: 13 Mar 2024

Peter DM, Schada von Borzyskowski L, Kiefer P, Christen P, Vorholt JA, Erb TJ (2015) Screening and engineering the synthetic potential of Carboxylating reductases from central metabolism and polyketide biosynthesis. Angew Chem 54(45):13457–13461. https://doi.org/10.1002/anie.201505282

Qin N, Li L, Wan X, Ji X, Chen Y, Li C, Liu P, Zhang Y, Yang W, Jiang J, Xia J, Shi S, Tan T, Nielsen J, Chen Y, Liu Z (2024) Increased CO_2 fixation enables high carbon-yield production of 3-hydroxypropionic acid in yeast. Nat Commun 15(1):1591. https://doi.org/10.1038/s41467-024-45557-9

Roberts JR, Lu WP, Ragsdale SW (1992) Acetyl-coenzyme A synthesis from methyltetrahydrofolate, CO, and coenzyme A by enzymes purified from *Clostridium thermoaceticum*: attainment of in vivo rates and identification of rate-limiting steps. J Bacteriol 174(14):4667–4676. https://doi.org/10.1128/jb.174.14.4667-4676.1992

Salehizadeh H, Yan N, Farnood R (2020) Recent advances in microbial CO_2 fixation and conversion to value-added products. Chem Eng J 390:124584. https://doi.org/10.1016/j.cej.2020.124584

Sánchez-Andrea I, Guedes IA, Hornung B, Boeren S, Lawson CE, Sousa DZ, Stams AJ (2020) The reductive glycine pathway allows autotrophic growth of Desulfovibrio desulfuricans. Nat Commun 11(1):5090. https://doi.org/10.1038/s41467-020-18906-7

Santos Correa S, Schultz J, Lauersen KJ, Soares Rosado A (2023) Natural carbon fixation and advances in synthetic engineering for redesigning and creating new fixation pathways. J Adv Res 47:75–92. https://doi.org/10.1016/j.jare.2022.07.011

Savage DF, Afonso B, Chen AH, Silver PA (2010) Spatially ordered dynamics of the bacterial carbon fixation machinery. Science (N.Y.) 327(5970):1258–1261. https://doi.org/10.1126/science.1186090

Schöne C, Poehlein A, Jehmlich N, Adlung N, Daniel R, von Bergen M, Scheller S, Rother M (2022) Deconstructing *Methanosarcina acetivorans* into an acetogenic archaeon. Proc Natl Acad Sci USA 119(2):e2113853119. https://doi.org/10.1073/pnas.2113853119

Schwander T, Schada von Borzyskowski L, Burgener S, Cortina NS, Erb TJ (2016) A synthetic pathway for the fixation of carbon dioxide in vitro. Science (N.Y.) 354(6314):900–904. https://doi.org/10.1126/science.aah5237

Shih PM, Ward LM, Fischer WW (2017) Evolution of the 3-hydroxypropionate bicycle and recent transfer of anoxygenic photosynthesis into the Chloroflexi. Proc Natl Acad Sci USA 114(40):10749–10754. https://doi.org/10.1073/pnas.1710798114

Siegel JB, Smith AL, Poust S, Wargacki AJ, Bar-Even A, Louw C, Shen BW, Eiben CB, Tran HM, Noor E, Gallaher JL, Bale J, Yoshikuni Y, Gelb MH, Keasling JD, Stoddard BL, Lidstrom ME, Baker D (2015) Computational protein design enables a novel one-carbon assimilation pathway. Proc Natl Acad Sci USA 112(12):3704–3709. https://doi.org/10.1073/pnas.1500545112

Strous M, Pelletier E, Mangenot S, Rattei T, Lehner A, Taylor MW, Horn M, Daims H, Bartol-Mavel D, Wincker P, Barbe V, Fonknechten N, Vallenet D, Segurens B, Schenowitz-Truong C, Médigue C, Collingro A, Snel B, Dutilh BE, Op den Camp HJ et al (2006) Deciphering the evolution and metabolism of an anammox bacterium from a community genome. Nature 440(7085):790–794. https://doi.org/10.1038/nature04647

Tabita FR, Hanson TE, Li H, Satagopan S, Singh J, Chan S (2007) Function, structure, and evolution of the RubisCO-like proteins and their RubisCO homologs. Microbiol Mol Biol Rev 71(4):576–599. https://doi.org/10.1128/MMBR.00015-07

Thauer RK (2007) Microbiology. A fifth pathway of carbon fixation. Science (N.Y.) 318(5857):1732–1733. https://doi.org/10.1126/science.1152209

Vasiliadou IA, Berná A, Manchon C, Melero JA, Martinez F, Esteve-Nuñez A, Puyol D (2018) Biological and bioelectrochemical systems for hydrogen production and carbon fixation using purple phototrophic bacteria. Front Ener Res 6:107. https://doi.org/10.3389/fenrg.2018.00107

Vorholt JA, Hafenbradl D, Stetter KO, Thauer RK (1997) Pathways of autotrophic CO_2 fixation and of dissimilatory nitrate reduction to N_2O in *Ferroglobus placidus*. Arch Microbiol 167(1):19–23. https://doi.org/10.1007/s002030050411

Wood HG (1991) Life with CO or CO_2 and H_2 as a source of carbon and energy. FASEB 5(2):156–163. https://doi.org/10.1096/fasebj.5.2.1900793

Yang J, Kang Y, Sakurai K, Ohnishi K (2017) Fixation of carbon dioxide by chemoautotrophic bacteria in grassland soil under dark conditions. Acta Agr Scand, Section B-S P Sci 67(4):362–371. https://doi.org/10.1080/09064710.2017.1281433

Yeates TO, Kerfeld CA, Heinhorst S, Cannon GC, Shively JM (2008) Protein-based organelles in bacteria: carboxysomes and related microcompartments. Nat Rev Microbiol 6(9):681–691. https://doi.org/10.1038/nrmicro1913

Yoshizawa Y, Toyoda K, Arai H, Ishii M, Igarashi Y (2004) CO_2-responsive expression and gene organization of three ribulose-1,5-bisphosphate carboxylase/oxygenase enzymes and carboxysomes in Hydrogenovibrio marinus strain MH-110. J Bacteriol 186(17):5685–5691. https://doi.org/10.1128/JB.186.17.5685-5691.2004

Youssef NH, Farag IF, Rudy S, Mulliner A, Walker K, Caldwell F, Miller M, Hoff W, Elshahed M (2019) The Wood-Ljungdahl pathway as a key component of metabolic versatility in candidate phylum Bipolaricaulota (Acetothermia, OP1). Environ Microbiol Rep 11(4):538–547. https://doi.org/10.1111/1758-2229.12753

Zarzycki J, Fuchs G (2011) Coassimilation of organic substrates via the autotrophic 3-hydroxypropionate bi-cycle in Chloroflexus aurantiacus. Appl Environ Microbiol 77(17):6181–6188. https://doi.org/10.1128/AEM.00705-1

Zhang T, Shi XC, Ding R, Xu K, Tremblay PL (2020) The hidden chemolithoautotrophic metabolism of Geobacter sulfurreducens uncovered by adaptation to formate. ISME J 14(8):2078–2089. https://doi.org/10.1038/s41396-020-0673-8

Climate-Driven Shift in Plant-Microbe Dynamics: A Comprehensive Review

3

Niraj Singh

Abstract

The impact of climate change on numerous ecological processes, such as the interactions between microbes and plants, has become a major global concern. The health of the entire plant community, the cycling of nutrients, and the operation of ecosystems depend on these relationships. The dynamics and distribution of these interactions can shift in response to changes in the climate. Wide-ranging effects on ecosystem functions result from these changes in plant-microbe dynamics brought on by climate change. Plant productivity and nutrient acquisition are affected by changes in symbiotic connections, including mycorrhizal associations and nitrogen-fixing partnerships, which are a result of climate change. Threats to plant health and agricultural output might arise from pathogenic interactions that intensify in response to shifting environmental circumstances. Determining future changes in ecosystem dynamics and developing practical plans for sustainable agriculture and ecosystem management require a thorough examination of how climate change affects interactions between plants and microbes. Forecasting how ecosystems will react to climate change and creating plans to lessen its negative effects on plant health and the ecosystem depend on having a thorough understanding of the precise mechanisms underlying the changes in plant-microbe interactions brought about by climate change. The present review explores the complex interplay between plant-microbe interaction dynamics and climate change, providing insight into the processes via which changing weather patterns influence these relationships.

N. Singh (✉)
Department of Microbiology, The Assam Royal Global University, Guwahati, Assam, India

P. Pandey et al. (eds.), *Microorganisms Resilience to Climate Change*,
Microorganisms for Sustainability 56, https://doi.org/10.1007/978-981-96-3748-5_3

Keywords
Climate change · Plant · Microbe · Interactions · Dynamics · Microbial Community

3.1 Introduction

Climate change has significantly impacted all living organisms on Earth, including plants, animals, and microbes, as well as processes such as biogeographical cycles and food chains (Pecl et al. 2017; Rojas-Downing et al. 2017; Kumar and Verma 2018; Caminade et al. 2019; Kasperson et al. 2022). This pressing global issue has profound implications for the natural world, particularly in the interactions between plants and their associated microbes. Recently, the impact of climate change on these associations has received substantial attention due to their vital role in ecosystem health. Rapid changes in climate patterns, including increasing temperatures and extreme weather events, have already begun to affect plant-microbe dynamics. These alterations can significantly impact plant growth, development, and adaptability to changing environmental conditions. Additionally, climate change influences the growth behaviour of various crop species, particularly affecting nutrient uptake in plants. Understanding the mechanisms behind these impacts is crucial for predicting and mitigating potential disruptions in ecosystem stability and productivity.

Plants are intricately linked with diverse microbial communities, and their interaction is unavoidable and inseparable. It is well known that plants and soil host a wide array of microbiomes, that resides within plant tissues and on their external surface (Khondoker et al. 2020). Furthermore, soil ecosystem serves as a reservoir for microorganisms, while plants that grows in soil associates with those microorganisms present at different regions in the soil (Trivedi et al. 2020). The interaction of the plants with the microorganisms are collectively known as the "holobiont," a term that signifies the interaction of the plant with microbial communities around it thereby creating an ecosystem. These microbes occupy diverse locations and play key roles in various ecological functions in both above-ground and below-ground environments. They include prokaryotes protozoa as well as viruses, all of which are crucial in regulating resistance to climate change (Trivedi et al. 2020). Microbial communities inhabiting the soil, rhizosphere, roots, and other plant tissues establish complex and dynamic interactions with their host plants. These interactions, heavily influenced by the environment, can enhance plant resilience to environmental stresses (Trivedi et al. 2020; Naylor et al. 2017) Despite the recognized importance of the microbiome and its role in plant health, the tolerance of plants against adverse climatic change is yet to combat. A profound molecular knowledge of the interaction of plant with the microbial community is crucial for developing techniques to alleviate the effects of varying conditions of climate on ecosystem productivity and plant species richness. The constant shift in the climate has placed significant pressure on plant health, intensifying disturbances in the microbe based ecosystem. This has led to higher costs for management strategies to tackle the varying climatic change. The relationship between the host plant and the microbes depends on the surrounding ecosystem, and any disturbance in the factors of the environment affecting both the plants and their associated microbes. This book chapter explores how environmental parameters influence molecular interactions between plants and microbes and their consequences.

3.2 Climate-Driven Influence on Microbial Community

The impact of change in climatic conditions and global warming on ecosystems, including microbes, are substantial and concerning, although measuring these impacts precisely remains difficult (Guo et al. 2019). Changes in environmental factors such as precipitation patterns, drought stress, and radiation notably influence soil microbial diversity by raising surface soil temperatures and reducing moisture levels (Guo et al. 2018; Zhou et al. 2016).

The influence of climate change on microbial communities in terrestrial and aquatic environments is complex and sometimes contradictory, shaped by both deterministic and stochastic factors. Temperature changes, in particular, significantly affect microbial biodiversity at various levels, including geographical range, phenology, and distribution (Guo et al. 2018). As temperatures rise, microbial metabolism, population growth rates, and species numbers increase exponentially. Conversely, bacterial and fungal diversity typically decreases with increasing latitude as temperatures drop (Zhou et al. 2016). Temperature variations impact microbial biodiversity through several mechanisms. A commonly reported mechanism is that higher temperatures increase metabolic rates, leading to faster population doubling times (Zhou et al. 2016) and accelerated evolutionary and other related ecological processes such as mutation, speciation, and interactions like parasitism, predation, competition, and mutualism (Zhou et al. 2016; Guo et al. 2019; Yuan et al. 2021). As a result, warming ecosystems, from tropical to alpine forests, may become more diverse and active, potentially enhancing processes like decomposition, nutrient cycling, carbon sequestration, etc.(Zhou et al. 2016). Moreover, higher temperatures can also impose a selective filter kind to things which favor more adapted microorganisms and limit random species dispersal, which may decrease the temporal scaling rates of soil microbial communities. Additionally, microbial community responses to climate change vary greatly depending on microbial lineages and functional groups (Zhou et al. 2016; Yuan et al. 2021).

Higher temperatures also influence the plant species, which helps to boost up the nutrient allocation to below-ground ecosystems given the well-known correlation between microbial taxa richness and plant species diversity (Zhou et al. 2016). Further, this shift towards nutrient allocation affects the plant-microbe interactions, specifically, it led to the population alteration of plant growth-promoting rhizobacteria (PGPR) that rely on rhizodeposition (Singh et al. 2019; Sharma et al. 2022). Over and above, changes in precipitation patterns and climate change induced drought stress and influence negatively the formation of the mycelium of the mycorrhiza within the plant roots (Singh et al. 2019; Zhang and Xi 2021). In the context of plant disease, these climatic changes increase environmental stresses, which may induce the pathogenic nature of microorganisms and the redistribution of beneficial microbes across diverse ecological niches (Singh et al. 2019; Sharma et al. 2022).

Apart from the known significant impacts of climate change on microbial communities, composition, and their interactions, most of the documented research has focused on specific regions and conditions, particularly terrestrial ecosystems. Given the complex interplay between microorganisms and their surroundings, data obtained from single points in time provide only a snapshot of microbial communities, making them unsuitable and insufficient for the comprehensive ecosystem model simulations and interpretations. Further in-depth research is needed to develop an accurate understanding of this dynamic relationship.

3.3 Plant-Microbe Interactions in the Face of Climate Change

Plant health and ecosystem stability are impacted by plant-microbe interactions, which are greatly impacted by climate change. Increased greenhouse gas concentrations in the atmosphere, rising temperatures, and drought are some of the main causes of these changes. The microbial community and diversity in the rhizosphere are impacted by dry conditions brought on by climate change, which puts plants under more stress, increases their susceptibility to diseases, and changes the composition of their root exudate. The amount and quality of root exudates, which are essential for microbial compounds, can change as a result of elevated CO_2 levels in the atmosphere encouraging plant development and root biomass. Climate change can also throw off the timing of microbial activity and plant phenology. Inappropriate interactions could arise, for instance, if springtime temperatures are too early and plants begin to leaf out before helpful microorganisms become active.

Climate change significantly impacts plant-microbe interactions, affecting ecosystem stability and plant health (Table 3.1). Key factors driving these changes include rising temperatures, drought, and increased greenhouse gas levels in the atmosphere. Climate change-induced drought conditions increase plant stress, making them more susceptible to infections and altering root exudate composition, which in turn affects the microbial community and diversity in the rhizosphere. Elevated CO_2 levels in the atmosphere can promote plant growth and root biomass, altering the quantity and quality of root exudates, which are vital for microbial substances. Furthermore, climate change can disrupt the synchrony between plant phenology and microbial activity. For example, earlier spring temperatures might cause plants to leaf out before beneficial microbes become active, resulting in mismatched interactions. These disruptions can reduce plant resilience and productivity, ultimately impacting food security and ecosystem services.

Addressing these challenges requires integrated strategies to manage plant-microbe interactions in the face of climate change.

Table 3.1 Plant-microbe interaction studies under unfavourable climatic conditions

Sl. No	Unfavorable climatic condition	Plant-microbe interaction studies	References
1	Heat stress tolerance	Soybean—*Bacillus aryabhatti*	Park et al. (2017)
2	Heat stress	Arabidopsis—Cyanobacteria	Chua et al. (2020)
3	Low-temperature stress	Cucumber—Seaweed extract	Sarhan et al. (2014)
4	Drought stress	Maize—Azotobacter	Shirinbayan et al. (2019)
5	Drought tolerance	Pepper—Bacillus subtilis	Vigani et al. (2019)
6	Chilling stress tolerance	Tomato—*Pseudomonas vancouverensis*	Subramanian et al. (2015)
7	Salt stress	Tomato—Microalgae/cyanobacteria	Mutale-Joan et al. (2021)
8	Salts tress	Redpepper—*Pseudomonas frederiksbergensis* OS261	Chatterjee et al. (2017)
9	Nutrient stress tolerance	Chickpea—Serratia species	Zaheer et al. (2016)
10	Flooding	Rice—*Pseudomonas flurorescence*	Etesami et al. (2014)

3.3.1 Influence of High Temperature on Plant-Microbe Interactions

The general health and morphology of plants are greatly impacted by rising air temperatures. Studies have indicated that high temperatures have an impact on a range of crops and decorative plants (Chen et al. 2017, 2021). Higher temperatures also have an impact on the makeup, activities, and interactions of microorganisms with plants. Increased temperature has the potential to change respiration rates and hasten microbial development. It has been discovered that rising temperatures and global warming impact the respiration rates of microbial communities residing in the rhizospheric area (Classen et al. 2015). Researchers has studied the temperature rise which boost the rate of respiration by the soil microbial communities significantly (Karhu et al. 2014). Furthermore, Frey et al. (2013) indicated that temperature affects how quickly soil bacteria break down organic materials. Additionally, it shows that cyanobacterial growth and development are significantly impacted by elevated temperature conditions (Frey et al. 2013; Deng et al. 2022).

In addition, higher temperatures have been shown to affect the incidence of illness. For example, *Pectobacterium atrosepticum* causes soft rot disease, and specific bacteria that are transferred from seeds and encourage the breakdown of cell wall enzymes cause the disease state in plants (Hasegawa et al. 2005). The loss of ETI (Effector-Triggered Immunity) due to global climate change, particularly elevated temperatures, is a serious concern as the flora of the environment depends on ETI for their maturation as well as for disease protection (Cheng et al. 2019; Velásquez et al. 2018). It also said that brief exposure to high temperatures alters the gene expression of the ETI route. In the published research paper of Huot et al. (2017), it was reported that the elevation in temperature supresses the salicylic acid mechanism in plants, leading to an increased vulnerability to infection by *Pseudomonas syringae.* High temperatures alters the expression of genes involved in the defence mechanism (Zhang et al. 2018; Wang et al. 2020). Oxid stress developed because of the overproduction of reactive and non-reactive oxygen species brought on by the declining rate of ETI, elevated the rate of photorespiration. To overcome the pressure brought on by high temperatures, plants therefore upregulate a variety of enzymatic and non-enzymatic antioxidant production processes (Awasthi et al. 2015).

Over and above, Microbes are also greatly impacted by global climate change because they require a specific period of temperature for development and growth If microorganisms don't promptly adjust to temperature changes, they become inactive or inhibited beyond the optimal temperature range, which suppresses their capabilities to promote plant growth (Velásquez et al. 2018). To offset the detrimental effects of high temperatures on plants, certain microorganisms, such as endophytic bacteria and arbuscular mycorrhizal fungus, alter their morphology. Bentonite seaweed (B. cereus SA1), a thermotolerant bacterial strain, increases the biomass and chlorophyll content of soybean plants under high-temperature stress by producing components including gibberellin, salicylic acid, and indole-3-acetic acid. *Klebsiella sp.*, a facultative thermophile with a temperature range of 18–65 °C, may simultaneously secrete substances that promote plant growth and adapt to fluctuating temperatures. According to (Mukherjee et al. 2018), the inoculated *Oryza sativa* plants outgrew the control plants by two times their length and eighteen times their total grain mass yield. Mycorrhizal fungi also experienced a change in hyphal structure as temperatures rose, going from a higher number of vesicles at low temperatures to an increased presence of hyphae at elevated temperatures. (Hawkes et al. 2008). Bilal et al. (2020) demonstrated that augmentation of *Paecilomyces formosus* LHL10 and *Penicillium funiculosum* LHL06 enhanced the tolerance level of plants in presence of multiple stress condition. The increase in the uptake of micronutrients, decreased lipid peroxidation, increased photosynthetic activity, and increased antioxidant activity in plants Better resistance and carbon allocation to the rhizosphere are the outcomes of these (Compant et al. 2010; Heinemeyer et al. 2006). The microbial population surrounding a plant may be able to mitigate the harmful impacts of extreme heat caused by change in climatic condition on crops by broadening the growth rate of plant at different extents of temperatures. As for example, both tropical grass (*Dichanthelium lanuginosum*) and the fungi (*Curvularia protuberate*) thrive symbiotically with each other at extreme heat conditions (Márquez et al. 2007). According to Rodriguez et al. (2008), tomato plants that had the *C. protuberate* fungus in the soil showed a similar pattern of heat tolerance, indicating the significance of microbial contact with plants to improve tolerance (Rodriguez et al. 2008). In the meantime, certain microbes assist plants in adjusting to a variety of stressors. For example, *Burkholderia phytofirmans* PsJN strain has demonstrated tolerance against heat, salinity, cold temperature, drought as well as other stressors (Miotto-Vilanova et al. 2016; Issa et al. 2018).

3.3.2 Influence of Greenhouse Gases on Plant-Microbe Interactions

The greenhouse gases (GHGs) that absorb infrared radiation reflected from the earth's surface include carbon dioxide, methane, nitrogen oxides, and ozone. In addition, the ozone absorbs Ultraviolet radiation and prevent it from causing harmful impact on the biosphere (Cassia et al. 2018). Under optimal greenhouse gas concentrations, the plant intake the gases and these harmful gases get incorporated into the plant's biochemical cycle, hence converting GHGs into biomolecules. This leads to the production of root exudates and plant sugars that benefit the rhizopheric bacteria, enhancing the association between plant and the bacteria (Weyens et al. 2015). The high GHG levels causes various natural calamities which creates negative impact on crops and microbes as well as the agronomic sector. Moreover, high GHG concentrations have serious repercussions for the flora and the related microorganisms (Cassia et al. 2018).

3.3.2.1 Carbon-di-Oxide (CO_2)

According to (Cassia et al. 2018), there are serious consequences for plants and related microbes from oxidative stress and ROS generation caused by rising GHG levels and climate change. Carbon dioxide accounts for 76% of global GHG emissions, and high CO_2 levels can affect how plants and microbes interact and how well they can withstand stress. Research has shown that high CO_2 levels are directly related to increased the generation of biomass, uptake of nutrition, rate of photosynthesis and concentration of carbohydrates in plants. Increased photosynthesis efficiency results in high accumulation of the photosynthetic product in the plant root, which can result in branched roots and change the composition of root exudate. Additionally, numerous research has indicated an increase in biomass content of soil microbes under elevated pressure of $CO_{2.}$ The atmospheric CO_2 can break down the moisture retained within soil particles where it is absorbed by microbes, thereby influencing their adaption to stress. These circumstances boosted the microbial community in the soil by selecting the microorganisms from the classes Alphaproteobacteria, Gamma Proteobacteria, and Bacteroidia that are resistant to metals. (Luo et al. 2020). Additionally, increased CO_2 may have an impact on soil microbes' capacity for biodegradation because of a drop in nitrogen brought on by plant biomass production and decreased ratio of carbon-to-nitrogen (Singh et al. 2019).

3.3.2.2 Nitrogen Oxides (NOx)

The forms of nitrogen oxides that contribute to greenhouse gas emissions are nitrous oxide ($_{O}$), nitric oxide (N_2O), and nitrogen dioxide (NO_2). Plants absorb NOx from the atmosphere through the interaction of rhizobia with the plant. These oxides serve as plant nutrition by acting as the building blocks of the compounds such as amines and amino acids during photochemical processes via the nitrate uptake mechanism (Weyens et al. 2015). Increased NOx in the atmosphere and plant stress brought on by climate change may cause an excess of reactive nitrogen species

(RNS) to be generated, which could result in oxidative stress that is harmful to the growth and photosynthetic activity of the plant (Cassia et al. 2018). Additionally, RNS serve as signalling molecules that advance the relationship between bacteria and legumes to promote the nodulation process in the roots and regulate the expression of genes responsible for various nitrogen fixation processes. (Signorelli et al. 2020). Additionally, NOx can be absorbed by microbes associated with plants as nitrite or nitrate, which is then used as a source of nitrogen for the growth and maintenance of microbial cells (Singh et al. 2019).

3.3.2.3 Methane (CH_4)

Twenty percent of global greenhouse gas emissions are attributed to methane, which is 25 times more effective in retaining heat than carbon dioxide (USEPA, 2022). Methanotrophic bacteria linked to plants convert ambient CH_4 to CO_2 and then enter the photosynthetic channel. According to (Stępniewska et al. 2018), methanotrophs that colonize *Sphagnum sp.* include *Methylocella palustris* and *Methylocapsa acidiphila*. Furthermore, a variety of compounds, including ectoine, glutamate, 5-oxoproline, methanol sucrose and lipids can be synthesized from CH_4 by several methanotrophs from the genera *Methylocystis*, *Methylomicrobium*, *Methylococcus*, and *Methylomonas*. According to (Sahoo et al. 2021), these bioactive compounds can help plants and related microorganisms adapt to the stresses brought on by climate change. Therefore, enhanced microbial stress tolerance results in enhanced plant tolerance to climate change impact.

3.3.3 Impact of Drought on Plant-Microbe Dynamics

High frequency fluctuation in precipitation patterns that affect soil and air moisture content and cause drought or flooding is one of the observable effects of climate change. Drought stress due to insufficient or absent precipitation is more dangerous for food security since different exposure levels have variable amplitude effects on crop productivity (Gupta et al. 2022). Water scarcity also impacts the availability and uptake of nutrients by roots, as well as their movement to different areas (Gupta et al. 2022). Furthermore, according to (Kumar and Verma 2018), drought stress impacts the integrity of cell walls, generates reactive oxygen species (ROS), encourages premature senescence of leaves, enhances the formation of ethylene, lowers the level of chlorophyll, and photosynthetic efficiency. Additionally, the salinity stress in the rhizosphere is also one of the impact (Gupta et al. 2022). Microbial diversity, cellular activities, and soil properties are all impacted by water scarcity or lack thereof (Manzanera 2021). To increase the bioavailability of water nearby the microbe, pseudomonads additionally secrete alginate and hygroscopic biosurfactants (Stevenson et al. 2015). After being subjected to drought stress, plant-associated microorganisms start their defence mechanism and restore growth. According to (Al-Karaki et al. 2004; He et al. 2021; Rawal et al. 2022), and other studies, the microorganisms releases compounds which enhances the normal growth of related crops and their reaction of stress by elevating the root/shoot ratio, enhancing the water absorption capacity and boosting the uptake of nutrients, along with the yield thereby alleviating drought impact. Furthermore, the root exudates provide further elements that affect these microorganisms' stress response (Gupta et al. 2022). (Melotto et al. 2006) reported that Arabidopsis closes its stomata to restrict microbe ingress when it detects the presence of bacteria like *P. syringae*, which have f lg22 as PAMPs. According to (Jiang et al. 2010), plants that have higher levels of ABA also have lower ETI and tolerance because it represses the salicylic acid pathway in plants after infection.(Xu et al. 2018). Because of the lower transcription of cell wall gene WAK16 and WAK3-2, that shows kinase activity a weakened immune system in ginger plants is noticed. In addition, *Ralstonia solanacearum* is found to play a significant role in determining the content of moisture present in the soil (Jiang et al. 2010). In Glycine max L, the production of biofilm by rhizobacteria on plant root and the release of secondary metabolites from the plant lessened the severe effects of drought stress while boosting the fresh weight of the roots and shoots (Braga et al. 2022). Adzuki bean plants' capacity to withstand drought was significantly altered by indole acetic acid that *Rhodobacter sphaeroides* KE149 drained into the rhizosphere. Because drought reduces the amount of iron and phytosiderophores available in the rhizosphere, it makes Actinobacteria more abundant. Actinobacteria can flourish in low-iron environments, which increases their fitness advantage and capacity to promote plant growth. In order to interact with insects, nematodes, and microorganisms, plants use volatile organic compounds (VOCs)(Weisskopf et al. 2021). The increase in greenhouse gas emissions due to climate change results in alteration in plant immunity or stress response pathways within the host, which may be linked to changes in the microbiome composition mediated by root exudate during drought and warmth. To enhance or maintain the current agroeconomic situation, these discoveries also need to be in line with changing climatic patterns.

3.4 Plant-Pathogen Interactions in a Changing Climate

Tripartite interactions between the environment, host, and pathogen significantly influence plant health and are key determinants of a plant's resistance to disease. Climate change and its associated factors can change pathogen prevalence and actions, impacting relation of the host and the pathogen and potentially leading to onset of novel pathogen (Cohen and Leach 2020). As global temperatures rise, it is predicted that the number of plant pathogens will increase (Delgado-Baquerizo et al. 2020), resulting in a higher incidence of plant diseases. Additionally, many widely used disease management approaches have been ineffective at extreme heat (Burdon and Zhan 2020). Disease causing agent may also develop contemporary approaches to infect plants by modifying their virulence systems, potentially overcoming resistance conferred by R genes. Increase in adverse environmental conditions have been observed to affect effector-triggered immunity (ETI), leading to infections in various plant ecosystems (Cheng et al. 2019). Numerous research on the effects of climate change on plant-pathogen dynamics relies on simplified models involving a single host plant and pathogen. But in innate environments plants engage with a wide variety of disease causing microorganisms. (Bartoli et al. 2018), and development of infectious agent relies on the collaboration or contestation among the microbiota present in plant and the pathogens. At present, there is no precise knowledge of the relationship between pathogenic microbes and plant associated microorganisms will be affected by prolonged climate change exposure.

3.5 Cutting-Edge Methods for Decoding Plant-Microbe Interactions

Microbial populations are essential for aiding the acclimitization of the host in accordance with the varying climatic conditions (Trivedi et al. 2022). Consequently, developing effective techniques to study the interactions of the plant with the microorganisms is crucial to understand the shift in climatic conditions. Currently, metagenomics is considered a state-of-the-art method for determining microbial diversity in soil, where plant and microbe interactions occur. This approach involves direct soil sampling and subsequent nucleotide/DNA sequencing analysis (Castro-Moretti et al. 2020; Fadiji et al. 2020). Numerous reports have examined the association of nematodes with soybean roots, nematode interactions with rice roots, and the interaction of the bacterium *Ralstonia solanacearum* with solanaceous crop roots using metagenomics (Hu et al. 2017; Kang et al. 2018; Zeiss et al. 2019).

Next Generation Sequencing (NGS) facilitates the analysis of a wide range of soil samples to identify various microbes and their potential associations with plants. Further studies using transcriptomics and metabolomics can deepen our understanding of these interactions (Sogin et al. 2006; Caporaso et al. 2012). For instance, shotgun sequencing of *Brevibacterium frigoritolerans* interaction with *Zea mays* with subject to different stress condition disclosed proteins that plays vital role against drought and salinity and increased the growth and development of the plant (Zhang et al. 2019). Additionally, whole genome studies aid in apprehending the numerous roles of genomic content present in the microbe, enabling researchers to create artificial environments that support plant health in presence of adverse circumstances (Brockhurst et al. 2019; Costa et al. 2020; Golicz et al. 2020). RNA sequencing is predominantly used to study plant-microbe associations under specific conditions by predicting the importance of each gene that plays a role in tolerance of plant under stress (Mirzaee et al. 2015; Yi et al. 2017). For example, RNA profiling of *Bacillus subtilis* revealed that *B. subtilis* affects cucumber roots by differentially expressing genes involved in signaling pathways (LRR, PR-4, ARG7, auxin response genes) under biotic stress (Samaras et al. 2022). Techniques such as GC/MS and LC/MS (Gas Chromatography and Liquid Chromatography-Mass Spectrometry) are also commonly used to identify and characterize chemical components released during plant-microbe interactions, such as flavonoids, polyamines, and spermidine (Rauha et al. 2000; Zeiss et al. 2019; Macoy et al. 2015). In context of microbiome, the proteomic approach, which investigates bacterial communities, provides detailed data on stress-related plant proteins responses (Novello et al.

2017; Bona et al. 2019). In addition to OMICS-based approaches, imaging techniques like fluorescence microscopy, X-ray crystallography, microscopic methods and nanoscale secondary ion mass spectroscopy (NanoSIMS) analysis are employed to gather substantial information on plant-microbe interactions under climatic changes (Vos et al. 2013; Young and Crawford 2004; Steffens et al. 2017).

3.6 Eco-Evolutionary Dynamics: Boosting Plant Resilience in a Changing Climate

Climate change can significantly influence the ecological interactions between the plant and its related microorganisms. Research indicates that drought-induced hormonal signalling and altered gene expression have shifted virus-host transitions from harmful to beneficial interactions (González et al. 2021). Additionally, adaptation encourages microrganisms to develop greater coordinate with the specific genotypes of their local hosts. (Batstone et al. 2020). The ecological and evolutionary relationships between host plants and their microbial assemblage are essential for enhancing plant resilience to climate change. Although plant environments significantly influence the microbial community, environmental dynamics and adaptive processes also has a critical role. According to Gao et al. (2020), arid conditions can reduce the host's impact on the composition of fungal organisms linked to host plants. Microbial species' adaptive relationships under stress show that microbes tend to be more competitive under moderate stress but more cooperative under severe conditions. Short-term plant resilience and productivity under climate change is governed by these microbiomes associated with the host. Moreover, recent studies suggest that the same plant characteristics can be obtained by either modifying the plant genome or adjusting its microbiome. (Ravanbakhsh et al. 2021). Extreme weather events such as droughts and heat waves are predicted to become more frequent and intense. Microbiomes adapt more rapidly to these events than plants, possibly offering instant resistance and adaptability to plants in response to climate change. Over time, plant communities are likely to reach comparable degrees of ecology and evolution-based adjustment. It is important to acknowledge that there are various routes for plant-microbe acclimatization to environmental changes.

3.7 Conclusion and Prospect

Research in this area is still ongoing, but it is now widely known that there are important factors that affect plant microbiomes and their changes in the environment. There is still much to learn about the behaviour and microbial communities in soil and plant surfaces as well as many factors influencing the plants. A global worry, climate change also brings with it a rise in climatic adversity. As a result of these changes to the current climatic patterns, which have a negative impact on plant and microbial growth, food security has become a global issue. The detrimental effects of climate change on the structure and function of microbes in ecological niches are also a topic of discussion. It is necessary to look into the extent to which we may modify the plant microbiome to boost sustainable agricultural productivity while preserving the environment. Host genetic factors have recently been shown to facilitate the research of microbiome diversity and structure, which may aid in the selection of microbiomes based on features. The relation between microbial populations and environmental processes can be linked through the use of advanced integrated innovative molecular methodologies such as ecological models, meta-omics, and bioinformatics, which are necessary to explore a deeper understanding of the host-specific plant-microbe interaction.

References

Al-Karaki G, McMichael B, Zak J (2004) Field response of wheat to arbuscular mycorrhizal fungi and drought stress. Mycorrhiza 14:263–269

Awasthi R, Bhandari K, Nayyar H (2015) Temperature stress and redox homeostasis in agricultural crops. Front Environ Sci 3:11

Bartoli C, Frachon L, Barret M, Rigal M, Huard-Chauveau C, Mayjonade B, Zanchetta C, Bouchez O, Roby D, Carrère S (2018) In situ relationships between microbiota and potential pathobiota in Arabidopsis Thaliana. ISME J 12(8):2024–2038

Batstone RT, O'Brien AM, Harrison TL, Frederickson ME (2020) Experimental evolution makes microbes more cooperative with their local host genotype. Science 370(6515):476–478

Bilal S, Shahzad R, Imran M, Jan R, Kim KM, Lee I-J (2020) Synergistic association of endophytic fungi enhances Glycine max L. resilience to combined abiotic stresses: heavy metals, high temperature and drought stress. Environ Sci Biol 143:111931

Bona E, Massa N, Novello G, Boatti L, Cesaro P, Todeschini V, Magnelli V, Manfredi M, Marengo E (2019) Metaproteomic characterization of Vitis vinifera rhizosphere. FEMS Microbiol Ecol 95(1):fiy204

Braga APA, Cruz JM, de Melo IS (2022) Rhizobacteria from Brazilian semiarid biome as growth promoters of soybean (Glycine max L.) under low water availability. Braz J Microbiol 53(2):873–883

Brockhurst MA, Harrison E, Hall JP, Richards T, McNally A, MacLean CJ (2019) The ecology and evolution of pangenomes. Curr Biol 29(20):R1094–R1103

Burdon JJ, Zhan JJ (2020) Climate change and disease in plant communities. PLoS Biol 18(11):e3000949

Caminade C, McIntyre KM, Jones AE (2019) Impact of recent and future climate change on vector-borne diseases. Ann N Y Acad Sci 1436(1):157–173

Caporaso JG, Lauber CL, Walters WA, Berg-Lyons D, Huntley J, Fierer N, Owens SM, Betley J, Fraser L, Bauer M (2012) Ultra-high-throughput microbial community analysis on the illumina HiSeq and MiSeq platforms. ISME J 6(8):1621–1624

Cassia R, Nocioni M, Correa-Aragunde N, Lamattina LJ (2018) Climate change and the impact of greenhouse gasses: CO2 and NO, friends and foes of plant oxidative stress. Front Plant Sci 9:273

Castro-Moretti FR, Gentzel IN, Mackey D, Alonso AP (2020) Metabolomics as an emerging tool for the study of plant–pathogen interactions. Metabolites 10(2):52

Chatterjee P, Samaddar S, Anandham R, Kang Y, Kim K, Selvakumar G, Sa TJ (2017) Beneficial soil bacterium Pseudomonas frederiksbergensis OS261 augments salt tolerance and promotes red pepper plant growth. Front Sci Plant 8:705

Chen C, Xin K, Liu H, Cheng J, Shen X, Wang Y, Zhang LJ (2017) Pantoea alhagi, a novel endophytic bacterium with ability to improve growth and drought tolerance in wheat. Sci Rep 7(1):41564

Chen J, Liu Q, Yu L, Korpelainen H, Niinemets Ü, Li C (2021) Elevated temperature and CO2 interactively modulate sexual competition and ecophysiological responses of dioecious Populus cathayana. Forest Ecol Manag 481:118747

Cheng YT, Zhang L, He SY (2019) Plant-microbe interactions facing environmental challenge. Cell Host Microbe 26:183–192

Chua A, Sherwood OL, Fitzhenry L, Ng CK-Y, McCabe PF, Daly CT (2020) Cyanobacteria-derived proline increases stress tolerance in Arabidopsis thaliana root hairs by suppressing programmed cell death. Front Plant Sci 11:490075

Classen AT, Sundqvist MK, Henning JA, Newman GS, Moore JA, Cregger MA, Moorhead LC, Patterson CM (2015) Direct and indirect effects of climate change on soil microbial and soil microbial-plant interactions: What lies ahead? Ecosphere 6(8):1–21

Cohen SP, Leach JE (2020) High temperature-induced plant disease susceptibility: more than the sum of its parts. Curr Opin Plant Biol 56:235–241

Compant S, Van Der Heijden MG, Sessitsch AJ (2010) Climate change effects on beneficial plant–microorganism interactions. FEMS Microbiol Ecol 73(2):197–214

Costa SS, Guimarães LC, Silva A, Soares SC, Baraúna RA (2020) First steps in the analysis of prokaryotic pan-genomes. Bioinform Biol Insights 14:1177932220938064

Delgado-Baquerizo M, Guerra CA, Cano-Díaz C, Egidi E, Wang J-T, Eisenhauer N, Singh BK, Maestre FT (2020) The proportion of soil-borne pathogens increases with warming at the global scale. Nat Clim Chang 10(6):550–554

Deng L, Cheung S, Kang CK, Liu K, Xia X, Liu H (2022) Elevated temperature relieves phosphorus limitation of marine unicellular diazotrophic cyanobacteria. Limnol Oceanogr 67(1):122–134

Etesami H, Mirsyed Hosseini H, Alikhani HA (2014) In planta selection of plant growth promoting endophytic bacteria for rice (Oryza sativa L.). J Soil Sci Plant Nutr 14(2):491–503

Fadiji AE, Babalola OO (2020) Elucidating mechanisms of endophytes used in plant protection and other bioactivities with multifunctional prospects. Front Bioeng Biotechnol 8:467

Frey SD, Lee J, Melillo JM, Six J (2013) The temperature response of soil microbial efficiency and its feedback to climate. Nat Clim Chang 3(4):395–398

Gao C, Montoya L, Xu L, Madera M, Hollingsworth J, Purdom E, Singan V, Vogel J, Hutmacher RB, Dahlberg JA (2020) Fungal community assembly in drought-stressed sorghum shows stochasticity, selection, and universal ecological dynamics. Nat Commun 11(1):34

Golicz AA, Bayer PE, Bhalla PL, Batley J, Edwards D (2020) Pangenomics comes of age: from bacteria to plant and animal applications. Trends Genet 36(2):132–145

González R, Butković A, Escaray FJ, Martínez-Latorre J, Melero Í, Pérez-Parets E, Gómez-Cadenas A, Carrasco P, Elena SF (2021) Plant virus evolution under strong drought conditions results in a transition from parasitism to mutualism. Proc Natl Acad Sci U S A 118(6):e2020990118

Guo X, Feng J, Shi Z, Zhou X, Yuan M, Tao X, Hale L, Yuan T, Wang J, Qin Y (2018) Climate warming leads to divergent succession of grassland microbial communities. Nat Clim Change 8(9):813–818

Guo X, Zhou X, Hale L, Yuan M, Ning D, Feng J, Shi Z, Li Z, Feng B, Gao Q (2019) Climate warming accelerates temporal scaling of grassland soil microbial biodiversity. Nat Ecol Evol 3(4):612–619

Gupta A, Bano A, Rai S, Mishra R, Singh M, Sharma S, Pathak N (2022) Mechanistic insights of plant-microbe interaction towards drought and salinity stress in plants for enhancing the agriculture productivity. Int J Mol Sci 4:100073

Hasegawa H, Chatterjee A, Cui Y (2005) Elevated temperature enhances virulence of Erwinia carotovora subsp. carotovora strain EC153 to plants and stimulates production of the quorum sensing signal, N-acyl homoserine lactone, and extracellular proteins. Appl Environ Microbiol 71(8):4655–4663

Hawkes CV, Hartley IP, Ineson P, Fitter AH (2008) Soil temperature affects carbon allocation within arbuscular mycorrhizal networks and carbon transport from plant to fungus. Glob Change Biol 14(5):1181–1190

He A, Niu S, Yang D, Ren W, Zhao L, Sun Y, Meng L, Zhao Q, Paré PW, Zhang J (2021) Two PGPR strains from the rhizosphere of Haloxylon ammodendron promoted growth and enhanced drought tolerance of ryegrass. Plant Physiol Biochem 161:74–85

Heinemeyer A, Ineson P, Ostle N, Fitter A (2006) Respiration of the external mycelium in the arbuscular mycorrhizal symbiosis shows strong dependence on recent photosynthates and acclimation to temperature. New Phytol 171(1):159–170

Hu Y, You J, Li C, Williamson VM, Wang C (2017) Ethylene response pathway modulates attractiveness of plant roots to soybean cyst nematode Heterodera glycines. Sci Rep 7(1):41282

Huot B, Castroverde CDM, Velásquez AC, Hubbard E, Pulman JA, Yao J, Childs KL, Tsuda K, Montgomery BL, He SY (2017) Dual impact of elevated temperature on plant defence and bacterial virulence in Arabidopsis. Nat Commun 8(1):1808

Issa A, Esmaeel Q, Sanchez L, Courteaux B, Guise J-F, Gibon Y, Ballias P, Clément C, Jacquard C, Vaillant-Gaveau N (2018) Impacts of Paraburkholderia phytofirmans strain PsJN on tomato (Lycopersicon esculentum L.) under high temperature. Front Plant Sci 9:1397

Jiang C-J, Shimono M, Sugano S, Kojima M, Yazawa K, Yoshida R, Inoue H, Hayashi N, Sakakibara H, Takatsuji H (2010) Abscisic acid interacts antagonistically with salicylic acid signaling pathway in rice–Magnaporthe grisea interaction. Mol Plant Microbe Interact 23(6):791–798

Kang W, Zhu X, Wang Y, Chen L, Duan Y (2018) Transcriptomic and metabolomic analyses reveal that bacteria promote plant defense during infection of soybean cyst nematode in soybean. BMC Plant Biol 18:1–14

Karhu K, Auffret MD, Dungait JA, Hopkins DW, Prosser JI, Singh BK, Subke J-A, Wookey PA, Ågren GI, Sebastià M-T (2014) Temperature sensitivity of soil respiration rates enhanced by microbial community response. Nature 513(7516):81–84

Kasperson JX, Kasperson RE, Turner B, Hsieh W, Schiller A (2022) Vulnerability to global environmental change. In: Social contours of risk. Routledge, pp 245–285

Khondoker NA, Uddin FJ, Sarker MAR, Rahman A (2020) Influence of nitrogen and phosphorus level for the performance of French Bean (Phaseolus Vulgaris L.). ASM 4(1):34–38

Kumar A, Verma J (2018) Does plant—microbe interaction confer stress tolerance in plants: a review? Microbiol Res 207:41–52

Luo J, Song Y, Liang J, Li J, Islam E, Li T (2020) Elevated CO2 mitigates the negative effect of CeO2 and Cr2O3 nanoparticles on soil bacterial communities by alteration of microbial carbon use. Environ Pollut 263:114456

Macoy DM, Kim W-Y, Lee SY, Kim MG (2015) Biotic stress related functions of hydroxycinnamic acid amide in plants. J Plant Biol 58:156–163

Manzanera M (2021) Dealing with water stress and microbial preservation. Environ Microbiol 23(7):3351–3359

Márquez LM, Redman RS, Rodriguez RJ, Roossinck MJ (2007) A virus in a fungus in a plant: three-way symbiosis required for thermal tolerance. Science 315(5811):513–515

Melotto M, Underwood W, Koczan J, Nomura K, He SY (2006) Plant stomata function in innate immunity against bacterial invasion. Cell 126(5):969–980

Miotto-Vilanova L, Jacquard C, Courteaux B, Wortham L, Michel J, Clément C, Barka EA, Sanchez L (2016) Burkholderia phytofirmans PsJN confers grapevine resistance against Botrytis cinerea via a direct antimicrobial effect combined with a better resource mobilization. Front Plant Sci 7:1236

Mirzaee H, Shuey L, Schenk PM (2015) Transcriptomics of plants interacting with pathogens and beneficial microbes. Wiley, pp 525–536

Mukherjee PK, Hurley JF, Taylor JT, Puckhaber L, Lehner S, Druzhinina I, Schumacher R, Kenerley CM (2018) Ferricrocin, the intracellular siderophore of Trichoderma virens, is involved in growth, conidiation, gliotoxin biosynthesis and induction of systemic resistance in maize. Biochem Biophys Res Commun 505(2):606–611

Mutale-Joan C, Rachidi F, Mohamed HA, Mernissi NE, Aasfar A, Barakate M, Mohammed D, Sbabou L, Arroussi HE (2021) Microalgae-cyanobacteria–based biostimulant effect on salinity tolerance mechanisms, nutrient uptake, and tomato plant growth under salt stress. J Appl Phycol 33:3779–3795

Naylor D, DeGraaf S, Purdom E, Coleman-Derr D (2017) Drought and host selection influence bacterial community dynamics in the grass root microbiome. ISME J 11(12):2691–2704

Novello G, Gamalero E, Bona E, Boatti L, Mignone F, Massa N, Cesaro P, Lingua G, Berta G (2017) The rhizosphere bacterial microbiota of Vitis vinifera cv. Pinot Noir in an integrated pest management vineyard. Front Microbiol 8:1528

Park Y-G, Mun B-G, Kang S-M, Hussain A, Shahzad R, Seo C-W, Kim A-Y, Lee S-U, Oh KY, Lee DY (2017) Bacillus aryabhattai SRB02 tolerates oxidative and nitrosative stress and promotes the growth of soybean by modulating the production of phytohormones. PLoS One 12(3):e0173203

Pecl GT, Araújo MB, Bell JD, Blanchard J, Bonebrake TC, Chen I-C, Clark TD, Colwell RK, Danielsen F, Evengård B (2017) Biodiversity redistribution under climate change: Impacts on ecosystems and human well-being. Science 355(6332):eaai9214

Rauha J-P, Remes S, Heinonen M, Hopia A, Kähkönen M, Kujala T, Pihlaja K, Vuorela H, Vuorela P (2000) Antimicrobial effects of Finnish plant extracts containing flavonoids and other phenolic compounds. Int J Food Microbiol 56(1):3–12

Ravanbakhsh M, Kowalchuk GA, Jousset A (2021) Targeted plant hologenome editing for plant trait enhancement. New Phytol 229(2):1067–1077

Rawal R, Scheerens JC, Fenstemaker SM, Francis DM, Miller SA, Benitez M-S (2022) Novel Trichoderma isolates alleviate water deficit stress in susceptible tomato genotypes. Front Plant Sci 13:869090

Rodriguez RJ, Henson J, Van Volkenburgh E, Hoy M, Wright L, Beckwith F, Kim Y-O, Redman RS (2008) Stress tolerance in plants via habitat-adapted symbiosis. ISME J 2(4):404–416

Rojas-Downing MM, Nejadhashemi AP, Harrigan T, Woznicki SA (2017) Climate change and livestock: Impacts, adaptation, and mitigation. Clim Risk Manag 16:145–163

Sahoo KK, Goswami G, Das D (2021) Biotransformation of methane and carbon dioxide into high-value products by methanotrophs: current state of art and future prospects. Front Microbiol 12:636486

Samaras A, Kamou N, Tzelepis G, Karamanoli K, Menkissoglu-Spiroudi U, Karaoglanidis GS (2022) Root transcriptional and metabolic dynamics induced by the plant growth promoting rhizobacterium (PGPR) Bacillus subtilis Mbi600 on cucumber plants. Plants (Basel) 11(9):1218

Sarhan TZ, Ismael SF (2014) Effect of low temperature and seaweed extracts on flowering and yield of two cucumber cultivars (Cucumis sativus L.). Agric Food Sci 3(1)

Sharma B, Singh BN, Dwivedi P, Rajawat MVS (2022) Interference of climate change on plant-microbe interaction: present and future prospects. Front Agron 3:725804

Shirinbayan S, Khosravi H, Malakouti M (2019) Alleviation of drought stress in maize (Zea mays) by inoculation with Azotobacter strains isolated from semi-arid regions. Appl Soil Ecol 133:138–145

Signorelli S, Sainz M, Tabares-da Rosa S, Monza J (2020) The role of nitric oxide in nitrogen fixation by legumes. Front Plant Sci 11:521

Singh VK, Shukla AK, Singh AK (2019) Impact of climate change on plant–microbe interactions under agroecosystems. Climate change and agricultural ecosystems. Elsevier, pp 153–179

Sogin ML, Morrison HG, Huber JA, Welch DM, Huse SM, Neal PR, Arrieta JM, Herndl G (2006) Microbial diversity in the deep sea and the underexplored 'rare biosphere'. Proc Natl Acad Sci U S A 103(32):12115–12120

Steffens M, Rogge DM, Mueller CW, Höschen C, Lugmeier J, Kölbl A, Kögel-Knabner I (2017) Identification of distinct functional microstructural domains controlling C storage in soil. Environ Sci Technol 51(21):12182–12189

Stępniewska Z, Goraj W, Kuźniar A, Szafranek-Nakonieczna A, Banach A, Górski A, Pytlak A, Urban D (2018) Methane oxidation by endophytic bacteria inhabiting Sphagnum sp. and some vascular plants. Wetlands 38:411–422

Stevenson A, Burkhardt J, Cockell CS, Cray JA, Dijksterhuis J, Fox-Powell M, Kee TP, Kminek G, McGenity TJ, Timmis KN (2015) Multiplication of microbes below 0690 water activity: implications for terrestrial and extraterrestrial life. Environ Microbiol 17(2):257–277

Subramanian P, Mageswari A, Kim K, Lee Y, Sa T (2015) Psychrotolerant endophytic Pseudomonas sp. strains OB155 and OS261 induced chilling resistance in tomato plants (Solanum lycopersicum Mill.) by activation of their antioxidant capacity. Mol Plant Microbe Interact 28(10):1073–1081

Trivedi P, Leach JE, Tringe SG, Sa T, Singh BK (2020) Plant–microbiome interactions: from community assembly to plant health. Nat Rev Microbiol 18(11):607–621

Trivedi P, Batista BD, Bazany KE, Singh BK (2022) Plant–microbiome interactions under a changing world: responses, consequences and perspectives. New Phytol 234(6):1951–1959

US EPA (2022) Importance of methane. https://www.epa.gov/gmi/importancemethane#:~:text=Methane%20is%20the%20second%20most,trapping%20heat%20in%20the%20atmosphere

Velásquez AC, Castroverde CDM, He SY (2018) Plant–pathogen warfare under changing climate conditions. Curr Biol 28(10):R619–R634

Vigani G, Rolli E, Marasco R, Dell'Orto M, Michoud G, Soussi A, Raddadi N, Borin S, Sorlini C, Zocchi G (2019) Root bacterial endophytes confer drought resistance and enhance expression and activity of a vacuolar H+-pumping pyrophosphatase in pepper plants. Environ Microbiol 21(9):3212–3228

Vos M, Wolf AB, Jennings SJ, Kowalchuk GA (2013) Micro-scale determinants of bacterial diversity in soil. FEMS Microbiol Rev 37(6):936–954

Wang X, Wang C, Cotrufo MF, Sun L, Jiang P, Liu Z, Bai E (2020) Elevated temperature increases the accumulation of microbial necromass nitrogen in soil via increasing microbial turnover. Glob Chang Biol 26(9):5277–5289

Weisskopf L, Schulz S, Garbeva PJNRM (2021) Microbial volatile organic compounds in intra-kingdom and inter-kingdom interactions. Nat Rev Microbiol 19(6):391–404

Weyens N, Thijs S, Popek R, Witters N, Przybysz A, Espenshade J, Gawronska H, Vangronsveld J, Gawronski SW (2015) The role of plant–microbe interactions and their exploitation for phytoremediation of air pollutants. Int J Mol Sci 16(10):25576–25604

Xu L, Naylor D, Dong Z, Simmons T, Pierroz G, Hixson KK, Kim Y-M, Zink EM, Engbrecht KM, Wang Y (2018) Drought delays development of the sorghum root microbiome and enriches for monoderm bacteria. Proc Natl Acad Sci U S A 115(18):E4284–E4293

Yi Y, de Jong A, Frenzel E, Kuipers OP (2017) Comparative transcriptomics of Bacillus mycoides strains in response to potato-root exudates reveals different genetic adaptation of endophytic and soil isolates. Front Microbiol 8:1487

Young IM, Crawford JW (2004) Interactions and self-organization in the soil-microbe complex. Science 304(5677):1634–1637

Yuan MM, Guo X, Wu L, Zhang Y, Xiao N, Ning D, Shi Z, Zhou X, Wu L, Yang Y (2021) Climate warming enhances microbial network complexity and stability. Nat Climate Change 11(4):343–348

Zaheer A, Mirza BS, Mclean JE, Yasmin S, Shah TM, Malik KA, Mirza MS (2016) Association of plant growth-promoting Serratia spp. with the root nodules of chickpea. Res Microbiol 167(6):510–520

Zeiss DR, Mhlongo MI, Tugizimana F, Steenkamp PA, Dubery IA (2019) Metabolomic profiling of the host response of tomato (Solanum lycopersicum) following infection by Ralstonia solanacearum. Int J Mol Sci 20(16):3945

Zhang C, Xi N (2021) Precipitation changes regulate plant and soil microbial biomass via plasticity in plant biomass allocation in grasslands: a meta-analysis. Front Plant Sci 12:614968

Zhang Y, Yan X, Guo H, Zhao F, Huang L (2018) A novel protein elicitor BAR11 from Saccharothrix yanglingensis Hhs 015 improves plant resistance to pathogens and interacts with catalases as targets. Front Microbiol 9:700

Zhang S, Lehmann A, Zheng W, You Z, Rillig MC (2019) Arbuscular mycorrhizal fungi increase grain yields: A meta-analysis. New Phytol 222(1):543–555

Zhou J, Deng Y, Shen L, Wen C, Yan Q, Ning D, Qin Y, Xue K, Wu L, He Z (2016) Temperature mediates continental-scale diversity of microbes in forest soils. Nat Commun 7(1):12083

Microbial Communities in Changing Climate: Challenges and Opportunities

4

Amar P. Garg and Afeefa Ateeq

Abstract

Climate of earth is changing very fast resulting in global warming including rising temperature and acidity of oceans, melting glaciers, depleting ground water level, increasing floods, changing pattern of rains, increasing events of extreme weather that are threatening all life forms including changing microbial communities. Climate change cause structural, physiological and ecological disruptions in microbial communities and may result in increased prevalence of plant and animal pathogens. Microbial communities are responsible for emissions of carbon dioxide, methane and nitrous oxide that contribute to green house gases. Methane is approximately 25 times more efficient at trapping heat than carbon dioxide and lowering of concentration of methane in atmosphere may slow down the global warming. Scientists have discovered that more than 50% of the methane emissions come from human and animal activities and if the physiology of rumen microbes is so regulated to produce lesser amounts of methane, it can reduce harmful impact of green house gases (GHG). The changing climate determine the pattern of microbial communities e.g. *Vibrio* spp. have expanded in warmer waters and increasing acidity of oceans has led to disappearance of large number of microbial taxa besides the evolution of new species. The change of climate has induced environmental shifts exhibiting huge impact on human, animal and plant pathogens. The microbes have the capabilities to mitigate the negative impact of climate change as they have fast growth rates, large population and ability to share genetic information with other organisms. The viruses have the capabilities to infect bacteria, cyanobacteria, fungi, algae and other life forms

A. P. Garg (✉) · A. Ateeq
Swami Vivekanand Subharti University, Meerut, India

P. Pandey et al. (eds.), *Microorganisms Resilience to Climate Change*,
Microorganisms for Sustainability 56, https://doi.org/10.1007/978-981-96-3748-5_4

and the infection by phages (viruses) may be used to shut down the carbon dioxide fixation during infection which can be exploited to reduce green house gases. Microbes can be exploited as solution for climate change and methanotrophs may be used to reduce methane emissions to mitigate the impact of global warming.

4.1 Introduction

Few decades many man-made and natural activities recovering the increase in carbon dioxide and temperature levels are cause many varies in climate conditions globally. These conditions of changing climate affect volatile events like rise in rainfall and snowfall over the world and the after effects of these volatile events cause global warming from green house effect, higher temperature, decrease in soil moisture content which cause soil dryness and formed drought like conditions. Climate change causes different types of pollutions like air pollution, CO_2 emission, aerosols, methane and increase in ground level ozone. Seasonal variation cause by the changing pattern of wind especially reverse pattern of climate and weather conditions, which can occur by the emission of greenhouse gases, cause wavering and extreme effects in the upcoming years (Coleine et al. 2024; Dastagir 2019; Allsup et al. 2023; Dutta and Dutta 2016; Hacquard et al. 2022).

The conditions of today's climate are affecting the food security through several reasons like: rising temperature, changing patterns of rainfall, drought, floods and heat waves. These conditions of climate may affect hygienic conditions with higher temperature, change in wind patterns, drastic events, and other. Studies recommended that the climate changing conditions affect the crop productions in different regions. In Ethiopia climate change cause crucial drought and due to the drought the locals faced the crop failure and food aid (Dastagir 2019; Koutroulis 2019; deMenocal et al. 2000). The productivity of crop species are also influenced by altering condition of climate and impact of climate on crop species include nutrient addition and mineral accumulation in the plant (Dastagir 2019; Hacquard et al. 2022).

Climate change is a critical threat to the nature and humanity in the twenty-first century. All dominant living organisms on earth, especially the microorganisms are considerabley affected by a changing climate (Allsup et al. 2023; Dutta and Dutta 2016) and they in turn affect circular ecosystem. Microbes are responsible for elemental cycles of carbon, nitrogen, sulphur, minerals and phosphorus, and are important producers and consumers of greenhouse gases. They are potential pathogens of humans, animals, and plants also. The threat of climate change is at present at alarming rate, hence the conversations and protection of relationship between different living species and the microorganisms are essential to maintain microbial community in such a balance that a sustainable future can be planned (Fig. 4.1).

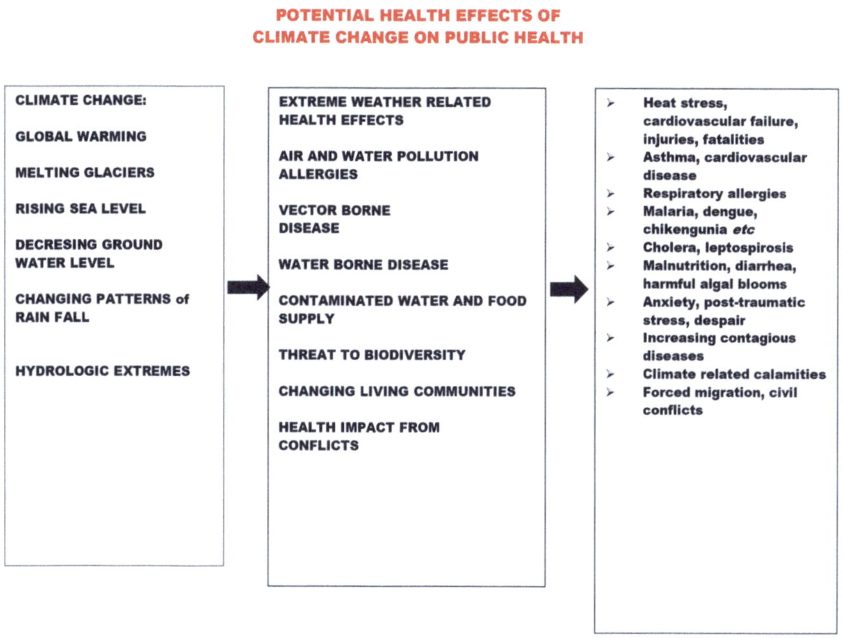

Fig. 4.1 Ill effects of climate change on public health and human life

4.1.1 Impact of Climatic Change on Plants and Soil Microorganisms

Climate change modify the relative growth, their physiology, survival, development and functions of soil microbes because the soil inhabitants different in their metabolic activities, thermo sensitive, morphological forms, growth and reproduction rates (Koutroulis 2019; Castro et al. 2010; Gray 2011; Lennon et al. 2012; Briones 2014; Manzoni et al. 2012; DeAngelis et al. 2015). The impact of climatic change cause direct impact on microbial diversity and function has been entirely discussed by various authors from time to time (Blankinship et al. 2011; Henry 2012; Manzoni et al. 2012). Temperature is raising by 58 °C in a temperate forest, for example, alters the relative density, frequency and increase soil bacteria growth in comparison to fungi, actinomycetes and nematodes. Iincreased number of the ratio of bacterial species to fungal species of the community has been observed (Dastagir 2019; Lu et al. 2018; Davila et al. 2021).

Microbial communities are highly sensitive to warming and other disturbance through resistance, and it has been observed that the microbial trait plasticity, or their resilience, returns to a formulation after the stress conditions have been over as the microbes are able to produce stress survival morphological forms (Koutroulis 2019). For instance, the recurrence of seasonal microbial communities every year after overcoming the high heat and high temperature conditions during summer in northern States of India, reveal that the microbial communities have greater flexibility and adaptability to produce dormant forms that helps them to overcome the adverse conditions. The origin of species and evolution confirm that archaeobacteria have not altered their structure and functions over a period of billions of years of organic evolution, though several climatic changes have occurred during this period. Dysbiosis compositions are likely to cause changes in the functions of ecosystem when soil microorganisms different in their functional characters and control a restricting step (Coleine et al. 2024; Koutroulis 2019). Some special group of microbe regulates ecosystem functions such as nitrogen fixation, nitrification (Koutroulis 2019; Fei et al. 2022), denitrification (Isobe et al. 2011; Ahemad and Kibret 2014), and plant microbe interactions (Garg and Sharma 1984).

Interchange in the diversity of organisms which can control specific processes than it can have a immediate effect on the growth speed. Although in some processes that happen at a bristly scale, like nitrogen mineralization, are match up with abiotic factors such as moisture and temperature than microbial community composition because a biodiversity drives these processes. Global changes especially rising temperature of earth and warming of atmosphere are directly responsible for alteration of soil microbial respiration rates, release of microbial enzymes, and plant microbe interactions (Garg and Sharma 1984) because soil microbial community, and the processes they moderate, are thermal sensitive. Microbial physiology, metabolism and their survival affect by the high temperature and it received more attention (Bradford 2013; Frey et al. 2013; Rinnan et al. 2007).

Due to increase in temperature activity of microbes increases. Q10 is a climate model which is used to study thermolabile microbes. As a result using only Q10 to account for temperature sensitivity in models may lead to bad signal. Moreover, if soil respiration decomposes than the organic matter and growth of microbes increases with temperature, these responses to experimental warming are often temporary in field studies (Chevrier et al. 2020; Bradford 2013).

The transient effects of alleviating on soil communities may be possible because the thermal active soil carbon substrates are depleted with the increased microbial activity and may also be because of the exchange of microbial communities either integrate, change in composition, or regulate their biomass to respond to convert conditions and the substrate availability (Bradford 2013; Adhya et al. 2015b). Experiments have revealed that rising temperature can initially alter the composition and diversity of microbial communities, changing in plenty of Gram-positive and Gram-negative bacteria both including actinomycetes (Davila et al. 2021; Kabir et al. 2023), or warming effects is a slow process and sometimes it may take many years to be visible response, within the microbial community (Rinnan et al. 2007). It has been found that the results from *in-vitro* and *in-vivo* studies often differ and may contradict one another and both long-term field experiments (Neill and Gignoux 2006) and short-term laboratory tests (Singh et al. 2010) compensation of temperature by microbial communities can support opposite conclusions. We have observed huge changes in microbial communities of soils of Meerut and their plant microbe interactions over the past 50 years (Garg and Sharma 1985; Garg 2023; Ruchika and Garg 2023) as the fungal species responsible for degradation of plant litter in the soils were 150+ while these have been reduced to less than 100 in the year 2023, obviously, because of the changing climate over the period of 50 years (Chevrier et al. 2020; Bradford 2013; Aanderud et al. 2011). The impacts of temperature on microbial physiology are complex that leads to adaptations of microbes, evolution, and interactions over time (Garg and Sharma 1985; Garg 2023; Ruchika and Garg 2023).

Microorganisms are the greatest part of the biodiversity on earth and they play an important role in ecosystem and issue the functions that at last sustain all of life (Ahemad and Kibret 2014; American Society for Microbiology 2008). Comprehensively the link between ecosystem biodiversity is required to predict ecosystem response to change the environment (Castro et al. 2010; Gray 2011; Almario et al. 2022; Fei et al. 2022). Molecular technique growing very fast so it develops a new interest in the field of microbiology and the distribution pattern of microbes in various habitat and their useful functions (Koutroulis 2019; Jeffries 1995; Huntington et al. 2017). Although, some studies have been focused on principal mechanisms and temperate regions in the controlling consortium change within the tropics are not discussed. Tropical ecosystems differ significantly from temperate regions in many aspects. They serve as crucial reservoirs of plant and animal biodiversity, which play essential roles in biogeochemical cycles (American Society for Microbiology 2008; Pajares et al. 2016). However, these ecosystems are under significant threat due to their conversion for alternative uses (Agler et al. 2012; Ceballos et al. 2015; Aanderud et al. 2011). Worldwide change is critical to understand how the factors of environmental and biogeography patterns. Pattern of utilizing land can change the interaction to influence the function and structure of community of microbes in the ecosystems. Possibly in tropical ecosystems different rules are applicable. Phosphorous deficiency increases with the high flow of nitrogen by human-induced activities in tropical regions; in such manner these situations are not common in temperate ecosystems (Abdel Fattah et al. 2022). Although it is not yet very clear that how availability of phosphorus affects soil microbes (Jeffries 1995; Berger 1988; Aanderud et al. 2013), or how process of microbes deal with nitrogen deposition in tropical ecosystems (Blankinship et al. 2011; Ceballos et al. 2017), however, the phosphorus availability is essential for the growth and survival of plant communities. The studies on interaction of various elements and mineral with microbial communities are limited to laboratory investigations only and their natural role is yet to be understood (Ruchika and Garg 2023). Microbial distribution is shaped by environmental factors as well as variations at regional scales and over time (Blankinship et al. 2011; Poloczanska et al. 2013; Aanderud et al. 2011; Young et al. 2008). While these factors primarily influence biodiversity patterns in larger organisms, their specific role in microbial diversity remains unclear (Singh et al. 2010; Isobe et al. 2011), particularly in tropical ecosystems. There is a lack of significant studies on microbial community responses to disturbances in tropical regions. Land-use changes, particularly for plant cultivation, are key drivers of biodiversity shifts in plant and animal communities. In tropical forest ecosystems, rapid transformations occur primarily due to the conversion of land for agriculture and other purposes (Agler et al. 2012). The type of vegetation govern and control the pattern and diversity of microbial communities. Land-use changes significantly impact soil microorganisms, making them a critical focus for conservation efforts. Temporal variations in phosphorus and nitrogen availability in tropical forest soils play a crucial role in shaping microbial community structure and activity, which are essential for the decomposition of biodegradable compounds. Tropical rainforests exhibit significant ecological variation in soil composition, influenced by soil properties and microbial communities, including bacteria, fungi, algae, nematodes, and actinomycetes. These communities differ notably across forest types. In tropical rice paddies, the distribution of methanogens and methanotrophs shows a gradual decline, with clear patterns linked to oxygen, methane, and organic carbon concentrations in the soil. Both natural and anthropogenic factors, such as nitrogen deposition and land-use changes, play pivotal roles in regulating the nitrogen cycle (Coleine et al. 2024; Hacquard et al. 2022; Lu et al. 2018; Falkowski et al. 2008).

4.1.2 Redundancy, Resilience and Resistance

Microbial communities are widely recognized for their high functional redundancy, enabling them to resist disturbances (resistance) and recover from disruptions (resilience) effectively (Chevrier et al. 2020; Martínez-Valderrama et al. 2020; Biggs et al. 2020). This adaptability is attributed to their high biomass, rapid growth potential, ubiquitous distribution, low extinction risk, and the prevalence of horizontal gene transfer (Wierzchos et al. 2012; Fei et al. 2022). Additionally, the sheer number of microbes in a biological community surpasses that of other living organisms, further reinforcing their ecological robustness and adaptability. The power of microbes vest is their high speed of growth and reproduction and most of the soil bacteria double in their number and biomass within 20 min to 6 h and they can survive under hardest and adverse conditions ranging from −120 °C to +150 °C temperature and over a wide range of p^H from 1 to 12 (Biggs et al. 2020; Xie 2022; Allison and Martiny 2008).

Existence of microbes in universe suggest that composition of microbial community is important for ecosystem's function (Leung et al. 2020; Erlandson et al. 2022). The success of 'black-box' models supports the notion that biogeochemical cycles are not constrained by the abundance of microbes. These models rely on simplified mathematical functions to represent microbially driven processes across diverse environmental conditions, highlighting the functional efficiency and adaptability of microbial communities in regulating ecosystem processes (Delgado-Baquerizo et al. 2018; Debray et al. 2022; Xie 2022; Allison and Martiny 2008).

The area of ecosystem modeling is growing fast but there is dispute about the generalized functional utility and these functions have broad and approximate change over medium term. The black-box models are better at pathway prediction than other models but sometime it is not able to categorize specific pathways but instead focus on general functions, such as gross nitrogen mineralization. The black-box model depends on three factors.

1. Microbial communities may be resistant, resilient and functionally redundant.
2. Function and diversity of microbial communities may be completely decoupled and deregulated.
3. Although function and diversity are linked, diversity remains constant across temporal and spatial scales (Santosh et al. 2024; Babalola et al. 2022; Allison and Martiny 2008).

These concepts differ from findings at higher levels of ecological organization, where the loss of any species is directly associated with functional changes (Jeffries 1995; Babalola et al. 2022). A recent study on resilience and resistance in soil microbial communities (Neill and Gignoux 2006; Almario et al. 2022; Allison and Martiny 2008) highlighted that soil microbes are influenced by various factors and can stabilize in response to disturbances. However, predicting their exact response to disturbances remains challenging (Huntington et al. 2017; Abdel Fattah et al. 2022). The engineering definition of resilience, which is most commonly used in microbial ecology, is often employed due to its simplicity and ease of measurement (Fig. 4.2).

Studies on microbes categorized ecosystems to investigate disturbance responses is presented in Fig. 4.2. Anaerobic, aerobic, and denitrifying microorganisms are commonly used to examine disturbance responses due to their well-understood functions and diverse phylogenetic diversity. Anthropogenic pollutants, such as hydrocarbons and metals, have been extensively studied for their effects on

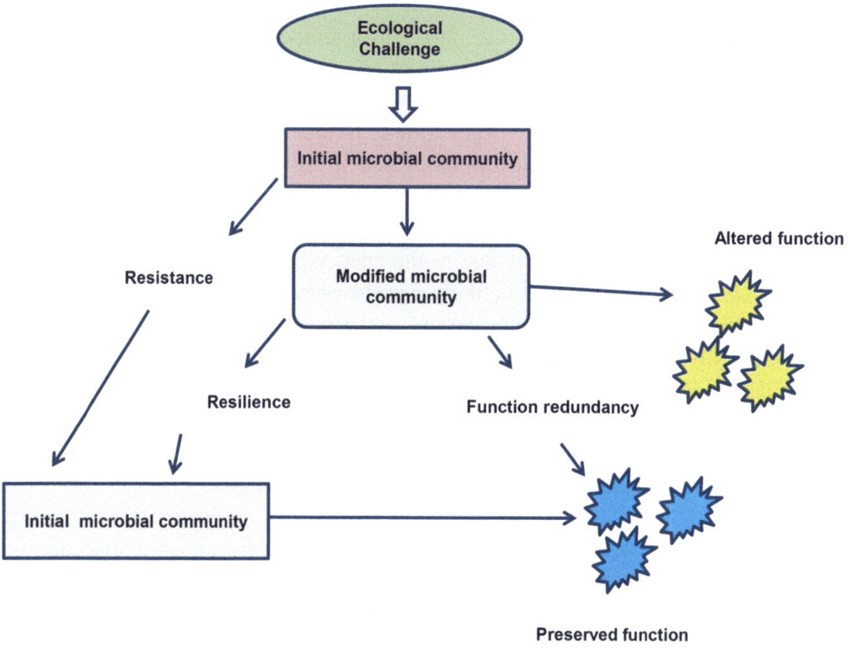

Fig. 4.2 Ecological challenges of climate change on microbial community

ammonia-oxidizing bacteria (AOB) and other microbes, particularly regarding their degradation potential and microbial activities (Sahrawat and Garg 2023; Chaudhary and Garg 2024; Ateeq and Garg 2023; Ji et al. 2015). For system recovery it is very important to obtain specific resistance for the effect of pollutants. Hydrocarbons may interact with ammonia at the active site of ammonia monooxygenase (AMO), potentially impacting microbial communities (Castro et al. 2010; Bardgett et al. 2008; Allison and Martiny 2008; Ji et al. 2015). The toxicants have ability and established mode of action that act differently on ammonia oxidizing bacteria and regulate their growth and survival rate. The mode of action shows high sensitivity of aerobic and anaerobic bacteria to hydrocarbon, aerobic and anaerobic bacteria are most sensitive organisms toward hydrocarbon in any ecosystem (Lennon et al. 2012; American Society for Microbiology 2008; Ji et al. 2015). In addition, hydrocarbons exert a non-specific toxic effect as hydrophobic narcotics, disrupting lipid-rich microbial membranes. Aerobic and anaerobic ammonia oxidizers (AOs) are also vulnerable to hydrocarbon spills resulting from disturbances in the ecosystem (Soliveres et al. 2014; American Society for Microbiology 2008) hydrocarbon degrading microbes rapidly consume oxygen, and essential nutrients. Soil microbial communities have directly affected by the hydrocarbons and it increase soil temperature and water retention by increasing soil hydrophobicity (Briones 2014; Ahemad and Kibret 2014). In our laboratories, a number of fungal taxa especially the species of *Aspergillus* have been isolated that actively degrade hydrocarbons in soils (Ruchika and Garg, 2023; Garg et al. 2023, 2024; Jabborova et al. 2024; Ateeq and Garg 2023; Ateeq and Garg 2023) under aerobic conditions. AO do not recover from the impact of hydrocarbon in extreme system like Antarctic (Whitaker et al. 2014; Agler et al. 2012; Adhya et al. 2015b). In temperate systems, the ammonia oxidizer (AO) community adjusts to indirect effects by altering enzyme activity, a process that is associated with levels of soil organic matter (Blankinship et al. 2011; Aanderud et al. 2011, 2013). This adaptation to indirect pollutant impacts is crucial for the system's response to disturbances and is influenced by various factors, including the system's energy status, other biota (Henry 2012), and the timing of pollutant exposure. In extreme environments, however, the AO community may not adapt to indirect pollutant effects due to the energy-poor nature of Antarctic soils (Agler et al. 2012). Therefore, it is likely that the ecosystem's energy status, which includes the soil's chemical environment and the interactions between distinct organisms, plays a key role in the resilience of the AO community (Manzoni et al. 2012; Garg et al. 2024).

4.1.3 Impact of Climate Change on Planet and Life

The most recent cooling phase ended approximately 20,000 years ago, marking the onset of the current warming phase. Around 8000 years ago, global temperatures increased by 3–4 °C, corresponding to a gradual rise of approximately 0.03 °C per century. For much of the period between 8000 years ago and the onset of the Industrial Revolution, temperatures remained relatively stable. However, since the Industrial

Revolution, global temperatures have been rising at an accelerated pace. Projections for the year 2100 suggest a temperature increase of 2.0–4.9 °C compared to pre-industrial levels. This implies that, in the coming century, the rate of temperature rise will be approximately 100 times faster than the pace observed over the past 20,000 years. Several factors are responsible for climate change like sun energy coming on earth change in concentration of atmospheric gases but for now only human activities are mainly responsible. This is the main reason current situation are called as human-induced climate change. Burning of fossil fuel species for the sake of industrial revolution to meet the need of energy for today's generation (Henry 2012; Ceballos et al. 2015). Over the past 200 years, human activities have led to a 46% increase in atmospheric carbon dioxide (CO_2) levels. However, CO_2 alone is not the sole driver of climate change. The burning of fossil fuels, agricultural practices, industrial activities, deforestation, and various other factors all contribute to the ongoing changes in the climate (Koutroulis 2019; DeAngelis et al. 2015; Barnosky et al. 2011). While the effects of anthropogenic climate change may not be immediately visible, they are expected to profoundly affect the daily lives of future generations. The frequency and intensity of natural disasters, such as droughts, floods, and storms, are likely to increase, with significant consequences for human populations (Koutroulis 2019; Ceballos et al. 2017; American Society for Microbiology 2008). Additionally, the melting of ice caps and rising sea levels are altering shorelines globally. Climate change will also impact all forms of life, including microbes, plants, and animals, many of which may face extinction. For example, the vast Sahara Desert, once a lush green landscape, was significantly wetter around 6000 years ago, a result of more humid climatic conditions at the time (Gibson et al. 2011; Schimel and Schaeffer 2012; Liu et al. 2014; Ateeq and Garg 2023).

4.1.4 Effect of Climate Change on Life Forms

Interaction between living being and plants and living being to living being also get affected by the climate changes in which they are living. After facing the climate change species have left only two option survival with the situation or migration, from algae to fish they move towards poles in searching of cold water. When species faced climate related challenges so they have only two options (Rinnan et al. 2007; Ahemad and Kibret 2014; Angulo et al. 2022; Garg et al. 2024). For species to migrate, they must be able to locate new habitats where they can survive and thrive. Large animals like whales and cougars can travel long distances relatively easily, but for smaller organisms like orchids and sloths, migration is more challenging, especially if they cannot disperse their seeds effectively. Evolution can occur on a small scale, as seen with alpine chipmunks at Yosemite National Park in California, USA. Over the course of 100 years, these chipmunks evolved longer faces in response to climate change (Koutroulis 2019; Isobe et al. 2011; Walsh et al. 2016) (Fig. 4.3). Around 480,000 to 340,000 years ago, brown bears and polar bears were the same species, but over time, as the two groups accumulated more differences, they eventually became distinct species. Unfortunately, when species are unable to

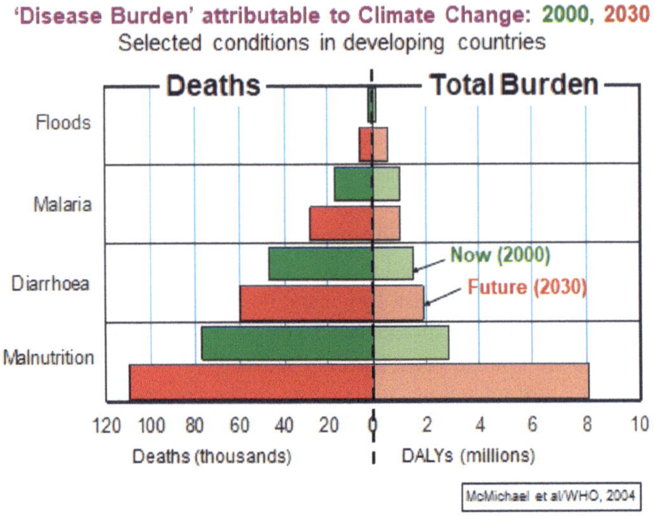

Fig. 4.3 Attributable disease burden in 2000–2030. (Mc Michael 2004)

migrate or evolve in response to environmental changes, they face extinction. The rapid growth of the human population has added significant challenges to the survival of many species. Modern lifestyles play a crucial role in altering climatic conditions, with two vertebrate species becoming extinct each year. Examples of harmful human activities include intense deforestation driven by agriculture and the pollution of air and water. The impact of anthropogenic climate change requires urgent attention, as it differs from other environmental changes. Unlike localized disturbances, anthropogenic climate change affects the global climate, even in areas that are rarely impacted by human activities, such as protected regions. The pace of anthropogenic climate change is accelerating, making it increasingly difficult for species to adapt or migrate in time (Chevrier et al. 2020; Bakken et al. 2012; Poloczanska et al. 2013; Debray et al. 2022; Ateeq and Garg 2023). This indicates that anthropogenic climate change poses a significant threat to species that are already struggling to survive in an Earth that has been extensively altered by human activity. Humans are causing dangerous impacts on earth which are long-lasting and disastrous impact. 75% of species which exist now but in few centuries these species will extinct. This extinction of life forms is dangerous because we depend on these species to survive. The air and food quality, + many things on which we directly depend for survival of these species will be disturbed (Koutroulis 2019; Schimel and Schaeffer 2012; Isobe et al. 2011; Annan and Hargreaves 2013; Raftery et al. 2017; deMenocal et al. 2000).

Earth is a dynamic planet, constantly undergoing change due to both internal and external forces. The movement of magma currents beneath the surface drives the movement of tectonic plates, which form the continental crust. Over time, this process leads to the creation of mountains, which can eventually erode into valleys. These valleys may later transform into lakes or seas (Koutroulis 2019; Briones

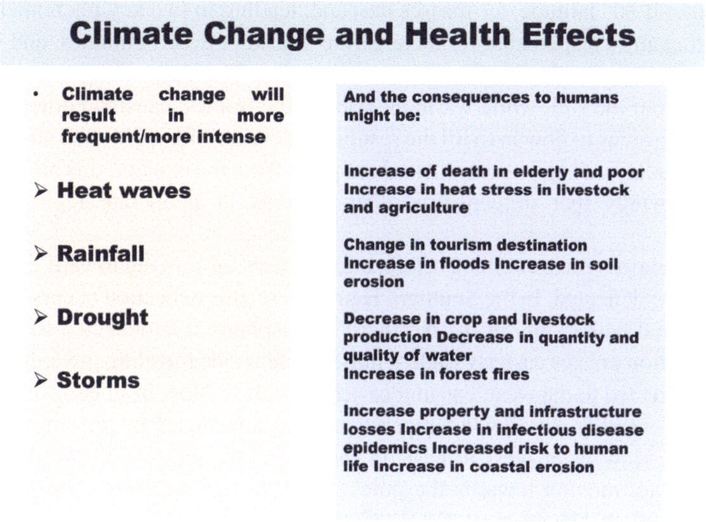

Fig. 4.4 Climate change and its health effects

2014; Annan and Hargreaves 2013; Aanderud et al. 2013). On the planet's surface, the most significant factor influencing Earth is sunlight. The Sun provides energy for living organisms and drives weather and climate patterns by creating temperature gradients in the atmosphere and oceans (Fig. 4.4).

4.2 Circulation of Climate

Earth receives light and heat from the sun, with some regions being exposed to these rays more intensely than others. And the tropics have less seasonal variations because of less sunlight exposure. Regions near the equator receive warm, moist air. As this air rises to higher levels in the atmosphere, it cools down, after that water molecules condense and formation of clouds takes place than fall as rain. On Earth surface warm air rises it pushes the air mass away from the tropical area, release moisture in the form of precipitation and travels towards poles (Singh et al. 2010; DeAngelis et al. 2015; Gildemyn et al. 2015; Salvioli et al. 2016).

Evaporation, condensation, and precipitation are influenced by Earth's rotation on its axis. Without this spin, water and air would not move along a north-south axis. The Earth's rotation creates three circulation belts. Air masses sink at approximately 30° north and south latitudes, and then air moves from the tropical regions toward the poles and regional areas. "This belt of air circulation is referred to as a Hadley cell, after George Hadley, who first described it (Gray 2011; Agler et al. 2012; Fierer and Jackson 2006). Two additional belts of circulating air exist in the temperate latitudes (between 30° and 60° latitude) and near the poles (between 60° and 90° latitude)".

At around 30° latitude, air masses descend, leading to two key phenomena: this descending air is responsible for the creation of arid or desert climates, and the circulation of air from the tropics towards the poles. The descending dry air draws moisture from the soil, while warm air rises in the tropics, causing cooler air from surrounding areas to flow in to fill the resulting vacuum. Although some air descends from the Hadley cell and moves toward the poles from the equator, this air mass also generates winds that influence weather patterns in temperate regions (Dang et al. 2019).

Due to Earth's rotation, the Coriolis effect causes air moving toward the Earth's surface to be deflected. In the Southern Hemisphere, this deflection occurs to the left of its original path, while in the Northern Hemisphere, it is deflected to the right. This deflection creates easterly trade winds, where winds traveling toward the equator are redirected to the west, causing easterly winds to blow from east to west. The majority of weather patterns in temperate regions are carried by prevailing westerlies, which blow from west to east. In these regions, the Coriolis force deflects winds that are moving towards the poles in an easterly direction (Dastagir 2019; Koutroulis 2019; Davila et al. 2021; Castro et al. 2010; Henry 2012; Davidson et al. 2006).

4.2.1 Global Climate

Redistribution of heat and moisture of global climate drives through circulation of oceanic and atmospheric combination. Tropical areas remain warm and wet throughout the year but on the other hands in temperate regions solar inputs drive seasonal change. Temperature may change in Northern Hemisphere because of the concentration of land masses (Blankinship et al. 2011; Basak and Biswas 2009; Church and White 2011; Jabborova et al. 2024). The Southern Hemisphere is largely covered by water, while significant landmasses are found near the equator. The presence or absence of precipitation plays a key role in temperature variations in these regions (Fig. 4.5).

Global climate patterns are constantly evolving due to changes in solar radiation, greenhouse gas concentrations, and climatic driving forces. The cyclical variations in solar radiation reaching the poles are predictable. "These cycles, first described by (Debray et al. 2022), involve Earth's orbit, tilt, and the precession of the equinoxes" (Whitaker et al. 2014; Basak and Biswas 2009).

The gravitational pull of other planets in our solar system causes variations in Earth's elliptical orbit around the Sun. The Earth is drawn farther away from its energy source as the orbit changes from being almost circular to being elongated throughout a 100,000-year cycle. Over a 41,000-year cycle, Earth's axial tilt varies between 21.5° and 24.5°, and currently, the tilt is at 23.5°. Additionally, the Earth's axis, or north-south direction, shifts over time. The 23,000-year precession of the equinoxes alters the planet's orientation relative to its orbit. These three Milankovitch cycles influence solar radiation and impact atmospheric and oceanic circulation patterns, often amplifying each other's effects (Soliveres et al. 2014; Andersen et al. 2016; deMenocal et al. 2000; Jabborova et al. 2024).

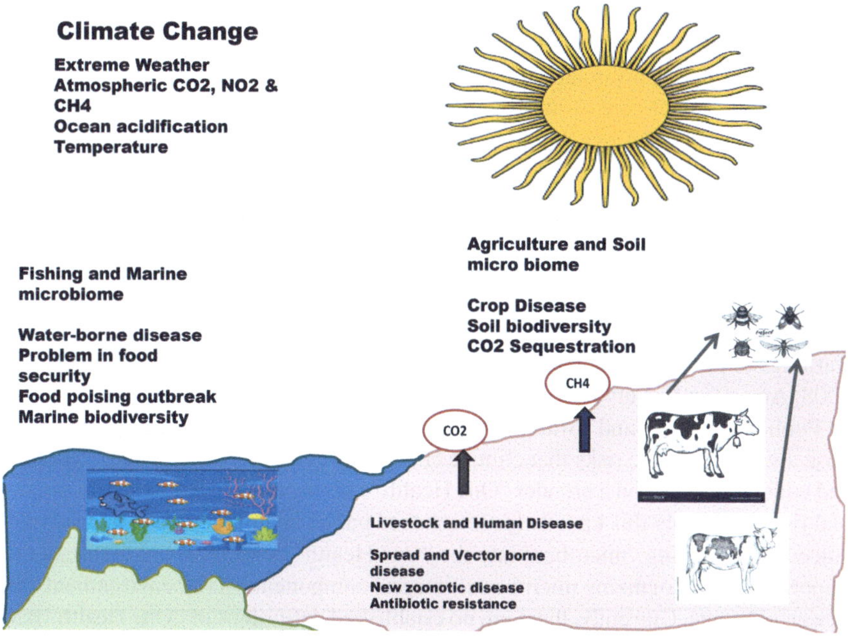

Fig. 4.5 Main effects of climate change on microorganisms: (i) expansion and increase in water-borne diseases affecting fish and shellfish and causing an increase in food poisoning outbreaks and loss of marine biodiversity; (ii) expansion and intensification of pathogenic fungal infections in crops all around the world; and (iii) outbreaks of livestock diseases, emergence of new zoonotic diseases and increase in antibiotic resistance

4.2.2 Influence of Climate Change on Microbial Communities

Humans are less adaptive and opportunistic than microbes, which are more versatile. The microbes change with climatic conditions as the climate changes their risk of infection also increase and they cause chronic diseases. Although becteria, fungi and viruses do not cause chronic diseases but in the warmer temperature and in extreme weather condition risk of new pathogens increases, these new pathogen can be thermal resistant and they can adapt in the human body. Some fungi, once non-pathogenic, are now linked to the rise in fungal diseases. These fungi have become antibiotic-resistant and highly invasive, posing a significant threat to human health (Huntington et al. 2017; Berger 1988; Adhya et al. 2015a).

Growth and distance transport of microbes also accelerate with increase in temperature; in human and animal physiology also alter with the higher temperature and environmental stress. With the warmer temperature gut microbiome and skin microbes may less protective which can cause serious problems. Zoonotic pathogens from wildlife carry an additional risk because they can mutate when transmitted from humans to animals, potentially leading to the emergence of new disease threats. When the microbes interact with weakened hosts than the microbes induce in to persister and became the antibiotic resistant or niche expansion (Neill and Gignoux 2006; Rinnan et al. 2007; Salvioli et al. 2016).

Public should be aware about the adaptability and opportunistic microbes and it needs more instance care to combat climate change (Carrell et al. 2022). Monitoring programs for air and water should be developed to integrate climate models capable of forecasting shifts in regional temperatures. Recent IPCC report, 2018 (Levy et al. 2018) states that "high confidence" that 'climate-sensitive aquatic pathogens like *Vibrio* spp. have increased regional risks of water and food borne disease. Data and documentation from surveillance efforts, which enabled the development of tools like the *Vibrio* Map Viewer in response to the northern expansion of *Vibrio* in Atlantic waters, will enable federal agencies to provide stronger guidance and early warnings of public health threats' (Jeffries 1995; Castro et al. 2010; Falkowski et al. 2008; American Society for Microbiology 2008).

Public education and awareness-raising on the human microbiome may help raise awareness of the risks that climate change poses to microbial health. Research and education based on a broader "One Health" approach could help identify microbial risks and apply this knowledge to enhance public health protection and surveillance. Incorporating "microbes" into the "One Health" concept would highlight the importance of recognizing microbes as integral components of our environment and powerful agents. Currently, there are no established definitions of "One Health" that explicitly include the term "microbes". Maybe they ought do, in order to maintain public health protection as the primary driving force behind the fight against climate change and to maintain microbiological awareness (Santosh et al. 2024; Neill and Gignoux 2006; Frey et al. 2013; Salvioli et al. 2016) (Fig. 4.6).

4.2.3 Microbes in Models: Reduce the Gap Through Innovation

For the climate models and pathogens surveillance metagenomics and other genomics technologies provide the potential data input for the efforts under global warming scenarios. However, this potential data must be translated into addressing the urgent impacts of climate change. To accelerate this process, bold and innovative actions are required, such as expanding high-throughput tools to measure microbial functions and metabolic activity rates in situ, and developing advanced mathematical models that consider microbial-scale factors (Kabir et al. 2023; Berger 1988).

Phenotypic properties and metabolic dynamics of microbes are obscured. Knowledge and the researches are very limited for the combine studies on microbial parameters and the climate models, from which we can understand how microbes actively participate in nutrient cycle, interact with species and response to disruption. Similarly, new statistical and experimental methods must be developed for isolating genomes from microbes to complement metagenomic data, helping to infer dynamic processes. These innovative techniques will enable studies that contribute to a predictive understanding of how microbes influence ecosystems (Leung et al. 2020; Delgado-Baquerizo et al. 2018; Bradford 2013; Bardgett et al. 2008).

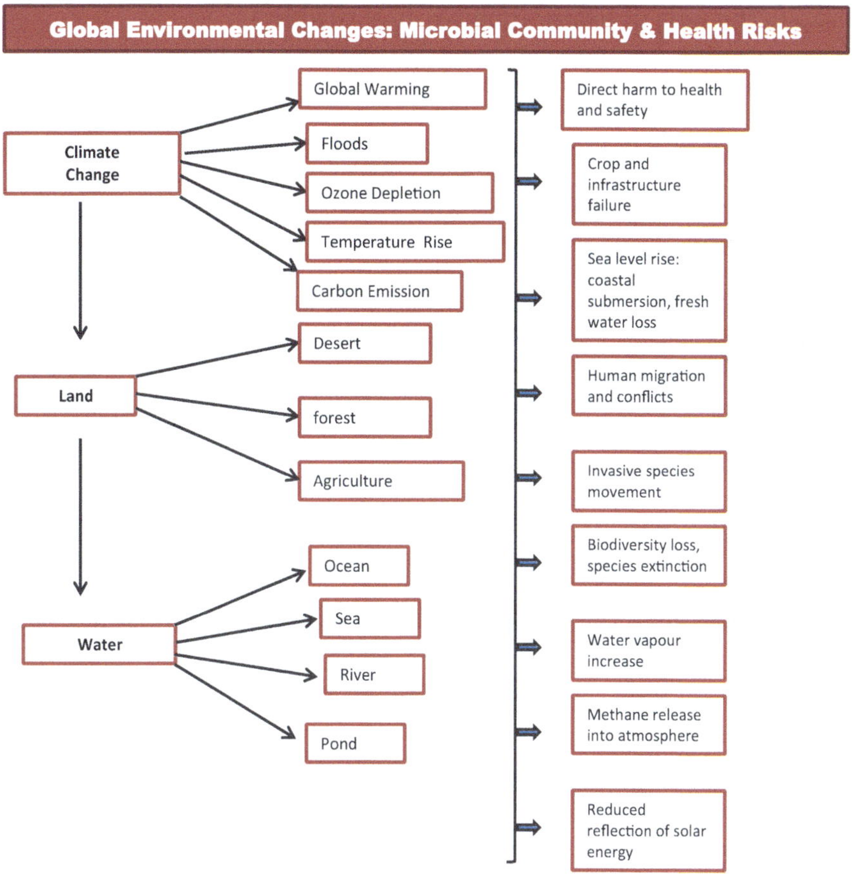

Fig. 4.6 Global environmental changes: microbial community and health risks

4.2.4 Microbes as Possible Solutions for Negative Impact of Climate Change

4.2.4.1 Microorganisms in Agriculture

Microorganisms are commonly found in almost all habitats. Microbes exist in single-celled or in the form of colony, microbes can be found in the extreme conditions like; desert, rocks, equator and in the deep water (Leung et al. 2020; Huntington et al. 2017; American Society for Microbiology 2008). Microbes are versatile organism, they can adapt high to low temperature and survive in extreme conditions. Microbes make up the active organic component of soil, playing a crucial role in the overall life of the soil. This small fraction of organic matter is responsible for facilitating nutrient absorption by plants, from the soil minerals to the rhizosphere (Wierzchos et al. 2012; Schimel and Schaeffer 2012; Isobe et al. 2011; Bakken et al. 2012).

4.2.5 The Role of Microorganism in Agriculture

4.2.5.1 Microbes for Plant Nutrition

Microorganisms like Cyanobacteria, *Azotobacter chroococcum, Azospirillum brasilense, Bacillus subtilis, Pseudomonas corrugata, Rhizobium* spp., mycorrihiza, and *Streptomyces nojiriensis* enhance plant growth and contribute to better plant development (Martínez-Valderrama et al. 2020). Plants absorb nutrients primarily from the rhizosphere, where microbes play a critical role in nutrient management (Koutroulis 2019; Schimel and Schaeffer 2012). Actinomycetes, which possess antagonistic properties, help mitigate pathogen effects in the soil and are effective in reducing plant pathogen impacts during growth (Manzoni et al. 2012; Salvioli et al. 2016; Bardgett et al. 2008). Microbes assist in supplying nitrogen, phosphorus, and potassium to plant roots. Arbuscular mycorrhizal fungi, present in the roots of higher plants, aid in the absorption of water, phosphorus, and other essential nutrients. Microorganisms such as *Nostoc, Anabaena, Aphanocapsa, Chroococcus, Oscillatoria, Erwinia herbicola*, and *Phormidium* contribute to the fixation of atmospheric nitrogen (Rinnan et al. 2007; Bardgett et al. 2008).

4.2.6 Microbes as Growth Regulating Agent

Microorganisms synthesize and release auxins, they regulate the plant growth (Koutroulis 2019; Davila et al. 2021). Plant growth regulator produced from soil microbes such as; bacteria, fungi and algae (Chevrier et al. 2020; DeAngelis et al. 2015). Phytoharmones like; indole acetic acid (IAA), gibberellic acid, cytokins and metabolites like; siderophores, HCN and antibiotic are produced by plant growth regulator rhizobacteria (PGPR). In rhizosphere many species like PGRPs, pathogenic, symbiotic and free-living rhizobacteria took part. Fungi controlled parasitic spores, sclerotia or hyphae of pathogenic microbes (Bardgett et al. 2008). Large quantity of enzymes like; chitinases, proteases, and glucanases are secrete by biocontrol process. Trichoderma strains are reported to inhabit with diverse plant roots (Saba et al. 2012). This advantageous association of fungi with plant growth is known as mycoparasitism (Manzoni et al. 2012).

4.2.7 Microbes for Phosphorus Solubilization

Among the macronutrients found in plants, phosphate is the least mobile element. Mineralisation and solubilisation are significantly influenced by phosphorus solubilising microorganisms (PSMs) (Koutroulis 2019; Wierzchos et al. 2012; Henry 2012; American Society for Microbiology 2008; Jabborova et al. 2024). Acid phosphatase releases organic phosphorus after the microbial communities produce organic acids, which lowers the pH of the soil and initiates the phosphate solubilisation process. PSM is most effective in solubilising phosphorus when it is co-inoculated with mycorrhizal fungi or other beneficial bacteria (Hacquard et al.

2022; Koutroulis 2019; Wierzchos et al. 2012; Almario et al. 2022). When it comes to phosphorous solubilisation, bacteria are more effective than fungi (Sharma et al. 2013). Rhizobium, Enterobacter, endosymbiotic rhizobia, ectorhizospheric strains of Pseudomonas and Bacillus, and bacterial populations in soils form effective microbial communities of phosphate solubilizers to enhance soils with P (American Society for Microbiology 2008; Debray et al. 2022). While phosphate-solubilizing fungi (PSF) have a population of only 0.1–0.5%, phosphate-solubilizing bacteria (PSB) have a population of 1–50% in typical soil (Isobe et al. 2011; Agler et al. 2012; Abdel Fattah et al. 2022). Potential phosphate-solubilizing species strains include Pseudomonas striata, Bacillus megaterium, Bacillus circulans, Bacillus subtilis, Bacillus polymyxa, and Bacillus sircalmous (Church and White 2011; Annan and Hargreaves 2013).

4.2.8 Microbes for Potash Mobilization

One of the most crucial elements for plants is potassium (K). Soils contain a lot of K. The top soil contains between 3000 and 1,000,000 kg of K/ha (Salvioli et al. 2016). Soil has four forms of K: exchangeable, nonexchangeable, water-soluble (solution K), and structural or mineral (Agler et al. 2012). A number of variables affect how much K is released into soils. The release of K is affected by changes in soil factors such as pH, moisture content, texture, oxygen content, temperature, soil tilling, topography, and biogeochemical characteristics (Barnosky et al. 2011; Aanderud et al. 2011). Microbes have a significant function in K mobilisation. Certain beneficial microbes, including *Acidithiobacillus ferrooxidans, Arthrobacter sp., Azotobacter sp., Bacillus mucilaginosus, Bacillus edaphicus, Frateuria sp., Klebsiella sp., Paenibacillus sp., Pseudomonas sp., and Rhizobium sp.* (Manzoni et al. 2012; Frey et al. 2013; Liu et al. 2014) play a very crucial role.

Microbial biofertilizers and biopesticides are best for sustainable agriculture (Wierzchos et al. 2012; Agler et al. 2012; Aanderud et al. 2011); microbes in the rhizosphere play an important role in the growth of plant for nutrients supply (Rinnan et al. 2007; Ahemad and Kibret 2014; Almario et al. 2022). And biopesticides play an important role in the production of antibiotics in plants.

4.3 Conclusion

The impact of industrial revolution and race for comfortable materialistic life in the world has resulted in visible impact of huge climate change on nature, human, animal and plant lives including microbial communities. Increased global warming has led to changing patterns of rain fall, melting glaciers, increasing sea water level, changing alkalinity leading to disturbed marine flora and fauna, diminishing forests, depleting ozone layer and increasing soil, water and air pollution. These factors have grossly affected the microbial communities, their diversity and quantity both leading in decreased degradation and accumulation of waste in the nature resulting in poisoned soil, water and air that have entered into the food web. The contaminated environment including poisonous gases and compounds have led to the emergence of new viral diseases like Corona virus in the year 2019. The enhanced use of systemic antimicrobial and unrestricted use of antibiotics have complicated the problem. The increasing carbon emissions and accumulated e-waste, decreasing carbon and nitrogen content of soils, decreasing ground water level have changed microbial communities that in turn have adversely affected human health and public life. We urgently need to address this issue at international level by organizing fruitful and meaningful deliberations with polity, business communities and scientists to evolve concrete strategies and planning to meet the ill effects of climate change.

Conflict of Interest The authors declare that they have no conflict of interest.

References

Aanderud ZT, Schoolmaster DR Jr, Lennon JT (2011) Plants mediate the sensitivity of soil respiration to rainfall variability. Ecosystems 14:156–167

Aanderud ZT, Jones SE, Schoolmaster DR Jr, Fierer N, Lennon JT (2013) Sensitivity of soil respiration and microbial communities to altered snowfall. Soil Biol Biochem 57:217–227

Abdel Fattah A, Tack AJM, Wasserman B, Liu J, Berg G, Norelli J, Droby S, Wisniewski M (2022) Evidenceforhost–microbiomeco-evolutioninapple. New Phytol 234:2088–2100

Adhya TK, Kumar N, Reddy G, Podile AR, Bee H, Samantaray B (2015a) Microbial mobilization of soil phosphorus and sustainable P management in agricultural soils. Curr Sci 12:1280–1287

Adhya TK, Kumar N, Reddy G, Podile AR, Bee H, Bindiya S (2015b) Microbial mobilization of soil phosphorus and sustainable P management in agricultural soils. Curr Sci 108:1280–1287

Afeefa A, Amar GP (2023) Biological degradation of low density polyethylene (LDPE). Asian Jr of Microbiol Biotech Env 25(4):664–669

Afeefa A, Prakash GA (2023) Plastic waste: current environmental pollution health Hazard and biodegradation strategies and its management, vol 3. The Challenges and Issues of Sustainability in the Hospitality and Tourism Industry, Iterative International Publishers, pp 59–85

Agler MT, Spirito CM, Usack JG, Werner JJ, Angenent LT (2012) Chain elongation with reactor microbiomes: upgrading dilute ethanol to medium-chain carboxylates. Energy Environ Sci 5:8189–8192

Ahemad M, Kibret M (2014) Mechanisms and applications of plant growth promoting rhizobacteria: current perspective. J King Saud Univ Sci 26:1–20

Allison SD, Martiny JB (2008) Resistance, resilience, and redundancy in microbial communities. Proc Natl Acad Sci 105(supplement_1):11512–11519

Allsup CM, George I, Lankau RA (2023) Shifting microbial communities can enhance tree tolerance to changing climates. Science 380(6647):835–840

Almario J, Fabianska I, Saridis G, Bucher M (2022) Unearthing the plant–microbe quid pro quo in root associations with beneficial fungi. New Phytol 234(1967):1976–1976

American Society for Microbiology (2008) Climate change could impact vital functions of microbes. Science Daily. www.sciencedaily.com/releases/2008/06/080603085922.htm. Accessed 15 Sept 2018

Aminov RI (2011) Horizontal gene exchange in environmental microbiota. Front Microbiol 2:158

Andersen SJ, Berton J, Naert P, Gildemyn S, Rabaey K, Stevens CV (2016) Extraction and esterification of low-titer short-chain volatile fatty acids from anaerobic fermentation with ionic liquids. ChemSusChem 9:2059–2063

Angulo V, Beriot N, Garcia-Hernandez E, Li E, Masteling R, Lau JA (2022) Plant microbe eco-evolutionary dynamics in a changing world. New Phytol 234:1919–1928
Annan JD, Hargreaves JC (2013) A new global reconstruction of temperature changes at the last glacial maximum. Clim Past 9:367–376. https://doi.org/10.5194/cp-9-367-20132
Babalola BJ, Li J, Willing CE, Zheng Y, Wang Y-L, Gan H-Y, Li X-C, Wang C, Adams CA (2022) Gao nitrogen fertilisation disrupts the temporal dynamics of arbuscular mycorrhizal fungal hyphae but not spore density and community composition in a wheat field. New Phytol 234:2057–2072
Bakken LR, Bergaust L, Liu B, Frostegård Å (2012) Regulation of denitrification at the cellular level: a clue to the understanding of N2O emissions from soils. Philosophical Transactions of the Royal Society B: Biological Sciences 367(1593):1226–1234
Bardgett RD, Freeman C, Ostle NJ (2008) Microbial contributions to climate change through carbon cycle feedbacks. ISME J 2:2805–2814
Barnosky AD, Matzke N, Tomiya S, Wogan GOU, Swartz B, Quental TB (2011) Has the earth's sixth mass extinction already arrived? Nature 471:51–57. https://doi.org/10.1038/nature096789
Basak BB, Biswas DR (2009) Influence of potassium solubilizing microorganism (Bacillus mucilaginosus) and waste mica on potassium uptake dynamics by Sudan grass (Sorghum vulgare Pers.) grown under two Alfisols. Plant Soil 317:235–255
Berger A (1988) Milankovitch theory and climate. Rev Geophys 26(4):624–657
Biggs CR, Yeager LA, Bolser DG, Bonsell C, Dichiera AM, Hou Z et al (2020) Does functional redundancy affect ecological stability and resilience? A review and meta-analysis. Ecosphere 11(7):e03184
Blankinship JC, Niklaus PA, Hungate BA (2011) A meta-analysis of responses of soil biota to global change. Oecologia 165:553–565
Bradford MA (2013) Thermal adaptation of decomposer communities in warming soils. Front Microbiol 4:333
Briones MJI (2014) Soil fauna and soil functions: a jigsaw puzzle. Front Environ Sci 2:7
Carrell AA, Lawrence TJ, Cabugao KGM, Carper DL, Pelletier DA, Lee JH, Jawdy SS, Grimwood J, Schmutz J, Hanson PJ (2022) Habitat-adapted microbial communities mediate *Sphagnumpeat mossresilience* to warming. New Phytologist 234:2111–2125
Castro HF, Classen AT, Austin EE, Norby RJ, Schadt CW (2010) Soil microbial community responses to multiple experimental climate change drivers. Appl Environ Microbiol 76(4):999–1007
Ceballos G, Ehrlich PR, Barnosky AD, Garcia A, Pringle RM, Palmer TM (2015) Accelerated modern human-induced species losses: entering the sixth mass extinction. Sci Adv 1:e1400253. https://doi.org/10.1126/sciadv.1400253
Ceballos G, Ehrlich PR, Dirzo R (2017) Biological annihilation via the ongoing sixth mass extinction signaled by vertebrate population losses and declines. Proc Natl Acad Sci 114:E6089–E6096. https://doi.org/10.1073/pnas.17049491148
Chaudhary M, Garg AP (2024) Global impact of plastic pollution and its management for sustainable development. In: Shukla, Kasturi, Patil Y, Estoque RC, de Haro PAL (eds) Quality of life and climate change. Chapter 6, IGI Global Publisher of Timely Knowledge, pp 122–152. https://doi.org/10.4018/978-1-6684-9863-7.ch006
Chevrier VF, Roy R, Meslin PY et al (2020) Geochemical and spectral characterization of an altered Antarctic dolerite: implications for recent weathering on Mars. Planet Space Sci 194:105–106
Church JA, White NJ (2011) Sea-level rise from the late 19th to the early 21st century. Surv Geophys 32:585–602
Coleine C, Delgado-Baquerizo M, DiRuggiero J, Guirado E, Harfouche AL, Perez-Fernandez C, Egidi E (2024) Dryland microbiomes reveal community adaptations to desertification and climate change. ISME J 18(1):wrae056
Dang HL, Li E, Nuberg I, Bruwer J (2019) Factors influencing the adaptation of farmers in response to climate change: A review. Clim Dev 11(9):765–774
Dastagir MR (2019) Role of microorganisms in managing climate change impacts. Microbial interventions in agriculture and environment. Soil Crop Health Manag 3:1–16

Davidson E, Janssens I, Luo Y (2006) On the variability of respiration in terrestrial ecosystems: moving beyond Q (10). Glob Chang Biol 12:154–164
Davila AF, Warren-Rhodes K, DiRuggiero J (2021) The Atacama Desert: a window into late Mars surface habitability. In: Soare S, Conway S, Williams J-P (eds) Mars geological enigmas. Elsevier, Gainesville, pp 333–355
DeAngelis KM, Pold G, Topçuoğlu BD, van Diepen LT, Varney RM, Blanchard JL et al (2015) Long-term forest soil warming alters microbial communities in temperate forest soils. Front Microbiol 6:104
Debray R, Socolar Y, Kaulbach G, Guzman A, Hernandez CA, Curley R, Dhond A, Bowles T, Koskella B (2022) Water stress and disruption of mycorrhizas induce parallel shifts in phyllosphere microbiome composition. New Phytol 234:2018–2031
Delgado-Baquerizo M, Oliverio AM, Brewer TE et al (2018) A global atlas of the dominant bacteria found in soil. Science 359:320–325. https://doi.org/10.1126/science.aap9516
deMenocal P, Ortiz J, Guilderson T, Adkins J, Sarnthein M, Baker L (2000) Abrupt onset and termination of the African humid period: rapid climate responses to gradual insolation forcing. Q Sci Rev 19:347–361. https://doi.org/10.1016/S0277-3791(99)00081-54
Dutta H, Dutta A (2016) The microbial aspect of climate change. Energy Ecol Environ 1:209–232
Erlandson SR, Margis R, Ramirez A, Nguyen N, Lofgren LA, Liao H-L, Vilgalys R, Kennedy PG, Peay KG (2022) Transcriptional acclimation and spatial differentiation characterize drought response by the ectomycorrhizal fungus Suillus pungens. New Phytol 234:1910–1913
Falkowski PG, Fenchel T, Delong EF (2008) The microbial engines that drive Earth's biogeochemical cycles. Science 320(5879):1034–1039
Fei S, Kivlin SN, Domke GM, Jo I, La Rue EA, Phillips RP (2022) Coupling of plant and mycorrhizal fungal diversity: its occurrence, relevance, and possible implications under global change. New Phytol 234:1960–1966
Fierer N, Jackson RB (2006) The diversity and biogeography of soil bacterial communities. Proc Natl Acad Sci 103:626–631
Frey SD, Lee J, Melillo JM, Six J (2013) The temperature response of soil microbial efficiency and its feedback to climate. Nat Clim Chang 3(4):395–398
Ruchika, Garg AP (2023) Isolation and identification from diesel oil contaminated soils and their physico-chemical analysis. Int J Environ Climate Change 13(10):3476–3487. Article no. IJECC 106072
Garg AP, Sharma PD (1984) Ecology of phylloplane and litter fungi of triticale. Nordic J Bot 4:707–715
Garg AP, Sharma PD (1985) Ecology of phylloplane and leaf-litter fungi of *Cyamopsis tetragonoloba* (L.) Taub. Rev Ecol Biol Sol 22:35–55
Garg AP, Chaudhary M, Garg C (2023) Global impact of carbon emissions and strategies for its management. In: Shukla K, Patil Y, Estoque RC, de Haro PAL (eds) Quality of life and climate change. Chapter 4, IGI Global Publisher of Timely Knowledge, pp 75–107. https://doi.org/10.4018/978-1-6684-9863-7.ch004
Garg AP, Ateeq A, Bisht N, Patil B (2024) Management of asthma using probiotics. Exploration of Asthma & Allergy 2(1):9–32
Gibson L, Lee TM, Koh LP, Brook BW, Gardner TA, Barlow J et al (2011) Primary forests are irreplaceable for sustaining tropical biodiversity. Nature 478(7369):378–381
Gildemyn S, Verbeeck K, Slabbinck R, Andersen SJ, Prevoteau A, Rabaey K (2015) Integrated production, extraction, and concentration of acetic acid from CO2 through microbial electrosynthesis. Environ Sci Technol Lett 2:325–328
Gray RM (2011) Entropy and information theory. Springer Science & Business Media
Hacquard S, Wang E, Slater H, Martin F (2022) Impact of global change on the plant microbiome. Special Issue 234(6):1907–1909
Henry KM (2012) Linking nitrogen biogeochemistry to different stages of wetland soil development in the Mississippi River delta. Louisiana State University and Agricultural & Mechanical College, Louisiana

Huntington JL, Hegewisch KC, Daudert B, Morton CG, Abatzoglou JT, McEvoy DJ, Erickson T (2017) Climate engine: cloud computing and visualization of climate and remote sensing data for advanced natural resource monitoring and process understanding. Bull Am Meteorol Soc 98(11):2397–2410

Isobe K, Koba K, Otsuka S, Senoo K (2011) Nitrification and nitrifying microbial communities in forest soils. J For Res 16:351–362

Jabborova D, Mamarasulov B, Davranov K, Enakiev Y, Bisht N, Singh S, Stoyanov S, Garg AP (2024) Diversity and plant growth properties of rhizospheric bacteria associated with medicinal plants. Indian J Microbiol 64:409–417. https://doi.org/10.1007/s12088-024-01275-w

Jeffries RD (1995) The kinematics of lithium-rich, active late-type stars: evidence for a low-mass local association. Mon Not R Astron Soc 273(3):559–572

Ji B, Yang K, Zhu L, Jiang Y, Wang H, Zhou J, Zhang H (2015) Aerobic denitrification: a review of important advances of the last 30 years. Biotechnol Bioprocess Eng 20:643–651

Kabir M, Habiba UE, Khan W, Shah A, Rahim S, Patricio R, Shafiq M (2023) Climate change due to increasing concentration of carbon dioxide and its impacts on environment in 21st century; a mini review. J King Saud Univ Sci 35(5):102693

Koutroulis AG (2019) Dryland changes under different levels of global warming. Sci Total Environ 655:482–511. https://doi.org/10.1016/j.scitotenv.2018.11.215

Lennon JT, Aanderud ZT, Lehmkuhl BK, Schoolmaster DR Jr (2012) Mapping the niche space of soil microorganisms using taxonomy and traits. Ecology 93(8):1867–1879

Leung PM, Bay SK, Meier DV et al (2020) Energetic basis of micro bial growth and persistence in desert ecosystems. mSystems 5:e00495–e00419. https://doi.org/10.1128/mSystems.00495-19

Levy K, Smith SM, Carlton EJ (2018) Climate change impacts on waterborne diseases: moving toward designing interventions. Curr Environ Health Rep 5(2):272–282. https://doi.org/10.1007/s40572-018-0199-7. PMID: 29721700; PMCID: PMC6119235

Liu S, Lorenzen ED, Fumagalli M, Li B, Harris K, Xiong Z (2014) Population genomics reveal recent speciation and rapid evolutionary adaptation in polar bears. Cell 157:785–794. https://doi.org/10.1016/j.cell.2014.03.0547

Lu N, Wang M, Ning B (2018) Research advances in ecosystem services in drylands under global environmental changes. Curr Opin Environ Sustain 33:92–98. https://doi.org/10.1016/j.cosust.2018.05.004

Manzoni S, Taylor P, Richter A, Porporato A, Ågren GI (2012) Environmental and stoichiometric controls on microbial carbon-use efficiency in soils. New Phytol 196(1):79–91

Martínez-Valderrama J, Guirado E, Maestre FT (2020) Decertifying deserts. Nat Sustain 3:572–575. https://doi.org/10.1038/s41893-020-0561-2

McMichael AJ (2004) Environmental and social influences on emerging infectious diseases: past, present and future. Philos Trans R Soc Lond Ser B Biol Sci 359(1447):1049–1058. https://doi.org/10.1098/rstb.2004.1480. PMID: 15306389; PMCID: PMC1693387

Neill C, Gignoux J (2006) Soil organic matter decomposition driven by microbial growth: a simple model for a complex network of interactions. Soil Biol Biochem 38(4):803–811

Pajares S, Bohannan BJ, Souza V (2016) Editorial: the role of microbial communities in tropical ecosystems. Front Microbiol 7:1805. https://doi.org/10.3389/fmicb.2016.01805. PMID: 27891123; PMCID: PMC5104756

Poloczanska ES, Brown CJ, Sydeman WJ, Kiessling W, Schoeman DS, Moore PJ (2013) Global imprint of climate change on marine life. Nat Clim Chang 3:919–925. https://doi.org/10.1038/nclimate19585

Raftery AE, Zimmer A, Frierson DMW, Startz R, Liu P (2017) Less than 2°C warming by 2100 unlikely. Nat Clim Chang 7:637–641. https://doi.org/10.1038/nclimate33523

Rinnan R, Michelsen A, Bååth E, Jonasson S (2007) Fifteen years of climate change manipulations alter soil microbial communities in a subarctic heath ecosystem. Glob Chang Biol 13(1):28–39

Ruchika, Amar P (2023) Garg Screening of diesel oil degrading species of *Aspergillus* for the production of different enzymes using agar plate assay method. Int J Environ Climate Change 13(11):1360–1371. Article no. IJECC.107868

Saba H, Vibhash D, Manisha M, Prashant KS, Farhan H, Tauseef A (2012) Trichoderma–a promising plant growth stimulator and biocontrol agent. Mycosphere 3(4):524–531

Sahrawat R, Garg AP (2023) Biodegradation of rice straw using mixed cultures of microfungi and their extracellular crude enzymes preparations under field conditions. Int J Environ Climate Change 13(8):2097–2104. https://doi.org/10.9734/IJECC/2023/v13i82167

Salvioli A, Ghignone S, Novero M, Navazio L, Venice F, Bagnaresi P, Bonfante P (2016) Symbiosis with an endobacterium increases the fitness of a mycorrhizal fungus, raising its bioenergetic potential. ISME J 10(1):130–144

Santosh M, Groves DI, Yang CX (2024) Habitable planet to sustainable civilization: global climate change with related clean energy transition reliant on declining critical metal resources. Gondwana Res 130:220–233

Schimel JP, Schaeffer SM (2012) Microbial control over carbon cycling in soil. Front Microbiol 3:348

Sharma SB, Sayyed RZ, Trivedi MH, Gobi TA (2013) Phosphate solubilizing microbes: sustainable approach for managing phosphorus deficiency in agricultural soils. SpringerPlus 2:1–14

Singh BK, Bardgett RD, Smith P, Reay DS (2010) Microorganisms and climate change: terrestrial feedbacks and mitigation options. Nat Rev Microbiol 8(11):779–790

Soliveres S, Maestre FT, Eldridge DJ, Delgado-Baquerizo M, Quero JL, Bowker MA, Gallardo A (2014) Plant diversity and ecosystem multifunctionality peak at intermediate levels of woody cover in global drylands. Glob Ecol Biogeogr 23(12):1408–1416

Walsh RE, Assis APA, Patton JL, Marroig G, dawson TE, Lacey EA (2016) Morphological and dietary responses of chipmunks to a century of climate change. Glob Change Biol 22:3233–3252. https://doi.org/10.1111/gcb.132166

Whitaker J, Ostle N, Nottingham AT, Ccahuana A, Salinas N, Bardgett RD et al (2014) Microbial community composition explains soil respiration responses to changing carbon inputs along an Andes-to-Amazon elevation gradient. J Ecol 102(4):1058–1071

Wierzchos J, de los Ríos A, Ascaso C (2012) Microorganisms in desert rocks: the edge of life on earth. Int Microbiol 5:173–183

Xie HW (2022) Methods for Bayesian inference and data assimilation of soil biogeochemical models. University of California, Irvine

Young IM, Crawford JW, Nunan N, Otten W, Spiers A (2008) Microbial distribution in soils: physics and scaling. Adv Agron 100:81–121

Part II

Climate Resilience in Agroecosystem

Significance of Plant Growth-Promoting Rhizobacteria in Alleviating Drought Stress in Crop Plants Under a Changing Climate for Sustainable Agricultural Production

5

Aditya Sharma, Pratyoosh Shukla, and Piyush Pandey

Abstract

The intensification of drought and climate change presents a significant risk to agricultural production, increasing the urgency of finding environmentally friendly solutions to ensure global food security. Drought stress altered the stomatal conductance, osmotic adjustment, antioxidant, root architecture and morphology, regulation of hormonal signaling, metabolic regulations, and molecular responses in crop plants. In this context, plant growth-promoting rhizobacteria (PGPR) has emerged as a crucial sustainable agricultural solution in a changing climate. These bacteria significantly alleviate drought stress in crop plants through various mechanisms, including modifying root morphology, regulating stress-responsive genes, influencing plant hormones and stress signalling pathways, and producing volatile organic compounds and exopolysaccharides. Furthermore, PGPR improves plant resilience to drought by enhancing water and nutrient uptake efficiency, modulating stress hormone levels, and inducing systemic resistance mechanisms. This chapter explores the sophisticated roles of PGPR in mitigating drought stress, highlighting recent advances in understanding their interactions with crop plants under changing climatic conditions. Additionally, it discusses practical applications of PGPR-based strategies for

A. Sharma · P. Shukla
Enzyme Technology and Protein Bioinformatics Laboratory, School of Biotechnology, Institute of Science, Banaras Hindu University, Varanasi, Uttar Pradesh, India
e-mail: shuklap@bhu.ac.in

P. Pandey (✉)
Department of Microbiology, Soil and Environmental Microbiology Laboratory, Assam University, Silchar, India

P. Pandey et al. (eds.), *Microorganisms Resilience to Climate Change*, Microorganisms for Sustainability 56, https://doi.org/10.1007/978-981-96-3748-5_5

sustainable agriculture production. It emphasizes their potential to enhance crop yields, reduce environmental impact, and promote resilience in agricultural systems vulnerable to drought.

5.1 Introduction

In the current climate change scenario, intense heat waves, longer fire seasons, shifting rainfall patterns, high temperatures, floods, and speeding drought are important severe climate changes all around the world (Dey and Lewis 2021). As the effects of global drought continue to rise, the worldwide population experiencing even more challenges to declining global crop yields such as maize, corn, wheat, rice, and soybeans (Neupane et al. 2022). Drought stress is one of the most devastating abiotic stresses that significantly affects an average of 25% of the agriculture sector worldwide (Santos-Medellín et al. 2021). It is worsened by climate change and water scarcity. According to a National Centres for Environmental Information (NCEI) report, drought-related United States agriculture lost approximately $9.4 billion annually over the past decade, varying based on regional severity. Another report by the Farm Credit Administration (FCA), USA (2017), indicated that drought stress led to an 89% reduction in corn productivity, an 85% decrease in wheat yield, and a 90% decline in soybean production, resulting in significant losses in the agriculture sector. The rising severity of drought results from climate change and human-caused warming, speeding up the duration, magnitude, and episodes of drought stress worldwide (Li et al. 2023a, b; Bloomfield et al. 2019). The impact of worldwide climate change and drought significantly affects crop productivity, causing changes in the morpho-physiological and biochemical aspects of crop plants (Fatema et al. 2023). The most significant strategies of stomatal closure, osmotic adjustment, root morphology changes and exudates, antioxidant regulation (catalase (CAT), superoxide dismutase (SOD), ascorbate peroxidase (APX), malonaldehyde (MDA), and peroxide dismutase (POD)), gene expression regulation, and metabolic adjustments collectively help plants withstand and recover from drought stress (Kang et al. 2022). An investigation by Dayoub et al. (2021) showed that *Glycine max* L. enhanced the root system architecture, root density, root hairs, depth, and length to mitigate the adverse effects of drought stress. Moreover, Sharma et al. (2021) advocated the mitigation strategy in *Oryza sativa* L. where drought stresses lead to enhance the signal transduction pathway such as ABA (Abscisic acid) synthesis, aquaporins via MAPKs (mitogen-activated protein kinase) and $Ca2^{+}$ mediated signaling upon drought stress perception, which is responsible for reduction in rice yield. Similarly, in response to drought stress, *Zea mays* L. showed a noticeable improvement in antioxidant defense regulation, with SOD increasing by 58.9%, POD (36%), and MDA content by 28.7% compared to non-stressed plants (Ren et al. 2024). Furthermore, *Zea mays* L. showed higher proline and relative water content (RWC) (Gopalakrishna et al. 2023), while *Triticum aestivum* L. exhibited increased expression levels of TaCAT, TaPOD, TaDREB1, and TaSOD under drought stress (Ge et al. 2024).

PGPR (Plant Growth Promoting Rhizobacteria) offers a cost-effective and eco-friendly method for improving crop resilience to drought stress. By utilizing specific PGPR strains for different crops and environments, we can maximize their benefits in sustainable agriculture (Sun et al. 2024). PGPR can enhance cell growth by upregulating gene expression, increasing water use efficiency, ABA levels, and the production of compatible solutes such as soluble sugar, proline, and organic acids (Kumar et al. 2024; Sharma et al. 2024; İpek and Mutluay 2022). Additionally, PGPR can promote photosynthesis, antioxidant enzyme activity (SOD, POD, CAT, and APX), chlorophyll levels, shoot length, relative water content (RWC) in drought conditions, and plant biomass (Hasan et al. 2024). Furthermore, PGPR also upregulates different phytochemicals such as indole3-acetic acid (IAA) production, ACC deaminase (1-aminocyclopropane-1-carboxylate) ammonia (NH3) production, hydrogen cyanide (HCN) production, catalase production under drought stress conditions (Baruah et al. 2021). As biofertilizers, PGPR is categorized into diverse groups of beneficial bacteria that colonize the rhizosphere of plants and provide various benefits, especially in mitigating the effects of drought stress on crop plants. Such group of PGPR encompasses several genera such as *Pseudomonas*, *Herbaspirillum*, *Azotobacter*, *Azospirillum*, *Bacillus*, *Paraburkholderia*, *Burkholderia*, *Enterobacter*, *Serratia*, *Klebsiella*, and *Gluconacetobacter* (Amaresan et al. 2020; Bittencourt et al. 2023). Therefore, understanding these mechanisms is pivotal, particularly in climate change and uncertain water availability, as it paves the way for developing drought-resistant crop varieties and implementing sustainable agricultural practices (Ojuederie et al. 2019).

5.2 Response Mechanisms by Crop Plants Under Drought Stress

Crop plants employ various mechanisms to tackle the negative consequences of drought stress. These mechanisms involve physiological, biochemical, molecular, and morphological changes that collectively enhance their resilience (Fatema et al. 2023). As a result, crop plants can withstand and maintain productivity even under drought conditions. Understanding these responses is crucial to developing strategies for breeding crops with enhanced drought tolerance, thus ensuring food security in water-deficit regions (Kang et al. 2022; Bali et al. 2023; Parmar et al. 2022). The significant adaptive mechanism employed by crop plants under abiotic stress (drought), such as stomatal regulation (stomatal conductance), osmotic adjustment (accumulation of osmoprotectants), antioxidant defense (activation of antioxidant enzymes i.e. SOD, POD, CAT and APX), root adaptations (root architecture and morphology), regulation of hormonal signaling (Abscisic acid, Jasmonic acid, Ethylene), metabolic regulations and molecular responses are briefly explained in further sections.

5.2.1 Stomatal Regulation

Stomatal regulation is one of the primary responses to drought stress. It involves closing stomata, which are tiny pores on the leaf surface (Jezek and Blatt 2017). This helps reduce water loss through transpiration and maintains water balance within the plant. Plants adjust the sensitivity of stomata to environmental cues, such as light and humidity, to optimize water use efficiency (Wang et al. 2018). Under drought stress conditions, crops such as *Hordeum vulgare* (Kaczmarek et al. 2017), *Zea mays* (Mansour et al. 2023), *Oryza sativa* (Chen et al. 2018), and *Triticum aestivum* (Li et al. 2023a, b) experience the reduction of turgor pressure in guard cells via ABA-mediated signaling response, which is responsible for the imbalance of K+ and Cl− ion channels. Krishna et al. (2022), advocated that stomatal regulation and conductance are phenomenal physiological processes of *solanum lycopersicum* L. under drought stress. Furthermore, Parasuraman et al. (2024) also observed that drought stress altered the pattern of stomatal closure and opening in *Glycine max* L., ultimately affecting ion effluxes, such as K+, $Ca2^+$, and H^+.

5.2.2 Osmotic Adjustment

Plants accumulate low molecular weight compatible solutes (organic and inorganic) such as proline, sugars, and betaines (glycine betaines) to uphold cell turgor pressure and osmotic balance (Krishna et al. 2022). At physiological pH, osmolytes maintain a neutral charge, which avoids disrupting the water molecules closely bound to macromolecules surfaces. This preservation enables the macromolecules to uphold their native structure and function even under osmotic stress conditions (Bhutto et al. 2023). These compounds aid in stabilizing proteins and membranes, preventing cells from damage caused by drought stress. (Parasuraman et al. 2024). An investigation reported that under drought stress conditions, *Glycine max* L. had shown enhanced osmotic potential (higher negative −1.82 Mpa) compared to non-stress (lower negative −1.16 Mpa) (Parasuraman et al. 2024). Additionally, Bhutto et al. (2023) found that under drought and salt stress a surge of osmolytes was observed which includes, organic solutes, a wide range of amino acids, sugars, osmolytes (proline) in vacuoles and cytosol of *Triticum aestivum*.

5.2.3 Production of Reactive Oxygen Species

Drought stress increases the production of reactive oxygen species (ROS) in plants, including free radicals (O_2^-), hydrogen peroxide (H_2O_2), and hydroxyl radicals (-OH), which are responsible for respiratory burst oxidase (RBO) in plant cells (Verma et al. 2019). Vijayaraghavareddy et al. (2022) have shown that under drought stress (60% filed capacity), a higher accumulation of free radicle (O_2^-) was observed in *Oryza sativa* L. genotype IR64 and APO compared to *Triticum aestivum* L. genotype Weebill. Furthermore, Rane et al. (2022), reviewed the combined effect of reactive oxygen species (ROS) and antioxidative system (AOS) such as SOD, POD, CAT, and APX, in response to drought stress in several crops such as *Glycine max*, *Pisum sativum*, *Gossypium*, *Arabidopsis thaliana* and *Nicotiana tabacum*. Sharma et al. (2017) also described the harmful effect of free radicle (O_2^-) in different crops. Similarly in *Brassica napus* L. drought stress induced the O_2^- leading to cell damage and injury (Lee et al. 2019).

5.2.4 Root Architecture and Hormone Signaling

Drought-tolerant crops often have deeper and more extensive root systems, allowing them to access water stored deeper in the soil profile, which remains beyond the reach of shallower-rooted plants. Crop plants change their root density and enhance root branching to improve nutrient and water uptake efficiency (Dash et al. 2023; Altaf et al. 2021). Additionally, phytohormones such as ABA significantly mediate plant responses to drought stress, regulate stomatal closure, and enhance the synthesis of osmoprotectants. Besides the ABA several phytohormones such as ethylene, cytokine, and jasmonic acid also contribute to the drought stress response by modulating root architecture and plant growth (Wahab et al. 2022). Gusain et al. (2024), summarized the root-mediated response and phytohormone signaling pathways under drought stress, such as ABA, Auxin, Cytokinins, Ethylene, and Jasmonic acid in several economically important crop plants i.e. *Populus*, *Zea mays*, *Oryza sativa*, *alfalfa*, *Medicago truncatula*, and *Arabidopsis thaliana*.

5.2.5 Relative Water Content

Relative water content (available for plant absorption) is the amount of water available for physiological processes in soil (Ojuederie et al. 2019). RWC is the midpoint of wilting potential and filed capacity. Abiotic stress (drought) significantly reduced the RWC and several physiological parameters, photosynthesis efficiency, stomal regulation, and hormonal signalling (Boussora et al. 2024). Ojuederie et al. (2019), highlighted the significance of drought on crop plants and several physiological parameters such as Lower RWC, leaf water potential (ψw), chlorophyll, and Stomata conductance. These factors can impair plant processes and reduce productivity in crop plants, namely, Barley (*Hordeum vulgare* L.), Chickpea (*Cicer arietinum* L.), *Camptotheca acuminate*, Cowpea (*Vigna unguiculata* L.), Faba bean (*Vicia faba* L.), Wheat (*Triticum aestivum* L.) and Maize (*Zea mays* L.) (Ojuederie et al. 2019).

5.3 PGPR-Mediated Alleviation of Drought Stress in Crop Plants

Although plants have their defense strategies which help to overcome and withstand environmental constraints but additional bonds with PGPR offer a promising approach in modern agriculture practices by significantly enhancing plant growth, improving nutrient uptake, effectively managing plant diseases, and increasing tolerance to environmental stresses (Sun et al. 2024; Hasan et al. 2024). Their use as biofertilizers and biocontrol agents promotes sustainable farming practices by reducing chemical inputs and supporting ecosystem health in agricultural soils (Ojuederie et al. 2019). PGPR, the beneficial soil bacteria, inhabits the rhizosphere, the region of soil influenced by plant roots, and supports plant growth through various mechanisms (Bittencourt et al. 2023). Throughout evolution, a wide variety of PGPRs have developed a natural, symbiotic relationship with plants. This relationship is evident in the interactions that take place under different environmental conditions, such as abiotic stress (drought, salinity, temperature, and cold) and biotic stress (pathogens, fungi, bacteria, nematodes, and viruses) (Amaresan et al. 2020). Mitra et al. (2024), highlighted the significant role of specific PGPR groups in various challenging environmental conditions, notably *Rhizobium*, *Burkholderia*, *Mesorhizobium*, *Achromobacter*, *Frankia*, *Enterobacter*, *Rhodococcus*, *Pseudomonas* sp., *Acidothiobacillus* sp., *Bacillus*, and *Serratia*. PGPR-mediated drought stress mitigation strategies are comprehensively discussed in the following sections.

5.3.1 Modification in the Root Morphology

As water and nutrient availability are crucial factors for plant growth and yield, even moderate drought stress can reduce plant productivity. Therefore, PGPR (*Pseudomonas fluorescens*) has been found to aid in the absorption of nutrients such as magnesium (Mg), calcium (Ca), potassium (K), phosphorus (P), and sulfur (S) under moderate to severe drought conditions by modifying the root system to be more intense and deeper in *Zea mays* L. (Zaib et al. 2023). PGPR as a biotechnological approach plays a crucial role in shaping root morphology by promoting root elongation, enhancing root density, stimulating root hair development, modifying root architecture, and influencing hormone signaling pathways, exopolysaccharides (EPSs) production, and enhancing production of microbial phytohormone (Kumar et al. 2024; Goswami and Suresh 2020). Harkhani and Sharma (2024), highlighted the significant mitigation approach in different crop plants under drought stress by root-associated beneficial rhizobacteria. Furthermore, Meenakshi et al. (2019) demonstrated the role of endophytic PGPR in improving the root-to-shoot ratio under drought stress in *Triticum aestivum*. Similarly, Abbasi et al. (2020) discovered that rhizobacteria (*Streptomyces* sp.) improved water use efficiency (WUE) and root and shoot biomass under limited water conditions, consequently mitigating the adverse effects of drought stress in *Solanum lycopersicum* L. Ahmad et al. (2022) summarised the role of PGPR, which can help to alleviate the impact of drought stress in crops by changing root structure, accumulating osmolytes, emitting volatile organic compounds (VOCs), activating stress-responsive genes, producing plant hormones like indole-3-acetic acid, creating iron-chelating compounds (siderophores) and EPS, and increasing 1-aminocyclopropane-1-carboxylate deaminase (ACC) activities.

5.3.2 Regulation of Stress-Responsive Genes Under Drought Stress

PGPR activates specific sets of genes (play metabolic and regulatory roles) in plants that are important for responding to drought stress. These genes help adjust to osmotic stress, regulate antioxidant response, and regulate hormone signaling (Ferrante et al. 2024) Fig. 5.1. These genes are responsible for osmotic adjustment and produce substances such as proline, glycine betaine, and sugars, which are crucial for maintaining the water balance in cells and preventing

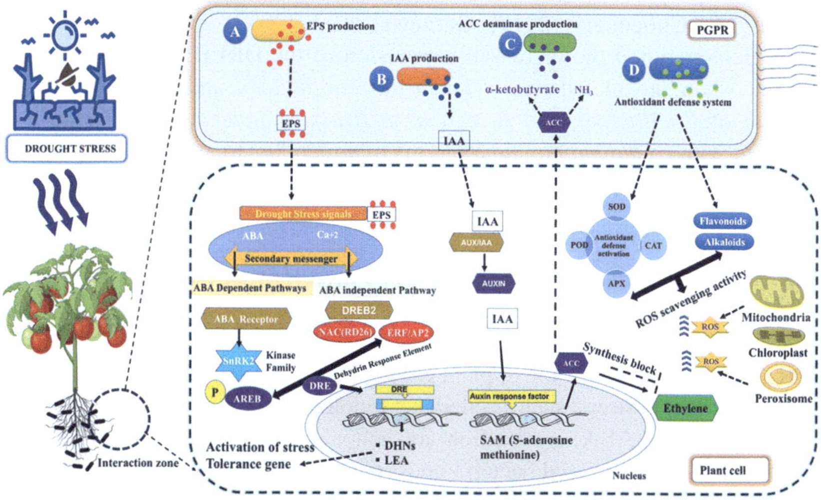

Fig. 5.1 Representing the PGPR-mediated stress response under drought stress. (**a**) Exopolysaccharide-producing PGPR enhances the endogenous level of ABA hormone in plant cells by recognizing the drought signal and activating the stress tolerance gene. (**b**) Bacterial IAA produced the ACC via activation of SAM (**c**). The ACC deaminase enzyme reduces an elevated level of ethylene, which breaks down the ACC into α-ketobutyrate and production. (**d**) The antioxidant defense system is activated to reduce plant ROS accumulation under drought stress. EPS (Exopolysaccharide), ABA (Abscisic acid), AREB (Arabidopsis response element binding), DRE (dehydrin response elements), LEA (Late embryogenesis abundant), IAA (Indole-3-acetic acid), SAM (S-adenosine methionine), ACC (1-Aminocyclopropane-1-Carboxylate), ROS (Reactive oxygen species), SOD (Superoxide dismutase), POD (Peroxide dismutase), CAT (Catalase), APX (Ascorbate peroxidase)

dehydration in plants (Lucas et al. 2023; Ansari and Ahmad 2019). An investigation by Lucas et al. (2023), showed that inoculation of PGPR (*Pseudomonas* strain N5.12 and N21.24) in *Solanum lycopersicum* L. under moderate drought stress, a significantly upregulated expression genes were observed, such as P5CS (pyrroline-5-carboxylate synthase), NCDE1 (9-cisepoxycarotenoid dioxygenase), and PM ATPase 1 (Plasma membrane ATPase 1), which play a crucial role in the synthesis of compatible solute proline, ABA-synthesis and stomatal conductance and turgidity of guard cell under drought stress respectively. Similarly, Kakar et al. (2016), reported that the introduction of three rhizospheric bacteria (*Brevibacillus laterosporus*, *Bacillus methylotrophicus*, and *B. amyloliquefaciens*) led to increased expression of the OsMYB3R-2 gene in rice (*Oryza sativa* L.) when exposed to drought stress. As a result, the plants exhibited enhanced height and an improved shoot-to-root ratio. In addition, Saikia et al. (2018), advocated that crop plants such as *Pisum sativum* L and *Vigna mungo* L. inoculated with P*seudomonas* sp. RJ15, *Ochrobactrum pseudogrignonense* RJ12 and *Bacillus subtilis* RJ46 (responsible for ACC deaminase-production) reduce the amount of ACC (1-aminocyclopropane-1-carboxylate) and ACO (1-aminocyclop ropane-1-carboxylate oxidase) genes and upregulation of ACS (Aminocyclopropane-1-carboxylate synthase) under drought stress. Furthermore, Vaishnav and Choudhary (2019), Barnawal et al. (2017) and Bandeppa et al. (2019), demonstrated the increased expression of the DREB2 gene in *Triticum aestivum* when inoculated with *Dietzia natronolimnaea* and *Bacillus subtilis*, treatment of *Bacillus* sp. and *B. cereus* in *Brassica juncea*, and *Glycine max* inoculated with *Pseudomonas simiae*. This treatment led to an increase in the dry weight of these crop plants.

5.3.3 PGPR-Mediated Regulation of Plant Hormones and Stress Signaling Pathways

PGPR enhances the production of plant growth regulators and stress signaling pathways under drought stress, such as ethylene, Indole-3-acetic acid (IAA) production, and Abscisic acid (ABA) modulation are major groups of phytohormones (Sati et al. 2023). These beneficial bacteria boost plant growth and tolerance to drought by modulating phytohormone levels and forming symbiotic relationships with plants (Sevim et al. 2023) Table 5.1. Ullah et al. (2019), showed that nearly 80% of PGPR species produced a significant amount of IAA to enhance the root exudation, root biomass, and architecture, and combat drought stress tolerance along with their colonization in plant roots. Similarly, two PGPR isolates, *Cupriavidus necator*, and *Pseudomonas fluorescens*, enhanced drought tolerance mechanisms by increasing nitrogen and phosphorus content availability under drought stress (Pereira et al. 2020). Furthermore, *Bacillus subtilis*, *Mesorhizobium*, and *B. mojavensis* successfully mitigate drought stress in *Cicer arietinum* by improving the root-to-shoot ratio by producing IAA (Mahmood Aulakh et al. 2020). Ambreetha et al. (2018), reported

Table 5.1 PGPR-mediated drought stress mitigation in various crop plants under drought stress

Sr. No.	PGPR isolates	Host plant	Drought stress mitigation strategy	References
1.	*Pseudomonas*, *Brevibacillus* and *Paenibacillus*	*Solanum lycopersicum* L.	Enhance the IAA production, phosphate solubilization, and exopolysaccharide production	Astorga-Eló et al. (2021)
2.	*Azospirillum brasilense* and *Stenotrophomonas maltophilia*	*Triticum aestivum* L.	Reduced the electrolyte leakage and accumulation of malondialdehyde and hydrogen peroxide	Kasim et al. (2021)
3.	*Bacillus subtilis*, *B. cereus* and *Brucella haematophilum*	*Oryza sativa* L.	Induced indole-3 acetic acid production, ACC deaminase, and antioxidant defense	Mahreen et al. (2023)
4.	*Bacillus rugosus*	*Triticum aestivum* L.	Increased the proline and chlorophyll content	Salem et al. (2024)
5.	*Achromobacter xylosoxidans*, *Pseudomonas fluorescence*, *Bacillus subtilis*	*Zea mays* L.	Enhanced the superoxide dismutase, glutathione, and catalase activity	Rai et al. (2024)
6.	*Pantoea agglomerans*, *B. subtilis*, *B. cereus* and *B. licheniformis*	*Glycine max* (L.) Merr.)	Reduced the ethylene stress via ACC deaminase production	Dubey et al. (2024)
7.	*Pantoea* and *Pseudomonas*	*Hordeum vulgare* L.	Improve the chlorophyll levels, photosynthetic rate, and stomatal conductance	Abideen et al. (2022)
8.	*Serratia odorifera*	*Hordeum vulgare* L.	PGPR isolates enhanced the antioxidant defense (superoxide dismutase (SOD), catalase (CAT), and peroxidase)	Gul et al. (2023)
9.	*Enterobacter sp.*, *Kluyvera sp.* and *Bacillus subtilis*	*Oryza sativa* L.	Enhanced the antioxidant properties	Singha and Pandey (2023)
10.	*Bacillus subtilis*	*Solanum tuberosum*	Enhanced the antioxidant defense (superoxide dismutase, catalase, and peroxidase)	Batool et al. (2020)
11.	*Glomus versiforme* and *Bacillus methylotrophicus*	*Nicotiana tabacum*	Increased the SOD, POD, CAT, and glutathione reductase (GR)	Begum et al. (2022)
12.	*Pseudomonas moraviensis*	*Triticum aestivum*	Increased the Ascorbate and catalase production	Yaseen et al. (2020)
13.	*Pseudomonas fluorescens* and *Pseudomonas putida*	*Hordeum vulgare* L.	Increase in the absorption of enhancement of Mg, K, S, P, Ca, Al, and Si	Zaib et al. (2023)

(continued)

Table 5.1 (continued)

Sr. No.	PGPR isolates	Host plant	Drought stress mitigation strategy	References
14.	*Bacillus amyloliquefaciens*	*Pennisetum glaucum*	Increase the production of ACC deaminase under drought stress	Murali et al. (2021)
15.	*Bacillus albus*	*Panicum sumatrense* L.	Enhanced the IAA, siderophore, nutrient solubilization	Rajkumar et al. (2024)

that *Bacillus altitudinis* altered the endogenous level of IAA which is responsible for changes in the root architecture and ensuring the rapid mitigation of drought stress in *Triticum aestivum* L. Some strains of PGPR can trigger the expression of ABA biosynthetic genes in plants, resulting in increased ABA (stress hormone) production. Elevated ABA levels assist plants in closing stomata to reduce water loss through transpiration. An investigation by Barnawal et al. (2017), reported that Bacillus subtilis produces the synthesis of IAA by synthesizing its precursor (abscisic acid and 1-aminocyclopropane-1-carboxylate) which triggers the pathways of TaCTR1 and STR1 under drought stress in wheat. It was observed that inoculation with PGPR significantly increased the level of the ABA hormone during drought stress, leading to a reduction in transpiration rate and an enhanced synthesis of the AQP transporter (aquaporin) in plants (Goswami and Suresh 2020; Bouremani et al. 2023). Additionally, *Bacillus subtilis* is also crucial in promoting plant growth and mitigating stress in *Solanum lycopersicum* L. by synthesizing the ACC deaminase enzyme. This enzyme helps reduce the endogenous level of ethylene hormone in plants under drought stress by breaking down ACC (a precursor of ethylene) into ammonia and α-ketobutyrate (Gowtham et al. 2020). In further context, PGPR also stimulates osmolyte production via the synthesis of volatile organic compounds (VOCs) to maintain stomatal turgidity under drought stress. A study by Yasmin et al. (2021) suggested that *Pseudomonas pseudoalcaligenes* enhances drought tolerance in *Zea mays* L. by reducing the level of malonaldehyde (MDA) through the secretion of dimethyl disulfide and 2,3-butanediol VOCs.

5.3.4 Production of Exopolysaccharide (EPS) Under Drought

EPS (hydrated polymers) produced by PGPR can bind soil particles together, participate in the microbial biofilm formation, improve water holding capacity, create higher moisture content, enhance soil aggregation, entrapped nutrients (N, P, and K), and induce plant hormones (auxins and cytokinins), improving soil porosity, and water retention (Mandal et al. 2022; Naseem et al. 2018). Approximately 15% to 40% of the rhizospheric surface is colonized by microorganisms that produce extracellular polymeric substances (EPS) (Carezzano et al. 2023). Latif et al. (2022), demonstrated that EPS-producing PGPR isolates *Klebsiella* sp., *Chryseobacterium* sp., and *Acinetobacter* sp. can alleviate drought stress in *Triticum aestivum*

L. Similarly, Ilyas et al. (2020), found that *Bacillus subtilis* and *Azospirillum brasilense* could produce higher amounts of EPS (sugar, protein and uronic acid) and osmolytes (proline) under drought stress in wheat plants. *Helianthus annuus* L. seedlings were able to withstand drought stress because the EPS-producing bacterium *Pseudomonas putida* formed a biofilm. Additionally, it improved soil aggregation, the relative water content of leaves, and soil adhesion to roots. Furthermore, *Bacillus cereus* and *Pseudomonas aeruginosa* also reduced the drought stress in *Zea mays* L. by EPS production, phosphate solubilization, and water retention (Naseem et al. 2024).

5.4 Conclusion and Future Perspective

Drought is a significant stress for crop plants in many parts of the world, severely affecting growth and yield. Plant growth-promoting rhizobacteria (PGPR) is a promising strategy to help crops withstand drought stress in a changing climate, promoting sustainable agricultural production. PGPR helps plants cope with water scarcity by improving nutrient availability, producing phytohormones, and modulating stress signals. They can also form symbiotic relationships with plants, offering a cost-effective and eco-friendly alternative to chemical use. While significant efforts have been taken to understand the benefits of PGPR, further study is needed to conduct and optimize their use across different crop types and environmental conditions. In the current global climate change situation, leveraging the beneficial interactions between plants and microbes is crucial to enhance food production and support the growing global population. Researchers are increasingly focusing on rhizosphere engineering to create specialized environments that promote enhanced crop yields, soil health, and sustainable agricultural practices.

Acknowledgements PS acknowledges the support from Science & Engineering Research Board (SERB) Department of Science and Technology, GoI (Grant No. CRG/2021/001206), and the 'Faculty Incentive Grant' by Institute of Eminence (IoE) Scheme by BHU, Varanasi (Letter No R/Dev/D/IoE/Seed & Incentive/2022-23/50024). PS also acknowledges, the BTIS-Net-Sub-Distributed Information Centre, funded by DBT, Govt. of India at the School of Biotechnology, Banaras Hindu University, Varanasi, India. AS acknowledges the Senior Research Fellowship from UGC, New Delhi, India (NTA Ref. No: 191620105037) and the Institute of Eminence-External Research Experience (IoE-ERE) scheme (Sanction No. SRICC/IoE-ERE/AS-SOB 6215). PP is grateful to the Department of Biotechnology, Ministry of Science and Technology Govt of India for financial support.

Declarations **Ethics Approval and Consent to Participate**

Not Applicable.
Consent for Publication
Not Applicable.
Availability of Data and Materials
Not Applicable.
Competing Interests
The authors declare that they have no competing interests.

Authors' Contributions PS and PP conceptualized the manuscript. AS wrote the first draft of the manuscript. PS and PP reviewed and edited the manuscript. All the authors read and approved the final manuscript.

Funding AS acknowledges the Senior Research Fellowship from UGC, New Delhi, India (NTA Ref. No: 191620105037) and the Institute of Eminence-External Research Experience (IoE-ERE) scheme (Sanction No. SRICC/IoE-ERE/AS-SOB 6215).

References

Abbasi S, Sadeghi A, Safaie N (2020) Streptomyces alleviate drought stress in tomato plants and modulate the expression of transcription factors ERF1 and WRKY70 genes. Sci Hortic 265:109206

Abideen Z, Cardinale M, Zulfiqar F, Koyro HW, Rasool SG, Hessini K, Darbali W, Zhao F, Siddique KH (2022) Seed endophyte bacteria enhance drought stress tolerance in Hordeum vulgare by regulating, physiological characteristics, antioxidants and minerals uptake. Front Plant Sci 13:980046

Ahmad HM, Fiaz S, Hafeez S, Zahra S, Shah AN, Gul B, Aziz O, Fakhar A, Rafique M, Chen Y, Yang SH, Wang X (2022) Plant growth-promoting rhizobacteria eliminate the effect of drought stress in plants: a review. Front Plant Sci 13:875774

Altaf MA, Shahid R, Ren MX, Khan LU, Altaf MM, Jahan MS, Nawaz MA, Naz S, Shahid S, Lal MK, Tiwari RK, Shahid MA (2021) Protective mechanisms of melatonin against vanadium phytotoxicity in tomato seedlings: insights into nutritional status, photosynthesis, root architecture system, and antioxidant machinery. J Plant Growth Regul:1–17

Amaresan N, Kumar MS, Annapurna K, Kumar K, Sankaranarayanan A (eds) (2020) Beneficial microbes in agro-ecology: bacteria and fungi. Academic

Ambreetha S, Chinnadurai C, Marimuthu P, Balachandar D (2018) Plant-associated Bacillus modulates the expression of auxin-responsive genes of rice and modifies the root architecture. Rhizosphere 5:57–66

Ansari FA, Ahmad I (2019) Alleviating drought stress of crops through PGPR: mechanism and application. In: Microbial interventions in agriculture and environment: volume 2: rhizosphere, microbiome and agro-ecology, pp 341–358

Astorga-Eló M, Gonzalez S, Acuña JJ, Sadowsky MJ, Jorquera MA (2021) Rhizobacteria from 'flowering desert' events contribute to the mitigation of water scarcity stress during tomato seedling germination and growth. Sci Rep 11(1):13745

Bali A, Rawal S, Singla K (2023) Challenges and strategies to improve drought tolerance in plants through agronomic managements. In: Salinity and drought tolerance in plants: physiological perspectives. Springer Nature Singapore, Singapore, pp 491–506

Bandeppa S, Paul S, Thakur JK, Chandrashekar N, Umesh DK, Aggarwal C, Asha AD (2019) Antioxidant, physiological and biochemical responses of drought susceptible and drought tolerant mustard (Brassica juncea L) genotypes to rhizobacterial inoculation under water deficit stress. Plant Physiology and Biochemistry, 143:19–28

Barnawal D, Bharti N, Pandey SS, Pandey A, Chanotiya CS, Kalra A (2017) Plant growth-promoting rhizobacteria enhance wheat salt and drought stress tolerance by altering endogenous phytohormone levels and TaCTR1/TaDREB2 expression. Physiol Plant 161(4):502–514

Baruah I, Sarki YN, Chikkaputtaiah C, Baruah G (2021) Role of PGPR in conferring drought stress tolerance in rice. In: Antioxidants in plant-microbe interaction, pp 425–448

Batool T, Ali S, Seleiman MF, Naveed NH, Ali A, Ahmed K, Abid M, Rizwan M, Shahid MR, Alotaibi M, Al-Ashkar I (2020) Plant growth promoting rhizobacteria alleviates drought stress in potato in response to suppressive oxidative stress and antioxidant enzymes activities. Sci Rep 10(1):16975

Begum N, Wang L, Ahmad H, Akhtar K, Roy R, Khan MI, Zhao T (2022) Co-inoculation of arbuscular mycorrhizal fungi and the plant growth-promoting rhizobacteria improve growth and photosynthesis in tobacco under drought stress by up-regulating antioxidant and mineral nutrition metabolism. Microb Ecol:1–18

Bhutto LA, Osborne CP, Quick WP (2023) Osmotic adjustment and metabolic changes under drought stress conditions in wheat (Triticum aestivum L.) genotypes. Pak J Bot 55:915–923

Bittencourt PP, Alves AF, Ferreira MB, da Silva Irineu LES, Pinto VB, Olivares FL (2023) Mechanisms and applications of bacterial inoculants in plant drought stress tolerance. Microorganisms 11(2):502

Bloomfield JP, Marchant BP, McKenzie AA (2019) Changes in groundwater drought associated with anthropogenic warming. Hydrol Earth Syst Sci 23(3):1393–1408

Bouremani N, Cherif-Silini H, Silini A, Bouket AC, Luptakova L, Alenezi FN, Baranov O, Belbahri L (2023) Plant growth-promoting rhizobacteria (PGPR): a rampart against the adverse effects of drought stress. Water 15(3):418

Boussora F, Triki T, Bennani L, Bagues M, Ben Ali S, Ferchichi A, Ngaz K, Guasmi F (2024) Mineral accumulation, relative water content and gas exchange are the main physiological regulating mechanisms to cope with salt stress in barley. Sci Rep 14(1):14931

Carezzano ME, Alvarez Strazzi FB, Pérez V, Bogino P, Giordano W (2023) Exopolysaccharides synthesized by Rhizospheric bacteria: a review focused on their roles in protecting plants against stress. Appl Microbiol 3(4):1249–1261

Chen T, Feng B, Fu W, Zhang C, Tao L, Fu G (2018) Nodes protect against drought stress in rice (Oryza sativa) by mediating hydraulic conductance. Environ Exp Bot 155:411–419

Crops. Federal Credit Administration: Office of Regulatory Policy, September 28

Dash GK, Purkaystha S, Sahoo SK, Kumar A, Jena J, Nayak L, Lal P, Kumar R, Tiwari RK, Lal MK (2023) Redesigning root architecture for improved performance of plants under stress. In: New Frontiers in plant-environment interactions: innovative technologies and developments. Springer Nature Switzerland, Cham, pp 515–534

Dayoub E, Lamichhane JR, Schoving C, Debaeke P, Maury P (2021) Early-stage phenotyping of root traits provides insights into the drought tolerance level of soybean cultivars. Agronomy 11(1):188

Dey R, Lewis SC (2021) Natural disasters linked to climate change. In: The impacts of climate change. Elsevier, pp 177–193

Dubey A, Malla MA, Kumar A, Khan ML, Kumari S (2024) Seed bio-priming with ACC deaminase-producing bacterial strains alleviates impact of drought stress in Soybean (Glycine max (L.) Merr.). Rhizosphere 30:100873

Fatema MK, Mamun MAA, Sarker U, Hossain MS, Mia MAB, Roychowdhury R, Ercisli S, Marc RA, Babalola OO, Karim MA (2023) Assessing morpho-physiological and biochemical markers of soybean for drought tolerance potential. Sustain For 15(2):1427

FCA (Farm Credit Administration) (2017) Crop insurance covers most major

Ferrante R, Campagni C, Vettori C, Checcucci A, Garosi C, Paffetti D (2024) Meta-analysis of plant growth-promoting rhizobacteria interaction with host plants: implications for drought stress response gene expression. Front Plant Sci 14:1282553

Ge M, Tang Y, Guan Y, Lv M, Zhou C, Ma H, Lv J (2024) TaWRKY31, a novel WRKY transcription factor in wheat, participates in regulation of plant drought stress tolerance. BMC Plant Biol 24(1):27

Gopalakrishna KN, Hugar R, Rajashekar MK, Jayant SB, Talekar SC, Virupaxi PC (2023) Simulated drought stress unravels differential response and different mechanisms of drought tolerance in newly developed tropical field corn inbreds. PLoS One 18(3):e0283528

Goswami M, Suresh DEKA (2020) Plant growth-promoting rhizobacteria—alleviators of abiotic stresses in soil: a review. Pedosphere 30(1):40–61

Gowtham HG, Singh B, Murali M, Shilpa N, Prasad M, Aiyaz M, Amruthesh KN, Niranjana SR (2020) Induction of drought tolerance in tomato upon the application of ACC deaminase producing plant growth promoting rhizobacterium Bacillus subtilis Rhizo SF 48. Microbiol Res 234:126422

Gul F, Khan IU, Rutherford S, Dai ZC, Li G, Du DL (2023) Plant growth promoting rhizobacteria and biochar production from Parthenium hysterophorus enhance seed germination and productivity in barley under drought stress. Front Plant Sci 14:1175097

Gusain S, Kumari K, Joshi R (2024) Physiological, hormonal and molecular dynamics of root system architectural response to drought stress signaling in crops. Rhizosphere 31:100922

Harkhani K, Sharma AK (2024) Alleviation of drought stress by plant growth-promoting Rhizobacteria (PGPR) in crop plants: a review. Commun Soil Sci Plant Anal 55(5):735–758

Hasan A, Tabassum B, Hashim M, Khan N (2024) Role of plant growth promoting rhizobacteria (PGPR) as a plant growth enhancer for sustainable agriculture: a review. Bacteria 3(2):59–75

Ilyas N, Mumtaz K, Akhtar N, Yasmin H, Sayyed RZ, Khan W, Enshasy HAE, Dailin DJ, Elsayed EA, Ali Z (2020) Exopolysaccharides producing bacteria for the amelioration of drought stress in wheat. Sustain For 12(21):8876

İpek M, Mutluay E (2022) Enhancing the physiological and molecular responses of horticultural plants to drought stress through plant growth–promoting rhizobacterias. In: Sustainable horticulture. Academic, pp 185–199

Jezek M, Blatt MR (2017) The membrane transport system of the guard cell and its integration for stomatal dynamics. Plant Physiol 174(2):487–519

Kaczmarek M, Fedorowicz-Strońska O, Głowacka K, Waśkiewicz A, Sadowski J (2017) CaCl 2 treatment improves drought stress tolerance in barley (Hordeum vulgare L.). Acta Physiol Plant 39:1–11

Kakar KU, Ren XL, Nawaz Z, Cui ZQ, Li B, Xie GL, Hassan MA, Ali E, Sun GC (2016) A consortium of rhizobacterial strains and biochemical growth elicitors improve cold and drought stress tolerance in rice (Oryza sativa L.). Plant Biol 18(3):471–483

Kang J, Peng Y, Xu W (2022) Crop root responses to drought stress: molecular mechanisms, nutrient regulations, and interactions with microorganisms in the rhizosphere. Int J Mol Sci 23(16):9310

Kasim WA, Osman ME, Omar MN, Salama S (2021) Enhancement of drought tolerance in Triticum aestivum L. seedlings using Azospirillum brasilense NO40 and Stenotrophomonas maltophilia B11. Bull Natl Res Cent 45(1):95

Krishna R, Ansari WA, Soumia PS, Yadav A, Jaiswal DK, Kumar S, Singh AK, Singh M, Verma JP (2022) Biotechnological interventions in tomato (Solanum lycopersicum) for drought stress tolerance: achievements and future prospects. BioTech (Basel) 11(4):48

Kumar M, Gupta A, Vandana P, Tiwari LD, Patel MK, Siddique KH (2024) Recent advances of plant growth-promoting rhizobacteria (PGPR)-mediated drought and waterlogging stress tolerance in plants for sustainable agriculture. In: Microbial biostimulants for plant growth and abiotic stress amelioration, pp 315–344

Latif M, Bukhari SAH, Alrajhi AA, Alotaibi FS, Ahmad M, Shahzad AN, Dewidar AZ, Mattar MA (2022) Inducing drought tolerance in wheat through exopolysaccharide-producing rhizobacteria. Agronomy 12(5):1140

Lee BR, Islam MT, Park SH, Jung HI, Bae DW, Kim TH (2019) Characterization of salicylic acid-mediated modulation of the drought stress responses: reactive oxygen species, proline, and redox state in Brassica napus. Environ Exp Bot 157:1–10

Li Q, Gao Y, Hamani AKM, Fu Y, Liu J, Wang H, Wang X (2023a) Effects of warming and drought stress on the coupling of photosynthesis and transpiration in winter wheat (Triticum aestivum L.). Appl Sci 13(5):2759

Li W, Reichstein M, May OSC, Destouni G, Migliavacca M, Kraft B, Weber U, Orth R (2023b) Contrasting drought propagation into the terrestrial water cycle between dry and wet regions. Earth's Future 11(7):e2022EF003441

Lucas JA, Garcia-Villaraco A, Montero-Palmero MB, Montalban B, Ramos Solano B, Gutierrez-Mañero FJ (2023) Physiological and genetic modifications induced by plant-growth-promoting rhizobacteria (PGPR) in tomato plants under moderate water stress. Biology 12(7):901

Mahmood Aulakh A, Qadir G, Hassan FU, Hayat R, Sultan T, Billah M, Hussain M, Khan N (2020) Desert soil microbes as a mineral nutrient acquisition tool for chickpea (Cicer arietinum L.) productivity at different moisture regimes. Plan Theory 9(12):1629

Mahreen N, Yasmin S, Asif M, Yahya M, Ejaz K, Yousaf S, Amin I, Zulfiqar S, Imran A, Khaliq S, Arif M (2023) Mitigation of water scarcity with sustained growth of Rice by plant growth promoting bacteria. Front Plant Sci 14:1081537

Mandal M, Chatterjee S, Majumdar S (2022) Outside the cell surface: encoding the role of exopolysaccharide producing rhizobacteria to boost the drought tolerance in plants. In: Plant stress: challenges and management in the new decade. Springer International Publishing, Cham, pp 295–310

Mansour E, El-Sobky ESE, Abdul-Hamid MI, Abdallah E, Zedan AM, Serag AM, Silvar C, El-Hendawy S, Desoky ESM (2023) Enhancing drought tolerance and water productivity of diverse maize hybrids (Zea mays) using exogenously applied biostimulants under varying irrigation levels. Agronomy 13(5):1320

Meenakshi AK, Govindasamy V, Ajit V, Choudhary DK (2019) Mitigation of drought stress in wheat crop by drought tolerant endophytic bacterial isolates. Vegetos 32(4):486–493

Mitra D, Pellegrini M, Guerra-Sierra BE (2024) Interaction between plants and growth-promoting Rhizobacteria (PGPR) for sustainable development. Bacteria 3(3):136–140

Murali M, Singh SB, Gowtham HG, Shilpa N, Prasad M, Aiyaz M, Amruthesh KN (2021) Induction of drought tolerance in Pennisetum glaucum by ACC deaminase producing PGPR-Bacillus amyloliquefaciens through Antioxidant defense system. Microbiol Res 253:126891

Naseem H, Ahsan M, Shahid MA, Khan N (2018) Exopolysaccharides producing rhizobacteria and their role in plant growth and drought tolerance. J Basic Microbiol 58(12):1009–1022

Naseem M, Chaudhry AN, Jilani G, Alam T, Naz F, Ullah R, Zahoor M, Zaman S (2024) Exopolysaccharide-producing bacterial cultures of Bacillus cereus and Pseudomonas aeruginosa in soil augment water retention and maize growth. Heliyon 10(4):e26104

Neupane D, Adhikari P, Bhattarai D, Rana B, Ahmed Z, Sharma U, Adhikari D (2022) Does climate change affect the yield of the top three cereals and food security in the world? Earth 3(1):45–71

Ojuederie OB, Olanrewaju OS, Babalola OO (2019) Plant growth promoting rhizobacterial mitigation of drought stress in crop plants: implications for sustainable agriculture. Agronomy 9(11):712

Parasuraman B, Rajamanickam V, Rathinavelu S, Geethanjali S, Alagarswamy S (2024) Interactive effect of drought and high temperature on physiological traits of soybean (Glycine max). Plant Physiol Rep 29(1):116–124

Parmar H, Venkatapuram AK, Rashid A, Sangeetha K, Mehta S, Reddy MK, Goel A, Achary VMM (2022) Portfolio of drought stress response and genetic enhancement strategies for development of future drought-tolerant crop. In: Plant stress Mitigators: action and application. Springer Nature Singapore, Singapore, pp 515–539

Pereira SIA, Abreu D, Moreira H, Vega A, Castro PML (2020) Plant growth-promoting rhizobacteria (PGPR) improve the growth and nutrient use efficiency in maize (Zea mays L.) under water deficit conditions. Heliyon 6(10):e05106

Rai PK, Sabharwal U, Singh S, Yadav AN, Choure K (2024) Exploration of drought tolerant PGPR and their role in regulating antioxidant enzymes in maize (Zea mays L.). J Soil Sci Plant Nutr:1–16

Rajkumar M, Narayanasamy S, Uthandi S (2024) A root-associated Bacillus albus LRS2 and its metabolites for plant growth promotion and drought stress tolerance in little millet (Panicum sumatrense L.). Plant Stress 12:100446

Rane J, Singh AK, Tiwari M, Prasad PV, Jagadish SK (2022) Effective use of water in crop plants in dryland agriculture: implications of reactive oxygen species and antioxidative system. Front Plant Sci 12:778270

Ren Z, Fu J, Abou-Elwafa SF, Ku L, Xie X, Liu Z, Shao J, Wen P, Al Aboud NM, Su H, Wang T, Wei L (2024) Analysis of the molecular mechanisms regulating how ZmEREB24 improves drought tolerance in maize (Zea mays) seedlings. Plant Physiol Biochem 207:108292

Saikia J, Sarma RK, Dhandia R, Yadav A, Bharali R, Gupta VK, Saikia R (2018) Alleviation of drought stress in pulse crops with ACC deaminase producing rhizobacteria isolated from acidic soil of Northeast India. Scientific reports 8(1):3560

Salem MA, Ismail MA, Radwan KH, Abd-Elhalim HM (2024) Unlocking the potential of plant growth-promoting Rhizobacteria to enhance drought tolerance in Egyptian wheat (Triticum aestivum). Sustain For 16(11):4605

Santos-Medellín C, Liechty Z, Edwards J, Nguyen B, Huang B, Weimer BC, Sundaresan V (2021) Prolonged drought imparts lasting compositional changes to the rice root microbiome. Nat Plants 7(8):1065–1077

Sati D, Pande V, Pandey SC, Samant M (2023) Recent advances in PGPR and molecular mechanisms involved in drought stress resistance. J Soil Sci Plant Nutr 23(1):106–124

Sevim G, Ozdemir-Kocak F, Unal D (2023) Hormonal signaling molecules triggered by plant growth-promoting bacteria. In: Phytohormones and stress responsive secondary metabolites. Academic, pp 187–196

Sharma M, Gupta SK, Deeba F, Pandey V (2017) Effects of reactive oxygen species on crop productivity: an overview. In: Reactive oxygen species in plants: boon or bane-revisiting the role of ROS, pp 117–136

Sharma R, Upadhyay S, Bhattacharya S, Singh A (2021) Abiotic stress-responsive miRNA and transcription factor-mediated gene regulatory network in Oryza sativa: construction and structural measure study. Frontiers in Genetics, 12:618089

Sharma A, Choudhary P, Chakdar H, Shukla P (2024) Molecular insights and omics-based understanding of plant–microbe interactions under drought stress. World Journal of Microbiology and Biotechnology, 40(2):42

Singha KM, Pandey P (2023) Endophytic bacteria with host-supportive genetic determinants in their genomes induce growth and antioxidant activity related gene functions in transcriptome of black rice (Oryza sativa L.). Environ Exp Bot 213:105396

Sun W, Shahrajabian MH, Soleymani A (2024) The roles of plant-growth-promoting Rhizobacteria (PGPR)-based biostimulants for agricultural production systems. Plan Theory 13(5):613

Ullah A, Nisar M, Ali H, Hazrat A, Hayat K, Keerio AA, Ihsan M, Laiq M, Ullah S, Fahad S, Khan A (2019) Drought tolerance improvement in plants: an endophytic bacterial approach. Appl Microbiol Biotechnol 103:7385–7397

Vaishnav A, Choudhary DK (2019) Regulation of drought-responsive gene expression in Glycine max l. Merrill is mediated through Pseudomonas simiae strain AU. Journal of Plant Growth Regulation, 38:333–342

Verma G, Srivastava D, Tiwari P, Chakrabarty D (2019) ROS modulation in crop plants under drought stress. In: Reactive oxygen, nitrogen and sulfur species in plants: Production, metabolism, signaling and defense mechanisms, pp 311–336

Vijayaraghavareddy P, Lekshmy SV, Struik PC, Makarla U, Yin X, Sreeman S (2022) Production and scavenging of reactive oxygen species confer to differential sensitivity of rice and wheat to drought stress. Crop Environ 1(1):15–23

Wahab A, Abdi G, Saleem MH, Ali B, Ullah S, Shah W, Wahab A, Abdi G, Saleem MH, Ali B, Ullah S, Shah W, Marc RA (2022) Plants' physio-biochemical and phyto-hormonal responses to alleviate the adverse effects of drought stress: a comprehensive review. Plan Theory 11(13):1620

Wang Y, Blatt MR, Chen ZH (2018) Ion transport at the plant plasma membrane. eLS:1–16

Yaseen R, Aziz O, Saleem MH, Riaz M, Zafar-ul-Hye M, Rehman M, Ali S, Rizwan M, Nasser Alyemeni M, El-Serehy HA, Al-Misned FA (2020) Ameliorating the drought stress for wheat growth through application of ACC-deaminase containing rhizobacteria along with biogas slurry. Sustain For 12(15):6022

Yasmin H, Rashid U, Hassan MN, Nosheen A, Naz R, Ilyas N, Sajjad M, Azmat A, Alyemeni MN (2021) Volatile organic compounds produced by Pseudomonas pseudoalcaligenes alleviated drought stress by modulating defense system in maize (Zea mays L.). Physiol Plant 172(2):896–911

Zaib S, Zubair A, Abbas S, Hussain J, Ahmad I, Shakeel SN (2023) Plant growth-promoting Rhizobacteria (PGPR) reduce adverse effects of salinity and drought stresses by regulating nutritional profile of barley. Appl Environ Soil Sci 2023(1):7261784

Disentangling the Contribution of Plant Genetics in Rhizomicrobiome Recruitment for Climate Resilient Agroecosystems

6

Izzah Shahid and Samina Mehnaz

Abstract

Plant roots are surrounded by a vibrant and ever-changing microbiome that plays an important role in the plant's health and growth. It contributes significantly to diverse functions such as nutrient cycling, mobilization, and mineralization, production of phytohormones, antimicrobials, enzymes, and augmentation of plant defense mechanisms against pathogens. In contrast, plants have an immense knowledge of what their microbiome can offer, therefore they manipulate these interactions to their benefit through a process called host selection by recruiting certain species while enhancing their performance under both internal and external stressors. Recent advances in plant breeding by using techniques such as gene editing, trait identification or genome characterization have enabled the production of improved crops that maximize the benefits of plant-microbe relationships. These advances have resulted in genetically modified crops with better crop productivity and increased resistance toward diseases. Different genotypes exhibit variations in root development patterns, rhizodeposition rates, immune responses, chemical signaling mechanisms, and nutrient uptake strategies. Because of these differences in traits among different plants, there is more exclusive and customized rhizomicrobiome arrangement due to this selective pressure that influences colonizing microorganisms largely. To get a lasting influence regarding microbial input and also for maintaining a durable specific microbiota

I. Shahid (✉)
Institute of Multidisciplinary Research in Applied Biology, Public University of Navarra, Pamplona, Spain
e-mail: izzah.shahid@unavarra.es

S. Mehnaz
Kauser Abdulla Malik School of Life Sciences, Forman Christian College (A Chartered University), Lahore, Pakistan

P. Pandey et al. (eds.), *Microorganisms Resilience to Climate Change*,
Microorganisms for Sustainability 56, https://doi.org/10.1007/978-981-96-3748-5_6

found on these organisms, it is on this basis that plants employ different strategies for microbial recruitment at the root surface.

Keywords

Rhizomicrobiome · Root exudates · Plant immunity · Plant-microbe interactions · Microbial community · Cultivars

6.1 Introduction

Global agricultural practices have reshaped the floral landscape of the world leaving a profound effect on earth's biodiversity and ecosystem (Gaut et al. 2015). The process of domesticating specific wild plant species involved the deliberate selection of traits such as enhanced seed size and quantity, the capacity to synthesize metabolites associated with phenology, seed dormancy, palatability, and reduced vulnerability to environmental pressures (Purugganan 2019). Both the artificial selection and adaptability to spatially variable local conditions by the selected species have come up with domestication syndrome (Denham et al. 2020). The emergence of new plant genotypes and the development of desirable plant cultivars have given rise to unprecedented characterization of plant genomes and the identification of metabolic activities. In addition, this has sparked a great interest in exploiting the nature of plant-microbe interactions under specific conditions (Porter and Sachs 2020). Because of the inextricable nature of these interactions, the plant microbiome is now regarded as the extended phenotype of the host plant (Henry et al. 2021).

Plant microbiome is very diverse and occupies roots, shoots, and ariel parts of the host plant. These microbes are vital for proper plant functioning and are influenced by certain abiotic and biotic factors and external stimuli. The most significant of plant-microbe interactions are the interactions in the rhizosphere region (Guo et al. 2021). These interactions get modulated in response to environmental changes to benefit both partners and specifically to enhance the resilience of host plants. For instance, plants can influence the spatial distribution and abundance of associated microbes by releasing certain chemical signals. However, the genetic mechanisms and environmental factors determining rhizosphere microbial dynamics and assembly are poorly understood (Wagner 2021). Numerous studies have nevertheless shown the variability of rhizomicrobiome among different species and plant genotypes suggesting the role of plant genetics in recruitment of specific microbes (Brown et al. 2020).

Overall, the domestication of hundreds of plant species and the introduction of new breeding lines are hypothesized to impact the rhizosphere microbiome because of the genomic modification and the loss of genetic diversity of host plants, maneuvering of the key traits controlling microbial assembly and functioning, and cultivation of plants in artificial habitats than their indigenous ones (Soldan et al. 2021). New plant genotypes developed focusing only on limited traits are more disease-prone and face continuous challenges from climate change, emerging pathogens, and soil degradation. A comprehensive understanding of the plant-microbe interactions especially how new cultivars shape their rhizomicrobiome can help them to better thrive in the global landscape by mitigation of external stressors. Some of the mechanisms summarizing effects of plant genotypes on rhizomicrobiome assembly and dynamics are outlined below in detail.

6.2 Significance of Plant-Microbe Interactions for Plant Health and Productivity

The rhizomicrobiome or the second genome of a plant is the plethora of trillions of microorganisms associated with plant roots and the surrounding soil called as rhizosphere, possesses a unique two-way relationship. The rhizosphere has more microbial concentration than the bulk soil due to its higher nutrient content, which is an attribute of the root zone (Finkel et al. 2017). These microorganisms are permanent residents of the rhizosphere ecosystem, and that explains why their underlying mechanisms for protection of plant counterparts create important impacts on well-being and health of plants (Bhat et al. 2023). The most remarkable among these is the relationship between bacteria and plants where many bacteria enhance plant growth leading to majority of microorganisms shaping the rhizomicrobiome (Khatoon et al. 2020). These bacteria also produce hydrolytic enzymes, siderophores, phytohormones, secondary metabolites and volatile compounds that promote plant growth by breaking down complex organic substances into easily absorbable forms (John et al. 2020). Sometimes these bacteria act in synergy causing multiple metabolic reactions that initiate systemic immunity in plants against dead or alive factors to sustain competition within ecological niches (Elnahal et al. 2022). This predominant property of rhizobacteria demonstrates their potential as biological control agents. Most plant-bacteria interactions are beneficial; however, some bacterial genera can also be plant pathogens, negatively affecting plant growth.

In addition to bacteria, the rhizomicrobiome comprises a considerable proportion of fungi and oomycetes, which can have both beneficial and damaging effects on the host plants. Some fungi can contribute to soil nutrient availability through nutrient cycling and protect plants through the biological control of phytopathogens (Baron and Rigobelo 2022). At the same time, most fungi are lethal and damaging phytopathogens, causing significant agricultural losses (Hariharan and Prasannath 2021). In addition to interacting with plants, these microbes also communicate and collaborate through inter-kingdom signaling. The coordination of these microbes broadens their capacity to fight external biotic and abiotic stressors, thereby ensuring plant protection (Checcucci and Marchetti 2020). Moreover, plants being fully aware of the diversity and dynamics of their associated partners can accordingly categorize them. Therefore, it is possible that root compartments such as rhizoplane, rhizosphere or endosphere could exhibit certain patterns of spatial arrangement of microorganisms depending on the types of the plant signals that are released to attract specific taxa of microbes into or away from them (Jamil et al. 2022). This initiates signaling that facilitates selective rooting involving distinct microbial communities so as to accomplish favorable exploitation by plant hosts. Consequently, plants learn how to monitor their associated microbes and tame them in response. Hence, the ability to shape the microbiome surrounding the roots is one of the central factors determining their health status and survival probability.

6.3 Plant Genetics as a Key Determinant in Shaping Rhizomicrobiome

The cultivation of existing crop varieties has experienced notable influences from adverse climatic circumstances, recurrent periods of drought, and the emergence of pathogens that present a significant challenge to the prospective food security (Deutsch et al. 2018). The rise in global temperatures and reduced precipitation levels have prompted plant breeders and scientists to create robust and adaptable plant varieties specifically designed to endure the deteriorating conditions of agro-ecosystems. Multiple breeding strategies and principles are currently being used to enhance genetic makeup of the plants to ensure nutritional intactness and smooth crop yields. Advancements in genome characterization techniques and tools to maneuver plant genomes have helped to develop a significant collection of genome assemblies (Zhang et al. 2019a, b). These genome assemblies have been exploited for trait discovery and to develop proteomic, metabolomic, and epigenomic plant maps, and gene-expression atlas (Varshney et al. 2019). Global gene banks hold the data of millions of alleles as plant genetic resource which have led to significant crop improvements (Langridge and Waugh 2019). A range of systems biology, trait-mapping approaches, and functional genomic strategies have helped in the characterization of candidate genes encoding for traits of interest (Lavarenne et al. 2018). Similarly, genomic breeding-assisted crop breeding programs have been extensively used for the development of commercially available improved breeding lines (Bevan et al. 2017). Gene editing has appeared as a powerful approach to establishing diverse biotic and abiotic stress-tolerant lines. All of these methods have set a trend to cultivate resilient and improved crop varieties shrinking the planting of low-yielding and disease-prone conventional cultivars. Genetic variations in plants overall impact plant traits and significantly influence the enrichment, recruitment, and diversity of their below-ground microbial friends.

6.4 Factors Influencing Rhizomicrobiome Assembly in Different Cultivars

The establishment and maintenance of rhizomicrobiome is a highly selective and complex process governed by an array of mechanisms including chemotaxis, external stimuli, antibiosis, biofilm formation, and motility (Chagas et al. 2018). Diverse biotic and abiotic factors greatly influence and shape the rhizomicrobiome or microbial communities associated with plants. Rhizodeposits are one of the major drivers shaping the structure and function of the plant microbiome. Rhizodeposits refer to assorted collection of chemicals secreted by plants in the root zone (Hirsch et al. 2013). Sloughed-off tissues and cells, soluble lysates, mucilage, root exudates, and volatile compounds are the major components of rhizodeposits. All of these constituents of rhizodeposits are largely influenced by plant genetics (Fig. 6.1 and Table 6.1).

6.4.1 Diversity of Root Exudates and Rhizosphere Microbiome

Root exudates are complex blends of phenolics, amino acids, saccharides, vitamins, organic acids, and bioactive molecules that serve as the nutrients or signaling molecules for plants (Semchenko et al. 2021). They primarily act as chemo-attractants and chemo-repellents and mediate the interactions between plants and other organisms (Vives-Peris et al. 2020). Plant genetics heavily influence the quantity and composition of the root exudates. Different plant genotypes are variably involved in the biosynthesis of root exudates which can selectively recruit favorable groups of microbes (Ling et al. 2022). For instance, different rice plant genotypes showed variable profiles of root exudation leading to discrete rhizosphere microbial assemblages (Lakshmanan et al. 2014). A comparison of rhizosphere microbial diversity of two barley genotypes, landrace (Chevallier) and modern (NFC Tipple) barley (*Hordeum vulgare*), showed the clear differences and the ability of *Pseudomonas* sp. to substantially colonize Tipple rhizosphere (Pacheco-Moreno et al. 2024). This cultivar-driven selection of microbiota has been shown to operate at both the molecular level and composition of root exudates. Tipple root exudates being rich in hexose sugars better attracted the *Pseudomonas* for root-colonization. Clear

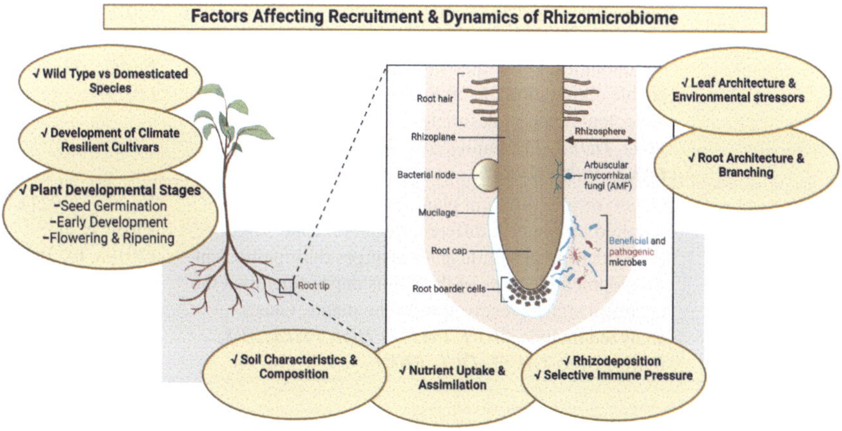

Fig. 6.1 Factors affecting rhizomicrobiome recruitment, dynamics, and assembly in in different plant genotypes. The major factors in shaping the rhizosphere microbiome include rhizodeposits, nutrient uptake capacities of the plants, root architecture and branching system, plant immune response against certain stressors, soil characteristics, and developmental stages. All of these determinants are variable in wild type, domesticated, and genetically engineered cultivars

Table 6.1 Variable factors in genotypically different plants influencing rhizomicrobiome recruitment

Traits/Factors	Functioning or mechanisms	References
Rhizodeposits	Complex blends of soluble lysates, root-exudates, polysaccharides, proteins, DNAs, lubricants, sloughed off cells and debris of detached tissues, phenols and flavones, antimicrobials and volatiles. Different plant genotypes display variable synthesis and production of rhizodeposits attracting distinct microbial communities in the rhizosphere. Rhizodeposits secrete diverse signals acting as attractants for specific microbial taxa or serve as carbon source. The difference in exudates composition is primarily attributed to plant domestication and introduction of new cultivars. Genes analyzed include *TPS25, TPS30, THAH, THAO, MYB72, CYP71A27, S8H*	Zhang et al. (2023), Ling et al. (2022), Pacheco-Moreno et al. (2024)
Root and leaf architecture	Root length, branching concentration, ability to navigate in soil and root hair formation capacities are different in various newly introduced plant cultivars. New breeding lines and domestication significantly impacted root architecture and subsequently rhizomicrobiome assembly and recruitment. Diversity in thickness and lengths of the rooting system highly influences attached microbial communities Trichome branching, cuticular wax, and cutin vary in different plant genotypes and also influence diversity and abundance of associated microbial communities Genes analyzed include SNPs in *YUC 1*, aquaporin *SlTIP2.3, FIT, CPR1, FHY3, MAX4, PAD4, BX1, C2, MATE 1*	Atkinson et al. (2014), Zhang et al. (2023), Zhang et al. (2018)
Immune system	Differential plant immunity and variable expression of immunity-related genes in diverse plant cultivars attract and influence specific groups of rhizosphere microbial communities. Difference in quorum suppressing signals enable elimination of certain microbial taxa and favors the colonization of desirables. Genes studied include *SID2, DDE3, JAZ* family, *SAV, RGL1, AXR1, LYM1, BAK1, JAR1, MYB51*	Zhalnina et al. (2018), Zhang et al. (2023), Smigielski et al. (2019)
Nutrient uptake capacities and development stage	Genotypically different plants show diversity of required nutrients at different developmental stages. Root microbiota significantly changes during different developmental stages of the plants depending upon the chemical signals received from the plants. Genes analyzed include *PHF1, PHT1, PHR1, PHL1, SPX, PGR5, CER, LACS2, DCL, RBK1*	Yu et al. (2021), Zhang et al. (2019a, b), Tang et al. (2022)

differences in root and rhizosphere microbiota of different barley genotypes have also been previously shown by Bulgarelli et al. (2015).

Likewise, several *Streptomyces* sp. have been seen enriched in the rhizosphere and endosphere of salicylate producing plants (Worsley et al. 2021). Rhizosphere spatial structure analysis of three maize genotypes showed that β-glucosidase released by root hairs facilitated microbial activity and recruitment, whereas, benzoxazinoids in root exudates led to the suppression of microbial activity and assembly (Bilyera et al. 2021). A study on *Populus trichocarpa* validated the contribution of plant genotypes and chemotypes in exerting selective pressure to recruit rhizomicrobiome. Particularly, the levels of salicylic acid varied significantly among different genotypes causing substantial shifts in dominant fungal and bacterial groups (Veach et al. 2019). An analysis of the different rice cultivars planted under drought stress demonstrated variability in root exudation and associated microbial communities. Levels of abscisic acid were significantly higher under drought stress, however jasmonic acid and L-cystine concentrations were low and differentially attracted different bacterial genera for stress mitigation (Li et al. 2023). The increased prevalence of specific bacterial taxa in the presence of particular root exudates at the level could be attributed to the presence of specific responsive proteins within the microbial community. For example, research has demonstrated that *Bacillus velezensis* SQR9 exhibits a chemotactic response to root exudates from cucumber plants (Zhang et al. 2014). Similarly, it has been observed that *Pseudomonas putida* strain KT2440 displays chemotaxis towards various amino acids, organic acids, and inorganic phosphate compounds (Ortega et al. 2017). Rhizobacteria's chemosensory system involves methyl-accepting chemotaxis proteins (MCP) and dCache domains, which undergo intricate chemical modifications in response to plant exudates (Flack and Parkinson 2018).

6.4.2 Differential Root Architecture and Microbial Assemblage

Root architecture depends considerably on the genetic composition of plants, and has a significant impact on the microbiome that colonizes them. Plant genotypes differ in their root characteristics such as hair development, length, density, branching pattern or network structure, spatial arrangement and soil distribution (Saleem et al. 2018). The variations in root architectural traits of diverse plant genotypes are associated with specific regulatory pathways and different genetic loci. Atkinson et al. showed that different *Arabidopsis thaliana* genetic varieties had distinct root lengths and densities because of genetic variations at specific loci such as Root Growth Factor 1 (Atkinson et al. 2014). Additionally, studies involving various barley cultivars have indicated that root hair formation and densities were regulated by the *RTH1* gene through genetic analysis (Kwasniewski et al. 2013). Finally, studies employing different rice cultivars revealed the significance of genes like Deeper Rooting 1 (DRO1) in determining gravitropic responses in roots leading to development of deeper rooting system (Uga et al. 2013).

The varying extent and structure of the root system create incredible niches within the soil shaping unique microbial composition and colonization. Plant genotypes with deeper root systems exude more compounds recruiting distinct microbial populations. Zhang and colleagues have shown enhanced microbial diversity in different rice genotypes with extensive root branching systems (Zhang et al. 2018). Similarly, barley genotypes with higher root hair densities have been shown to have abundant plant growth promoting rhizobacteria as compared to the ones with less root hair (Haling et al. 2014). A comparison of wild and modern bean accessions showed clear differences in root phenology between modern and wild beans with thick and thin roots, respectively (Pérez-Jaramillo et al. 2017). This led to a differential abundance of microbial OTUs and phyla in both cultivars with modern bean accessions showing enrichment in *Actinobacteria* whereas, *Bacteroidetes* in wild beans. Likewise, two commercial *Nicotiana* sp. (burley tobacco) cultivars showed the dominance of *Rhizobium* and *Pantoea* sp. species in the microbiome across all phenotypic root traits (Saleem et al. 2016). Interestingly, the abundance of *Actinobacteria* was observed to decrease from root hair to primary roots. The abundance of *Spartobacteria* and *Verrucomicrobia*, nevertheless, was seen as variable across the root phenotypic gradient (Saleem et al. 2016).

Some studies have also demonstrated the development of root phenome-specific microbiome showing the preferential colonization of inside roots (Mousa et al. 2016). Multiple ecological, molecular, and biochemical traits including quorum-sensing signals, species competition, cell wall degradation, and metabolite consumption have been shown to govern these root-microbiome interactions (Massalha et al. 2017). In addition to this, a complex interplay between microbiome and roots has also been witnessed in varying soil compaction. Soil compaction and texture have great influence on the patterns of root growth and variably affect different plant genotypes (Rogers et al. 2016). This in turn has a reciprocal effect on the recruitment and abundance of microbial communities (Saleem et al. 2018). A more recent study on maize cultivars showed a significant effect of plant genotype and planting density on rhizomicrobiome (Zhang et al. 2023). Specific establishment of microbial communities was observed in root endocompartments when subjected to differential planting. Similarly, higher densities of microbial populations were observed in maize genotypes with extensive lateral root system (Peiffer et al. 2013). Higher microbial biomass and diversity were noted in wheat genotypes with longer and denser root systems due to increased root-soil contact (Fan et al. 2018). Overall, root architecture diversity among different plant genotypes creates unique microenvironments and directly influences microbial community composition and selection together with root exudates.

6.4.3 Selective Pressure by Diverse Plant Genotypes

Plant genotypes are capable of imposing selection pressure in their surroundings through the synthesis of several bioactive compounds and modulation of soil pH. Diverse plant genotypes and communities exert diversity-driven favorable selection pressures to avoid competition and loss of diversity. Species-rich plant communities can maintain complex and diverse microbial communities simultaneously diluting the specie-specific pathogens (Milcu et al. 2013). A complex mutual selection pressure in specie-rich plant community was observed in below-ground microbial composition and diversity and their assemblage (Schmid et al. 2019).

Likewise, differential plant diversity chooses to select altered plant phenotypes in association with rhizosphere microbial communities (Hahl et al. 2020).

The Zigzag model of plant-triggered immune responses is another key driver in shaping the microbial community. Different plant genotypes can directly and differentially select microbiome through the initiation of cascade reactions, thereby suppressing and favoring the growth of desirable microbes (Zhalnina et al. 2018). There are two widely characterized immune responses across different genotypic backgrounds in plants: 1) pattern-triggered immunity and; 2) effector-triggered immunity (ETI). As a first line of defense, PTI comprises protein recognition receptors (PRRs) which can subsequently recognize and bind to microbe-associated molecular patterns (MAMPs). This in turn initiates a cascade fashion defense response directly suppressing microbial growth (Berrabah et al. 2019). On the contrary, beneficial plant-associated bacteria such as *Pseudomonas* sp., *Frankia* sp., and *Rhizobia* evade these defense pathways and establish symbiosis with host plants. This is yet another demonstration of the durability that strengthens plant-symbiont bond endogenously triggered to withstand pathogens (Smigielski et al. 2019). These well-coordinated defenses can be differentially modulated by distinct plant genotypes and are quite fine-tuned both at temporal and spatial levels.

Conversely, effector-triggered immunity (ETI) is a more specific type of defense comprising intracellular resistance genes containing nucleotide-binding leucine-rich repeats. The diversity of intracellular resistance genes (R genes) among different plant genotypes can shape rhizomicrobiome by selective suppression (Cui et al. 2015). This immune response leads to the death of the target microbes or pathogenic organisms. Additionally, several phytohormones produced by the plants also influence the recruitment and abundance of rhizosphere microbes. Phytohormones such as jasmonic acid, salicylic acid, and ethylene constitute the systematic acquired resistance (SAR) of plants and exert a notable influence on the associated microbiome (Carvalhais et al. 2014). The differential levels and diverse expression of these phytohormones have been identified in various plant varieties, subsequently influencing the dynamics of hormone-microbiome interactions. For example, it is well established that salicylic acid impacts the plant microbiome through selective pressure, enhancing colonization and community diversity (Vincent et al. 2022). Furthermore, increased levels of salicylic acid have been linked to diminished community diversity and function (Lebeis et al. 2015). Similarly, the deficient biosynthesis of jasmonic acid in mutant tomato plants has been associated with greater endophytic fungal diversity and community richness compared to wild types (Manzotti et al. 2020). Likewise, plants incapable of producing tryptophan-derived metabolites exhibited a significant rise in fungal communities (Wolinska et al. 2021). An altered composition of endophytic communities and dysbiosis were observed in quadruple mutants with impaired pattern-triggered immunity (Chen et al. 2020). Furthermore, numerous studies have explored the intricate relationship between phytohormone pathways and wild plant varieties, revealing distinctive microbial diversity compared to cultivated ones (Gutierrez and Grillo 2022). Several allelopathic compounds produced by plants are genetically controlled and also influence structure of plant microbial communities. These allelopathic compounds can reduce the abundance and diversity of several bacterial taxa by selective repression. For instance, rice plants producing higher levels of allelopathic compounds displayed an overall low microbial diversity, nonetheless, a higher abundance of specific beneficial bacteria (Li et al. 2016).

6.4.4 Nutrient Uptake Capacities Modulating Rhizomicrobiome

Over the years, different plant cultivars have been developed with the focus on regulating nutrient patterns and uptake. Microbes in the soil also assist plants in mobilization of soil nutrients by converting them into more bioavailable forms. Plants formulate their microbiome by recruiting specific groups of more helpful microbes enabling smooth nutrient acquisition for them and have been genetically modified for certain response genes (Trivedi et al. 2020). Plants can sense specific nutritional changes in their surroundings through certain transporters such as Nitrate Transporter 1.1B to sense the nitrate levels in soil. This modulates related gene expression and root metabolites which subsequently changes root microbiome recruitment and dynamics (Zhang et al. 2019a, b). Similarly, maize plants defective in chalcone synthase gene displayed significantly reduced abundance of nitrogen assimilating and plant growth promoting *Oxalobacteraceae*, when cultivated in nitrogen poor soils (Yu et al. 2021). This study also demonstrated that root-derived flavones favored the dynamics and assembly of microbes facilitating nitrogen acquisition in plants.

Phosphorus is another essential nutrient vital for the maintenance of metabolism and growth in plants. To meet their need, plants initiate phosphate starvation response mechanisms which activate and modulate associated microbiome (Fabiańska et al. 2019). Under phosphorus-starved conditions, certain plant genotypes respond using Phosphate Starvation Response 1 transcriptional regulator to alleviate phosphorous starvation by modulation and recruitment of certain bacterial taxa (Tang et al. 2022). In the same way, plants can release iron-mobilizing phenols coumarins under iron-limited conditions (Harbort et al. 2020). Iron is a significant micronutrient and is needed as a co-factor in several metabolic reactions. Iron deficiency in plants can lead to stunted and retarded growth. Plant cultivars with disrupted coumarin pathway and iron scarce conditions have shown notable changes in microbiome assembly and recruitment modulating them to enhance iron uptake. Likewise, MYB72 gene has been documented to regulate scopoletin release and induction of enhanced root colonization by iron-mobilizing microbiota to mitigate iron deficiency in plants (Stringlis et al. 2018). Like other nutrients, sulfur is also significant in maintaining plant health and is one of the four major nutrients needed by the plants. It is essential for the formation of chlorophyll, activation of certain enzymes, and is part of the several vital amino acids. Certain plant cultivars have been developed for the improved uptake and metabolism of sulphur.

6.4.5 Plant Development Tailors Microbiome

A general perception is that plant chemical signals in addition to root exudates are vital to shape the microbial communities tailored to the needs of host. These chemical signals are changed over the different plant developmental stages and across different genotypes. Different developmental stages in plants have been documented to significantly influence microbiome composition and recruitment, and interkingdom dynamics of microbiota not only in rhizosphere but also in the aeriel parts (Xiong et al. 2021). Genotypically different rice cultivars with varying developmental stages were shown to host distinct root microbiota (Edwards et al. 2018). Similarly, wheat rhizomicrobiome and different developmental stages showed great plant-bacteria interaction as compared to plant-fungal interactions and jointing of these microbes more at ripening stage than tillering stage (Chen et al. 2019). Moreover, the genes involved in plant growth phase regulation such as the ones involved in flowering and root development were examined for their direct association with rhizosphere microbiome (Lu et al. 2018). In another study on legumes, plant symbiosis genes were cordially involved in structuring and regulating root-associated microbiomes (Thiergart et al. 2019). Epigenetic changes including DNA methylation, histone methylation, and demethylation significantly impact microbial assembly and co-occurrence of microbial networks in root zone (Lv et al. 2021). Moreover, signature microbiota of each plant tends to co-evolve with plants during different developmental stages. Such as legume plants have been shown to display co-evolutionary dynamics indicating the richness of specific rhizobia during plant life (Afkhami et al. 2014). Most of these genotypic variations in plants lead to an overall re-shaping and clustering of microbiome for the best exploitation of their association for optimal plant growth.

6.5 Conclusion

Plant genetics is central in shaping the rhizosphere microbiome through diverse mechanisms involving plant immune responses, rhizodeposition, root architecture, and symbiotic relationships. All of these mechanisms intricately influence the recruitment, diversity, and assemblage of rhizomicrobiome, and select the microbial taxa to best fit the needs of the host. Advances in multi-omics technologies and synthetic ecology are deepening our understanding of these complex interactions. Comprehensive mechanistic insights into these mechanisms can lead to the development of customized and tailored cropping systems to address food security challenges in face of the global climatic changes.

References

Afkhami ME, Rudgers JA, Stachowicz JJ (2014) Multiple mutualist effects: conflict and synergy in multispecies mutualisms. Ecology 95(4):833–844

Atkinson JA, Rasmussen A, Traini R, Voß U, Sturrock C, Mooney SJ et al (2014) Branching out in roots: uncovering form, function, and regulation. Plant Physiol 166(2):538–550

Baron NC, Rigobelo EC (2022) Endophytic fungi: a tool for plant growth promotion and sustainable agriculture. Mycology 13(1):39–55

Berrabah F, Ratet P, Gourion B (2019) Legume nodules: massive infection in the absence of defense induction. Mol Plant-Microbe Interact 32(1):35–44

Bevan MW, Uauy C, Wulff BB, Zhou J, Krasileva K, Clark MD (2017) Genomic innovation for crop improvement. Nature 543(7645):346–354

Bhat MA, Mishra AK, Jan S, Bhat MA, Kamal MA, Rahman S et al (2023) Plant growth promoting rhizobacteria in plant health: a perspective study of the underground interaction. Plan Theory 12(3):629

Bilyera N, Zhang X, Duddek P, Fan L, Banfield CC, Schlüter S et al (2021) Maize genotype-specific exudation strategies: an adaptive mechanism to increase microbial activity in the rhizosphere. Soil Biol Biochem 162:108426

Brown SP, Grillo MA, Podowski JC, Heath KD (2020) Soil origin and plant genotype structure distinct microbiome compartments in the model legume Medicago truncatula. Microbiome 8:1–17

Bulgarelli D, Garrido-Oter R, Münch PC, Weiman A, Dröge J, Pan Y et al (2015) Structure and function of the bacterial root microbiota in wild and domesticated barley. Cell Host Microbe 17(3):392–403

Carvalhais LC, Dennis PG, Schenk PM (2014) Plant defence inducers rapidly influence the diversity of bacterial communities in a potting mix. Appl Soil Ecol 84:1–5

Chagas FO, de Cassia Pessotti R, Caraballo-Rodríguez AM, Pupo MT (2018) Chemical signaling involved in plant–microbe interactions. Chem Soc Rev 47(5):1652–1704

Checcucci A, Marchetti M (2020) The rhizosphere talk show: the rhizobia on stage. Front Agron 2:591494

Chen S, Waghmode TR, Sun R, Kuramae EE, Hu C, Liu B (2019) Root-associated microbiomes of wheat under the combined effect of plant development and nitrogen fertilization. Microbiome 7:1–13

Chen T, Nomura K, Wang X, Sohrabi R, Xu J, Yao L et al (2020) A plant genetic network for preventing dysbiosis in the phyllosphere. Nature 580(7805):653–657

Cui H, Tsuda K, Parker JE (2015) Effector-triggered immunity: from pathogen perception to robust defense. Annu Rev Plant Biol 66(1):487–511

Denham T, Barton H, Castillo C, Crowther A, Dotte-Sarout E, Florin SA et al (2020) The domestication syndrome in vegetatively propagated field crops. Ann Bot 125(4):581–597

Deutsch CA, Tewksbury JJ, Tigchelaar M, Battisti DS, Merrill SC, Huey RB, Naylor RL (2018) Increase in crop losses to insect pests in a warming climate. Science 361(6405):916–919

Edwards JA, Santos-Medellín CM, Liechty ZS, Nguyen B, Lurie E, Eason S et al (2018) Compositional shifts in root-associated bacterial and archaeal microbiota track the plant life cycle in field-grown rice. PLoS Biol 16(2):e2003862

Elnahal AS, El-Saadony MT, Saad AM, Desoky ESM, El-Tahan AM, Rady MM et al (2022) The use of microbial inoculants for biological control, plant growth promotion, and sustainable agriculture: a review. Eur J Plant Pathol 162(4):759–792

Fabiańska I, Gerlach N, Almario J, Bucher M (2019) Plant-mediated effects of soil phosphorus on the root-associated fungal microbiota in Arabidopsis thaliana. New Phytol 221(4):2123–2137

Fan K, Weisenhorn P, Gilbert JA, Chu H (2018) Wheat rhizosphere harbors a less complex and more stable microbial co-occurrence pattern than bulk soil. Soil Biol Biochem 125:251–260

Finkel OM, Castrillo G, Paredes SH, González IS, Dangl JL (2017) Understanding and exploiting plant beneficial microbes. Curr Opin Plant Biol 38:155–163

Flack CE, Parkinson JS (2018) A zipped-helix cap potentiates HAMP domain control of chemoreceptor signaling. Proc Natl Acad Sci 115(15):E3519–E3528

Gaut BS, Díez CM, Morrell PL (2015) Genomics and the contrasting dynamics of annual and perennial domestication. Trends Genet 31(12):709–719

Guo J, Ling N, Li Y, Li K, Ning H, Shen Q et al (2021) Seed-borne, endospheric and rhizospheric core microbiota as predictors of plant functional traits across rice cultivars are dominated by deterministic processes. New Phytol 230(5):2047–2060

Gutierrez A, Grillo MA (2022) Effects of domestication on plant–microbiome interactions. Plant Cell Physiol 63(11):1654–1666

Hahl T, van Moorsel SJ, Schmid MW, Zuppinger-Dingley D, Schmid B, Wagg C (2020) Plant responses to diversity-driven selection and associated rhizosphere microbial communities. Funct Ecol 34(3):707–722

Haling RE, Yang Z, Shadwell N, Culvenor RA, Stefanski A, Ryan MH, Sandral GA, Kidd DR (2014) Growth and root morphology of herbaceous perennial pasture species influence rhizosphere microbial communities and soil phosphorus dynamics. Plant Soil 383:231–244

Harbort CJ, Hashimoto M, Inoue H, Niu Y, Guan R, Rombolà AD et al (2020) Root-secreted coumarins and the microbiota interact to improve iron nutrition in Arabidopsis. Cell Host Microbe 28(6):825–837

Hariharan G, Prasannath K (2021) Recent advances in molecular diagnostics of fungal plant pathogens: a mini review. Front Cell Infect Microbiol 10:600234

Henry LP, Bruijning M, Forsberg SK, Ayroles JF (2021) The microbiome extends host evolutionary potential. Nat Commun 12(1):5141

Hirsch PR, Miller AJ, Dennis PG (2013) Do root exudates exert more influence on rhizosphere bacterial community structure than other rhizodeposits? In: Molecular microbial ecology of the rhizosphere, vol 1. Wiley, pp 229–242

Jamil F, Mukhtar H, Fouillaud M, Dufossé L (2022) Rhizosphere signaling: insights into plant–rhizomicrobiome interactions for sustainable agronomy. Microorganisms 10(5):899

John CJ, Kumar S, Ge M (2020) Probiotic prospects of PGPR for green and sustainable agriculture. Arch Phytopathol Plant Protect 53(19–20):899–914

Khatoon Z, Huang S, Rafique M, Fakhar A, Kamran MA, Santoyo G (2020) Unlocking the potential of plant growth-promoting rhizobacteria on soil health and the sustainability of agricultural systems. J Environ Manag 273:111118

Kwasniewski M, Szymanowska-Pulka J, Szarejko I (2013) Molecular mechanisms of root hair formation in barley. J Exp Botany 64:3601–3614

Lakshmanan V, Selvaraj G, Bais HP (2014) Functional soil microbiome: belowground solutions to an aboveground problem. Plant Physiol 166(2):689–700

Langridge P, Waugh R (2019) Harnessing the potential of germplasm collections. Nat Genet 51(2):200–201

Lavarenne J, Guyomarc'h S, Sallaud C, Gantet P, Lucas M (2018) The spring of systems biology-driven breeding. Trends Plant Sci 23(8):706–720

Lebeis SL, Paredes SH, Lundberg DS, Breakfield N, Gehring J, McDonald M et al (2015) Salicylic acid modulates colonization of the root microbiome by specific bacterial taxa. Science 349(6250):860–864

Li G, Wang K, Qin Q, Li Q, Mo F, Nangia V, Liu Y (2023) Integrated microbiome and metabolomic analysis reveal responses of rhizosphere bacterial communities and root exudate composition to drought and genotype in Rice (Oryza sativa L.). Rice 16(1):19

Li X, Shen J, Zhang F (2016) Allocation of allelopathic rice root exudates and their effects on rhizosphere microbes. Plant Soil 399:373–384

Ling N, Wang T, Kuzyakov Y (2022) Rhizosphere bacteriome structure and functions. Nat Commun 13(1):836

Lu T, Ke M, Lavoie M, Jin Y, Fan X, Zhang Z et al (2018) Rhizosphere microorganisms can influence the timing of plant flowering. Microbiome 6:1–12

Lv Z, Dai R, Xu H, Liu Y, Bai B, Meng Y et al (2021) The rice histone methylation regulates hub species of the root microbiota. J Genet Genomics 48(9):836–843

Manzotti A, Bergna A, Burow M, Jørgensen HJ, Cernava T, Berg G et al (2020) Insights into the community structure and lifestyle of the fungal root endophytes of tomato by combining amplicon sequencing and isolation approaches with phytohormone profiling. FEMS Microbiol Ecol 96(5):fiaa052

Massalha H, Korenblum E, Malitsky S, Shapiro OH, Aharoni A (2017) Live imaging of root–bacteria interactions in a microfluidics setup. Proc Natl Acad Sci 114(17):4549–4554

Milcu A, Allan E, Roscher C, Jenkins T, Meyer ST, Flynn D et al (2013) Functionally and phylogenetically diverse plant communities key to soil biota. Ecology 94(8):1878–1885

Mousa WK, Shearer C, Limay-Rios V, Ettinger CL, Eisen JA, Raizada MN (2016) Root-hair endophyte stacking in finger millet creates a physicochemical barrier to trap the fungal pathogen Fusarium graminearum. Nat Microbiol 1(12):1–12

Ortega Á, Zhulin IB, Krell T (2017) Sensory repertoire of bacterial chemoreceptors. Microbiol Mol Biol Rev 81(4):10–1128

Pacheco-Moreno A, Bollmann-Giolai A, Chandra G, Brett P, Davies J, Thornton O et al (2024) The genotype of barley cultivars influences multiple aspects of their associated microbiota via differential root exudate secretion. PLoS Biol 22(4):e3002232

Peiffer JA, Spor A, Koren O, Jin Z, Tringe SG, Dangl JL et al (2013) Diversity and heritability of the maize rhizosphere microbiome under field conditions. Proc Natl Acad Sci 110(16):6548–6553

Pérez-Jaramillo JE, Carrión VJ, Bosse M, Ferrão LF, De Hollander M, Garcia AA et al (2017) Linking rhizosphere microbiome composition of wild and domesticated Phaseolus vulgaris to genotypic and root phenotypic traits. ISME J 11(10):2244–2257

Porter SS, Sachs JL (2020) Agriculture and the disruption of plant–microbial symbiosis. Trends Ecol Evol 35(5):426–439

Purugganan MD (2019) Evolutionary insights into the nature of plant domestication. Curr Biol 29(14):R705–R714

Rogers ED, Monaenkova D, Mijar M, Nori A, Goldman DI, Benfey PN (2016) X-ray computed tomography reveals the response of root system architecture to soil texture. Plant Physiol 171(3):2028–2040

Saleem M, Law AD, Moe LA (2016) Nicotiana roots recruit rare rhizosphere taxa as major root-inhabiting microbes. Microb Ecol 71:469–472

Saleem M, Law AD, Sahib MR, Pervaiz ZH, Zhang Q (2018) Impact of root system architecture on rhizosphere and root microbiome. Rhizosphere 6:47–51

Schmid MW, Hahl T, van Moorsel SJ, Wagg C, De Deyn GB, Schmid B (2019) Feedbacks of plant identity and diversity on the diversity and community composition of rhizosphere microbiomes from a long-term biodiversity experiment. Mol Ecol 28(4):863–878

Semchenko M, Xue P, Leigh T (2021) Functional diversity and identity of plant genotypes regulate rhizodeposition and soil microbial activity. New Phytol 232(2):776–787

Smigielski L, Laubach EM, Pesch L, Glock JML, Albrecht F, Slusarenko A et al (2019) Nodulation induces systemic resistance of Medicago truncatula and Pisum sativum against Erysiphe pisi and primes for powdery mildew-triggered salicylic acid accumulation. Mol Plant-Microbe Interact 32(9):1243–1255

Soldan R, Fusi M, Cardinale M, Daffonchio D, Preston GM (2021) The effect of plant domestication on host control of the microbiota. Commun Biol 4(1):1–9

Stringlis IA, Yu KE, Feussner K, de Jonge R, Van Bentum S, Van Verk MC et al (2018) MYB72-dependent coumarin exudation shapes root microbiome assembly to promote plant health. Proc Natl Acad Sci 115(22):E5213–E5222

Tang J, Wu D, Li X, Wang L, Xu L, Zhang Y et al (2022) Plant immunity suppression via PHR1-RALF-FERONIA shapes the root microbiome to alleviate phosphate starvation. EMBO J 41(6):e109102

Thiergart T, Zgadzaj R, Bozsóki Z, Garrido-Oter R, Radutoiu S, Schulze-Lefert P (2019) *Lotus japonicus* Symbiosis Genes Impact Microbial Interactions between Symbionts and Multikingdom Commensal Communities. MBio 10:10–128

Trivedi P, Leach JE, Tringe SG, Sa T, Singh BK (2020) Plant–microbiome interactions: from community assembly to plant health. Nat Rev Microbiol 18(11):607–621

Uga Y, Sugimoto K, Ogawa S, Rane J, Ishitani M, Hara N et al (2013) Control of root system architecture by DEEPER ROOTING 1 increases rice yield under drought conditions. Nat Genet 45(9):1097–1102

Varshney RK, Thudi M, Roorkiwal M, He W, Upadhyaya HD, Yang W et al (2019) Resequencing of 429 chickpea accessions from 45 countries provides insights into genome diversity, domestication and agronomic traits. Nat Genet 51(5):857–864

Veach AM, Morris R, Yip DZ, Yang ZK, Engle NL, Cregger MA et al (2019) Rhizosphere microbiomes diverge among Populus trichocarpa plant-host genotypes and chemotypes, but it depends on soil origin. Microbiome 7:1–15

Vincent SA, Ebertz A, Spanu PD, Devlin PF (2022) Salicylic acid-mediated disturbance increases bacterial diversity in the phyllosphere but is overcome by a dominant core community. Front Microbiol 13:809940

Vives-Peris V, De Ollas C, Gómez-Cadenas A, Pérez-Clemente RM (2020) Root exudates: from plant to rhizosphere and beyond. Plant Cell Rep 39(1):3–17

Wagner MR (2021) Prioritizing host phenotype to understand microbiome heritability in plants. New Phytol 232(2):502–509

Wolinska KW, Vannier N, Thiergart T, Pickel B, Gremmen S, Piasecka A et al (2021) Tryptophan metabolism and bacterial commensals prevent fungal dysbiosis in Arabidopsis roots. Proc Natl Acad Sci 118(49):e2111521118

Worsley SF, Macey MC, Prudence SM, Wilkinson B, Murrell JC, Hutchings MI (2021) Investigating the role of root exudates in recruiting Streptomyces bacteria to the Arabidopsis thaliana microbiome. Front Mol Biosci 8:686110

Xiong C, Singh BK, He JZ, Han YL, Li PP, Wan LH et al (2021) Plant developmental stage drives the differentiation in ecological role of the maize microbiome. Microbiome 9:1–15

Yu P, He X, Baer M, Beirinckx S, Tian T, Moya YA et al (2021) Plant flavones enrich rhizosphere Oxalobacteraceae to improve maize performance under nitrogen deprivation. Nat Plants 7(4):481–499

Zhalnina K, Louie KB, Hao Z, Mansoori N, Da Rocha UN, Shi S et al (2018) Dynamic root exudate chemistry and microbial substrate preferences drive patterns in rhizosphere microbial community assembly. Nat Microbiol 3(4):470–480

Zhang N, Wang D, Liu Y, Li S, Shen Q, Zhang R (2014) Effects of different plant root exudates and their organic acid components on chemotaxis, biofilm formation and colonization by beneficial rhizosphere-associated bacterial strains. Plant Soil 374:689–700

Zhang J, Zhang N, Liu YX, Zhang X, Hu B, Qin Y et al (2018) Root microbiota shift in rice correlates with resident time in the field and developmental stage. Sci China Life Sci 61:613–621

Zhang J, Liu YX, Zhang N, Hu B, Jin T, Xu H et al (2019a) NRT1. 1B is associated with root microbiota composition and nitrogen use in field-grown rice. Nat Biotechnol 37(6):676–684

Zhang X, Zhang S, Zhao Q, Ming R, Tang H (2019b) Assembly of allele-aware, chromosomal-scale autopolyploid genomes based on Hi-C data. Nat Plants 5(8):833–845

Zhang Y, Lin J, Chen S, Lu H, Liao C (2023) The Influence of the Genotype and Planting Density on the Structure and Composition of Root and Rhizosphere Microbial Communities in Maize. Microorganisms 11(10):2443

Plant Growth-Promoting Bacteria-Mediated Climatic Stress Tolerance in Plants

7

Ying Ma and Xiang Li

Abstract

The flourishing of crops and their subsequent productivity are significantly impaired by the strain of climatic stress. Various abiotic stresses, including drought, extreme temperatures, freezing conditions, and increased salinity, have made cultivated plants susceptible to the ongoing global climate change. The precarious survival of the plants and agricultural output is significantly reduced due to these stresses. Reactive oxygen species (ROS) are produced and accumulate in plants as a result of abiotic stressors, which damages essential biomolecules such as proteins, lipids, and nucleic acids that are found within the cells. Plant growth-promoting bacteria (PGPB) are resilient microbes that boost plant growth and crop productivity, strengthening food security under abiotic stress. They enhance plant growth through mechanisms including the synthesis of ACC deaminase, exopolysaccharides, HCN, phytohormones, extracellular polymers, siderophores, abscisic acid, and release ofvolatile organic compounds (VOCs). PGPB also enhance the antioxidant pathways in plants, supporting the prevention of oxidative stress, regulation of ion homeostasis, and maintenance of osmotic balance. They also trigger stress response genes, reprogram metabolism, and supply transcription factors. This review explores how domesticated PGPB strains can aid sustainable agriculture by mitigating abiotic stress, while also examining the potential of multi-omics to uncover new strains and mechanisms.

Keywords

Climatic stress · Plant growth-promoting bacteria · Plant-PGPB interaction · Abiotic stress tolerance · Sustainable agriculture

Y. Ma (✉) · X. Li
College of Resources and Environment, Southwest University, Chongqing, China

P. Pandey et al. (eds.), *Microorganisms Resilience to Climate Change*,
Microorganisms for Sustainability 56, https://doi.org/10.1007/978-981-96-3748-5_7

7.1 Introduction

7.1.1 The Stresses Caused by Global Climate Change

Both natural and human-caused factors, particularly the emission of greenhouse gases (GHGs) such as methane (CH_4), carbon dioxide (CO_2), water vapor and nitrous oxide (N_2O), has been largely attributed to the industrial revolution as the main cause of the dramatic global climate shifts we're experiencing today (Zhang et al. 2011; Tian et al. 2016). The 2014 Intergovernmental Panel on Climate Change (IPCC) report emphasizes how the greenhouse effect, primarily caused by these GHGs, results in global warming and can trigger drought, posing a significant threat to ecosystems (Pachauri et al. 2014; Cassia et al. 2018) (Fig. 7.1).

Simultaneously, climate change induces various environmental changes, including Rising sea levels, altered weather patterns, melting ice caps, ecosystem shifts, increased frequency and intensity of extreme weather events, disruptions in biodiversity, glaciers, rising global temperatures, and other ecological impacts (Sala et al. 2000; Traill et al. 2010; Kumar et al. 2012; Lehmann and Rillig 2014; Mazdiyasni and AghaKouchak 2015).

The global agriculture industry, which is dealing with significant difficulties as a result of climate change, is the subject of focus for the researchers. Modern agriculture's increasing susceptibility to climate change, which is a result of both natural and human-caused variables, has garnered more attention in recent decades (Altieri et al. 2015) (e.g., the carbon cycle, depicted in Fig. 7.2). In rural Pakistan, in recent years, crops like wheat, rice, cotton, and sugarcane are severely impacted by climate change, according to a report by Abid et al. 2015. Additionally, research also

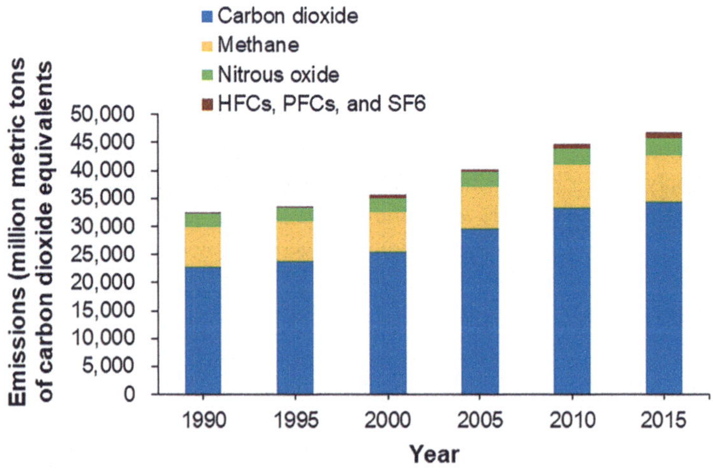

Fig. 7.1 Global greenhouse gas emissions, 1990–2015
Note: PFCs: perfluoro-carbons, HFCs: hydrofluorocarbons, SF6: Sulphur hexafluoride
Data source: EPA's Climate Change Indicators in the United States: www.epa.gov/climate-indicators

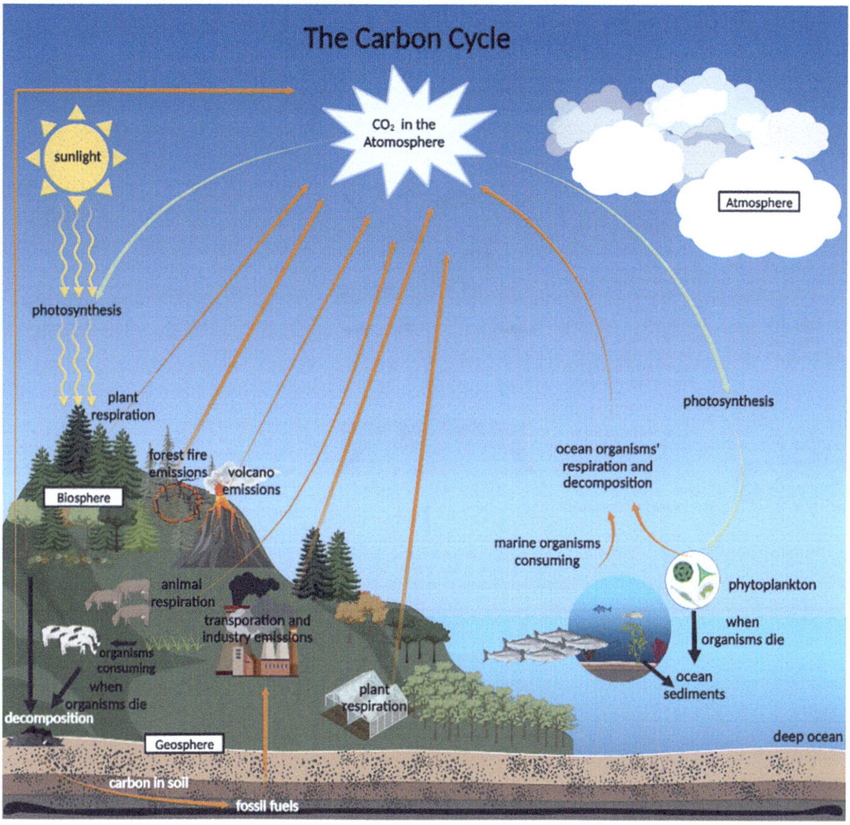

Fig. 7.2 Schematic diagram of the global carbon cycle
Note: Plants use photosynthesis to absorb carbon dioxide from the air and produce sugars with the help of water and sunlight. Animals that eat plants acquire the carbon from them, and the carbon also transfers when animals eat other animals. Through respiration, both plants and animals emit carbon dioxide into the atmosphere. Upon their death, their bodies decompose, transferring carbon into the soil. Some of this carbon has been converted into fossil fuels over millions of years. When fossil fuels are burned in factories, power plants, and automobiles, stored carbon is released as carbon dioxide (CO_2) back into the atmosphere, a portion of the carbon from the atmosphere is absorbed by various water bodies, including rivers, lakes, and oceans, where it dissolves

indicates that plant biomass and metabolic activities are greatly impacted by increasing atmospheric CO_2 concentrations (Nogia et al. 2016). Global food production will undoubtedly continue to be impacted by climate change, with significant consequences for crop yields (Rosenzweig and Parry 1994).

Global climate change renders plants vulnerable to a spectrum of climatic stresses, including not only temperature extremes (both high and low) but also salinity stress, flooding, dehydration, excessive light, ultraviolet radiation, and pollution from various sources (such as heavy metals and ozone (O_3)), among other factors (Lal et al. 2018) as depicted in Fig. 7.3.

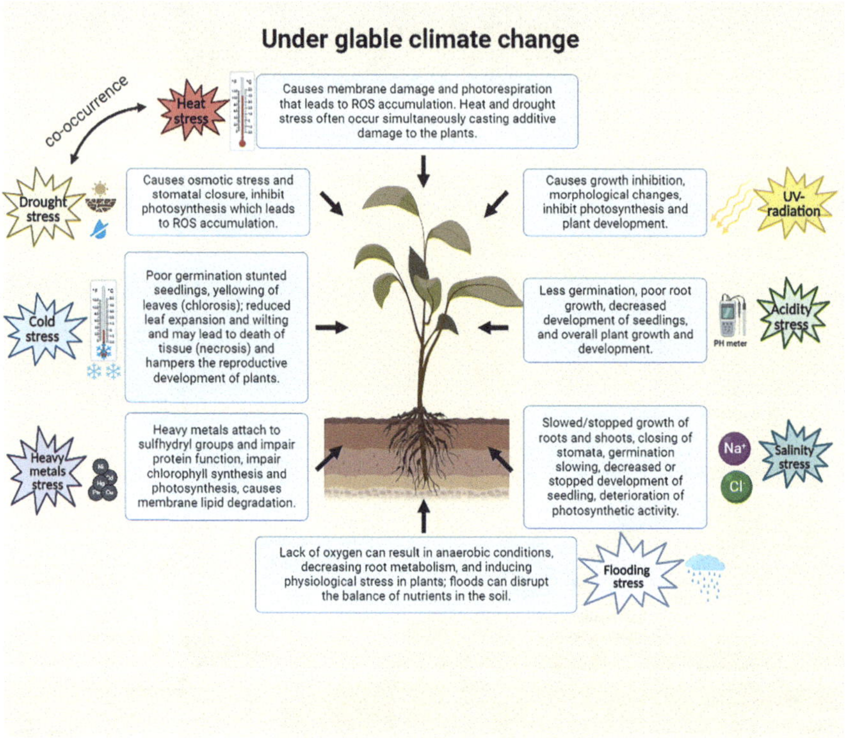

Fig. 7.3 Evaluating climatic stress indicators in plants under global climate change

7.1.2 Effects of Climate-Induced Stresses on Plant Development

It is essential to note that the primary factors affecting grain yield are drought and high temperature (Barnabás et al. 2008). High-temperature stress, coupled with drought stress, can induce cellular and physiological changes, including protein denaturation, loss of membrane integrity, induced cellular function, and slowed plant growth. Plant growth and development are negatively impacted by these alterations, and the effects of dryness and high temperatures alone or in combination may cause oxidative and osmotic stress (Moriondo et al. 2011; Quint et al. 2016). In plants, stomatal closure—where tiny holes on the leaf surface control gas exchange with the atmosphere—is a vital physiological reaction to dryness. This reaction reduces transpirational water loss (Grzesiak et al. 1997; Jones et al. 2003).

In a study examining the impact of climate change on economic growth in Malaysia, researchers identified a direct correlation between economic growth and temperature fluctuations (Rahim and Guan Puay 2020).

Simultaneously, exposure to cold and low temperatures can have a significant influence on the development, growth, and overall productivity (Guy 1990; Yadav

2010). Freezing injury occurs at temperatures below freezing (0 °C) and can cause ice crystal formation, cell wall rupture, and tissue damage (Devireddy et al. 2021).

Furthermore, salinity stress constitutes a substantial abiotic challenge that detrimentally impacts plant development and growth (Gul et al. 2022). When subjected to elevated salinity levels in the soil, plants undergo osmotic stress and experience an imbalance in ion concentrations, leading to reduced water uptake (Steudle 1994). This water deficit adversely affects multiple physiological processes in plants, including photosynthesis, nutrient uptake, and hormonal regulation (Hao et al. 2021).

Excess moisture in the soil affects various aspects of plant growth and physiology. One significant impact is the alteration of root growth and architecture. Excess moisture refers to elevated water content in the soil, often resulting from heavy rainfall, poor drainage, or waterlogging (Kaur et al. 2020). In waterlogged conditions, plants often exhibit reduced root growth due to limited oxygen availability in the rhizosphere (Toral-Juárez et al. 2021). This reduced root growth can decrease nutrient uptake, hampering overall plant productivity.

Certainly, excessive light and ultraviolet (UV) radiation can profoundly affect plants, affecting various physiological and molecular processes. Elevated light intensities beyond the plant's optimal range can lead to photoinhibition, where damage to the photosynthetic apparatus surpasses the repair rate (Mackey et al. 2008). Ultraviolet radiation, particularly UV-B and UV-A, can cause harm to plant cells and DNA. UV-B radiation can trigger the formation of thymine dimers, leading to mutations and impairments in normal growth and development (Jenkins 2014). Moreover, prolonged UV-B exposure can disrupt cellular processes such as nutrient uptake and inhibit plant reproduction (Paul and Gwynn-Jones 2003).

Heavy metal pollution is a significant environmental concern affecting plant health and ecosystem functioning. Plants can accumulate heavy metals from contaminated soils, water sources, and atmospheric deposition, leading to physiological and biochemical disruptions. One critical effect of heavy metals on plants is the disruption of essential nutrient uptake and utilization. Heavy metals such as mercury (Hg), lead (Pb), and cadmium (Cd), can interfere with the absorption and transport of essential nutrients, impairing plant growth and development (Parmar et al. 2013). This disruption often leads to nutrient imbalances and deficiencies, affecting photosynthesis and protein synthesis. In addition, ozone (O^3) is a highly reactive gas that can significantly affect plant physiology and growth. Ozone-induced oxidative stress destroys chloroplasts, reduces chlorophyll content, and impairs electron transport, decreasing photosynthetic efficiency and reducing plant growth (Bohler et al. 2013).

Taken together, these studies collectively demonstrate that, firstly, abiotic stresses exert a more pronounced influence on plants worldwide when compared to biotic stresses. The impact of abiotic stress on plant growth is inevitable, and, secondly, processes such as plant development, germination, and photosynthesis, are all significantly reduced (Gao et al. 2007; Khan et al. 2019).

7.1.3 Contemporary Technologies for Mitigating Stress in Plants

Technological advancements have ushered in various innovative solutions for alleviating plant stress caused by climatic factors. Initially, researchers employ diverse crop management strategies to enhance plant tolerance against heat and drought stress, including soil management, tillage practices, irrigation, and the selection of appropriate crop varieties (Lamaoui et al. 2018). Additionally, extensive research has been conducted to combat abiotic stresses induced by climate change, focusing on augmenting the expression of desirable crop traits through gene editing techniques (Hamdan et al. 2022). Notably, nanotechnology, a cutting-edge technology, has gained considerable attention across various disciplines in recent years. Nanoparticles (NPs) have proven effective in aiding plants in mitigating abiotic stress by establishing a dependency on NPs throughout their development and growth (Rastogi et al. 2019). One study demonstrated that the application of silica NPs improved the growth and physiological response of hawthorn seedlings to drought, resulting in enhanced drought tolerance (Ashkavand et al. 2015). Furthermore, NPs not only enhance plant drought tolerance but also alleviate stress induced by salinity, flood, and ultraviolet radiation stress (Siddiqui et al. 2014; Elsheery et al. 2020; Faraji and Sepehri 2020).

Currently, various agricultural strategies are being employed to promote sustainable development. These strategies encompass agricultural intensification, the adoption of sustainable agricultural technology management practices, an exploration of microbial transgenic technology or utilization of plant growth promoters (Passeri et al. 2016), and the advancement of biofertilizers (Kamkar 2016; Suhag 2016). Using microorganisms, such as Plant growth-promoting bacteria (PGPB), is vital in enhancing plant growth and augmenting agricultural product yields (Vejan et al. 2016; Trivedi et al. 2020). PGPB are recognized as a beneficial microbial community within the soil. Besides their ability to stimulate plant growth (Beneduzi et al. 2012), Plants can gain increased resilience to abiotic stresses such as salinity, extreme temperatures, drought, and toxic metals through the activity of plant growth-promoting bacteria (PGPB). (Rajkumar et al. 2008; Khan et al. 2017; Numan et al. 2018). Moreover, the efficacy of a specific PGPB strain may vary due to various factors, including plant species, geographic environment, ecological conditions, and plant characteristics (Volis et al. 2015; Lastochkina et al. 2017). PGPB plays a crucial role in mitigating constraints to plant growth under climatic stress. This knowledge aids in fully harnessing the potential of these microorganisms as vital components of organic agriculture, thereby improving crop productivity and ensuring soil fertility and sustainability amidst climatic stresses.

This chapter presents the latest development in our understanding how PGPB contributes to enhancing plant tolerance to various abiotic stresses. Moreover, it comprehensively analyzes the precise mechanisms by which PGPB alleviates climatic stress in plants.

7.2 PGPB and Their Role in Mitigating Climatic Stresses in Plants

7.2.1 PGPB-Induced Soil Amelioration Alleviates Plant Climatic Stresses

In agriculture, soil stands as a crucial living medium that sustains crop production (Farrell and Hart 1998). Consequently, the preservation and upkeep of soil health today are of paramount importance for ensuring sustained crop productivity and the stability of the cultivation environment (Szczepanek et al. 2021). As integral elements of the soil ecosystem, microorganisms contribute significantly to processes such as nutrient cycling, the decomposition of organic matter, and maintaining soil structure (Mumtaz et al. 2017). Consequently, they are anticipated to represent environmentally friendly alternatives for preserving soil health and enhancing crop productivity (Varma and Choudhary 2019). PGPB, intimately intertwined with plant survival in the soil, has been harnessed as the primary reservoir of beneficial microorganisms for the development of organic biofertilizers and biopesticides. The primary merit lies in their ability to sustain soil environment effectiveness and abundance, while simultaneously enhancing plant tolerance to abiotic stresses (Dhir 2017) (Table 7.1). Furthermore, PGPB can also function as an antagonist against various pathogens (Velten et al. 2015).

7.2.1.1 Enhancing Soil Structural Framework

Research findings have demonstrated that the dynamic alterations in soil's physical, chemical, and structural properties stem from intricate interplays among rhizosphere microorganisms, roots, soil, and water (Tisdall and Oades 1980; Haynes and Swift 1990). Polysaccharides produced by certain organisms in the soil facilitate the adhesion of soil particles, leading to the formation of micro-network aggregates (including microaggregates with a diameter < 250 μm and macroaggregates with a diameter > 250 μm). These aggregates, in turn, promote the formation of soil aggregates and soil pores (Tisdall and Oades 1982; Oades 1993; Rashid et al. 2016). Ensuring the stability of soil structure is beneficial for maintaining the sustainability of agricultural development.

A previous study has reported that exopolysaccharides (EPS) produced by PGPB play a role in enhancing soil aggregation and structure, subsequently enhancing maize tolerance to drought stress (Naseem and Bano 2014). Moreover, the application of EPS produced by *Pseudomonas putida* strain GAP-P45 to sunflower seedlings has promoted soil structure stability by enhancing the proportion of soil aggregates (Sandhya et al. 2009). A well-structured soil offers numerous advantages for soil health, such as the preservation of natural soil tilth, heightened water use efficiency in plants, improved soil aeration, facilitation of average root growth, and promotion of organic matter accumulation (Miller and Jastrow 2000; Roesti et al. 2006).

It has been reported that the salinity accumulation in the soil leads to an increase in soil particle dispersion and swelling capacity, consequently reducing soil water

Table 7.1 A summary of findings demonstrating the PGPB strains applied to soil amelioration

PGPB species	Host plant species	Soil type	Effects	Mode of Action	References
Bacillus cereus, *Pseudomonas moraviensis*, and *Stenotrophomonas maltophilia*	Wheat (*Triticum aestivum* L.)	Saline soils	Increase adhesion of soils and indirectly improve soil structure and aggregation, resulted in better texture, EC, pH and organic matter in soil.	Production of extracellular polymeric substances (EPSs), formation of cohesive EPS-soil complexes to resist salinity.	Siebers and Kruse (2019) Creus et al. (2005), Hassan et al. (2018)
Halomonas variabilis and *Planococcus rifietoensis*	Chickpea (*Cicer arietinum* Var.)	Saline soils	Formation of soil aggregates have water retaining capacity around roots.	Biofilm formation and exopolysaccharide accumulation.	Qurashi and Sabri (2012), Ghosh et al. (2019)
Bacillus cereus	Mungbean (*Vigna radiata* (L.) Wilczek)	Saline soils	Increase dehydrogenase activity and microbial biomass carbon in soil.	EPS-mediated biofilms act as a carbon source and condenses soil particles to form humus, resulting in stable organic carbon.	Islam et al. (2016a)
Pseudomonas putida	Sunflower (*Helianthus annuus* L.)	Dry soils	Improving water retention in soils by increasing the proportion of stable soil aggregates leads to enhanced soil aggregation and greater aggregate stability.	Formation of biofilms by inoculated bacteria on root surfaces involves the release of EPS into the soil as capsular and slime materials.	Sandhya et al. (2009) Sandhya and Ali (2015)

Rhizobium sp.	Sunflower (*Helianthus annuus* L.)	Dry soils	Enhancing permeability by promoting soil aggregation helps maintain a higher water potential around the roots.	Production of EPS contributes to an increased ratio of root-adhering soil (RAS) mass to root tissue (RT) mass, known as the RAS/RT ratio.	Alami et al. (2000)
Pseudomonas sp.	Tall fescue (*Festuca arundinacea*) and Perennial ryegrass (*Lolium perenne*)	Soils contaminated with dichlorodiphenyltrichloroethane (DDT)	Increase the bioavailability of DDT, enhance microbial degradation of DDT in soil, and assist plants effectively in removing DDT from the soil.	Production of biosurfactants.	Wang et al. (2017)
P. putida	Mungbean (*Vigna radiata* (L.) Wilczek)	Heavy metal (lead and cadmium) soils	Reduced the free metal concentration in soils.	Production of siderophores and iron–siderophore complex.	Tripathi et al. (2005)
Microbacterium oxydans JYC17, *Burkholderia cepacia* J62 and *Pseudomonas thivervalensis* Y1–3-9	Rape plants	Heavy metal (copper) soils	Promote the phytoremediation of copper pollution in soil, enhance the copper enrichment efficiency of rape plants.	Oorganic acids and siderophores production, mitigation of the oxidative stress caused by copper in rape plants, and within the endosphere and rhizosphere of rape, variations occur in the composition of the bacterial community.	Ren et al. (2019)

(continued)

Table 7.1 (continued)

PGPB species	Host plant species	Soil type	Effects	Mode of Action	References
P. putida and *P. fluorescens*	Barley seeds (*Hordeum vulgare* L.) and Canola (*Brassica napus* L.)	Heavy metal (cadmium and lead ions) soils	Promote plant growth and absorb heavy metals in soils.	Production of siderophores, IAA and ACCD.	Yancheshmeh (2011)
Rhodococcus sp.	Reed canary grass (*Phalaris arundinacea* L.)	Soils contaminated with polychlorinated biphenyls (PCBs)	Degradation of PCB congeners, promotion of lateral root emergence, and increased plant biomass in PCB contaminated soil.	Production of ACCD, which help alleviate plant stress.	Vergani et al. (2019)
Klebsiella pneumoniae	*Scirpus triqueter* L.	Pyrene-Ni co-contaminated soils	Increase the efficiency of phytoremediation of pyrene and Ni in soil.	Increases fluorescein diacetate (FDA) activity in pyrene-Ni co-contaminated soil.	Chen et al. (2017)

permeability (Aslam and Ali 2018). Notably, the EPS secreted by PGPB in response to salinity stress plays a significant role in biofilm formation, aiding plants in adapting to salinity stress. This unique biofilm structure formed in both roots and soil particles contributes to strengthening soil structure and enhancing soil physical and chemical properties (Chan et al. 1984; Ishii et al. 2004; Batool and Hasnain 2005; Banerjee et al. 2019). In soils contaminated with heavy metals, EPS can also adsorb heavy metals and reduce their concentration in the soil (Harrison et al. 2006). Human activities have significantly exacerbated soil degradation in semi-humid, semi-arid, and arid regions, resulting in land desertification (Wang et al. 2004). A perspective on land desertification control highlights that when soil fertility and microbial communities cannot be replenished solely through water, the crucial aspect lies in artificial intervention and the introduction of specific microorganisms that facilitate plant growth (de-Bashan et al. 2012). Preliminary studies have revealed that inoculation of the PGPB *Azospirillum brasilense* into three species of cacti during a field trial in Mexico resulted in a significant increase in the cacti's survival rate (76%) compared to the controls; in contrast, less than 2% of the control group managed to survive beyond 3.5 years. Additionally, the experimental groups exhibited substantial improvements in plant biomass, soil properties, and a notable reduction in soil erosion and reclamation of topsoil (Bashan et al. 1999).

7.2.1.2 Regulating Acidity/Alkalinity, Moisture, and Nutrient Composition in Soils

An optimal soil pH fosters a favorable environment for microorganisms and plant growth (Wungrampha et al. 2018; Goswami and Suresh 2020). The soil acidity and alkalinity are governed by the concentration of hydrogen ions (H^+) (Zhang et al. 2019). Elevated levels of H^+ result in soil acidity, indicated by a pH value below 5.5–5.0, whereas reduced H^+ concentration leads to soil alkalinity (Zhang and Mu 2009; Shavrukov and Hirai 2016). Moreover, PGPB can enhance the capacity to utilize phosphorus (P) and potassium (K) in plants by lowering the soil pH (Adnan et al. 2017; Liu et al. 2021). Scholarly research has demonstrated that acidic soils, characterized by a lower pH value, tend to accumulate higher concentrations of metal ions, which can negatively impact the growth and development of plants (Mohite 2013). In a particular study, the inoculation of PGPBs into acidic sulfate soil showcased their ability to secrete organic acids that effectively chelate aluminum (Al) (Panhwar et al. 2015). This mechanism subsequently reduces the toxicity of Al towards plant roots (Gupta et al. 2013).

The application of PGPBs provides an environmentally friendly solution for mitigating high pH stress in plants. Recent research has unveiled the capabilities of two PGPBs, specifically the endophyte *Bacillus licheniformis* strain P2.3 and the rhizobacteria *Bacillus aerius* strain S2.14, in enhancing the absorption and utilization of iron (Fe) by soybean plants when grown in alkaline soil conditions (Roriz et al. 2023). This promotion of Fe uptake subsequently leads to improved Fe nutrition for plants, increasing biological yield (Zhou et al. 2016; Patel et al. 2018).

Moisture plays a pivotal role in agricultural development, with excess moisture accumulating in the soil during floods resulting from climate change. Waterlogging in farmland leads to the saturation of soil structure pores by water, which adversely affects the respiration of plant roots (Dong et al. 1983). Decreased oxygen availability to the roots can significantly impact crop quality and yield (Linkemer et al. 1998). PGPBs can improve soil drainage and permeability by facilitating the formation of soil particle structure (Mącik et al. 2020). This, in turn, contributes to enhanced soil infiltration capacity, promoting water distribution within the soil and reducing the adverse effects of water accumulation during floods. In arid regions, drought stress due to water scarcity is a critical constraint on agricultural production. Furthermore, the EPS secreted by PGPBs can enhance soil structure, retain soil moisture, and augment their colonization in roots (Morcillo and Manzanera 2021). It is worth noting that although PGPBs cannot directly regulate the pH and water distribution within the soil, their mechanisms described above contribute to improving the overall soil environment.

The loss of soil nutrients due to climatic stress underscores the crucial role of PGPB in regulating essential elements like nitrogen (N), P, sulfur (S), K, and micronutrients in the soil. For example, PGPBs, such as *Azotobacter*, *Azospirillum*, *Pseudomonas*, *Agrobacterium*, *Erwinia*, *Bacillus*, *Serratia*, *Klebsiella*, and *Burkholderia*, possessing nitrogen-fixing abilities can convert atmospheric nitrogen into forms, such as ammonia or nitrate, that are readily available for plants (Shilev 2020). These strains establish symbiotic relationships with plant roots, facilitating nitrogen fixation within nodules and supplying plants with the necessary nitrogen, ultimately enhancing nitrogen availability in the soil. Simultaneously, PGPBs can enhance soil nutrient availability and plant uptake, increasing crop yields (Berendsen et al. 2018; Etesami 2018; Majeed et al. 2018).

7.2.1.3 Optimizing Soil Microbial Community

The soil microbiome plays a pivotal role in biogeochemical cycles by regulating the cycling of macronutrients, micronutrients, and other essential elements crucial for the existence of plants and animals. Climate change will impact the regulatory capacity of individual species within the soil ecosystem, altering the interactions among microbial populations within soil communities (Berg et al. 2010). The regulation of soil function is intricately linked to the composition and physiology of the soil microbial community. During periods of drought stress, the abundance, structure, and activity of soil microorganisms are profoundly affected. For instance, a PGPB strain, *Sphingomonas* sp. strain Hbc-6, isolated from the leaves of *Nitraria tangutorum*, a desert plant, was introduced into corn root soil. The results demonstrated that strain Hbc-6 reshaped the corn rhizosphere bacterial community, enhancing its complexity and resilience. Moreover, it was observed that strain Hbc-6 facilitated the recruitment of a more significant number of potentially beneficial bacteria in the rhizosphere such as producing nitrogen-fixing bacteria and phosphorus-dissolving bacteria, thereby promoting synergistic plant growth together (Wang et al. 2022). Under salinity stress in tomato plants, researchers found that *Bacillus mesonae* strain H20–5 promoted tomato plant growth while concurrently influencing the soil bacterial community (Lee et al. 2021). The inoculation of PGPBs into the soil led to significant alterations in the structure of the pre-existing microbial community, driven by potential interactions with native microbial communities, resulting in both synergistic and antagonistic dynamics. Several strains of PGPB have demonstrated promise as biocontrol agents against diverse soil pathogens responsible for plant diseases (Liu et al. 2017a). PGPBs can mitigate the abundance of detrimental bacteria through competitive interactions and the synthesis of antibacterial compounds. These deleterious microorganisms can significantly impede plant health under stressful conditions, leading to the onset of plant diseases and compromised plant resistance. By diminishing the prevalence of harmful microbes, PGPBs effectively enhance plant resilience and adaptation to various stresses.

7.2.2 PGPB-Mediated Biological Processes in Plants for Stress Alleviation

It is vital for us to develop a comprehensive understanding of the mechanisms underlying the ability of PGPBs to alleviate stresses in plants (Table 7.2). This knowledge is essential for harnessing the potential of PGPBs to optimize plant productivity under climatic stresses (Cao et al. 2023). Drawing from prior investigations into the growth-promoting mechanisms of PGPB, these mechanisms can be classified into direct and indirect mechanisms (Goswami et al. 2016). PGPBs contribute to plant growth directly by facilitating improved nutrient accessibility, including N, Fe, and P, while concurrently enhancing the biosynthesis of plant hormones (Rastegari et al. 2020). Moreover, PGPBs employ direct strategies to enhance plant growth, encompassing the secretion of siderophores, phytohormones, phosphorus solubilization, and nitrogen fixation (Singh et al. 2020). PGPBs play a pivotal role in sustaining plant root growth by facilitating water uptake and improving nutrient use efficiency. The primary indirect strategies associated with this phenomenon encompass the upregulation of defense-related enzyme activity in plants, exemplified by zeatin, chitinase, indole-3-acetic acid (IAA), ethylene, gibberellic acid (GA_3), and abscisic acid (ABA) (Fadiji et al. 2022). Additionally, PGPBs enhance the antioxidant capacity in plants to cope with climatic stresses through direct and indirect mechanisms. These mechanisms are highly relevant to the discussion within this section, shedding light on how PGPBs employ these strategies to ameliorate the challenges posed by climatic stress in plants (Fig. 7.4).

7.2.2.1 Direct Mechanisms

As widely recognized, plants rely primarily on their root systems to absorb the essential water and nutrients required for growth and development. PGPB can possess the ability to regulate root structure, including the promotion of primary root elongation, lateral root development, and root hair growth (Wang et al. 2015). Impaired plant growth under drought stress is primarily attributed to inhibited cell elongation due to water scarcity. However, when treated with PGPB, plant root development and lateral root formation are stimulated, leading to increased root water absorption capacity (Wu et al. 2020). Research has revealed that PGPB

Table 7.2 Mechanisms of PGPBs alleviating climatic stresses in plants

Types of climatic stresses	PGPB species	Plants	Specific mechanisms	Beneficial effects	References
Salt stress	*Bacillus megaterium*	Alfafa (*Medicago sativa* L.)	Produced IAA and ACC deaminase.	Observed plant growth promotion and nodulation in alfalfa roots	Chinnaswamy et al. (2018)
Salt stress	*Lysinibacillus fusiformis*, *B. subtilis*, *Brevibacterium halotolerans*, *Bacillus licheniformis*, *Bacillus pumilus*, *Achromobacter xylosoxidans*, and *Pseudomonas putida*	Argentine screwbean (*Prosopis strombulifera*)	Produced ACC deaminase.	Promoted germination and early growth	Sgroy et al. (2009)
Heavy metal (Zn, Cd and Pb) stress	*Variovorax paradoxus*	Pasture legume species (*Astragalus hamosus* L., *Lotus edulis* L., *Lotus ornithopodioides* L.)	Produced ACC deaminase	Improved nodulation, root growth, and uptake of nutrients (N, P, Ca, Mg, Na, Mn, Zn, and Pb)	Safronova et al. (2012)
Heat stress	*Bacillus cereus*	Tomato (*Solanum lycopersicum* L.)	Produced ACC deaminase and exopolysaccharide; reduced extracellular polymeric substances production in plants	Promoted growth of shoot, root length, leaf surface area, fresh and dry weight.	Mukhtar et al. (2020)
Salt stress	*Pseudomonas* sp., *Agromyces tropicus*, *L. fusiformis*, *Bacillus flexus*, *Lysinibacillus* sp.,*Bacillus tequilensis*,*B. licheniformis*, and*Pseudomonas mendocina*	Wheat (*Triticum aestivum* var.)	Produced IAA, siderophores and catalase; inhibited phytopathogenic fungi; dissolved phosphate.	Enhanced seed germination percent, shoot length, root length， fresh and dry weigh.	Damodaran et al. (2019)

Salt stress	*Stenotrophomonas* sp.	Spinach and soybean	Induced synthesis of novel proteins (unknown to be identified) which help the plants in tolerating salts under salt stress conditions.	Improved plant growth, biomass and yield through counteracting the effects of soil salinity.	Nigam et al. (2022)
Drought stress	*Azospirillum* sp. and *Herbaspirillum* sp.	Corn (Zea mays L.)	Suppressed ZmVP14 gene expression (related to abscisic acid (ABA) biosynthesis) in plants.	Increased the carbon, nitrogen and chlorophyll levels of corn and crop yield, and suppress the content of abscisic acid and ethylene in corn.	Curá et al. (2017)
Drought stress	*Azospirillum* sp.	Wheat (*Triticum aestivum* L.) and tomato (*Lycopersicon esculentum* L.)	Production of IAA, siderophore, N and auxin, with P solubilizing and ACC deaminase activities.	Promoted the formation of lateral roots and alleviate the adverse effects of growth in plants.	Arzanesh et al. (2011) Creus et al. (2005)
Drought stress / osmotic-stress	*Pseudomonas aeruginosa*, *Bacillus endophyticus*, and *Bacillus tequilensis*	*Arabidopsis thaliana*	Production of auxin, EPS, and cytokinin (tZ); and regulation of ABA-responsive genes expression, ROS metabolism, and metabolic homeostasis.	Modulated root architecture especially the lateral root number, and increased in fresh weight and dry weight.	Ghosh et al. (2019) Shi et al. (2014)
Drought stress	*Bacillus* spp. and *Pseudomonas* spp.	Sorghum [*Sorghum bicolor* (L.) Moench]	Upregulation of the hormones gibberellic acid, indole acetic acid and cytokinin in plants.	Promoted growth (Including fresh and dry weight) and change root structure in plants	Carlson et al. (2020)
Flooding stress	*Azospirillum* spp.	Forage grasses	Biological nitrogen fixation (BNF) and indole-3-acetic acid (IAA) production.	Increased plant height, shoot dry matter and root volume.	Souza et al. (2017)

(continued)

Table 7.2 (continued)

Types of climatic stresses	PGPB species	Plants	Specific mechanisms	Beneficial effects	References
Drought stress	*Bacillus subtilis*	Timothy (*Phleum pratense* L.)	Accumulation of osmotic protective substances in roots and shoots; increase content of sucrose, fructans, asparagine, glutamic acid and glutamine in plants.	Increased in shoot and root biomass, photosynthesis, and stomatal conductance in plants	Gagné-Bourque et al. (2016)
Drought stress	*Bacillus spp.* and *Pseudomonas spp.*	Sorghum [*Sorghum bicolor* (L.) Moench]	Induced genes related to antioxidant (AOX) for enhancing the plant's ability to maintain cellular redox balance.	Metabolites such as SA, brassinolide, JA, sphingosine in roots and shoots of plants are significantly induced.	Jones et al. (2003), Carlson et al. (2020)
Heavy metal (Cd) stress	*Pseudomonas fluorescens*	*Sedum alfredii*	Increased expression levels of phytohormones response genes and phytohormones concentrations in plants.	Promoted lateral root formation and Cd uptake in hyperaccumulator.	Wu et al. (2020)
Saline-alkali stress	*Pseudomonas aeruginosa* and *Enterobacter aerogenes*	Alfafa (*Medicago sativa* L.)	Induced activity of antioxidant enzymes (SOD, POD, and CAT) andaccumulation of antioxidants in plants.	Enhanced plant biomass, removed harmful substances, increased in crude protein but decrease in fiber content.	Liu et al. (2019)
Drought stress	*Ochrobactrum pseudogrignonense*, *Pseudomonas* sp. and *B. subtilis*	Black gram (*Vigna mungo* L.) and the garden pea (*Pisum sativum* L.)	Elevated ROS scavenging enzymes (e.g., CAT, POD) and osmolytes (as proline and total phenolics) in plants.	Increased seed germination percentage, root and shoot length, and dry weight; and higher chlorophyll content, relative water content, and root recovery intention.	Saikia et al. (2018)

Drought stress	*Pseudomonas putida*	Chickpea (*Cicer arietinum* L.)	Modulated membrane integrity, osmolyte accumulation (proline, glycine betaine),and induced ROS scavenging enzymes (e.g., CAT, APX, and GST)	Improved and restored plant growth, water status, membrane integrity, osmoprotectant accumulation, and antioxidant enzyme activity.	Tiwari et al. (2016)
Salt stress	*B. subtilis*	Chickpea (*Cicer arietinum* L.)	Reduced ROS level and lipid peroxidation and increased activities of ROS scavenging antioxidants (superoxide dismutase, peroxidase, catalase, and glutathione reductase), the levels of non-enzymatic antioxidants (ascorbic acid and glutathione), and the total phenol content in plants.	Enhanced plant biomass and the synthesis of photosynthetic pigments.	Abd_Allah et al. (2018)
Salt stress	*Aneurinibacillus aneurinilyticus* and *Paenibacillus* sp.	Common bean (*Phaseolus vulgaris*)	Increased activities of antioxidative defensive enzymes (POD, SOD, and CAT) and phenolics content for scavenging salt-induced ROS	Enhanced seed germination, root and shoot length, fresh and dry weight of root and shoot, relative water content, chlorophyll, and carotenoid content.	Gupta and Pandey (2020)

(continued)

Table 7.2 (continued)

Types of climatic stresses	PGPB species	Plants	Specific mechanisms	Beneficial effects	References
Heat stress	*Bacillus cereus*	Tomato (*Solanum lycopersicum* L.)	Increased ascorbate peroxidase (APX) and SOD Decreased glutathione (GSH) in tomato plants.	Increased biomass and chlorophyll fluorescence of tomato plants.	Khan et al. (2020a)
Heat stress	*Bacillus cereus*	Tomato (*S. lycopersicum* L.)	Upregulated heat stress-responsive transcription factor (*SIHsfA1a*) and high-affinity potassium transporter (*SIHKT1*) in plants	Increased amino acid content in plants and reduced heat stress damage in tomato plants.	Khan et al. (2020a)
Heavy metal (Cd) stress	*Pseudomonas grimontii*, *Pantoea vagans*, *Pseudomonas veronii* and *Pseudomonas fluorescens.*	Switchgrass (*Panicum virgatum* L.)	Elevated the gene expression of *HSP70* and *HMA3* in switchgrass plants.	Reduced Cd accumulation and increased IAA production and biomass (including fresh weight and dry weight) in plants.	Begum et al. (2019)
Heat stress	*Bacillus cereus*	Soybean	Overexpression of stress-responsive *GMLAX3* and *GmAKT2*	Enhanced the chlorophyll fluorescence and photosynthetic pigments in soybean plants	Khan et al. (2020b)
Salt stress	*Enterobacter sp.*	*Arabidopsis thaliana*	Produced IAA and ACC deaminase;up-regulated the expression of salt stress responsive genes such as *DREB2b*, *RD29A*, *RD29B*, and *RAB18* in plants.	Increased fresh weight, dry weight, and plant height in plants.	Kim et al. (2014)
Heat stress	*Bacillus cereus*	Soybean	Increased heat shock protein (HSP) expression.	Increased essential amino acids for plant growth.	Khan et al. (2020b)

Heat stress	*Bacillus cereus*	Soybean	Enhanced levels of ascorbic acid peroxidase, superoxide dismutase, andglutathione in soybean plants.	Improved plant biomass, chlorophyll content and fluorescence.	Khan et al. (2020b)
Salt stress	*Pseudomonas spp.*	Tomato (*S. lycopersicum* L.)	Improved in shoot uptake of macronutrients (P, K, and Mg) and micronutrients (Mn, Fe, Cu, and Zn); and produced ACC deaminase.	Reduced stress-related ethylene levels, improved plant growth and root development.	Win et al. (2018)
CO_2, temperature, and salt stress	*Vibrio neocaledonicus, Thalassospira australica* and *Pseudarthrobacter oxydans.*	Slender glasswort (*Salicornia ramosissima*)	Enhanced seed nitrogen content and consequent increase in seedling vigor.	Improved seed germination and seedling biomass.	Mesa-Marín et al. (2019)

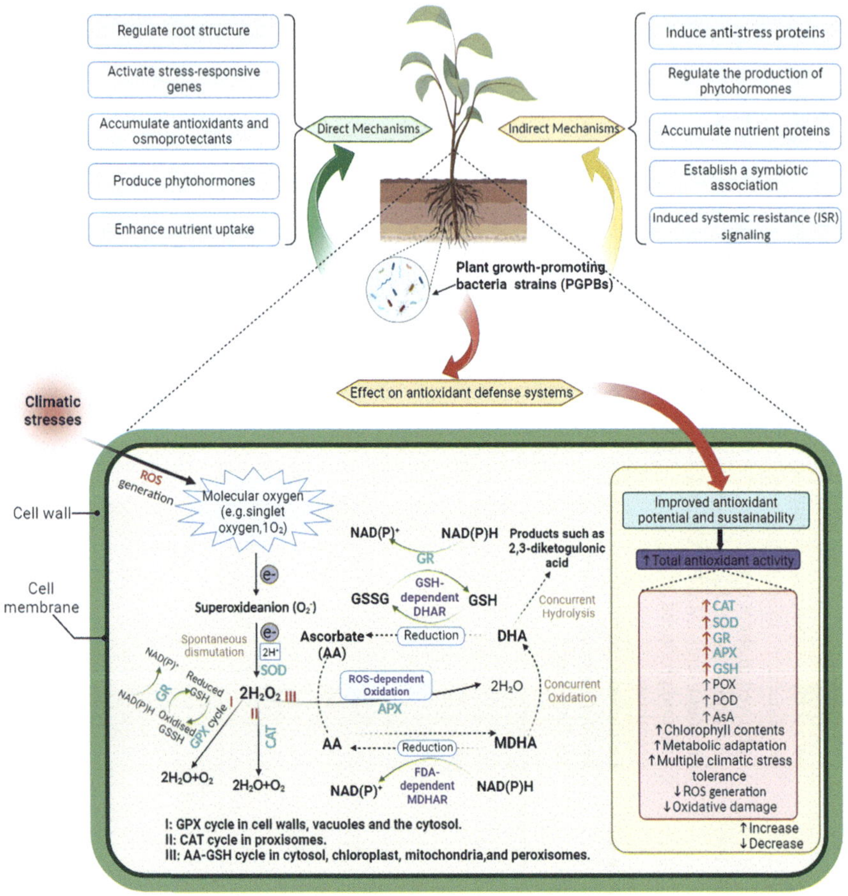

Fig. 7.4 Schematic diagram illustrated the major pathways of ROS generation and scavenging in plant cells during climatic stresses as well as the mechanisms of PGPBs in alleviating stresses in plants**Note:** When plant cells are subjected to climatic stresses, the most direct damage mechanism is the accelerated induction of production and accumulation of ROS. One pathway in plant cells is that molecular oxygen (e.g., singlet oxygen, $1O_2$) obtains electrons (e^-) and converts them into superoxide ion (O_2^-). Then, under the action of SOD enzyme, spontaneous dismutation occurs and is synthesized into H_2O_2, in order to reduce the accumulation of H_2O_2, plant cells will initiate three metabolic pathways (I, guaiacol peroxidase (GPX) cycle, II, catalase (CAT) cycle, and III, the ascorbate-glutathione (AA-GSH) cycle), among which the AA-GSH cycle mediated by ascorbate peroxidase (APX) is the main pathway. In this chemical reaction, ascorbate undergoes a transformation into monodehydroascorbate (MDHA) through APX enzyme. Additionally, MDHA is converted to dehydroascorbate (DHA) via a non-enzymatic pathway. In the AA-GSH cycle, monodehydroascorbate reductase (MDHAR) utilizes NADPH to convert MDHA to AA. Furthermore, the conversion of DHA to AA is catalyzed by dehydroascorbate reductase (DHAR), which involves the oxidation of redox glutathione (GSH). PGPBs could participate in various mechanisms and promote the degradation of ROS. Other abbreviations: reactive oxygen species (ROS), superoxide dismutase (SOD), glutathione reductase (GR), glutathione disulfide (GSSG), peroxidase (POX or POD), and acetylsalicylic acid (AsA)

positively influences root development and the uptake of nutrients and water in plants under salinity stress. This remarkable effect is achieved through the secretion of auxin by PGPBs. In a specific study, the inoculation of PGPBs producing auxin significantly augmented the proline content in barley and stimulated rhizosphere growth, ameliorating the impact of salinity stress (Mahmoud et al. 2020). In the context of heavy metal chromium (Cr) in the root system of maize (*Zea mays* L.), inoculation with two PGPBs (*Proteus mirabilis* T2Cr and CrP450) has been observed to effectively reduce the Cr uptake and translocation in the root system. However, the underlying mechanism behind this phenomenon remains unknown (Islam et al. 2016b). Simultaneously, under nutrient-deficient stress, inoculation with *Bacillus* sp. has been found to significantly enhance nutrient uptake in maize by expanding the surface area of the root system (de Sousa et al. 2021).

When plants confront abiotic stress induced by climate change, they activate specific stress-responsive genes (Ma et al. 2012). For instance, a phylogenetic analysis of the *GmNAC* transcription factor (TF) family in soybean predicted the presence of 58 dehydration-responsive genes (Le et al. 2011). Previous studies have convincingly shown that PGPBs play a crucial role in augmenting the expression of stress-related genes in plants, a factor of significant importance in alleviating climatic stress in various plant species (Nautiyal et al. 2013; Jan et al. 2019; Li et al. 2020). One study reported that *Bacillus amyloliquefaciens* strain SQR9 can enhance salinity tolerance and ameliorate ion toxicity in maize. This beneficial effect is achieved through the inhibition of the expression of genes involved in abscisic acid (ABA) biosynthesis, namely *NCED*, and the promotion of critical players in ion balance regulation, including genes such as *NHX1*, *H ± PPase*, and *HKT1* (Chen et al. 2016).

The impact of water on plants under climatic stress and the remarkable contribution of PGPB in regulating soil water have been discussed previously. However, exploring how PGPB directly contributes to maintaining water balance in plants under stress is vital. While adjusting the transpiration process through stomatal control on the inner leaf surface of plants under stress is essential for maintaining leaf water balance, it also indirectly contributes to a reduction in plant photosynthetic efficiency. Notably, the inoculation of PGPB helps to re-establish the dynamic water balance in durum wheat (*Triticum durum* Desf.) by enhancing stomatal conductance and transpiration efficiency (Yaghoubi Khanghahi et al. 2021). As a result, this PGPB-mediated approach improves drought and salinity tolerance in durum wheat. Under salinity stress, the inoculation of sweet acacia (*Acacia gerrardii*) with *Bacillus subtilis* resulted in enhanced salinity tolerance and improved water balance, achieved through the elevation of proline levels in plant tissues (Hashem et al. 2016).

PGPBs can accumulate adaptive substances to facilitate plant recovery following stress. They play a crucial role in supporting plants by providing essential nutrients and growth factors, promoting the repair of damaged tissues, stimulating the growth of new tissues, and expediting the restoration of plants to their normal physiological state. For instance, PGPBs can expedite the post-drought recovery and development of plants. The rhizosphere soil serves as the primary natural habitat for PGPBs, such as *Rhizobium*, *Azospirillum*, *Bacillus*, *Azotobacter*, *Paenibacillus*, *Pseudomonas*, *Serratia*, and *Klebsiella*. These strains play a crucial role in helping plants withstand abiotic stresses, particularly drought, through mechanisms such as nitrogen fixation, amino acid production, phytohormone synthesis, and enhancement of plant nutrient availability (ALKahtani et al. 2020a, b). Consequently, PGPBs have the capacity to transform infertile soil into a fertile substrate while alleviating stress in plants resulting from factors such as disease, drought, salinity, extreme temperatures (both high and low), and excessive light irradiation. For instance, a study report observed that under Zn and Cd stress, IAA produced by *Pseudomonas fluorescens* exhibited a growth-promoting effect on *Sedum alfredii* by elevating chlorophyll content and enhancing enzyme activity. Moreover, ABA produced by *Bacillus subtilis* can also reduce the accumulation of cadmium (Cd) in pakchoi (Pan et al. 2019). By down-regulating the transcriptional expression of *HKT1*, volatile organic compounds (VOCs) synthesized by *Bacillus subtilis* serve to alleviate salinity stress in *Arabidopsis* roots (Qin et al. 2016). *Bacillus* could synthesize exopolysaccharides and produce siderophores in response to water stress, salinity, and heavy metal stress. The combination of two compounds is critical for maintaining ion balance and facilitating efficient water transport in plants.

7.2.2.2 Indirect Mechanisms

A substantial body of research has provided compelling evidence demonstrating that PGPB could induce the production of anti-stress proteins, thereby exerting a protective role within plant cells and enhancing their resistance against abiotic stress conditions. For instance, when plants face salinity stress, they synthesize a diverse array of multifunctional proteins, such as chaperones, proteases, inhibitors, as well as low-molecular-weight organic compounds, including amino acids, betaines, and sugars, all of which collectively serve to alleviate cell damage (Szabados and Savouré 2010; Shakirova et al. 2012). Underwater deficiency conditions, *Bacillus licheniformis* can induce the overexpression of a diverse array of stress response proteins in pepper plants, effectively mitigating the adverse impacts of water shortage on plant physiology (Gupta et al. 2022). Furthermore, PGPB can elicit the production of heat shock proteins (HSPs) in plants, counteracting protein denaturation induced by high temperatures (Park et al. 2017; Tiwari et al. 2017). The mechanism enhances the resilience of plant roots against high-temperature stress and mitigates the detrimental effects of heat damage on root systems. PGPB can enhance the accumulation of nutrient proteins, soluble sugars, starch, and lipids in plants to mitigate the detrimental effects of drought and salinity stress (Yaghoubi Khanghahi et al. 2022). During drought and salinity stress, a common response in plants is the depletion of intracellular moisture. To mitigate stress in plants, compatible osmolytes such as proline, glycine, betaine, or trehalose accumulate within the cytoplasm, sustaining the cells' physiological osmotic pressure (Pattanagul and Thitisaksakul 2008). Remarkably, the synthesis of these compatible osmolytes occurs more rapidly in bacteria than in plants (Ilangumaran and Smith 2017).

Under drought and salinity stress, proline production in plants alleviates oxidative stress by directly scavenging reactive oxygen species (ROS) and stabilizing intracellular proteins and cell membranes (Gill and Tuteja 2010b). The inoculation of PGPB can either stimulate or suppress the synthesis of proline in plants, consequently modulating the stress tolerance in plants (Bharti et al. 2014; Ma et al. 2017; Singh and Jha 2017). Interestingly, results obtained through real-time PCR revealed that under prolonged drought stress, *P. putida* strain GAP-P45 can upregulate genes associated with proline synthesis and degradation in *Arabidopsis thaliana*. These findings, which correlate with the phenotypic variations observed in different treatment groups, suggest that PGPB has the potential to alleviate drought stress in *A. thaliana* (Ghosh et al. 2017). However, the precise mechanism by which PGPB could regulate the proline content in plants to mitigate stress remains unclear.

Trehalose (α-D-glucopyranosyl-(1 → 1)-α-D-glucopyranoside), a monosaccharide molecule composed of glucose units connected by α-(1 → 3) glycosidic bonds, is naturally occurring within plant cells (Eastmond and Graham 2003). Trehalose serves various essential functions in plants, including shielding plant cells from adverse stressors. When plants encounter environmental challenges such as low temperature, drought, or salinity stress, trehalose can accumulate within cells. This accumulation enables trehalose to contribute to the preservation of cell osmotic balance and the prevention of cellular dehydration (Redillas et al. 2012; Kosar et al. 2019). Remarkably, researchers have observed that the inoculation of *A. brasilense* imparts drought tolerance and sustains osmotic pressure stability in *Z. mays*. Additionally, this inoculation leads to an increase in trehalose accumulation by stimulating the overexpression of genes related to trehalose biosynthesis in *Z. mays* (Rodríguez-Salazar et al. 2009).

Polyamines (PAs) constitute essential biological components in plants, performing many crucial functions. These include their involvement in cell growth and development, the regulation of stress adaptation, the facilitation of immune responses, the modulation of metabolic processes, the regulation of plant hormone synthesis and signaling, and their participation in plant interaction and defense mechanisms (Martin-Tanguy 2001). The extensive and intricate functions of PAs in plants render them pivotal in regulating plant adaptability and viability (Alcázar et al. 2006; Gill and Tuteja 2010a; Chen et al. 2019). Notably, arginine serves as a significant precursor for the biosynthesis of PAs, including spermidine, spermine, and putrescine. Under osmotic stress conditions, when tomatoes were subjected to inoculation with *Bacillus megaterium* strain TG1-E1, there was an accumulation of elevated arginine levels within the plants (Morcillo et al. 2021). Furthermore, chickpea plants inoculated with *Azosprillum lipoferum* under salinity exhibited increased deposition of soluble sugars, glycine betaine, proline, and other salt-induced proteins.

PGPBs can facilitate plant adaptation to abiotic stress by establishing a symbiotic association with plants. Notably, PGPBs can improve plant tolerance to abiotic stress through a significant mechanism that regulates plant hormone homeostasis. Recently, PGPBs have shown the potential to stimulate the synthesis of phytohormones in plants, mainly including IAA, ABA, and GA_3. These phytohormones are essential in enhancing plant adaptation to abiotic stresses by regulating plant growth and development (Gupta et al. 2022). IAA plays a pivotal role in mitigating abiotic stress in plants. It exerts diverse effects on plant physiology, encompassing the promotion of growth and development, enhancement of root growth and branching, facilitation of water absorption and retention, and facilitation of water balance maintenance during drought and salinity stress. Furthermore, IAA can improve antioxidant capacity, mitigate oxidative damage, and enhance plant resistance to abiotic stress (Gupta et al. 2022). Research indicates that PGPBs could induce upregulation of endogenous IAA levels in plants, facilitate root formation and promote overall plant growth. Moreover, under stress conditions, IAA has been observed to enhance the absorption of essential nutrients such as N, P, and K in plants (Singh and Singh 1993; Grichko and Glick 2001; Martin and Stutz 2004; Turan et al. 2012). ABA plays a crucial role in protecting plants under stress. It could regulate stomatal closure, minimize water loss, and induce the expression of stress response genes to mitigate the harmful effects of stress on plants (Zhang et al. 2006). Interestingly, the regulation of the plant hormone ABA by PGPBs exhibits diversity under abiotic stresses, reflecting the complex mechanism of action of ABA on plants. Studies demonstrated that regardless of drought or salinity stress, inoculation of PGPB can influence the synthesis of ABA in plants, ultimately alleviating stress in plants (María José 2007). Under salinity stress, it was observed that higher levels of ABA within plants were associated with the inhibition of leaf extension and shoot development (Kang et al. 2010). However, plants inoculated with PGPB demonstrated resistance to stress by significantly increasing leaf area and exhibiting lower levels of ABA production compared to the control (Kang et al. 2014a). Gibberellins, commonly referred to as GA_3, contribute to plant tolerance responses to abiotic stress through diverse mechanisms, including enhancing ion homeostasis, regulating membrane permeability, bolstering antioxidant systems, promoting osmolyte accumulation, and downregulating the expression of stress-related genes (Emamverdian et al. 2020). Under salinity stress, plants are known to suppress the synthesis of gibberellins (GA) to mitigate damage caused by salinity (Shahzad et al. 2016). However, when plants were inoculated with PGPB, the endogenous levels of GA were triggered, therefore activating strategies to alleviate salinity stress and prevent damage (Kang et al. 2014b). For instance, when soybeans were exposed to salinity, the *P. putida* strain H-2-3 alleviated salinity stress by producing GA (Kang et al. 2014c).

Global climate change is widely recognized as an ongoing threat that is progressively deteriorating the living environment of plants. This underscores the importance of understanding and fortifying plant defense mechanisms. These defense mechanisms are primarily initiated through two inducible resistance signaling pathways, namely systemic acquired resistance (SAR) and induced systemic resistance (ISR) signaling in plants. ISR, in particular, represents a stable defense mechanism mediated by PGPBs (Choudhary and Johri 2009). Moreover, plants utilize SAR to develop hypersensitivity responses to infected pathogens, which means non-infected whole-body parts in plants can become resistant to local infection. This reduction in susceptibility to diseases can be crucial for plants under abiotic stress (Walters et al. 2013). During periods of stress, PGPBs activate the jasmonic acid (JA) and ethylene (ET) signaling pathway in plants to initiate ISR, while SA prompts systemic acquired resistance (SAR). These mechanisms help mitigate growth inhibition under stress (Kumar et al. 2022). Moreover, PGPB-mediated physiological adaptation in plants under abiotic stress conditions can enhance their tolerance, a phenomenon known as induced systemic tolerance (IST). Studies have demonstrated that PGPBs can induce induced systemic tolerance (IST) in response to various abiotic stress, such as drought stress. For instance, inoculating common beans with PGPR strains has been shown to enhance plant growth during drought stress (Figueiredo et al. 2008).

7.2.2.3 Strengthening Antioxidant Capacity in Plants

A review of previous studies has demonstrated that strengthening antioxidant capacity in plants can be achieved through two pathways: direct and indirect mechanisms. The direct mechanisms enhance antioxidant capacity in plants by regulating their antioxidant system. This system includes various antioxidant enzymes and substances, such as superoxide dismutase (SOD), peroxidase (POD), ascorbate peroxidase (APX), catalase (CAT), and glutathione reductase (GR), which effectively neutralize free radicals and mitigate cellular damage (Agarwal et al. 2005). The modulation of gene expression or provision of optimal nutrients facilitates the augmentation of the synthesis of these antioxidant enzymes and substances, ultimately leading to an overall improvement in the antioxidant capacity (Gamalero and Glick 2022). It is worth mentioning that PGPBs can increase the synthesis of antioxidant enzymes and non-enzymatic antioxidant substances (e.g., flavonoid and phenolic compounds) in plants under abiotic stress. For instance, in myrtle seedlings (*Myrtus communis* L.) subjected to drought stress, PGPBs were found to enhance the content of non-enzymatic antioxidant substances, including total phenols and total flavonoids in the leaves. This contribution helps the plants cope with oxidative stress more effectively compared to those not inoculated with PGPBs (Azizi et al. 2021).

The indirect mechanism operates by enhancing the antioxidant capacity of plants through the regulation of other metabolic pathways. For instance, augmenting the nutrient content of plants with substances such as vitamins C, E, and flavonoids can substantially bolster their antioxidant activity. This, in turn, enables the effective neutralization of free radicals within cells, affording protection against oxidative damage (Shen et al. 2022).

In conditions of severe drought stress, plants can experience the harmful accumulation of oxygen-free radicals, causing damage to organelles and disrupting cell metabolism (Munns 2002). However, the presence of antioxidant enzymes such as CAT and POD can alleviate cell damage by effectively scavenging free radicals in plants. Certain PGPBs can scavenge reactive oxygen species (ROS) by upregulating the activity of antioxidant enzymes in plants (Saravanakumar et al. 2011; Sarma and Saikia 2014). A study reported that *Lactuca sativa* L., when inoculated with *Pseudomonas mendocina* and *Glomus intraradices*, exhibited elevated levels of antioxidant enzymes, contributing to enhanced drought tolerance (Kohler et al. 2008). Similarly, in another investigation involving *Sorghum bicolor* exposed to heavy metal (Cr) and heat stress, the inoculation with *Providencia rettgeri* strain TCR21 resulted in the overexpression of genes associated with ROS scavenging activity (*sod*, *apx1*, and *cat*) as determined by quantitative real-time PCR (Bruno et al. 2020). Moreover, some PGPBs can produce a variety of antioxidant substances, such as SOD, AOD, catalase, glutathione, and tocopherols, which aid in ROS scavenging and improve plant tolerance to oxidative stress (Miller et al. 2010; Petrov et al. 2015; Shabaan et al. 2022). As mentioned above, PGPBs can induce proline accumulation in plants, serving as an osmotic regulator. Additionally, under conditions of drought, heavy metal, and salinity stress, researchers have observed that this mechanism indirectly assists PGPBs in activating antioxidant pathways in plants (e.g., the ratio of $NADP^+$/NADPH and GSH/GSSG) in plants (Mahmud et al. 2020; Ali and Khan 2021).

7.3 Genome Engineering in PGPBs for Climatic Stress Alleviation in Plants

Leveraging genetic engineering technology to harness a diverse range of microorganisms for improved crop growth in agriculture is a promising avenue. Scientists have the ability to edit the genetic material within these microorganisms, thereby regulating the biosynthetic pathways of various biochemical substances produced by these microorganisms and their by-products (Ali and Khan 2021). Genetic modification of PGPBs amplifies their capacity to mitigate various climatic stresses, including drought, salinity, high temperatures, and cold, in plants. In essence, the application of genome engineering in PGPBs opens up new possibilities for enhancing plants' adaptability and resistance to changing environmental conditions. With further scientific research, customized PGPBs could emerge as a crucial tool in future agriculture, aiding in addressing evolving climate challenges while safeguarding food security and promoting sustainable development.

7.3.1 Achieved Environmental Improvement in Soil by Engineering PGPB

Climate change has the potential to exert a substantial influence on soil health, leading to environmental degradation. Engineered PGPBs hold promise in playing a vital role in mitigating these effects by enhancing crucial aspects of soil health. Specifically, engineered PGPBs can contribute to improving soil fertility, a particularly critical concern in the context of climate change. The shifting temperature and precipitation patterns associated with climate change can disrupt nutrient availability and soil moisture levels. By facilitating nutrient cycling, fixing atmospheric nitrogen, improving plant nutrient use efficiency, and fostering overall soil fertility, engineered PGPBs present viable strategies for mitigating the adverse consequences of climate change on soil health.

Engineered PGPBs exert a significant influence on nutrient cycling in plants, with a particular emphasis on soil phosphorus dynamics. These impacts include phosphorus dissolution, phosphorus transformation, and direct provision of phosphorus sources (Rodríguez et al. 2006; Sharma et al. 2013). To illustrate this, a noteworthy milestone was reached in 1987 when researchers successfully cloned the genes responsible for phosphorus dissolution in *Erwinia herbicola* for the first time (Goldstein and Liu 1987). Furthermore, the *napA* phosphatase gene from *Morganella morganii* was introduced into the universal vector pRK293, which was subsequently transferred into *Burkholderia cepacia* strain IS-16, a plant growth-promoting bacterium used for biofertilizer. This genetic modification resulted in an augmentation of the activity of the recombinant extracellular phosphate enzyme (Fraga et al. 2001).

PGPBs can play a crucial role in facilitating atmospheric nitrogen fixation. PGPBs can also provide plants with an additional nitrogen source. They enhance plant growth and development by producing nitrogenase, therefore establishing symbiotic interactions with plants and augmenting soil organic nitrogen. According to the reports, genome sequencing of *Pseudomonas stutzeri* strain A1501, isolated from rice fields, revealed the presence of *nif* genes associated with nitrogen fixation (You and Zhou 1989; Yan et al. 2008). In the context of nitrogen fixation, researchers achieved the successful creation of an ammonium-excreting variant of *Pseudomonas stutzeri* strain A1501. This was accomplished by crafting a mutant with deleted ammonium transporter genes (*amtB*) and introducing a plasmid expressing the *nifA* gene (Zhang et al. 2012). Subsequently, the engineered ammonium-excreting *P. stutzeri* mutant strain 1568/pVA3 underwent further investigation. Under conditions of water stress, corn plants inoculated with the mutant strain 1568/pVA3 demonstrated a noteworthy increase in both shoot length and fresh weight when compared to groups inoculated with the wild-type strain (Jiang et al. 2022).

The application of engineered PGPBs in soil remediation has been explored extensively in recent studies. These engineered strains are designed with specific traits to enhance their ability to restore soil health and mitigate environmental contaminants. By harnessing the natural plant-microbe interactions, engineered PGPBs can play a crucial role in remediation processes such as organic pollutant degradation and heavy metal immobilization (Ullah et al. 2015). The most extensively researched approach involves using genetic engineering technology to develop engineered PGPBs that collaborate with hyperaccumulator plants for the remediation of heavy metal-contaminated soils (Singh et al. 2011). Potential bioremediation strategies utilizing engineered PGPBs for remediation of heavy metal-contaminated soil include: (i) chelation of toxic metals through the production of compounds such as metallothionein (MT), phytochelatins, and bacterial siderophores; (ii) modification of biochemical pathways to mitigate the toxic effects of metals on cells; (iii) utilization of secreted enzymes for the absorption and utilization of heavy metals; (iv) regulation of cellular homeostasis to lower intracellular metal concentrations (Jan et al. 2014). For example, MT constitutes a family of low-molecular-weight proteins known for their high metal ion content, typically zinc, copper, and sulfur atoms at their core. MT plays a crucial role in maintaining the balance of intracellular metal ions, aiding in detoxification, responding to oxidative stress, and regulating various cellular physiological processes. Intriguingly, the gene responsible for the synthetic tetrameric metallothionein (MTL4) found in *A. thaliana* was successfully transferred into *Mesorhizobium huakuii* subsp. *rengei* B3. This genetic modification resulted in the creation of a PGPB strain, which was demonstrated to enhance the accumulation of Cd in *Astragalus sinicus* (Ike et al. 2007).

In recent years, there has been a growing focus on soil carbon depletion as a consequence of climate change. A comprehensive study revealed the activation of the *cstA* gene, responsible for encoding a carbon storage regulator protein in *Klebsiella* PGPB under extreme soil carbon deficiency conditions (Sharma et al. 2021). This activation presents a promising avenue for genetic engineering of specialized PGPBs aimed at bolstering soil carbon sequestration efforts.

Engineered PGPBs have exhibited considerable potential in mitigating the effects of extreme weather events on soil erosion. The key to engineering PGPBs for soil erosion prevention lies in the manipulation of specific functional genes. These genes are typically involved in the production of extracellular polymeric substances (EPS), biosurfactants, and root adhesion factors, all of which contribute to soil aggregation and the reinforcement of soil structure (de-Bashan et al. 2012; Lehman et al. 2015; Fiodor et al. 2021). In addition, some PGPBs possess genes responsible for producing specific enzymes, such as cellulase, chitinase, and protease. These enzymes aid in the decomposition of organic matter and recycling nutrients, ultimately enhancing the overall stability of the soil.

However, further evidence is required to substantiate the promotion of plant growth by this recombinant strain. Additionally, there is a need to ascertain how this plant-microbe interaction responds to environmental stresses triggered by climate change, notably drought conditions. Engineered PGPBs indeed exhibit significant potential in mitigating various aspects of soil environmental degradation from climate change. The ongoing research and development in this realm are imperative for addressing the challenges posed by climate change and for securing the long-term sustainability of our soils.

7.3.2 PGPBs with Genes for Maintaining Basal Metabolism Under Stresses in Plants

In the context of climate stress induced by global climate change, it is imperative to elucidate the genes responsible for maintaining the normal basal metabolism of plants when introduced into PGPBs. These identified genes, such as anti-stress genes, phytohormone-related genes, and nutrient uptake genes, play a significant role in enhancing plants' adaptability to environmental stress, resulting in improved growth and survival.

In an intriguing discovery, researchers successfully cloned and transferred a cytokinin biosynthesis (*ipt*) gene from *Agrobacterium* into *Sinorhizobium meliloti*, a PGPB strain (Xu et al. 2012). Following weeks of drought stress, the genetically modified strain was inoculated into alfalfa. Remarkably, upon rehydration, the alfalfa plants exhibited a remarkable resurgence in their original vitality. This outcome strongly indicates that the heightened cytokinin production by the engineered strain significantly bolstered the drought resilience in alfalfa. Amidst ongoing global climate warming, methane has emerged as an additional greenhouse gas (GHG) that warrants attention. Furthermore, in addition to its direct effects, the engineered PGPB strain indirectly modulates IAA synthesis by excessively producing tryptophan within its cells. In a groundbreaking experiment, feedback-resistant enzymes were employed for the first time in a type I methanotroph, specifically *Methylotuvimicrobium alcaliphilum* 20Z. This innovation resulted in a significant enhancement in IAA levels, achieved through the overexpression of tryptophan and the optimization of the IAA production pathway by enhancing promoter efficiency (Pham et al. 2022). Subsequent inoculation with an IAA-producing strain under salinity conditions lead to remarkable increases in shoot and root length during the early growth stage of wheat seeds. Another intriguing study showed that the bacterial mutant *Pseudomonas fluorescens* strain BSP53a, characterized by IAA overproduction, had a stimulating effect on roots from blackcurrant cuttings but inhibited root development in cherry cuttings (Dubeikovsky et al. 1993). These findings suggest that the impact of IAA on plant growth varies depending on the plant species.

To recapitulate, in the face of drought and other abiotic stresses, the inoculation of plants with PGPB plays an indispensable role in enhancing stress tolerance. Trehalose functions as a signaling molecule in plants, activating plant defense systems in response to drought stress and mitigating the damage caused by such conditions. In a study, trehalose synthase-related genes, *ots*AB, isolated and identified from the drought-tolerant *Microbacterium* sp. strain 3 J1, were introduced into the drought-sensitive *P. putida* strain KT2440 (Vílchez et al. 2016). Consequently, under water stress conditions, the transformed strain KT2440 effectively mitigated the growth inhibition of pepper plants resulting from water deficiency. The use of genetic engineering technology to develop engineered PGPBs engineered for trehalose overproduction is frequently more straightforward than acquiring transgenic plants capable of trehalose production. Moreover, these engineered strains can confer protective benefits on various crops. Additionally, an engineered PGPB strain enhances stress tolerance in plants through the overproduction of trehalose (Suárez et al. 2008; Rodríguez-Salazar et al. 2009). Improvement in drought recovery was observed by engineering the genome of *Rhizobium etli* to enhance trehalose production, especially when inoculated into legume plants to stimulate nodule formation (Suárez et al. 2008). Likewise, maize plants inoculated with genetically engineered *A. brasilense* demonstrated enhanced trehalose production under drought stress. It was observed that these treated plants exhibited greater drought tolerance compared to the wild-type *A. brasilense* -treated plants, resulting in increased crop yield (Rodríguez-Salazar et al. 2009).

Moreover, the inoculation of genetically modified PGPBs has proven to be advantageous in enhancing plant tolerance to heavy metal stress. In previous studies, the phytase gene, originally isolated and identified from *Aspergillus niger*, was artificially introduced into *Bacillus* spp. strain, resulting in the creation of an engineered strain (Li et al. 2007). When this engineered strain was inoculated into plants growing in heavy metal-contaminated soil, it demonstrated the potential to enhance growth and increase tolerance to heavy metal stress (Sharma et al. 2013; Lin et al.

2023). Additionally, the transfer of the gene encoding MT into *Ralstonia eutropha* can enhance both plant growth and development and improve the degradation of pollutants (Singh et al. 2011).

The presence of PGPBs in the environment is widely recognized, and they can enhance crop yield under stress conditions, whether individually or when combined with multiple strains (Gray and Smith 2005). In situations where the concentrated effect of a single strain falls short compared to the synergistic inoculation of various strains, it becomes essential to utilize genetic engineering techniques to cultivate a consortium of multifunctional microorganisms that facilitates the compatibility among beneficial soil microorganisms (Reddy and Saravanan 2013). These findings carry significant implications for promoting plant growth and sustaining essential metabolic processes, such as nitrogen fixation, in plants facing climatic stresses.

7.3.3 Deliberately Activate Stress-Responsive Genes in Plants

Against the backdrop of a changing global climate and the continuous evolution of agricultural practices, the impact of stress on plant life has attained unparalleled significance. Indeed, the stress-responsive genes inherent to plants have emerged as pivotal players in challenging environments, orchestrating intricate regulatory mechanisms that govern the physiological and biochemical processes necessary to bolster their adaptability to adversities (Rodrigues et al. 2013; Bharti et al. 2016; Wetzler et al. 2018; Zhou et al. 2018). Through skillful utilization of genetic engineering, the manipulation of strains associated with the genus of PGPBs possesses remarkable potential to deliberately activate stress-responsive genes in plants, leading to a substantive augmentation in their resilience and productivity. Nevertheless, the deliberate and purposeful induction of stress-responsive genes in plants subjected to climatic stress through engineered PGPBs have received limited attention, necessitating more comprehensive and systematic investigations. Without compelling scientific evidence, including comprehensive long-term environmental assessments that evaluate potential ecological disturbances resulting from recombinant events between artificially engineered strains and coexisting microorganisms, researchers are prohibited from directly releasing such strains into the natural milieu (Urgun-Demirtas et al. 2006; Qiao et al. 2014). Conversely, the exploration of genetic engineering techniques to extract functional genes from PGPBs for creating transgenic plants has undergone extensive scrutiny (Tilgam et al. 2021). Nonetheless, the commercialization of transgenic plants uniquely incorporating microbial genes remains relatively limited, signifying that this technology has significant progress yet to be achieved. Amongst the few pioneering investigations, one study, in particular, unveiled the remarkable attributes of a cold shock protein gene, *cspB*, procured from the worthy bacteria *B. subtilis*, and this gene, which encodes an RNA chaperone, imparts remarkable resilience to maize against the detrimental effects of drought stress (Castiglioni et al. 2008).

The intentional genetic manipulation of PGPBs to induce the activation of stress-response genes in plants holds great promise for augmenting plant resilience in the presence of stressors. Nevertheless, to confirm its effectiveness and safety in practical applications, more research and real-world validation are required. Future studies should place paramount emphasis on refining the intricate development of genetically engineered strains and unraveling the underlying regulatory mechanisms that govern their functionality.

7.4 Domestication of PGPB Under Climatic Stresses

The ongoing escalation of climate change is increasingly exacerbating challenges in crop cultivation and production. In response to this pressing issue, the domestication of PGPBs has arisen as a promising approach to augment plant resilience in the face of climatic stresses. Leveraging the capabilities of PGPBs through domestication efforts hold the potential to ameliorate the detrimental impacts imposed by the ever-heightening climatic stressors. The process of domesticating PGPBs involves the judicious selection and intricate engineering, equipping them with an exquisite repertoire of attributes to flourish amidst the rigors of inhospitable surroundings while concurrently forging harmonious symbiotic alliances with the botanical realm (Rojas-Tapias et al. 2012). The ecological milieu facilitating the taming and cultivation of PGPBs, as explicated within this scholarly treatise, predominantly encompasses abiotic stresses (including temperature, pH, humidity, nutrient availability, oxygen levels, and light intensity), all of which are influenced by climate change (Islam et al. 2016b; Ahemad 2019; Ansari et al. 2019; Moreno-Galván et al. 2020; Nookongbut et al. 2020). The focal focus of future efforts lies in domesticating PGPBs that exhibit resilience to a multitude of extreme environments to meet the escalating demands of an ever-changing environment. Additionally, researchers must maintain a heightened level of vigilance to ensure meticulous oversight of the domestication process of PGPBs across various ecological settings, thereby effectively mitigating the risks of environmental contamination and the potential for cross-infection.

7.4.1 Screening and Optimization of Stress-Tolerant PGPBs

The screening and application of tolerant PGPBs have significant importance in agriculture and plant health management (Viscardi et al. 2016; Tirry et al. 2018; Mahmud-Ur-Rahman et al. 2022). Meanwhile, in the context of climate change, the meticulous screening and enhancement of stress-tolerant strains can offer numerous advantages for agriculture. By harnessing the potential of these beneficial bacteria, crop productivity can be optimized, reducing the reliance on chemical inputs and fostering environmentally friendly agricultural practices. However, the research on developing tolerance in PGPBs to mitigate the impacts of climate change remains ongoing and requires further extensive investigations. In recent years, a substantial body of literature has been published on the tolerance of PGPBs under abiotic stresses. Most studies have primarily focused on existing PGPBs that exhibit resistance to heavy metals, salinity, and extreme temperatures. It is noteworthy that researchers usually isolate tolerant PGPBs from abiotic stress environments. In laboratory settings, the initial steps typically involve optimizing the tolerance of the target strains, specifically characterizing the extent of tolerance.

According to a report, around 50% of the Earth's arable land is projected to become saline by the mid-century (Hertel 2011). Given the pressing need to address extreme conditions such as saline and alkaline conditions, there is an urgent requirement to identify and utilize extremotolerant PGPBs in agricultural settings (Egamberdieva et al. 2019). In nature, researchers have discovered the halotolerant and halophilic microorganisms thriving abundantly in saline environments (Margesin and Schinner 2001; Edbeib et al. 2016). Thus, it is essential to selectively isolate and cultivate salt-tolerant PGPBs microorganisms from these two distinct categories of microorganisms (Talaat 2018). For instance, the strains *Kocuria flava* AB402 and *Bacillus vietnamensis* AB403, isolated from the mangrove rhizosphere of the Indian Sundarbans, not only exhibit halophilic tendencies but also possess characteristics of heavy metal fixation and adsorption. Results from the potted experiment indicate that the inoculation of these two bacterial strains into rice plants improves their tolerance to salinity and arsenic (As) stress (Mallick et al. 2018). Similarly, in recent years, a study reported the isolation of a *Bacillus megaterium* strain, named STB1 from polluted estuary soil in Portugal. This strain demonstrates tolerance to salinity and heavy metals, while also promoting the germination and early growth of various plants (Nascimento et al. 2020).

The escalating challenges posed by global climate change, characterized by increasingly severe temperature extremes, present significant hurdles for agriculture on a global scale. These challenges primarily manifest as compromised plant immunity and reduced productivity from extreme temperature fluctuations. Climate change has given rise to a growing frequency of extreme weather events, such as frost (< -0 °C), further exacerbating the obstacle of cold stress in modern agriculture (Duman and Wisniewski 2014; Barlow et al. 2015). Similarly, an endophyte *Serratia marcescens* strain SRM (MTCC 8708) isolated from flowers of summer squash (*Cucurbita pepo*), exhibits persistent growth-promoting properties even at 4 °C. Furthermore, it possesses the capability to produce hydrogen cyanide (HCN) for pathogen suppression and IAA to stimulate plant growth and development (Selvakumar et al. 2008). In subsequent studies conducted in 2015, psychrotolerant PGPB were successfully isolated and characterized from three different areas of the Andes Mountains and one in Patagonia, Chile. Notably, *Pseudomonas* sp. TmR5a and *Curtobacterium* sp. BmP22c demonstrated a remarkable ability to achieve a bacterial cell survival percentage (BCS %) exceeding 50% at −20 °C (Vega-Celedón et al. 2021). Additionally, the two strains exhibited diverse growth-promoting properties, such as auxin and ACC deaminase production. As global warming engulfs the entire planet, the significance of thermotolerant PGPBs in agricultural ecosystems becomes increasingly prominent. *Brevibacterium linens* strain RS16, isolated from the coastal area of the Yellow Sea in South Korea, displays high tolerance to both salt and high temperature. Studies have revealed its capacity to enhance salt stress tolerance in rice by reducing ACC accumulation in salt-treated rice leaves and suppressing apocarotenoid oxygenase (ACO) activity (Chatterjee et al. 2018). Additionally, strain RS16 has been observed to alleviate thermo-oxidative stress in *Eucalyptus grandis* exposed to high-temperature stress (37–41 °C), thereby promoting increased plant's tolerance (Chatterjee et al. 2020). As the arid conditions resulting from high temperatures persist, the utilization of PGPB remains a sustainable strategy for mitigating the adverse impact on plant growth and productivity. In a potted experiment conducted under non-stressful and drought-stressful conditions, an analysis revealed that drought-tolerant *Pseudomonas* spp. isolated and identified from a rain-fed agriculture field in India enhances the growth performance of finger millet plants and significantly augments their antioxidant activity (Chandra et al. 2018).

Concerning the primary control of heavy metal pollution in the environment, climate change indirectly influences these dynamics. For instance, increased atmospheric CO2 concentrations can contribute to the formation of acid rain, subsequently raising the effective concentration of heavy metals in the soil. Another example involves intensified rainfall in humid climates, which enhances the migration and leaching of heavy metals in the soil, leading to their accumulation. The ability of heavy metals to trigger excessive generation of ROS within plant cells, causing cellular damage or even death, presents a substantial threat to agricultural productivity (Khalid et al. 2017; Kacholi and Sahu 2018). To address this challenge, environmentally friendly and well-established heavy metal-tolerant PGPBs have been extensively identified and screened. This plays a pivotal role in supporting plant growth and contributing significantly to soil remediation through various mechanisms.

It is well-established that when the concentration of heavy metal ions in the environment remains below the threshold of microbial tolerance, microbial cell metabolism can appropriately incorporate heavy metal cations as essential trace elements to sustain their vital functions (Rajkumar et al. 2010). However, as the environment's heavy metal concentrations steadily rise, researchers are motivated to selectively isolate heavy metal-tolerant PGPBs from persistently polluted environments. This effort serves as the foundation for their potential future application as bioinoculants or biofertilizers (Madhaiyan et al. 2007; Wani and Khan 2010).

Reports indicate that a cluster of heavy metal-tolerant PGPBs, including species such as *Pseudomonas*, *Arthrobacter*, *Agrobacterium*, *Bacillus*, *Azoarcus*, *Azospirillum*, *Azotobacter*, *Burkholderia*, *Klebsiella*, *Alcaligenes*, *Serratia*, *Rhizobium*, and *Enterobacter* species, have demonstrated immense potential in fostering diverse plant growth in metal-contaminated environments (Enebe and Babalola 2018). Consequently, these heavy metal-tolerant PGPBs can be considered as suitable candidate microorganisms for heavy metal remediation, with the goal of minimizing the detrimental impact of heavy metals on plants to the utmost extent.

7.4.2 Genome-Engineered PGPBs Used for Adapting to Stresses

Advancements in synthetic biology have introduced innovative tools and methods for tackling the challenge of climate change. Leveraging the innate abilities and precisely engineered traits of microorganisms enables the creation of environmentally friendly and sustainable solutions (Voigt 2006; Cambray et al. 2011; Schwander et al. 2016; DeLisi 2019; Miller et al. 2020).

In response to unstable and inhospitable environmental conditions, the formation of biofilms stands as a crucial defense mechanism. Consequently, PGPBs can thrive within rhizosphere biofilms, as colonies adhering to the biofilm matrix have demonstrated enhanced resistance against antimicrobial agents, drought, and ultraviolet radiation (Amaya-Gómez et al. 2020). Environmental pressures prompt free-floating PGPB to form biofilms, ensuring their presence in the plant rhizosphere. The ability to generate biofilms in *B. amyloliquefaciens* strain 54 has enhanced its colonization within tomato plants and improved drought tolerance in these plants. Mutants (Δ*abrB* and Δ*ywcC*), engineered through gene knockout techniques, can produce exceptionally robust biofilms, thereby aiding plants to better withstand drought stress (Wang et al. 2019a). Similarly, there are thermotolerant PGPBs that alleviate oxidative stress induced by extreme temperatures through the production of biofilms (Dastogeer et al. 2022).

When climatic variations directly result in abiotic stress, they have a dual impact. Firstly, the deficiency of nutrients in the environment hampers the stability of plant-associated PGPB communities in the soil. Secondly, under stress pressure, these variations disrupt the interactions between PGPB and plants. Researchers have also ingeniously engineered bacteria capable of remediating heavy metal pollution by introducing exogenous genes into PGPBs, offering a solution to the challenge of heavy metal contamination in the environment (Ullah et al. 2015). By conducting a comprehensive analysis of the metal resistance mechanisms in *P. putida*, a prominent PGPB, it is conceivable to augment its Cd tolerance through genetic modification (Gouda et al. 2018; Alka et al. 2020). A study reported here illustrates that the engineered strains resulting from genetic mutations (e.g., increased phosphate solubilization capacity) have been obtained to stabilize nutrient metabolism in both plants and microbial communities (Rodríguez et al. 2006). Another example of acquiring tolerant PGPB involves integrating exogenous *HSP* genes into *B. subtilis* strain 446. This genetic modification enhances tolerance to heat and osmotic stress while augmenting its riboflavin synthesis and secretion capacity in the strain (Wang et al. 2019b).

Apart from the straightforward studies mentioned above, the introduction of PGPB can also indirectly protect plants. It reduces the impact of biotic stress and reduces the susceptibility of plants under abiotic stress. Moreover, genetically engineered PGPBs can exhibit robust antagonistic effects, potentially inhibiting or even eradicating harmful organisms (including agricultural pathogenic microorganisms and harmful insects). This capacity is advantageous for PGPB in mounting more effective responses to abiotic stressors (Abbas 2018).

7.4.3 Application of Domesticated PGPBs for Combating Climatic Stresses

As the impact of climate change on agricultural yield and ecosystems becomes increasingly apparent, there is a pressing need to prioritize the development of effective domesticated strains of PGPB. Initially, it is essential to engineer domesticated PGPBs to serve as a bioinoculants or biofertilizer aimed at enhancing plant tolerance to climatic stresses (Orozco-Mosqueda et al. 2020). Numerous studies have shown that the widespread use of PGPBs (including *Azotobacter tropicalis*, *Burkholderia unamae*, and *B. subtilis*, etc.), as biofertilizers due to their advantages of improving growth in plants (Dadarwall et al. 1997; Leaungvutiviroj et al. 2010). An often-overlooked aspect involves introducing domesticated PGPBs into plants resilient to abiotic stressors, wherein PGPB not only manifests growth-promoting traits but also indirectly modulates the plants' resilience against climatic stresses (de-Bashan et al. 2012; Ma et al. 2020). To thoroughly assess the safety and efficacy of PGPBs in mitigating climatic stresses, it is essential to maintain consistency in assessing them alongside commercialized biofertilizers and chemical fertilizers. For example, one study evaluated the PGPB as a biofertilizer to improve the nutritional content of wheat grains under drought and salt stress, aiming to prevent grain yield losses. Most notably, the results of the consistency assessment were on par with the effectiveness of chemical fertilizers (Yaghoubi Khanghahi et al. 2022).

Although most studies on abiotic stress-tolerant PGPBs are currently conducted in laboratories, several resources offer valuable insights for future Research directions (Kumar et al. 2017; Niu et al. 2018). A recent advanced study has emphasized the significance of developing drought-tolerant PGPB communities as a prominent avenue for future research. Experimental findings from this study demonstrate that the application of *B. amyloliquefaciens* strains RHF6, LMG9814, and *Bacillus* sp. AGS84, either individually or in combination, promotes seed germination and enhances the seedling growth rate of *Spinacia oleracea*, even under drought stress conditions (Petrillo et al. 2022). Furthermore, co-inoculation of *Bradyrhizobium japonicum* USDA 110 and salt-tolerant *P. putida* TSAU1 in soybeans under salt stress was found to improve the agronomic traits of the plants (Egamberdieva et al. 2017). Another study examined the tolerance of PGPBs, including rhizobacteria and endophytic bacteria, isolated from arid regions where peppers were cultivated. These strains underwent testing for their tolerance to drought, extreme temperatures, salinity, and heavy metals as well as their ability to produce biofilms, auxin, and EPS (Admassie et al. 2022). The findings indicated a strong correlation between the growth of pepper plants in arid regions and the potential application of multiple PGPBs. Additional research findings revealed a noteworthy outcome regarding wheat under cold stress. In comparison to the non-inoculated group of PGPBs, the presence of cold-tolerant PGPB strain (*Pseudomonas* sp. PPERs23) led to a significant 13.4% increase in the wheat yield in comparison to the control group (Mishra et al. 2011; He et al. 2019). However, it's important to note that utilizing PGPBs to mitigate the effects of climate change on plants is not foolproof. In light of this, one study highlighted the advantages of bacterial fertilizer utilization and suggested that the effectiveness of PGPB inoculants is greatly influenced by factors such as plant age, biological characteristics, soil properties, and physicochemical properties (Kundan et al. 2015).

The utilization of stress-tolerant PGPBs raises numerous questions that require further investigation. Currently, the release of PGPBs is confined to a limited and artificially controllable range. The advancement and application of stress-tolerant PGPBs can meet the demand for transgenic-resistant crops while also alleviating the burden on farmland in the face of climate change. The exploration and identification of stress-tolerant PGPBs in soil have introduced novel perspectives, suggesting the adoption of a pyramid thinking approach to combat climate stress.

There is no direct evidence to suggest that the release of domesticated PGPBs, engineered PGPBs, or other microorganisms poses a severe threat to the environment or human health. However, in the context of decreasing available farmland worldwide due to climate change, ongoing research on PGPBs shows that their benefits significantly outweigh potential risks. This trend may lead to widespread adoption of biological agents in the future (Timmusk et al. 2017). However, it is worth noting that concerns similar to those associated with the historical use of chemical insecticides, such as DDT, have emerged regarding the release of PGPBs into the environment (Ferreira et al. 2019; Glick 2020).

Once gaining a comprehensive understanding of domesticated PGPBs' tolerance to extreme environmental conditions, recognizing their unique genetic diversity, and ensuring their well-balanced distribution, we can establish guidelines for their optimal utilization in agricultural production. This enables us to fully harness the advantages of their adaptation to extreme climates. Subsequently, the application model offers a practical framework for enhancing our comprehension of PGPBs' stress tolerance mechanisms, enabling their continued advancement and application. In light of this, the pursuit of safe, stable, and efficient PGPB-related bioagents or biofertilizers tailored to address climate stress conditions emerges as promising approaches to achieving sustainability in agriculture.

7.5 Unraveling Mechanism of PGPB for Stress Alleviation in Plants: The Omics Strategies

Taken together, there are multiple connections between PGPB and host plants, including biochemical, physiological, and molecular interactions. PGPBs largely influence and actively participate in plant responses to climatic stress (Farrar et al. 2014). Currently, there is a need to comprehend the underlying mechanisms through which PGPBs mitigate climatic stress in plants. To delve deeper into the evolving dynamics of microbe-plant-environment interactions mediated by PGPB and establish links between PGPB-mediated plant tolerance in plants under climatic stresses, the advancement of bioinformatics and omics research methods becomes imperative. These methods primarily encompass genomics, proteomics, transcriptomics and metabolomics (Meena et al. 2017; Córdova-Albores et al. 2021).

In accordance with the progressive tide of natural sciences, the integration of diverse omics analysis methods has emerged as a systematic and comprehensive strategy for unraveling the intricate and dynamic interrelationships within ecological systems (Atkinson and Urwin 2012; Baillo et al. 2019; Pinu et al. 2019). In this chapter, our aim is to provide a comprehensive overview of the biochemical and molecular mechanisms underpinning the PGPB-mediated alleviation of abiotic stresses in plants, particularly under climate stresses. Ultimately, by establishing a holistic understanding of the functional mechanisms of PGPBs within the environmental context via the application of multi-omics, we aspire to furnish a reference framework for future research endeavors focused on alleviating climatic stresses in plants.

7.5.1 Comprehensive Analysis of PGPB-Mediated Climatic Stress Tolerance in Plants: Transcriptomics, Proteomics and Metabolomics

7.5.1.1 Transcriptomics

Transcriptomics is a discipline dedicated to exploring the expression and functionality of RNA molecules within cells or tissues, offering valuable insights into the regulatory mechanisms and interaction networks of genes across diverse conditions (Guo et al. 2022). Molecular genetic profiles derived from transcriptome sequencing data are referred to as transcriptome profiles, serving as a reflection of the expression and regulatory patterns of genes in various cells or tissues. Frequently, when addressing scientific inquiries, the comparison of transcriptome profiles is usually used to analyze the disparities in gene expression levels among different transcripts (Bräutigam and Gowik 2010). Currently, transcriptome profiles generated through methods such as microarray, RNA sequencing (RNA-seq) technology, and emerging third-generation sequencing technology serve as the primary means for studying the interaction between PGPB and plants under climatic stress (Gut 2013; Lowe et al. 2017). Among these methods, RNA-Seq stands out as a technique for quantifying and qualitatively analyzing RNA using high-throughput sequencing technology. It boasts advantages such as high efficiency and cost-effectiveness, making it widely adopted for research purposes (Wang et al. 2009; Haas and Zody 2010; Conesa et al. 2016).

For example, to unravel the molecular mechanisms behind PGPB-mediated induction of salt tolerance in plants facing salinity stress, researchers conducted transcriptomic analyses. They explored the impact of inoculating *Arabidopsis* plants with *B. amyloliquefaciens* strain FZB42. The study revealed a significant induction of several essential resistance genes associated with JA-responsive defense, ACC synthase, and brassinosteroid regulation (Liu et al. 2017b). Furthermore, the comparative analysis of transcriptomic data from *Stenotrophomonas rhizophila* exposed to salt and root exudates from oilseed rape revealed significant similarities. The simultaneous analysis of transcriptomic data and phenotypes led to the hypothesis that spermidine, produced by *S. rhizophila*, serves as an essential stress-protecting agent (SPA) that regulates osmotic stress in plants (Alavi et al. 2013). In another comprehensive transcriptomics study delving into the molecular mechanisms underlying plant-PGPB interactions under salinity, RNA-Seq analysis was employed. The findings demonstrated that under salt stress, *B. amyloliquefaciens* strain SN13 induced significant expression changes in numerous genes associated with photosynthesis, hormone signaling, stress response, cell wall composition, and lipid metabolism in the model crop, rice (Chauhan et al. 2019). Furthermore, it's essential to mention another example of research focused on PGPB alleviating drought stress in plants. In 2014, the first transcriptomic study of its kind was published, stating that *Gluconacetobacter diazotrophicus* activates the hormone synthesis pathway, specifically upregulating ABA-dependent signaling genes, to confer drought tolerance in sugarcane (Vargas et al. 2014).

In contrast to the aforementioned transcriptome analysis methods, cDNA-AFLP technology presents an advantageous approach for obtaining comprehensive expression information of the transcriptome (Bachem et al. 1998). With the gradual completion of genome sequencing of model plants such as Arabidopsis and rice, the focus of gene expression studies has shifted from individual genes at specific developmental stages to a more comprehensive analysis of gene expression. In a previous study, cDNA-AFLP transcriptome analysis was employed to identify differentially expressed genes encoding proteins involved in the antioxidant system, HSPs, carbon metabolism, and temperature-sensitive transcription factors in wheat seedling leaves treated with *B. amyloliquefaciens* strain 5113 under heat, cold, or drought stress (Abd El-Daim et al. 2018).

Transcriptomic data alone are insufficient for a comprehensive understanding of the mechanism through which PGPB alleviates abiotic stress in plants. Numerous unknown issues await resolution in the future. Besides systematically and extensively investigating the changes in gene transcription levels in plants inoculated with PGPB under climate stress, researchers have also explored new research areas that pose numerous new challenges and opportunities.

7.5.1.2 Proteomics

Proteomics has significantly enhanced understanding of how plants interact with their environment. Previous studies have analyzed the protein expression profiles in plants and their functions in various aspects of bacterial interactions, such as pathogen resistance and symbiotic associations related to plant nutrition (Jorrín-Novo et al. 2009). Many studies have highlighted the importance of functional proteins, which are expressed through transcription and translation, in supporting essential phenotypes in plants when confronted with abiotic stresses (Xiong and Zhu 2001; Forrest and Bhave 2007; Luhua et al. 2013). Consequently, proteomics, which allows for the investigation of protein-protein interactions between PGPB and plants, stands as a potent tool for exploring how PGPB mediates the alleviation of climatic stresses in plants.

Currently, plant proteomics research primarily focuses on the following areas: (i) examining proteomic profiles in response to abiotic stresses; (ii) identifying differentially expressed proteins during plant-pathogen interactions; and (iii) studying proteomic expression of nitrogen-fixing rhizobia symbiosis in plants (Chen and Pachter 2005; Mehta et al. 2008; Mathesius 2009; Afroz et al. 2013; Kosová et al. 2015; Liu et al. 2015). While this review has discussed how PGPB can help plants cope with climatic stress, our understanding of the proteomic profiles of plants benefiting from PGPB under abiotic stresses remains limited. Quantitative proteomics analysis has revealed that wheat plants inoculated with PGPB strain *Enterobacter cloacae* SBP-8 exhibit improved salt stress tolerance. This inoculation induced the expression of transcription factors, antioxidant enzymes, defense enzymes, and lignin in wheat, ultimately enhancing the plant's ability to tolerate salt stress (Singh et al. 2017).

Recent studies investigated how plant protein expression changes following inoculation with PGPB, shedding light on the potential relevance of altered protein profiles to abiotic stress tolerance in plants. Shotgun proteomics was used to analyze tomato plants that were inoculated with the PGPB strain. The analysis revealed that inoculation induced the expression of proteins associated with oxidative stress, such as glutathione S-transferase (GST), heat shock protein 70 (Hsp70), and neutral ceramidase, as well as the phosphate transporter protein (SlPT1) (Ibort et al. 2018). Several proteome studies have explored various aspects, such as the effects of *Paenibacillus polymyxa* E681 inoculation on *Arabidopsis* (Kwon et al. 2016), the impact of *Herbaspirillum seropedicae* SmR1 (Alberton et al. 2013) and *Azoarcus* sp. (Miché et al. 2006) inoculation on rice, and the effects of *Azospirillum brasilense* FP2 inoculation on maize (Cangahuala-Inocente et al. 2013; Faleiro et al. 2015). However, these studies have revealed different proteome profiles and action mechanisms for different PGPBs, indicating the need for further research (Ryu et al. 2005; Long et al. 2008).

Plant protein expression plays a pivotal role in conferring tolerance to abiotic stresses. Proteome analysis serves as a valuable tool for elucidating alterations in the levels of various responsive proteins and the functions, post-translational modifications, interactions, and localization of these proteins in cells or tissues. Moreover, proteomics can validate data obtained from genomics and transcriptomics, although it involves greater complexity. To characterize the proteome profile in plants, it is necessary to analyze the expression changes induced by known abiotic stresses, such as salinity, alkalinity, temperature, drought, toxic metals, and UV radiation. This foundational knowledge can significantly contribute to a deeper comprehension of the mechanisms through which PGPB exert their effects.

7.5.1.3 Metabolomics

As previously described, PGPB can stimulate the expression of genes involved in plant responses to abiotic stress. Metabolomics is an emerging field that enables the comprehensive analysis of metabolites in plants. This technology, in conjunction with other -omics technologies, has facilitated an in-depth analysis of the dynamic processes with plant stress by monitoring alterations in metabolites and gene expressions (Anzano et al. 2022). PGPBs can alleviate environmental stress and elicit plant tolerance, thereby promoting their growth. Several PGPBs induce physical and chemical changes that contribute to plant defense through mechanisms like induced systemic resistance (ISR), which is effective against both abiotic and biotic stresses (Choudhary et al. 2016). The application of metabolomics to study the interactions between PGPB and plants under abiotic stress provides valuable insights into how PGPB can alleviate stress and promote plant growth.

Metabolomics studies the complete set of metabolites generated by organisms in specific environmental conditions; it unravels direct connections between various metabolic pathways within cells and their environments. Moreover, alterations in the metabolomics of organisms also reflect the most immediate physiological responses to environmental fluctuations (Bundy et al. 2005). By comparing the metabolome data between control and treated groups, researchers can pinpoint the specific metabolites and stress-related genes activated by abiotic stress in plants. Metabolomics offers valuable insight into how plants react to and cope with abiotic stress (Bundy et al. 2005; Jia et al. 2016).

Few studies have investigated how PGPBs affect the plant metabolome to alleviate abiotic stress. However, a recent review has shed light on the metabolite framework associated with abiotic stress tolerance and plant growth enhancement mediated by PGPBs for sustainable agriculture. The review aims to consolidate and establish a novel metabolite framework that links potential metabolite markers synthesized in response to salinity or drought stress induced by PGPBs. This framework is constructed by examining a range of PGPB-induced metabolite markers that have been modulated in various plants to alleviate stress and promote growth (Sabki et al. 2021). As discussed in the previous sections, PGPBs can reshape the plant metabolome under abiotic stress conditions, resulting in the accumulation of various metabolites, such as trehalose, betaine, and IAA. This advancement contributes to the field of plant metabolomics research in the context of climate stress, facilitating the exploration and elucidation of the intricate metabolomic mechanisms involving PGPBs in plants under climate stresses.

7.5.2 Soil Microbial Communities Related to PGPB Alleviating Climatic Stresses in Plants: A Metagenomic Perspective

Environmental factors such as climate, vegetation, soil type, and organic matter addition, exert significant influence over microbial communities and their functions. These factors cause variations and diversity within soil microbiota across regions and ecosystems. Among these soil microorganisms, there exists a group known as PGPBs, which play a crucial role in enhancing plant growth and health. PGPBs employ various mechanisms to mitigate the adverse impacts of climatic stress, such as drought, salinity, and extreme temperature on plants. It is widely recognized that 99% of soil microorganisms cannot be cultured using artificial media. To dissect the diversity of these elusive microorganisms and comprehend their interactions with plants, metagenomic sequencing technology assumes a pivotal role. Metagenomics is an approach to studying the genomic information of intricate microbial communities. This technique unveils insights into the composition, structure, function, and dynamics (Jing et al. 2022). Comprehending the metaphenome, which represents the collective phenotype exhibited by all microorganisms and their host organisms within a given environment, presents a formidable challenge when studying soil due to the intricate and ever-changing nature of interactions among plants, microbes, and the surrounding environment (Jansson and Hofmockel 2018).

Endophytes isolated from potatoes were subjected to genome analysis, revealing genes highly similar to the ACC-deaminase genes (*acdS*) found in *P. fluorescens*, a bacterium known for its stress-mitigating functions (Cho et al. 2015). Among the uncultured endophytes colonizing potato, a metagenomic library identified 195 bacterial genomes carrying the *acdS* gene, with a majority belonging to the *Pseudomonas* sp. Additionally, the metagenomic library highlighted the role of the upstream transcriptional regulator gene *acdR* in influencing *acdS* gene expression, a phenomenon observed in several *Burkholderia* strains (Nikolic et al. 2011). To enhance the genetic diversity available for the development of engineered salt-tolerant recombinant bacteria and transgenic plants, metagenomic sequencing was employed to analyze uncultured microorganisms derived from pond water. This analysis led to the identification of two cloned genes, specifically Enoyl-CoA hydratase (*EchM*), as putative stress proteins associated with the biological mechanism of salt tolerance (Kapardar et al. 2010). Researchers used metagenomic analysis to investigate the mechanisms underlying the acid and drought tolerance of microbial communities (potential PGPBs). This investigation revealed genes encoding proteins involved in fitness and survival, such as antifreeze proteins, cold shock proteins, compatible solute production, and pH homeostasis in low-temperature (6–10 °C) acid mine drainage (AMD) systems (Liljeqvist et al. 2015). Understanding the functional characteristics of endophytes is a critical aspect of comprehending rice's resistance to climate stress. In 2011, a groundbreaking study analyzed endophytes believed to be involved in various metabolic pathways, including the nitrogen cycle, iron acquisition and storage, quorum sensing, and detoxification of reactive oxygen species in rice roots, using metagenomic data (Sessitsch et al. 2012). These findings underscore the potential of metagenomics as a valuable research tool for identifying candidate PGPBs capable of alleviating climatic stress in plants. However, in the context of global climate change, comprehensive analyses that integrate metagenomics, metatranscriptomics, and metaproteomics are essential for a more profound understanding of the potential functions of PGPB microbiome interactions with plants, including nutrient metabolism and stress alleviation (Turner et al. 2013).

7.5.3 Integrating Multi-omics Platforms

The integration of data from multiple omics platforms, known as multi-omics integration (MOI) represents a powerful research tool that facilitates the comprehensive analysis, annotation, and modeling of genomic, transcriptomic, proteomic, and metabolomic data (Jamil et al. 2020). Combining data from multiple omics sources provides a holistic understanding of the interactions between PGPB and plants, shedding light on the underlying mechanisms by which they mitigate the impact of climate stress. By integrating genomic, transcriptomic, and proteomic data, researchers can investigate how PGPBs modulate gene expression and metabolic pathways in plants when confronted with environmental stress. The integration of a multi-omics platform enables the identification of biochemical and genetic markers associated with PGPB-mediated climate stress alleviation in plants. These markers, in turn, facilitate the assessment of stress responses and resistance in plants and enable the screening of stress-tolerant crop varieties. Moreover, integrating multi-omics platforms addresses the challenges inherent in the practical application of PGPB, optimizing the selection and application strategies of PGPB in physiological processes and metabolic pathways, thereby enhancing the sustainability of plant growth under climate stress conditions. Overall, the integration of multi-omics platforms plays a pivotal role in comprehending how PGPB mitigates effects of climate stress on plant growth.

7.6 Conclusions and Future Directions

Abiotic stresses, such as drought, extreme temperature, and salinity, impact crop growth and yield in the context of global climate change. To mitigate the effects of abiotic stress, plants undergo various physiological and metabolic changes. These adaptations involve the production of protective substances, such as proline and antioxidants to minimize cell damage, adjustments in hormone synthesis and signaling to adapt to changing environmental conditions, and the regulation of photosynthesis and respiration rates to balance energy supply and carbohydrate metabolism. PGPBs are beneficial bacteria that alleviate abiotic stresses associated with global climate change. PGPBs enhance plant growth, development, resistance, and adaptability by providing nutrients, generating biologically active substances, such as plant growth regulators and antioxidants, and establishing symbiotic relationships with plants to improve root structure and function. A comprehensive understanding of the mechanisms by which PGPBs operate under climatic stresses is essential for their practical application in agriculture, contributing to the sustainable development of the agriculture sector. It is imperative to investigate the interaction between PGPB and plants in hostile environments, refine PGPB application strategies such as seed treatment and soil fertilization, and anticipate the long-term impacts of PGPBs on crops subjected to various climatic stresses.

References

Abbas MST (2018) Genetically engineered (modified) crops (*Bacillus thuringiensis* crops) and the world controversy on their safety. Egypt J Biol Pest Control 28(1):52. https://doi.org/10.1186/s41938-018-0051-2

Abd El-Daim IA, Bejai S, Fridborg I, Meijer J (2018) Identifying potential molecular factors involved in *Bacillus amyloliquefaciens* 5113 mediated abiotic stress tolerance in wheat. Plant Biol 20(2):271–279. https://doi.org/10.1111/plb.12680

Abd_Allah EF, Alqarawi AA, Hashem A et al (2018) Endophytic bacterium *Bacillus subtilis* (BERA 71) improves salt tolerance in chickpea plants by regulating the plant defense mechanisms. J Plant Interact 13 (1):37–44. https://doi.org/10.1080/17429145.2017.1414321

Abid O, AghaKouchak A (2015) Substantial increase in concurrent droughts and heatwaves in the United States. Proceedings of the National Academy of Sciences 112(37):11484–11489. https://doi.org/10.1073/pnas.1422945112

Admassie M, Woldehawariat Y, Alemu T, Gonzalez E, Jimenez JF (2022) The role of plant growth-promoting bacteria in alleviating drought stress on pepper plants. Agric Water Manag 272:107831. https://doi.org/10.1016/j.agwat.2022.107831

Adnan M, Shah Z, Fahad S et al (2017) Phosphate-solubilizing bacteria nullify the antagonistic effect of soil calcification on bioavailability of phosphorus in alkaline soils. Sci Rep 7(1):16131. https://doi.org/10.1038/s41598-017-16537-5

Afroz A, Zahur M, Zeeshan N, Komatsu S (2013) Plant-bacterium interactions analyzed by proteomics. Front Plant Sci 4. https://doi.org/10.3389/fpls.2013.00021

Agarwal S, Sairam R, Srivastava G, Meena R (2005) Changes in antioxidant enzymes activity and oxidative stress by abscisic acid and salicylic acid in wheat genotypes. Biol Plant 49:541–550. https://doi.org/10.1007/s10535-005-0048-z

Ahemad M (2019) Remediation of metalliferous soils through the heavy metal resistant plant growth promoting bacteria: paradigms and prospects. Arab J Chem 12(7):1365–1377. https://doi.org/10.1016/j.arabjc.2014.11.020

Alami Y, Achouak W, Marol C, Heulin T (2000) Rhizosphere soil aggregation and plant growth promotion of sunflowers by an exopolysaccharide-producing *Rhizobium* sp. strain isolated from sunflower roots. Appl Environ Microbiol 66(8):3393–3398. https://doi.org/10.1128/AEM.66.8.3393-3398.2000

Alavi P, Starcher M, Zachow C, Mueller H, Berg G (2013) Root-microbe systems: the effect and mode of interaction of Stress Protecting Agent (SPA) Stenotrophomonas rhizophila DSM14405T. Front Plant Sci 4. https://doi.org/10.3389/fpls.2013.00141

Alberton D, Müller-Santos M, Brusamarello-Santos LCC et al (2013) Comparative proteomics analysis of the rice roots colonized by *Herbaspirillum seropedicae* strain SmR1 reveals induction of the methionine recycling in the plant host. J Proteome Res 12(11):4757–4768. https://doi.org/10.1021/pr400425f

Alcázar R, Marco F, Cuevas JC et al (2006) Involvement of polyamines in plant response to abiotic stress. Biotechnol Lett 28:1867–1876. https://doi.org/10.1007/s10529-006-9179-3

Ali S, Khan N (2021) Delineation of mechanistic approaches employed by plant growth promoting microorganisms for improving drought stress tolerance in plants. Microbiol Res 249:126771. https://doi.org/10.1016/j.micres.2021.126771

Alka S, Shahir S, Ibrahim N, Chai T-T, Mohd Bahari Z, Abd Manan F (2020) The role of plant growth promoting bacteria on arsenic removal: a review of existing perspectives. Environ Technol Innov 17:100602. https://doi.org/10.1016/j.eti.2020.100602

ALKahtani MD, Attia KA, Hafez YM, Khan N, Eid AM, Ali MA, Abdelaal KA (2020a) Chlorophyll fluorescence parameters and antioxidant defense system can display salt tolerance of salt acclimated sweet pepper plants treated with chitosan and plant growth promoting rhizobacteria. Agronomy 10(8):1180. https://doi.org/10.3390/agronomy10081180

ALKahtani MD, Fouda A, Attia KA et al (2020b) Isolation and characterization of plant growth promoting endophytic bacteria from desert plants and their application as bioinoculants for sustainable agriculture. Agronomy 10(9):1325. https://doi.org/10.3390/agronomy10091325

Altieri MA, Nicholls CI, Henao A, Lana MA (2015) Agroecology and the design of climate change-resilient farming systems. Agron Sustain Dev 35(3):869–890. https://doi.org/10.1007/s13593-015-0285-2

Amaya-Gómez CV, Porcel M, Mesa-Garriga L, Gómez-Álvarez MI (2020) A framework for the selection of plant growth-promoting rhizobacteria based on bacterial competence mechanisms. Appl Environ Microbiol 86(14):e00760–e00720. https://doi.org/10.1128/AEM.00760-20

Ansari M, Shekari F, Mohammadi MH, Juhos K, Végvári G, Biró B (2019) Salt-tolerant plant growth-promoting bacteria enhanced salinity tolerance of salt-tolerant alfalfa (*Medicago*

sativa L.) cultivars at high salinity. Acta Physiol Plant 41(12):195. https://doi.org/10.1007/s11738-019-2988-5
Anzano A, Bonanomi G, Mazzoleni S, Lanzotti V (2022) Plant metabolomics in biotic and abiotic stress: a critical overview. Phytochem Rev 21(2):503–524. https://doi.org/10.1007/s11101-021-09786-w
Arzanesh MH, Alikhani HA, Khavazi K, Rahimian HA, Miransari M (2011) Wheat (*Triticum aestivum* L.) growth enhancement by *Azospirillum* sp. under drought stress. World J Microbiol Biotechnol 27(2):197–205. https://doi.org/10.1007/s11274-010-0444-1
Ashkavand P, Tabari M, Zarafshar M, Tomásková I, Struve D (2015) Effect of SiO_2 nanoparticles on drought resistance in hawthorn seedlings. For Res Paper 76(4). https://doi.org/10.1515/frp-2015-0034
Aslam F, Ali B (2018) Halotolerant bacterial diversity associated with *Suaeda fruticosa* (L.) forssk. improved growth of maize under salinity stress. Agronomy 8(8):131. https://doi.org/10.3390/agronomy8080131
Atkinson NJ, Urwin PE (2012) The interaction of plant biotic and abiotic stresses: from genes to the field. J Exp Bot 63(10):3523–3543. https://doi.org/10.1093/jxb/ers100
Azizi S, Tabari Kouchaksaraei M, Hadian J, Fallah Nosrat Abad AR, Modarres Sanavi SAM, Ammer C, Bader MKF (2021) Dual inoculations of arbuscular mycorrhizal fungi and plant growth-promoting rhizobacteria boost drought resistance and essential oil yield of common myrtle. For Ecol Manag 497:119478. https://doi.org/10.1016/j.foreco.2021.119478
Bachem CWB, Oomen RJFJ, Visser RGF (1998) Transcript imaging with cDNA-AFLP: a step-by-step protocol. Plant Mol Biol Report 16(2):157–157. https://doi.org/10.1023/A:1007468801806
Baillo EH, Kimotho RN, Zhang Z, Xu P (2019) Transcription factors associated with abiotic and biotic stress tolerance and their potential for crops improvement. Genes 10(10):771. https://doi.org/10.3390/genes10100771
Banerjee A, Sarkar S, Cuadros-Orellana S, Bandopadhyay R (2019) Exopolysaccharides and biofilms in mitigating salinity stress: the biotechnological potential of halophilic and soil-inhabiting PGPR microorganisms. In: Microorganisms in saline environments: strategies and functions, pp 133–153
Barlow KM, Christy BP, O'Leary GJ, Riffkin PA, Nuttall JG (2015) Simulating the impact of extreme heat and frost events on wheat crop production: a review. Field Crop Res 171:109–119. https://doi.org/10.1016/j.fcr.2014.11.010
Barnabás B, Jäger K, Fehér A (2008) The effect of drought and heat stress on reproductive processes in cereals. Plant Cell Environ 31(1):11–38. https://doi.org/10.1111/j.1365-3040.2007.01727.x
Bashan Y, Rojas A, Puente ME (1999) Improved establishment and development of three cactus species inoculated with *Azospirillum brasilense* transplanted into disturbed urban desert soil. Can J Microbiol 45(6):441–451. https://doi.org/10.1139/w99-043
Batool R, Hasnain S (2005) Growth stimulatory effects of Enterobacter and Serratia isolated from biofilms on plant growth and soil aggregation. Biotech 4(4):347–353
Begum N, Hu Z, Cai Q, Lou L (2019) Influence of PGPB inoculation on HSP70 and HMA3 gene expression in Switchgrass under cadmium stress. Plants 8(11):504. https://doi.org/10.3390/plants8110504
Beneduzi A, Ambrosini A, Passaglia LM (2012) Plant growth-promoting rhizobacteria (PGPR): their potential as antagonists and biocontrol agents. Genet Mol Biol 35:1044–1051. https://doi.org/10.1590/S1415-47572012000600020
Berendsen RL, Vismans G, Yu K et al (2018) Disease-induced assemblage of a plant-beneficial bacterial consortium. ISME J 12(6):1496–1507. https://doi.org/10.1038/s41396-018-0093-1
Berg MP, Kiers ET, Driessen G et al (2010) Adapt or disperse: understanding species persistence in a changing world. Glob Chang Biol 16(2):587–598. https://doi.org/10.1111/j.1365-2486.2009.02014.x
Bharti N, Barnawal D, Awasthi A, Yadav A, Kalra A (2014) Plant growth promoting rhizobacteria alleviate salinity induced negative effects on growth, oil content and physiological status in *Mentha arvensis*. Acta Physiol Plant 36:45–60. https://doi.org/10.1007/s11738-013-1385-8

Bharti N, Pandey SS, Barnawal D, Patel VK, Kalra A (2016) Plant growth promoting rhizobacteria *Dietzia* natronolimnaea modulates the expression of stress responsive genes providing protection of wheat from salinity stress. Sci Rep 6(1):34768. https://doi.org/10.1038/srep34768

Bohler S, Sergeant K, Jolivet Y, Hoffmann L, Hausman J-F, Dizengremel P, Renaut J (2013) A physiological and proteomic study of poplar leaves during ozone exposure combined with mild drought. Proteomics 13(10–11):1737–1754. https://doi.org/10.1002/pmic.201200193

Bräutigam A, Gowik U (2010) What can next generation sequencing do for you? Next generation sequencing as a valuable tool in plant research. Plant Biol 12(6):831–841. https://doi.org/10.1111/j.1438-8677.2010.00373.x

Bruno LB, Karthik C, Ma Y, Kadirvelu K, Freitas H, Rajkumar M (2020) Amelioration of chromium and heat stresses in Sorghum bicolor by Cr^{6+} reducing-thermotolerant plant growth promoting bacteria. Chemosphere 244:125521. https://doi.org/10.1016/j.chemosphere.2019.125521

Bundy JG, Willey TL, Castell RS, Ellar DJ, Brindle KM (2005) Discrimination of pathogenic clinical isolates and laboratory strains of *Bacillus cereus* by NMR-based metabolomic profiling. FEMS Microbiol Lett 242(1):127–136. https://doi.org/10.1016/j.femsle.2004.10.048

Cambray G, Mutalik VK, Arkin AP (2011) Toward rational design of bacterial genomes. Curr Opin Microbiol 14(5):624–630. https://doi.org/10.1016/j.mib.2011.08.001

Cangahuala-Inocente GC, Plucani do Amaral F, Faleiro AC, Huergo LF, Maisonnave Arisi AC (2013) Identification of six differentially accumulated proteins of *Zea mays* seedlings (DKB240 variety) inoculated with *Azospirillum brasilense* strain FP2. Eur J Soil Biol 58:45–50. https://doi.org/10.1016/j.ejsobi.2013.06.002

Cao M, Narayanan M, Shi X, Chen X, Li Z, Ma Y (2023) Optimistic contributions of plant growth-promoting bacteria for sustainable agriculture and climate stress alleviation. Environ Res 217:114924. https://doi.org/10.1016/j.envres.2022.114924

Carlson R, Tugizimana F, Steenkamp PA, Dubery IA, Hassen AI, Labuschagne N (2020) Rhizobacteria-induced systemic tolerance against drought stress in *Sorghum bicolor* (L.) Moench. Microbiol Res 232:126388. https://doi.org/10.1016/j.micres.2019.126388

Cassia R, Nocioni M, Correa-Aragunde N, Lamattina L (2018) Climate change and the impact of greenhouse gasses: CO2 and NO, friends and foes of plant oxidative stress. Front Plant Sci 9:273. https://doi.org/10.3389/fpls.2018.00273

Castiglioni P, Warner D, Bensen RJ et al (2008) Bacterial RNA chaperones confer abiotic stress tolerance in plants and improved grain yield in maize under water-limited conditions. Plant Physiol 147(2):446–455. https://doi.org/10.1104/pp.108.118828

Chan R, Lam J, Lam K, Costerton J (1984) Influence of culture conditions on expression of the mucoid mode of growth of *Pseudomonas aeruginosa*. J Clin Microbiol 19(1):8–16. https://doi.org/10.1128/jcm.19.1.8-16.1984

Chandra D, Srivastava R, Glick BR, Sharma AK (2018) Drought-tolerant *Pseudomonas* spp. improve the growth performance of finger millet (*Eleusine coracana* (L.) *Gaertn.*) under non-stressed and drought-stressed conditions. Pedosphere 28(2):227–240. https://doi.org/10.1016/S1002-0160(18)60013-X

Chatterjee P, Kanagendran A, Samaddar S, Pazouki L, Sa T-M, Niinemets Ü (2018) Inoculation of *Brevibacterium linens* RS16 in *Oryza sativa* genotypes enhanced salinity resistance: impacts on photosynthetic traits and foliar volatile emissions. Sci Total Environ 645:721–732. https://doi.org/10.1016/j.scitotenv.2018.07.187

Chatterjee P, Kanagendran A, Samaddar S, Pazouki L, Sa T-M, Niinemets Ü (2020) Influence of *Brevibacterium linens* RS16 on foliage photosynthetic and volatile emission characteristics upon heat stress in *Eucalyptus grandis*. Sci Total Environ 700:134453. https://doi.org/10.1016/j.scitotenv.2019.134453

Chauhan PS, Lata C, Tiwari S et al (2019) Transcriptional alterations reveal *Bacillus amyloliquefaciens*-rice cooperation under salt stress. Sci Rep 9(1):11912. https://doi.org/10.1038/s41598-019-48309-8

Chen K, Pachter L (2005) Bioinformatics for whole-genome shotgun sequencing of microbial communities. PLoS Comput Biol 1(2):e24. https://doi.org/10.1371/journal.pcbi.0010024

Chen L, Liu Y, Wu G, Veronican Njeri K, Shen Q, Zhang N, Zhang R (2016) Induced maize salt tolerance by rhizosphere inoculation of *Bacillus amyloliquefaciens* SQR9. Physiol Plant 158(1):34–44. https://doi.org/10.1111/ppl.12441

Chen X, Liu X, Zhang X, Cao L, Hu X (2017) Phytoremediation effect of *Scirpus triqueter* inoculated plant-growth-promoting bacteria (PGPB) on different fractions of pyrene and Ni in co-contaminated soils. J Hazard Mater 325:319–326. https://doi.org/10.1016/j.jhazmat.2016.12.009

Chen D, Shao Q, Yin L, Younis A, Zheng B (2019) Polyamine function in plants: metabolism, regulation on development, and roles in abiotic stress responses. Front Plant Sci 9. https://doi.org/10.3389/fpls.2018.01945

Chinnaswamy A, Coba de la Peña T, Stoll A et al (2018) A nodule endophytic *Bacillus megaterium* strain isolated from *Medicago polymorpha* enhances growth, promotes nodulation by *Ensifer medicae* and alleviates salt stress in alfalfa plants. Ann Appl Biol 172(3):295–308. https://doi.org/10.1111/aab.12420

Cho ST, Chang HH, Egamberdieva D, Kamilova F, Lugtenberg B, Kuo CH (2015) Genome analysis of *Pseudomonas fluorescens* PCL1751: a rhizobacterium that controls root diseases and alleviates salt stress for its plant host. PLoS One 10(10):e0140231. https://doi.org/10.1371/journal.pone.0140231

Choudhary DK, Johri BN (2009) Interactions of *Bacillus* spp. and plants–with special reference to induced systemic resistance (ISR). Microbiol Res 164(5):493–513. https://doi.org/10.1016/j.micres.2008.08.007

Choudhary DK, Kasotia A, Jain S, Vaishnav A, Kumari S, Sharma KP, Varma A (2016) Bacterial-mediated tolerance and resistance to plants under abiotic and biotic stresses. J Plant Growth Regul 35(1):276–300. https://doi.org/10.1007/s00344-015-9521-x

Conesa A, Madrigal P, Tarazona S et al (2016) A survey of best practices for RNA-seq data analysis. Genome Biol 17(1):13. https://doi.org/10.1186/s13059-016-0881-8

Córdova-Albores LC, Zelaya-Molina LX, Ávila-Alistac N et al (2021) Omics sciences potential on bioprospecting of biological control microbial agents: the case of the Mexican agro-biotechnology. Revista Mex fitopatol 39(1):147–184. https://doi.org/10.18781/r.mex.fit.2009-3

Creus CM, Graziano M, Casanovas EM et al (2005) Nitric oxide is involved in the *Azospirillum brasilense*-induced lateral root formation in tomato. Planta 221(2):297–303. https://doi.org/10.1007/s00425-005-1523-7

Curá JA, Franz DR, Filosofía JE, Balestrasse KB, Burgueño LE (2017) Inoculation with *Azospirillum* sp. and *Herbaspirillum* sp. bacteria increases the tolerance of maize to drought stress. Microorganisms 5(3):41. https://doi.org/10.3390/microorganisms5030041

Dadarwall L, Yadv L, Sindhu S (1997) Biofertilizer production: technology: prospects in biotechnological approach in soil microorganisms for sustainable crop production. Scientific Publishers, Jodhpur, pp 323–337

Damodaran T, Mishra VK, Jha SK, Pankaj U, Gupta G, Gopal R (2019) Identification of rhizosphere bacterial diversity with promising salt tolerance, PGP traits and their exploitation for seed germination enhancement in sodic soil. Agric Res 8(1):36–43. https://doi.org/10.1007/s40003-018-0343-5

Dastogeer KMG, Zahan MI, Rhaman MS, Sarker MSA, Chakraborty A (2022) Microbe-mediated thermotolerance in plants and pertinent mechanisms- a meta-analysis and review. Front Microbiol 13. https://doi.org/10.3389/fmicb.2022.833566

de Sousa SM, de Oliveira CA, Andrade DL et al (2021) Tropical *Bacillus* strains inoculation enhances maize root surface area, dry weight, nutrient uptake and grain yield. J Plant Growth Regul 40(2):867–877. https://doi.org/10.1007/s00344-020-10146-9

de-Bashan LE, Hernandez J-P, Bashan Y (2012) The potential contribution of plant growth-promoting bacteria to reduce environmental degradation – a comprehensive evaluation. Appl Soil Ecol 61:171–189. https://doi.org/10.1016/j.apsoil.2011.09.003

DeLisi C (2019) The role of synthetic biology in climate change mitigation. Biol Direct 14(1):14. https://doi.org/10.1186/s13062-019-0247-8

Devireddy AR, Tschaplinski TJ, Tuskan GA, Muchero W, Chen J-G (2021) Role of reactive oxygen species and hormones in plant responses to temperature changes. Int J Mol Sci 22(16):8843. https://doi.org/10.3390/ijms22168843

Dhir B (2017) Biofertilizers and biopesticides: eco-friendly biological agents. Adv Environ Biotechnol:167–188

Dong J, Yu Z, Yu S (1983) Effect of increased ethylene production during different periods on the resistance of wheat plants to waterlogging. Acta Phytophysiol Sinica 9:383–389

Dubeikovsky AN, Mordukhova EA, Kochetkov VV, Polikarpova FY, Boronin AM (1993) Growth promotion of blackcurrant softwood cuttings by recombinant strain *Pseudomonas fluorescens* BSP53a synthesizing an increased amount of indole-3-acetic acid. Soil Biol Biochem 25(9):1277–1281. https://doi.org/10.1016/0038-0717(93)90225-Z

Duman JG, Wisniewski MJ (2014) The use of antifreeze proteins for frost protection in sensitive crop plants. Environ Exp Bot 106:60–69. https://doi.org/10.1016/j.envexpbot.2014.01.001

Eastmond PJ, Graham IA (2003) Trehalose metabolism: a regulatory role for trehalose-6-phosphate? Curr Opin Plant Biol 6(3):231–235. https://doi.org/10.1016/S1369-5266(03)00037-2

Edbeib MF, Wahab RA, Huyop F (2016) Halophiles: biology, adaptation, and their role in decontamination of hypersaline environments. World J Microbiol Biotechnol 32(8):135. https://doi.org/10.1007/s11274-016-2081-9

Egamberdieva D, Wirth S, Jabborova D, Räsänen LA, Liao H (2017) Coordination between *Bradyrhizobium* and *Pseudomonas* alleviates salt stress in soybean through altering root system architecture. J Plant Interact 12(1):100–107. https://doi.org/10.1080/17429145.2017.1294212

Egamberdieva D, Wirth S, Bellingrath-Kimura SD, Mishra J, Arora NK (2019) Salt-tolerant plant growth promoting rhizobacteria for enhancing crop productivity of saline soils. Front Microbiol 10. https://doi.org/10.3389/fmicb.2019.02791

Elsheery NI, Helaly MN, El-Hoseiny HM, Alam-Eldein SM (2020) Zinc oxide and silicone nanoparticles to improve the resistance mechanism and annual productivity of salt-stressed mango trees. Agronomy 10(4):558. https://doi.org/10.3390/agronomy10040558

Emamverdian A, Ding Y, Mokhberdoran F (2020) The role of salicylic acid and gibberellin signaling in plant responses to abiotic stress with an emphasis on heavy metals. Plant Signal Behav 15(7):1777372. https://doi.org/10.1080/15592324.2020.1777372

Enebe MC, Babalola OO (2018) The influence of plant growth-promoting rhizobacteria in plant tolerance to abiotic stress: a survival strategy. Appl Microbiol Biotechnol 102(18):7821–7835. https://doi.org/10.1007/s00253-018-9214-z

Etesami H (2018) Can interaction between silicon and plant growth promoting rhizobacteria benefit in alleviating abiotic and biotic stresses in crop plants? Agric Ecosyst Environ 253:98–112. https://doi.org/10.1016/j.agee.2017.11.007

Fadiji AE, Santoyo G, Yadav AN, Babalola OO (2022) Efforts towards overcoming drought stress in crops: revisiting the mechanisms employed by plant growth-promoting bacteria. Front Microbiol 13:962427. https://doi.org/10.3389/fmicb.2022.962427

Faleiro AC, Neto PAV, de Souza TV, Santos M, Maisonnave Arisi AC (2015) Microscopic and proteomic analysis of *Zea mays* roots (P30F53 variety) inoculated with *Azospirillum brasilense* strain FP2. J Crop Sci Biotechnol 18(2):63–71. https://doi.org/10.1007/s12892-014-0061-x

Faraji J, Sepehri A (2020) Exogenous nitric oxide improves the protective effects of TiO_2 nanoparticles on growth, antioxidant system, and photosynthetic performance of wheat seedlings under drought stress. J Soil Sci Plant Nutr 20:703–714. https://doi.org/10.1007/s42729-019-00158-0

Farrar K, Bryant D, Cope-Selby N (2014) Understanding and engineering beneficial plant–microbe interactions: plant growth promotion in energy crops. Plant Biotechnol J 12(9):1193–1206. https://doi.org/10.1111/pbi.12279

Farrell A, Hart M (1998) What does sustainability really mean?: the search for useful indicators. Environ: Sci Policy Sustain Dev 40(9):4–31

Ferreira CMH, Soares HMVM, Soares EV (2019) Promising bacterial genera for agricultural practices: an insight on plant growth-promoting properties and microbial safety aspects. Sci Total Environ 682:779–799. https://doi.org/10.1016/j.scitotenv.2019.04.225

Figueiredo MV, Burity HA, Martinez CR, Chanway CP (2008) Alleviation of drought stress in the common bean (*Phaseolus vulgaris* L.) by co-inoculation with *Paenibacillus polymyxa* and *Rhizobium tropici*. Appl Soil Ecol 40(1):182–188. https://doi.org/10.1016/j.apsoil.2008.04.005

Fiodor A, Singh S, Pranaw K (2021) The contrivance of plant growth promoting microbes to mitigate climate change impact in agriculture. Microorganisms 9(9):1841. https://doi.org/10.3390/microorganisms9091841

Forrest KL, Bhave M (2007) Major intrinsic proteins (MIPs) in plants: a complex gene family with major impacts on plant phenotype. Funct Integr Genomics 7(4):263–289. https://doi.org/10.1007/s10142-007-0049-4

Fraga R, Rodríguez H, González T (2001) Transfer of the gene encoding the NapA acid phosphatase of *Morganella morganii* to a *Burkholderia cepacia* Strain. Acta Biotechnol 21(4):359–369. https://doi.org/10.1002/1521-3846(200111)21:4<359::AID-ABIO359>3.0.CO;2-B

Gagné-Bourque F, Bertrand A, Claessens A, Aliferis KA, Jabaji S (2016) Alleviation of drought stress and metabolic changes in timothy (*Phleum pratense* L.) colonized with *Bacillus subtilis* B26. Front. Plant Sci 7. https://doi.org/10.3389/fpls.2016.00584

Gamalero E, Glick BR (2022) Recent advances in bacterial amelioration of plant drought and salt stress. Biology 11(3):437. https://doi.org/10.3390/biology11030437

Gao JP, Chao DY, Lin HX (2007) Understanding abiotic stress tolerance mechanisms: recent studies on stress response in rice. J Integr Plant Biol 49(6):742–750. https://doi.org/10.1111/j.1744-7909.2007.00495.x

Ghosh D, Sen S, Mohapatra S (2017) Modulation of proline metabolic gene expression in *Arabidopsis thaliana* under water-stressed conditions by a drought-mitigating *Pseudomonas putida* strain. Ann Microbiol 67:655–668. https://doi.org/10.1007/s13213-017-1294-y

Ghosh D, Gupta A, Mohapatra S (2019) A comparative analysis of exopolysaccharide and phytohormone secretions by four drought-tolerant rhizobacterial strains and their impact on osmotic-stress mitigation in *Arabidopsis thaliana*. World J Microbiol Biotechnol 35(6):90. https://doi.org/10.1007/s11274-019-2659-0

Gill SS, Tuteja N (2010a) Polyamines and abiotic stress tolerance in plants. Plant Signal Behav 5(1):26–33. https://doi.org/10.4161/psb.5.1.10291

Gill SS, Tuteja N (2010b) Reactive oxygen species and antioxidant machinery in abiotic stress tolerance in crop plants. Plant Physiol Biochem 48(12):909–930. https://doi.org/10.1016/j.plaphy.2010.08.016

Glick BR (2020) Issues regarding the use of PGPB. In: Beneficial plant-bacterial interactions. Springer International Publishing, Cham, pp 361–383

Goldstein AH, Liu ST (1987) Molecular cloning and regulation of a mineral phosphate solubilizing gene from *Erwinia Herbicola*. Biotechnology 5(1):72–74. https://doi.org/10.1038/nbt0187-72

Goswami M, Suresh D (2020) Plant growth-promoting rhizobacteria—alleviators of abiotic stresses in soil: a review. Pedosphere 30(1):40–61. https://doi.org/10.1016/S1002-0160(19)60839-8

Goswami D, Thakker JN, Dhandhukia PC (2016) Portraying mechanics of plant growth promoting rhizobacteria (PGPR): a review. Cogent Food Agric 2(1):1127500. https://doi.org/10.1080/23311932.2015.1127500

Gouda S, Kerry RG, Das G, Paramithiotis S, Shin H-S, Patra JK (2018) Revitalization of plant growth promoting rhizobacteria for sustainable development in agriculture. Microbiol Res 206:131–140. https://doi.org/10.1016/j.micres.2017.08.016

Gray EJ, Smith DL (2005) Intracellular and extracellular PGPR: commonalities and distinctions in the plant–bacterium signaling processes. Soil Biol Biochem 37(3):395–412. https://doi.org/10.1016/j.soilbio.2004.08.030

Grichko VP, Glick BR (2001) Amelioration of flooding stress by ACC deaminase-containing plant growth-promoting bacteria. Plant Physiol Biochem 39(1):11–17. https://doi.org/10.1016/S0981-9428(00)01212-2

Grzesiak S, Iijima M, Kono Y, Yamauchi A (1997) Differences in drought tolerance between cultivars of field bean and field pea. A comparison of drought-resistant and drought-sensitive cultivars. Acta Physiol Plant 19(3):349–357. https://doi.org/10.1007/s11738-997-0012-y

Gul Z, Tang Z-H, Arif M, Ye Z (2022) An insight into abiotic stress and influx tolerance mechanisms in plants to cope in saline environments. Biology 11(4):597. https://doi.org/10.3390/biology11040597

Guo Y, Sun Y, Zhang H (2022) Advances in microbial transcriptomics techniques. Chin J Biotechnol 38(10):3606–3615. https://doi.org/10.13345/j.cjb.220527

Gupta S, Pandey S (2020) Enhanced salinity tolerance in the common bean (*Phaseolus vulgaris*) plants using twin ACC deaminase producing rhizobacterial inoculation. Rhizosphere 16:100241. https://doi.org/10.1016/j.rhisph.2020.100241

Gupta N, Gaurav SS, Kumar A (2013) Molecular basis of aluminium toxicity in plants: a review. Am J Plant Sci 2013. https://doi.org/10.4236/ajps.2013.412A3004

Gupta A, Mishra R, Rai S et al (2022) Mechanistic insights of plant growth promoting bacteria mediated drought and salt stress tolerance in plants for sustainable agriculture. Int J Mol Sci 23(7):3741. https://doi.org/10.3390/ijms23073741

Gut IG (2013) New sequencing technologies. Clin Transl Oncol 15(11):879–881. https://doi.org/10.1007/s12094-013-1073-6

Guy CL (1990) Cold acclimation and freezing stress tolerance: role of protein metabolism. Annu Rev Plant Biol 41(1):187–223. https://doi.org/10.1146/annurev.pp.41.060190.001155

Haas BJ, Zody MC (2010) Advancing RNA-Seq analysis. Nat Biotechnol 28(5):421–423. https://doi.org/10.1038/nbt0510-421

Hamdan MF, Karlson CKS, Teoh EY, Lau S-E, Tan BC (2022) Genome editing for sustainable crop improvement and mitigation of biotic and abiotic stresses. Plants 11(19):2625. https://doi.org/10.3390/plants11192625

Hao S, Wang Y, Yan Y, Liu Y, Wang J, Chen S (2021) A review on plant responses to salt stress and their mechanisms of salt resistance. Horticulturae 7(6):132. https://doi.org/10.3390/horticulturae7060132

Harrison JJ, Rabiei M, Turner RJ, Badry EA, Sproule KM, Ceri H (2006) Metal resistance in *Candida* biofilms. FEMS Microbiol Ecol 55(3):479–491. https://doi.org/10.1111/j.1574-6941.2005.00045.x

Hashem A, Abd_Allah EF, Alqarawi AA, Al-Huqail AA, Wirth S, Egamberdieva D (2016) The interaction between arbuscular mycorrhizal fungi and endophytic bacteria enhances plant growth of *Acacia gerrardii* under salt stress. Front Microbiol 7:1089. https://doi.org/10.3389/fmicb.2016.01089

Hassan TU, Bano A, Naz I (2018) Halophyte root powder: an alternative biofertilizer and carrier for saline land. Soil Sci Plant Nutr 64(5):653–661. https://doi.org/10.1080/00380768.2018.1509676

Haynes R, Swift R (1990) Stability of soil aggregates in relation to organic constituents and soil water content. J Soil Sci 41(1):73–83. https://doi.org/10.1111/j.1365-2389.1990.tb00046.x

He Y, Pantigoso HA, Wu Z, Vivanco JM (2019) Co-inoculation of *Bacillus* sp. and *Pseudomonas putida* at different development stages acts as a biostimulant to promote growth, yield and nutrient uptake of tomato. J Appl Microbiol 127(1):196–207. https://doi.org/10.1111/jam.14273

Hertel TW (2011) The global supply and demand for agricultural land in 2050: a perfect storm in the making? Am J Agric Econ 93(2):259–275. https://doi.org/10.1093/ajae/aaq189

Ibort P, Imai H, Uemura M, Aroca R (2018) Proteomic analysis reveals that tomato interaction with plant growth promoting bacteria is highly determined by ethylene perception. J Plant Physiol 220:43–59. https://doi.org/10.1016/j.jplph.2017.10.008

Ike A, Sriprang R, Ono H, Murooka Y, Yamashita M (2007) Bioremediation of cadmium contaminated soil using symbiosis between leguminous plant and recombinant rhizobia with the *MTL4* and the *PCS* genes. Chemosphere 66(9):1670–1676. https://doi.org/10.1016/j.chemosphere.2006.07.058

Ilangumaran G, Smith DL (2017) Plant growth promoting rhizobacteria in amelioration of salinity stress: a systems biology perspective. Front Plant Sci 8:1768. https://doi.org/10.3389/fpls.2017.01768

Ishii S, Koki J, Unno H, Hori K (2004) Two morphological types of cell appendages on a strongly adhesive bacterium, *Acinetobacter* sp. strain Tol 5. Appl Environ Microbiol 70(8):5026–5029. https://doi.org/10.1128/AEM.70.8.5026-5029.2004

Islam F, Yasmeen T, Arif MS, Ali S, Ali B, Hameed S, Zhou W (2016a) Plant growth promoting bacteria confer salt tolerance in *Vigna radiata* by up-regulating antioxidant defense and biological soil fertility. Plant Growth Regul 80(1):23–36. https://doi.org/10.1007/s10725-015-0142-y

Islam F, Yasmeen T, Arif MS, Riaz M, Shahzad SM, Imran Q, Ali I (2016b) Combined ability of chromium (Cr) tolerant plant growth promoting bacteria (PGPB) and salicylic acid (SA) in attenuation of chromium stress in maize plants. Plant Physiol Biochem 108:456–467. https://doi.org/10.1016/j.plaphy.2016.08.014

Jamil IN, Remali J, Azizan KA, Nor Muhammad NA, Arita M, Goh H-H, Aizat WM (2020) Systematic multi-omics integration (MOI) approach in plant systems biology. Front Plant Sci 11. https://doi.org/10.3389/fpls.2020.00944

Jan AT, Azam M, Ali A, Haq QMR (2014) Prospects for exploiting bacteria for bioremediation of metal pollution. Crit Rev Environ Sci Technol 44(5):519–560. https://doi.org/10.1080/10643389.2012.728811

Jan R, Khan MA, Asaf S, Lubna LI-J, Kim KM (2019) Metal resistant endophytic bacteria reduces cadmium, nickel toxicity, and enhances expression of metal stress related genes with improved growth of *Oryza sativa*, via regulating its antioxidant machinery and endogenous hormones. Plants 8(10):363. https://doi.org/10.3390/plants8100363

Jansson JK, Hofmockel KS (2018) The soil microbiome—from metagenomics to metaphenomics. Curr Opin Microbiol 43:162–168. https://doi.org/10.1016/j.mib.2018.01.013

Jenkins GI (2014) The UV-B photoreceptor UVR8: from structure to physiology. Plant Cell 26(1):21–37. https://doi.org/10.1105/tpc.113.119446

Jia X, Sun C, Zuo Y, Li G, Li G, Ren L, Chen G (2016) Integrating transcriptomics and metabolomics to characterize the response of *Astragalus membranaceus* Bge. var. *mongolicus* (Bge.) to progressive drought stress. BMC Genomics 17(1):188. https://doi.org/10.1186/s12864-016-2554-0

Jiang S, Li J, Wang Q et al (2022) Maize growth promotion by inoculation with an engineered ammonium-excreting strain of nitrogen-fixing *Pseudomonasstutzeri*. Microorganisms 10(10):1986. https://doi.org/10.3390/microorganisms10101986

Jing J, Cong W-F, Bezemer TM (2022) Legacies at work: plant-soil-microbiome interactions underpinning agricultural sustainability. Trends Plant Sci 27(8):781–792. https://doi.org/10.1016/j.tplants.2022.05.007

Jones L, Milne JL, Ashford D, McQueen-Mason SJ (2003) Cell wall arabinan is essential for guard cell function. Proc Natl Acad Sci 100(20):11783–11788. https://doi.org/10.1073/pnas.1832434100

Jorrín-Novo JV, Maldonado AM, Echevarría-Zomeño S et al (2009) Plant proteomics update (2007–2008): second-generation proteomic techniques, an appropriate experimental design, and data analysis to fulfill MIAPE standards, increase plant proteome coverage and expand biological knowledge. J Proteome 72(3):285–314. https://doi.org/10.1016/j.jprot.2009.01.026

Kacholi DS, Sahu M (2018) Levels and health risk assessment of heavy metals in soil, water, and vegetables of Dar es Salaam, Tanzania. J Chem 2018:1402674. https://doi.org/10.1155/2018/1402674

Kamkar B (2016) Sustainable development principles for agricultural activities. Adv Plant Agric Res 3(5):1–2

Kang S-M, Hamayun M, Joo G-J et al (2010) Effect of *Burkholderia* sp. KCTC 11096BP on some physiochemical attributes of cucumber. Eur J Soil Biol 46(3):264–268. https://doi.org/10.1016/j.ejsobi.2010.03.002

Kang S-M, Khan AL, Waqas M et al (2014a) Plant growth-promoting rhizobacteria reduce adverse effects of salinity and osmotic stress by regulating phytohormones and antioxidants in *Cucumis sativus*. J Plant Interact 9(1):673–682. https://doi.org/10.1080/17429145.2014.894587

Kang S-M, Khan AL, You Y-H, Kim J-G, Kamran M, Lee I-J (2014b) Gibberellin production by newly isolated strain *Leifsonia soli* SE134 and its potential to promote plant growth. J Microbiol Biotechnol 24(1):106–112

Kang S-M, Radhakrishnan R, Khan AL et al (2014c) Gibberellin secreting rhizobacterium, *Pseudomonas putida* H-2-3 modulates the hormonal and stress physiology of soybean to improve the plant growth under saline and drought conditions. Plant Physiol Biochem 84:115–124. https://doi.org/10.1016/j.plaphy.2014.09.001

Kapardar RK, Ranjan R, Grover A, Puri M, Sharma R (2010) Identification and characterization of genes conferring salt tolerance to *Escherichia coli* from pond water metagenome. Bioresour Technol 101(11):3917–3924. https://doi.org/10.1016/j.biortech.2010.01.017

Kaur G, Singh G, Motavalli PP, Nelson KA, Orlowski JM, Golden BR (2020) Impacts and management strategies for crop production in waterlogged or flooded soils: a review. Agron J 112(3):1475–1501. https://doi.org/10.1002/agj2.20093

Khalid S, Shahid M, Niazi NK, Murtaza B, Bibi I, Dumat C (2017) A comparison of technologies for remediation of heavy metal contaminated soils. J Geochem Explor 182:247–268. https://doi.org/10.1016/j.gexplo.2016.11.021

Khan MY, Zahir ZA, Asghar HN, Waraich EA (2017) Preliminary investigations on selection of synergistic halotolerant plant growth promoting rhizobacteria for inducing salinity tolerance in wheat. Pak J Bot 49(4):1541–1551

Khan TA, Yusuf M, Ahmad A et al (2019) Proteomic and physiological assessment of stress sensitive and tolerant variety of tomato treated with brassinosteroids and hydrogen peroxide under low-temperature stress. Food Chem 289:500–511. https://doi.org/10.1016/j.foodchem.2019.03.029

Khan MA, Asaf S, Khan AL, Jan R, Kang S-M, Kim K-M, Lee I-J (2020a) Extending thermotolerance to tomato seedlings by inoculation with SA1 isolate of *Bacillus cereus* and comparison with exogenous humic acid application. PLoS One 15(4):e0232228. https://doi.org/10.1371/journal.pone.0232228

Khan MA, Asaf S, Khan AL, Jan R, Kang S-M, Kim K-M, Lee I-J (2020b) Thermotolerance effect of plant growth-promoting *Bacillus cereus* SA1 on soybean during heat stress. BMC Microbiol 20(1):175. https://doi.org/10.1186/s12866-020-01822-7

Kim K, Jang YJ, Lee SM, Oh BT, Chae JC, Lee KJ (2014) Alleviation of salt stress by *Enterobacter* sp. EJ01 in tomato and *Arabidopsis* is accompanied by up-regulation of conserved salinity responsive factors in plants. Mol Cell 37(2):109–117. https://doi.org/10.14348/molcells.2014.2239

Kohler J, Hernández JA, Caravaca F, Roldán A (2008) Plant-growth-promoting rhizobacteria and arbuscular mycorrhizal fungi modify alleviation biochemical mechanisms in water-stressed plants. Funct Plant Biol 35(2):141–151. https://doi.org/10.1071/FP07218

Kosar F, Akram NA, Sadiq M, Al-Qurainy F, Ashraf M (2019) Trehalose: a key organic osmolyte effectively involved in plant abiotic stress tolerance. J Plant Growth Regul 38(2):606–618. https://doi.org/10.1007/s00344-018-9876-x

Kosová K, Vítámvás P, Urban MO, Klíma M, Roy A, Prášil IT (2015) Biological networks underlying abiotic stress tolerance in temperate crops—a proteomic perspective. Int J Mol Sci 16(9):20913–20942. https://doi.org/10.3390/ijms160920913

Kumar S, Himanshu S, Gupta K (2012) Effect of global warming on mankind-a review. Int Res J Environ Sci 1(4):56–59

Kumar K, Amaresan N, Madhuri K (2017) Alleviation of the adverse effect of salinity stress by inoculation of plant growth promoting rhizobacteria isolated from hot humid tropical climate. Ecol Eng 102:361–366. https://doi.org/10.1016/j.ecoleng.2017.02.023

Kumar R, Swapnil P, Meena M, Selpair S, Yadav BG (2022) Plant growth-promoting rhizobacteria (PGPR): approaches to alleviate abiotic stresses for enhancement of growth and development of medicinal plants. Sustainability 14(23):15514. https://doi.org/10.3390/su142315514

Kundan R, Pant G, Jadon N, Agrawal PK (2015) Plant growth promoting rhizobacteria: mechanism and current prospective. J Fertil Pestic 06(02). https://doi.org/10.4172/2471-2728.1000155

Kwon YS, Lee DY, Rakwal R et al (2016) Proteomic analyses of the interaction between the plant-growth promoting rhizobacterium *Paenibacillus polymyxa* E681 and *Arabidopsis thaliana*. Proteomics 16(1):122–135. https://doi.org/10.1002/pmic.201500196

Lal SK, Kumar S, Sheri V et al (2018) Seed priming: an emerging technology to impart abiotic stress tolerance in crop plants. In: Advances in seed priming, pp 41–50

Lamaoui M, Jemo M, Datla R, Bekkaoui F (2018) Heat and drought stresses in crops and approaches for their mitigation. Front Chem 6:26. https://doi.org/10.3389/fchem.2018.00026

Lastochkina O, Yuldashev R, Pusenkova L (2017) The effect of bacteria *Bacillus subtilis* 26D on drought resistance of soft spring wheat cultivars of the West Siberian forest-steppe and Volga steppe ecotypes in the early stages of ontogenesis. Proc RAS Ufa Sci Centre 3(1):99–102

Le DT, Nishiyama R, Watanabe Y, Mochida K, Yamaguchi-Shinozaki K, Shinozaki K, Tran L-SP (2011) Genome-wide survey and expression analysis of the plant-specific NAC transcription factor family in soybean during development and dehydration stress. DNA Res 18(4):263–276. https://doi.org/10.1093/dnares/dsr015

Leaungvutiviroj C, Ruangphisarn P, Hansanimitkul P, Shinkawa H, Sasaki K (2010) Development of a new biofertilizer with a high capacity for N_2 fixation, phosphate and potassium solubilization and auxin production. Biosci Biotechnol Biochem 74(5):1098–1101. https://doi.org/10.1271/bbb.90898

Lee SA, Kim HS, Sang MK, Song J, Weon H-Y (2021) Effect of *Bacillus mesonae* H20–5 treatment on rhizospheric bacterial community of tomato plants under salinity stress. Plant Pathol J 37(6):662. https://doi.org/10.5423/PPJ.FT.10.2021.0156

Lehmann J, Rillig M (2014) Distinguishing variability from uncertainty. Nat Clim Change 4(3):153. https://doi.org/10.1038/nclimate2133

Lehman RM, Cambardella CA, Stott DE et al (2015) Understanding and enhancing soil biological health: the solution for reversing soil degradation. Sustainability 7(1):988–1027. https://doi.org/10.3390/su7010988

Li X, Wu Z, Li W et al (2007) Growth promoting effect of a transgenic *Bacillus mucilaginosus* on tobacco planting. Appl Microbiol Biotechnol 74(5):1120–1125. https://doi.org/10.1007/s00253-006-0750-6

Li X, Sun P, Zhang Y, Jin C, Guan C (2020) A novel PGPR strain *Kocuria rhizophila* Y1 enhances salt stress tolerance in maize by regulating phytohormone levels, nutrient acquisition, redox potential, ion homeostasis, photosynthetic capacity and stress-responsive genes expression. Environ Exp Bot 174:104023. https://doi.org/10.1016/j.envexpbot.2020.104023

Liljeqvist M, Ossandon FJ, González C et al (2015) Metagenomic analysis reveals adaptations to a cold-adapted lifestyle in a low-temperature acid mine drainage stream. FEMS Microbiol Ecol 91(4). https://doi.org/10.1093/femsec/fiv011

Lin H, Zhou M, Li B, Dong Y (2023) Mechanisms, application advances and future perspectives of microbial-induced heavy metal precipitation: a review. Int Biodeterior Biodegrad 178:105544. https://doi.org/10.1016/j.ibiod.2022.105544

Linkemer G, Board JE, Musgrave ME (1998) Waterlogging effects on growth and yield components in late-planted soybean. Crop Sci 38(6):1576–1584. https://doi.org/10.2135/cropsci1998.0011183X003800060028x

Liu Z, Li Y, Cao H, Ren D (2015) Comparative phospho-proteomics analysis of salt-responsive phosphoproteins regulated by the MKK9-MPK6 cascade in *Arabidopsis*. Plant Sci 241:138–150. https://doi.org/10.1016/j.plantsci.2015.10.005

Liu K, Newman M, McInroy JA, Hu C-H, Kloepper JW (2017a) Selection and assessment of plant growth-promoting rhizobacteria for biological control of multiple plant diseases. Phytopathology 107(8):928–936. https://doi.org/10.1094/PHYTO-02-17-0051-R

Liu S, Hao H, Lu X et al (2017b) Transcriptome profiling of genes involved in induced systemic salt tolerance conferred by *Bacillus amyloliquefaciens* FZB42 in *Arabidopsis thaliana*. Sci Rep 7(1):10795. https://doi.org/10.1038/s41598-017-11308-8

Liu J, Tang L, Gao H, Zhang M, Guo C (2019) Enhancement of alfalfa yield and quality by plant growth-promoting rhizobacteria under saline-alkali conditions. J Sci Food Agric 99(1):281–289. https://doi.org/10.1002/jsfa.9185

Liu C, Song Y, Dong X, Wang X, Ma X, Zhao G, Zang S (2021) Soil enzyme activities and their relationships with soil C, N, and P in peatlands from different types of permafrost regions, Northeast China. Front Environ Sci 9:143. https://doi.org/10.3389/fenvs.2021.670769

Long HH, Schmidt DD, Baldwin IT (2008) Native bacterial endophytes promote host growth in a species-specific manner; phytohormone manipulations do not result in common growth responses. PLoS One 3(7):e2702. https://doi.org/10.1371/journal.pone.0002702

Lowe R, Shirley N, Bleackley M, Dolan S, Shafee T (2017) Transcriptomics technologies. PLoS Comput Biol 13(5):e1005457. https://doi.org/10.1371/journal.pcbi.1005457

Luhua S, Hegie A, Suzuki N et al (2013) Linking genes of unknown function with abiotic stress responses by high-throughput phenotype screening. Physiol Plant 148(3):322–333. https://doi.org/10.1111/ppl.12013

Ma Y, Qin F, Tran L-SP (2012) Contribution of genomics to gene discovery in plant abiotic stress responses. Mol Plant 5(6):1176–1178. https://doi.org/10.1093/mp/sss085

Ma Y, Rajkumar M, Moreno A, Zhang C, Freitas H (2017) Serpentine endophytic bacterium *Pseudomonas azotoformans* ASS1 accelerates phytoremediation of soil metals under drought stress. Chemosphere 185:75–85. https://doi.org/10.1016/j.chemosphere.2017.06.135

Ma H, Wei M, Wang Z, Hou S, Li X, Xu H (2020) Bioremediation of cadmium polluted soil using a novel cadmium immobilizing plant growth promotion strain *Bacillus* sp. TZ5 loaded on biochar. J Hazard Mater 388:122065. https://doi.org/10.1016/j.jhazmat.2020.122065

Mącik M, Gryta A, Frąc M (2020) Biofertilizers in agriculture: an overview on concepts, strategies and effects on soil microorganisms. Adv Agron 162:31–87. https://doi.org/10.1016/bs.agron.2020.02.001

Mackey KR, Paytan A, Grossman AR, Bailey S (2008) A photosynthetic strategy for coping in a high-light, low-nutrient environment. Limnol Oceanogr 53(3):900–913. https://doi.org/10.4319/lo.2008.53.3.0900

Madhaiyan M, Poonguzhali S, Sa T (2007) Metal tolerating methylotrophic bacteria reduces nickel and cadmium toxicity and promotes plant growth of tomato (*Lycopersicon esculentum* L.). Chemosphere 69(2):220–228. https://doi.org/10.1016/j.chemosphere.2007.04.017

Mahmoud OMB, Hidri R, Talbi-Zribi O, Taamalli W, Abdelly C, Djébali N (2020) Auxin and proline producing rhizobacteria mitigate salt-induced growth inhibition of barley plants by enhancing water and nutrient status. S Afr J Bot 128:209–217. https://doi.org/10.1016/j.sajb.2019.10.023

Mahmud JA, Borhannuddin Bhuyan MHM, Nahar K, Parvin K, Hasanuzzaman M (2020) Response and tolerance of Fabaceae plants to metal/metalloid toxicity. In: Hasanuzzaman M, Araújo S, Gill SS (eds) The plant family Fabaceae: biology and physiological responses to environmental stresses. Springer Singapore, Singapore, pp 435–482

Mahmud-Ur-Rahman NIB, Mahmud NU, Sarker A, Hoque MN, Islam T (2022) A highly salt-tolerant bacterium *Brevibacterium sediminis* promotes the growth of rice (*Oryza sativa* L.) seedlings. Stresses 2(3):275–289. https://doi.org/10.3390/stresses2030020

Majeed A, Muhammad Z, Ahmad H (2018) Plant growth promoting bacteria: role in soil improvement, abiotic and biotic stress management of crops. Plant Cell Rep 37:1599–1609. https://doi.org/10.1007/s00299-018-2341-2

Mallick I, Bhattacharyya C, Mukherji S, Dey D, Sarkar SC, Mukhopadhyay UK, Ghosh A (2018) Effective rhizoinoculation and biofilm formation by arsenic immobilizing halophilic plant growth promoting bacteria (PGPB) isolated from mangrove rhizosphere: a step towards arsenic rhizoremediation. Sci Total Environ 610–611:1239–1250. https://doi.org/10.1016/j.scitotenv.2017.07.234

Margesin R, Schinner F (2001) Potential of halotolerant and halophilic microorganisms for biotechnology. Extremophiles 5(2):73–83. https://doi.org/10.1007/s007920100184

María José HM (2007) Abscisic acid determines arbuscule development and functionality in the tomato arbuscular mycorrhiza. New Phytologist 175(3):554–564. v.2175 no.2003. https://doi.org/10.1111/j.1469-8137.2007.02107.x

Martin CA, Stutz JC (2004) Interactive effects of temperature and arbuscular mycorrhizal fungi on growth, P uptake and root respiration of *Capsicum annuum* L. Mycorrhiza 14(4):241–244. https://doi.org/10.1007/s00572-003-0261-6

Martin-Tanguy J (2001) Metabolism and function of polyamines in plants: recent development (new approaches). Plant Growth Regul 34:135–148. https://doi.org/10.1023/A:1013343106574

Mathesius U (2009) Comparative proteomic studies of root–microbe interactions. J Proteome 72(3):353–366. https://doi.org/10.1016/j.jprot.2008.12.006

Mazdiyasni O, AghaKouchak A (2015) Substantial increase in concurrent droughts and heatwaves in the United States. Proceedings of the National Academy of Sciences 112 (37):11484–11489. https://doi.org/10.1073/pnas.1422945112

Meena KK, Sorty AM, Bitla UM et al (2017) Abiotic stress responses and microbe-mediated mitigation in plants: the omics strategies. Front Plant Sci 8. https://doi.org/10.3389/fpls.2017.00172

Mehta A, Brasileiro ACM, Souza DSL et al (2008) Plant–pathogen interactions: what is proteomics telling us? FEBS J 275(15):3731–3746. https://doi.org/10.1111/j.1742-4658.2008.06528.x

Mesa-Marín J, Pérez-Romero JA, Mateos-Naranjo E, Bernabeu-Meana M, Pajuelo E, Rodríguez-Llorente ID, Redondo-Gómez S (2019) Effect of plant growth-promoting rhizobacteria on *Salicornia ramosissima* seed germination under salinity, CO_2 and temperature stress. Agronomy 9(10):655. https://doi.org/10.3390/agronomy9100655

Miché L, Battistoni F, Gemmer S, Belghazi M, Reinhold-Hurek B (2006) Upregulation of jasmonate-inducible defense proteins and differential colonization of roots of *Oryza sativa* cultivars with the endophyte *Azoarcus* sp. Mol Plant-Microbe Interact 19(5):502–511. https://doi.org/10.1094/mpmi-19-0502

Miller R, Jastrow J (2000) Mycorrhizal fungi influence soil structure. Arbuscular mycorrhizas: molecular biology and physiology. Kluwer Academic Press, Dordrecht, Netherlands

Miller G, Suzuki N, Ciftci-Yilmaz S, Mittler R (2010) Reactive oxygen species homeostasis and signalling during drought and salinity stresses. Plant Cell Environ 33(4):453–467. https://doi.org/10.1094/MPMI-19-0502

Miller TE, Beneyton T, Schwander T et al (2020) Light-powered CO_2 fixation in a chloroplast mimic with natural and synthetic parts. Science 368(6491):649–654. https://doi.org/10.1126/science.aaz6802

Mishra PK, Bisht SC, Ruwari P et al (2011) Alleviation of cold stress in inoculated wheat (*Triticum aestivum* L.) seedlings with psychrotolerant Pseudomonads from NW Himalayas. Arch Microbiol 193(7):497–513. https://doi.org/10.1007/s00203-011-0693-x

Mohite B (2013) Isolation and characterization of indole acetic acid (IAA) producing bacteria from rhizospheric soil and its effect on plant growth. J Soil Sci Plant Nutr 13(3):638–649. https://doi.org/10.4067/S0718-95162013005000051

Morcillo RJ, Manzanera M (2021) The effects of plant-associated bacterial exopolysaccharides on plant abiotic stress tolerance. Metabolites 11(6):337. https://doi.org/10.3390/metabo11060337

Morcillo RJ, Vílchez JI, Zhang S et al (2021) Plant transcriptome reprograming and bacterial extracellular metabolites underlying tomato drought resistance triggered by a beneficial soil bacteria. Metabolites 11(6):369. https://doi.org/10.3390/metabo11060369

Moreno-Galván AE, Cortés-Patiño S, Romero-Perdomo F, Uribe-Vélez D, Bashan Y, Bonilla RR (2020) Proline accumulation and glutathione reductase activity induced by drought-tolerant rhizobacteria as potential mechanisms to alleviate drought stress in Guinea grass. Appl Soil Ecol 147:103367. https://doi.org/10.1016/j.apsoil.2019.103367

Moriondo M, Giannakopoulos C, Bindi M (2011) Climate change impact assessment: the role of climate extremes in crop yield simulation. Clim Chang 104(3–4):679–701. https://doi.org/10.1007/s10584-010-9871-0

Mukhtar T, Rehman S, Smith D et al (2020) Mitigation of heat stress in *Solanum lycopersicum* L. by ACC-deaminase and exopolysaccharide producing *Bacillus cereus*: effects on biochemical profiling. Sustainability 12(6):2159. https://doi.org/10.3390/su12062159

Mumtaz MZ, Ahmad M, Jamil M, Hussain T (2017) Zinc solubilizing *Bacillus* spp. potential candidates for biofortification in maize. Microbiol Res 202:51–60. https://doi.org/10.1016/j.micres.2017.06.001

Munns R (2002) Comparative physiology of salt and water stress. Plant Cell Environ 25(2):239–250. https://doi.org/10.1046/j.0016-8025.2001.00808.x

Nascimento FX, Hernández AG, Glick BR, Rossi MJ (2020) Plant growth-promoting activities and genomic analysis of the stress-resistant *Bacillus megaterium* STB1, a bacterium of agricultural and biotechnological interest. Biotechnol Rep 25:e00406. https://doi.org/10.1016/j.btre.2019.e00406

Naseem H, Bano A (2014) Role of plant growth-promoting rhizobacteria and their exopolysaccharide in drought tolerance of maize. J Plant Interact 9(1):689–701. https://doi.org/10.1080/17429145.2014.902125

Nautiyal CS, Srivastava S, Chauhan PS, Seem K, Mishra A, Sopory SK (2013) Plant growth-promoting bacteria *Bacillus amyloliquefaciens* NBRISN13 modulates gene expression profile of leaf and rhizosphere community in rice during salt stress. Plant Physiol Biochem 66:1–9. https://doi.org/10.1016/j.plaphy.2013.01.020

Nigam B, Dubey RS, Rathore D (2022) Protective role of exogenously supplied salicylic acid and PGPB (*Stenotrophomonas* sp.) on spinach and soybean cultivars grown under salt stress. Sci Hortic 293:110654. https://doi.org/10.1016/j.scienta.2021.110654

Nikolic B, Schwab H, Sessitsch A (2011) Metagenomic analysis of the 1-aminocyclopropane-1-carboxylate deaminase gene (*acdS*) operon of an uncultured bacterial endophyte colonizing *Solanum tuberosum* L. Arch Microbiol 193(9):665–676. https://doi.org/10.1007/s00203-011-0703-z

Niu X, Song L, Xiao Y, Ge W (2018) Drought-tolerant plant growth-promoting rhizobacteria associated with foxtail millet in a semi-arid agroecosystem and their potential in alleviating drought stress. Front Microbiol 8. https://doi.org/10.3389/fmicb.2017.02580

Nogia P, Sidhu GK, Mehrotra R, Mehrotra S (2016) Capturing atmospheric carbon: biological and nonbiological methods. Int J Low-Carbon Technol 11(2):266–274. https://doi.org/10.1093/ijlct/ctt077

Nookongbut P, Kantachote D, Khuong NQ, Tantirungkij M (2020) The biocontrol potential of acid-resistant *Rhodopseudomonas palustris* KTSSR54 and its exopolymeric substances against rice fungal pathogens to enhance rice growth and yield. Biol Control 150:104354. https://doi.org/10.1016/j.biocontrol.2020.104354

Numan M, Bashir S, Khan Y et al (2018) Plant growth promoting bacteria as an alternative strategy for salt tolerance in plants: a review. Microbiol Res 209:21–32. https://doi.org/10.1016/j.micres.2018.02.003

Oades JM (1993) The role of biology in the formation, stabilization and degradation of soil structure. In: Soil structure/soil biota interrelationships. Elsevier, pp 377–400

Orozco-Mosqueda MC, Glick BR, Santoyo G (2020) ACC deaminase in plant growth-promoting bacteria (PGPB): an efficient mechanism to counter salt stress in crops. Microbiol Res 235:126439. https://doi.org/10.1016/j.micres.2020.126439

Pachauri RK, Allen MR, Barros VR et al. (2014) Climate change 2014: synthesis report. Contribution of Working Groups I, II and III to the fifth assessment report of the Intergovernmental Panel on Climate Change: Ipcc

Pan W, Lu Q, Xu Q-R et al (2019) Abscisic acid-generating bacteria can reduce Cd concentration in pakchoi grown in Cd-contaminated soil. Ecotoxicol Environ Saf 177:100–107. https://doi.org/10.1016/j.ecoenv.2019.04.010

Panhwar QA, Naher UA, Radziah O, Shamshuddin J, Razi IM (2015) Eliminating aluminum toxicity in an acid sulfate soil for rice cultivation using plant growth promoting bacteria. Molecules 20(3):3628–3646. https://doi.org/10.3390/molecules20033628

Park Y-G, Mun B-G, Kang S-M et al (2017) *Bacillus aryabhattai* SRB02 tolerates oxidative and nitrosative stress and promotes the growth of soybean by modulating the production of phytohormones. PLoS One 12(3):e0173203. https://doi.org/10.1371/journal.pone.0173203

Parmar P, Dave B, Sudhir A, Panchal K, Subramanian R (2013) Physiological, biochemical and molecular response of plants against heavy metals stress. Int J Curr Res 5(1):80–89

Passeri V, Koes R, Quattrocchio FM (2016) New challenges for the design of high value plant products: stabilization of anthocyanins in plant vacuoles. Front Plant Sci 7:153. https://doi.org/10.3389/fpls.2016.00153

Patel P, Trivedi G, Saraf M (2018) Iron biofortification in mungbean using siderophore producing plant growth promoting bacteria. Environ Sustain 1:357–365. https://doi.org/10.1007/s42398-018-00031-3

Pattanagul W, Thitisaksakul M (2008) Effect of salinity stress on growth and carbohydrate metabolism in three rice (*Oryza sativa* L.) cultivars differing in salinity tolerance. Indian J Exp Biol 46(10):736–742

Paul ND, Gwynn-Jones D (2003) Ecological roles of solar UV radiation: towards an integrated approach. Trends Ecol Evol 18(1):48–55. https://doi.org/10.1016/S0169-5347(02)00014-9

Petrillo C, Vitale E, Ambrosino P, Arena C, Isticato R (2022) Plant growth-promoting bacterial consortia as a strategy to alleviate drought stress in *Spinacia oleracea*. Microorganisms 10(9):1798. https://doi.org/10.3390/microorganisms10091798

Petrov V, Hille J, Mueller-Roeber B, Gechev TS (2015) ROS-mediated abiotic stress-induced programmed cell death in plants. Front Plant Sci 6. https://doi.org/10.3389/fpls.2015.00069

Pham DN, Mai DHA, Nguyen AD, Chau THT, Lee EY (2022) Development of an engineered methanotroph-based microbial platform for biocatalytic conversion of methane to phytohormone for sustainable agriculture. Chem Eng J 429:132522. https://doi.org/10.1016/j.cej.2021.132522

Pinu FR, Beale DJ, Paten AM, Kouremenos K, Swarup S, Schirra HJ, Wishart D (2019) Systems biology and multi-omics integration: viewpoints from the metabolomics research community. Metabolites 9(4):76. https://doi.org/10.3390/metabo9040076

Qiao J-Q, Wu H-J, Huo R, Gao X-W, Borriss R (2014) Stimulation of plant growth and biocontrol by *Bacillus amyloliquefaciens* subsp. *plantarum* FZB42 engineered for improved action. Chem Biol Technol Agric 1(1):12. https://doi.org/10.1186/s40538-014-0012-2

Qin Y, Druzhinina IS, Pan X, Yuan Z (2016) Microbially mediated plant salt tolerance and microbiome-based solutions for saline agriculture. Biotechnol Adv 34(7):1245–1259. https://doi.org/10.1016/j.biotechadv.2016.08.005

Quint M, Delker C, Franklin KA, Wigge PA, Halliday KJ, Van Zanten M (2016) Molecular and genetic control of plant thermomorphogenesis. Nat Plants 2(1):1–9. https://doi.org/10.1038/nplants.2015.190

Qurashi AW, Sabri AN (2012) Bacterial exopolysaccharide and biofilm formation stimulate chickpea growth and soil aggregation under salt stress. Braz J Microbiol 43(3):1183–1191. https://doi.org/10.1590/s1517-83822012000300046

Rahim S, Guan Puay T (2020) The impact of climate on economic growth in Malaysia. J Adv Res Bus Manag Stud 6(2):108–119

Rajkumar M, Ma Y, Freitas H (2008) Characterization of metal-resistant plant-growth promoting *Bacillus weihenstephanensis* isolated from serpentine soil in Portugal. J Basic Microbiol 48(6):500–508. https://doi.org/10.1002/jobm.200800073

Rajkumar M, Ae N, Prasad MNV, Freitas H (2010) Potential of siderophore-producing bacteria for improving heavy metal phytoextraction. Trends Biotechnol 28(3):142–149. https://doi.org/10.1016/j.tibtech.2009.12.002

Rashid MI, Mujawar LH, Shahzad T, Almeelbi T, Ismail IM, Oves M (2016) Bacteria and fungi can contribute to nutrients bioavailability and aggregate formation in degraded soils. Microbiol Res 183:26–41. https://doi.org/10.1016/j.micres.2015.11.007

Rastegari AA, Yadav AN, Yadav N (2020) New and future developments in microbial biotechnology and bioengineering: trends of microbial biotechnology for sustainable agriculture and biomedicine systems: diversity and functional perspectives. Elsevier

Rastogi A, Tripathi DK, Yadav S et al (2019) Application of silicon nanoparticles in agriculture. 3 Biotech 9:1–11. https://doi.org/10.1007/s13205-019-1626-7

Reddy CA, Saravanan RS (2013) Chapter Three – Polymicrobial multi-functional approach for enhancement of crop productivity. In: Sariaslani S, Gadd GM (eds) Advances in applied microbiology, vol 82. Academic Press, pp 53–113

Redillas MC, Park S-H, Lee JW et al (2012) Accumulation of trehalose increases soluble sugar contents in rice plants conferring tolerance to drought and salt stress. Plant Biotechnol Rep 6:89–96. https://doi.org/10.1007/s11816-011-0210-3

Ren X-M, Guo S-J, Tian W et al (2019) Effects of plant growth-promoting bacteria (PGPB) inoculation on the growth, antioxidant activity, Cu uptake, and bacterial community structure of rape (*Brassica napus* L.) grown in Cu-contaminated agricultural soil. Front Microbiol 10. https://doi.org/10.3389/fmicb.2019.01455

Rodrigues MI, Bravo JP, Sassaki FT, Severino FE, Maia IG (2013) The tonoplast intrinsic aquaporin (TIP) subfamily of *Eucalyptus grandis*: characterization of *EgTIP2*, a root-specific and osmotic stress-responsive gene. Plant Sci 213:106–113. https://doi.org/10.1016/j.plantsci.2013.09.005

Rodríguez H, Fraga R, Gonzalez T, Bashan Y (2006) Genetics of phosphate solubilization and its potential applications for improving plant growth-promoting bacteria. Plant Soil 287(1):15–21. https://doi.org/10.1007/s11104-006-9056-9

Rodríguez-Salazar J, Suárez R, Caballero-Mellado J, Iturriaga G (2009) Trehalose accumulation in *Azospirillum brasilense* improves drought tolerance and biomass in maize plants. FEMS Microbiol Lett 296(1):52–59. https://doi.org/10.1111/j.1574-6968.2009.01614.x

Roesti D, Gaur R, Johri B, Imfeld G, Sharma S, Kawaljeet K, Aragno M (2006) Plant growth stage, fertiliser management and bio-inoculation of arbuscular mycorrhizal fungi and plant growth promoting rhizobacteria affect the rhizobacterial community structure in rain-fed wheat fields. Soil Biol Biochem 38(5):1111–1120. https://doi.org/10.1016/j.soilbio.2005.09.010

Rojas-Tapias D, Moreno-Galván A, Pardo-Díaz S, Obando M, Rivera D, Bonilla R (2012) Effect of inoculation with plant growth-promoting bacteria (PGPB) on amelioration of saline stress in maize (*Zea mays*). Appl Soil Ecol 61:264–272. https://doi.org/10.1016/j.apsoil.2012.01.006

Roriz M, Pereira SI, Castro PM, Carvalho SM, Vasconcelos MW (2023) Impact of soybean-associated plant growth-promoting bacteria on plant growth modulation under alkaline soil conditions. Heliyon 9(5):e14620. https://doi.org/10.1016/j.heliyon.2023.e14620

Rosenzweig C, Parry ML (1994) Potential impact of climate change on world food supply. Nature 367(6459):133–138. https://doi.org/10.1038/367133a0

Ryu C-M, Hu C-H, Locy RD, Kloepper JW (2005) Study of mechanisms for plant growth promotion elicited by rhizobacteria in *Arabidopsis thaliana*. Plant Soil 268(1):285–292. https://doi.org/10.1007/s11104-004-0301-9

Sabki MH, Ong PY, Ibrahim N, Lee CT, Klemeš JJ, Li C, Gao Y (2021) A review on abiotic stress tolerance and plant growth metabolite framework by plant growth-promoting bacteria for sustainable agriculture. Chem Eng Trans 83:367–372. https://doi.org/10.3303/CET2183062

Safronova VI, Piluzza G, Zinovkina NY, Kimeklis AK, Belimov AA, Bullitta S (2012) Relationships between pasture legumes, rhizobacteria and nodule bacteria in heavy metal polluted mine waste of SW Sardinia. Symbiosis 58(1):149–159. https://doi.org/10.1007/s13199-012-0207-x

Saikia J, Sarma RK, Dhandia R, Yadav A, Bharali R, Gupta VK, Saikia R (2018) Alleviation of drought stress in pulse crops with ACC deaminase producing rhizobacteria isolated from acidic soil of Northeast India. Sci Rep 8(1):3560. https://doi.org/10.1038/s41598-018-21921-w

Sala OE, Stuart Chapin F, Armesto JJ et al. (2000) Global biodiversity scenarios for the year 2100. science 287(5459):1770–1774. https://doi.org/10.1126/science.287.5459.1770

Sandhya V, Ali SZ (2015) The production of exopolysaccharide by *Pseudomonas putida* GAP-P45 under various abiotic stress conditions and its role in soil aggregation. Microbiology 84(4):512–519. https://doi.org/10.1134/S0026261715040153

Sandhya V, Sk ZA, Grover M, Reddy G, Venkateswarlu B (2009) Alleviation of drought stress effects in sunflower seedlings by the exopolysaccharides producing *Pseudomonas putida* strain GAP-P45. Biol Fertil Soils 46(1):17–26. https://doi.org/10.1007/s00374-009-0401-z

Saravanakumar D, Kavino M, Raguchander T, Subbian P, Samiyappan R (2011) Plant growth promoting bacteria enhance water stress resistance in green gram plants. Acta Physiol Plant 33(1):203–209. https://doi.org/10.1007/s11738-010-0539-1

Sarma RK, Saikia R (2014) Alleviation of drought stress in mung bean by strain *Pseudomonas aeruginosa* GGRJ21. Plant Soil 377(1):111–126. https://doi.org/10.1007/s11104-013-1981-9

Schwander T, Schada von Borzyskowski L, Burgener S, Cortina NS, Erb TJ (2016) A synthetic pathway for the fixation of carbon dioxide in vitro. Science 354(6314):900–904. https://doi.org/10.1126/science.aah5237

Selvakumar G, Mohan M, Kundu S, Gupta AD, Joshi P, Nazim S, Gupta HS (2008) Cold tolerance and plant growth promotion potential of *Serratia marcescens* strain SRM (MTCC 8708) isolated from flowers of summer squash (*Cucurbita pepo*). Lett Appl Microbiol 46(2):171–175. https://doi.org/10.1111/j.1472-765X.2007.02282.x

Sessitsch A, Hardoim P, Döring J et al. (2012) Functional characteristics of an endophyte community colonizing rice roots as revealed by metagenomic analysis. Mol Plant-Microbe Interact 25(1):28–36. https://doi.org/10.1094/mpmi-08-11-0204

Sgroy V, Cassán F, Masciarelli O, Del Papa MF, Lagares A, Luna V (2009) Isolation and characterization of endophytic plant growth-promoting (PGPB) or stress homeostasis-regulating (PSHB) bacteria associated to the halophyte *Prosopis strombulifera*. Appl Microbiol Biotechnol 85(2):371–381. https://doi.org/10.1007/s00253-009-2116-3

Shabaan M, Asghar HN, Zahir ZA, Zhang X, Sardar MF, Li H (2022) Salt-tolerant PGPR confer salt tolerance to maize through enhanced soil biological health, enzymatic activities, nutrient uptake and antioxidant defense. Front Microbiol 13. https://doi.org/10.3389/fmicb.2022.901865

Shahzad R, Waqas M, Khan AL et al (2016) Seed-borne endophytic *Bacillus amyloliquefaciens* RWL-1 produces gibberellins and regulates endogenous phytohormones of *Oryza sativa*. Plant Physiol Biochem 106:236–243. https://doi.org/10.1016/j.plaphy.2016.05.006

Shakirova FM, Avalbaev AM, Bezrukova MV et al (2012) Hormonal intermediates in the protective action of exogenous phytohormones in wheat plants under salinity. In: Phytohormones and abiotic stress tolerance in plants, pp 185–228

Sharma SB, Sayyed RZ, Trivedi MH, Gobi TA (2013) Phosphate solubilizing microbes: sustainable approach for managing phosphorus deficiency in agricultural soils. Springerplus 2(1):587. https://doi.org/10.1186/2193-1801-2-587

Sharma S, Gang S, Schumacher J, Buck M, Saraf M (2021) Genomic appraisal of *Klebsiella* PGPB isolated from soil to enhance the growth of barley. Genes & Genomics 43(8):869–883. https://doi.org/10.1007/s13258-021-01099-8

Shavrukov Y, Hirai Y (2016) Good and bad protons: genetic aspects of acidity stress responses in plants. J Exp Bot 67(1):15–30. https://doi.org/10.1093/jxb/erv437

Shen N, Wang T, Gan Q, Liu S, Wang L, Jin B (2022) Plant flavonoids: classification, distribution, biosynthesis, and antioxidant activity. Food Chem 383:132531. https://doi.org/10.1016/j.foodchem.2022.132531

Shi H, Chen L, Ye T, Liu X, Ding K, Chan Z (2014) Modulation of auxin content in *Arabidopsis* confers improved drought stress resistance. Plant Physiol Biochem 82:209–217. https://doi.org/10.1016/j.plaphy.2014.06.008

Shilev S (2020) Plant-growth-promoting bacteria mitigating soil salinity stress in plants. Appl Sci 10(20):7326. https://doi.org/10.3390/app10207326

Siddiqui MH, Al-Whaibi MH, Faisal M, Al Sahli AA (2014) Nano-silicon dioxide mitigates the adverse effects of salt stress on *Cucurbita pepo* L. Environ Toxicol Chem 33(11):2429–2437. https://doi.org/10.1002/etc.2697

Siebers N, Kruse J (2019) Short-term impacts of forest clear-cut on soil structure and consequences for organic matter composition and nutrient speciation: a case study. PLoS One 14(8):e0220476. https://doi.org/10.1371/journal.pone.0220476

Singh RP, Jha PN (2017) The PGPR *Stenotrophomonas maltophilia* SBP-9 augments resistance against biotic and abiotic stress in wheat plants. Front Microbiol 8:1945. https://doi.org/10.3389/fmicb.2017.01945

Singh HP, Singh TA (1993) The interaction of rock phosphate, *Bradyrhizobium*, vesicular-arbuscular mycorrhizae and phosphate-solubilizing microbes on soybean grown in a sub-Himalayan mollisol. Mycorrhiza 4(1):37–43. https://doi.org/10.1007/BF00203249

Singh JS, Abhilash P, Singh H, Singh RP, Singh D (2011) Genetically engineered bacteria: an emerging tool for environmental remediation and future research perspectives. Gene 480(1–2):1–9. https://doi.org/10.1016/j.gene.2011.03.001

Singh RP, Runthala A, Khan S, Jha PN (2017) Quantitative proteomics analysis reveals the tolerance of wheat to salt stress in response to *Enterobacter cloacae* SBP-8. PLoS One 12(9):e0183513. https://doi.org/10.1371/journal.pone.0183513

Singh A, Kumari R, Yadav AN, Mishra S, Sachan A, Sachan SG (2020) Tiny microbes, big yields: microorganisms for enhancing food crop production for sustainable development. In: New and future developments in microbial biotechnology and bioengineering. Elsevier, pp 1–15

Souza MST, de Baura VA, Santos SA et al (2017) *Azospirillum* spp. from native forage grasses in Brazilian Pantanal floodplain: biodiversity and plant growth promotion potential. World J Microbiol Biotechnol 33(4):81. https://doi.org/10.1007/s11274-017-2251-4

Suárez R, Wong A, Ramírez M et al (2008) Improvement of drought tolerance and grain yield in common bean by overexpressing trehalose-6-phosphate synthase in rhizobia. Mol Plant-Microbe Interact 21(7):958–966. https://doi.org/10.1094/mpmi-21-7-0958

Steudle E (1994) Water transport across roots. Plant Soil 167:79–90. https://doi.org/10.1007/BF01587602

Suhag M (2016) Potential of biofertilizers to replace chemical fertilizers. Int Adv Res J Sci Eng Technol 3(5):163–167. https://doi.org/10.17148/IARJSET.2016.3534

Szabados L, Savouré A (2010) Proline: a multifunctional amino acid. Trends Plant Sci 15(2):89–97. https://doi.org/10.1016/j.tplants.2009.11.009

Szczepanek M, Piotrowska-Długosz A, Konopka I (2021) Sustainable crop production protects the quality of soil and plant raw materials. Agronomy 11(6):1178. https://doi.org/10.3390/agronomy11061178

Talaat NB (2018) Exploring halotolerant rhizomicrobes as a pool of potent genes for engineering salt stress tolerance in crops. In: Kumar V, Wani SH, Suprasanna P, Tran L-SP (eds) Salinity responses and tolerance in plants, Volume 2: exploring RNAi, genome editing and systems biology. Springer International Publishing, Cham, pp 49–76

Tian H, Lu C, Ciais P et al. (2016) The terrestrial biosphere as a net source of greenhouse gases to the atmosphere. Nature 531(7593):225–228. https://doi.org/10.1038/nature16946

Tilgam J, Kumar K, Jayaswal D, Choudhury S, Kumar A, Jayaswall K, Saxena AK (2021) Success of microbial genes based transgenic crops: Bt and beyond Bt. Mol Biol Rep 48(12):8111–8122. https://doi.org/10.1007/s11033-021-06760-9

Timmusk S, Behers L, Muthoni J, Muraya A, Aronsson A-C (2017) Perspectives and challenges of microbial application for crop improvement. Front Plant Sci 8. https://doi.org/10.3389/fpls.2017.00049

Tirry N, Tahri Joutey N, Sayel H, Kouchou A, Bahafid W, Asri M, El Ghachtouli N (2018) Screening of plant growth promoting traits in heavy metals resistant bacteria: prospects in phytoremediation. J Genet Eng Biotechnol 16(2):613–619. https://doi.org/10.1016/j.jgeb.2018.06.004

Tisdall J, Oades J (1980) The effect of crop rotation on aggregation in a red-brown earth. Soil Res 18(4):423–433. https://doi.org/10.1071/SR9800423

Tisdall JM, Oades JM (1982) Organic matter and water-stable aggregates in soils. J Soil Sci 33(2):141–163. https://doi.org/10.1111/j.1365-2389.1982.tb01755.x

Tiwari S, Lata C, Chauhan PS, Nautiyal CS (2016) *Pseudomonas putida* attunes morphophysiological, biochemical and molecular responses in *Cicer arietinum* L. during drought stress and recovery. Plant Physiol Biochem 99:108–117. https://doi.org/10.1016/j.plaphy.2015.11.001

Tiwari S, Prasad V, Chauhan PS, Lata C (2017) *Bacillus amyloliquefaciens* confers tolerance to various abiotic stresses and modulates plant response to phytohormones through osmoprotection and gene expression regulation in rice. Front Plant Sci 8:1510. https://doi.org/10.3389/fpls.2017.01510

Toral-Juárez MA, Avila RT, Cardoso AA et al (2021) Drought-tolerant coffee plants display increased tolerance to waterlogging and post-waterlogging reoxygenation. Environ Exp Bot 182:104311. https://doi.org/10.1016/j.envexpbot.2020.104311

Traill LW, Lim MLM, Sodhi NS, Bradshaw CJA (2010) Mechanisms driving change: altered species interactions and ecosystem function through global warming. J Anim Ecol 79(5):937–947. https://doi.org/10.1111/j.1365-2656.2010.01695.x

Tripathi M, Munot HP, Shouche Y, Meyer JM, Goel R (2005) Isolation and functional characterization of siderophore-producing lead- and Cadmium-resistant *Pseudomonas putida* KNP9. Curr Microbiol 50(5):233–237. https://doi.org/10.1007/s00284-004-4459-4

Trivedi P, Leach JE, Tringe SG, Sa T, Singh BK (2020) Plant-microbiome interactions: from community assembly to plant health. Nat Rev Microbiol 18(11):607–621. https://doi.org/10.1038/s41579-020-0412-1

Turan M, Gulluce M, von Wirén N, Sahin F (2012) Yield promotion and phosphorus solubilization by plant growth-promoting rhizobacteria in extensive wheat production in Turkey. J Plant Nutr Soil Sci 175(6):818–826. https://doi.org/10.1002/jpln.201200054

Turner TR, Ramakrishnan K, Walshaw J et al (2013) Comparative metatranscriptomics reveals kingdom level changes in the rhizosphere microbiome of plants. ISME J 7(12):2248–2258. https://doi.org/10.1038/ismej.2013.119

Ullah A, Heng S, Munis MFH, Fahad S, Yang X (2015) Phytoremediation of heavy metals assisted by plant growth promoting (PGP) bacteria: a review. Environ Exp Bot 117:28–40. https://doi.org/10.1016/j.envexpbot.2015.05.001

Urgun-Demirtas M, Stark B, Pagilla K (2006) Use of genetically engineered microorganisms (GEMs) for the bioremediation of contaminants. Crit Rev Biotechnol 26(3):145–164. https://doi.org/10.1080/07388550600842794

Vargas L, Santa Brígida AB, Mota Filho JP et al (2014) Drought tolerance conferred to sugarcane by association with *Gluconacetobacter diazotrophicus*: a transcriptomic view of hormone pathways. PLoS One 9(12):e114744. https://doi.org/10.1371/journal.pone.0114744

Varma A, Choudhary DK (2019) Mycorrhizosphere and pedogenesis. Springer

Vega-Celedón P, Bravo G, Velásquez A et al (2021) Microbial diversity of psychrotolerant bacteria isolated from wild flora of andes mountains and patagonia of chile towards the selection of plant growth-promoting bacterial consortia to alleviate cold stress in plants. Microorganisms 9(3):538. https://doi.org/10.3390/microorganisms9030538

Vejan P, Abdullah R, Khadiran T, Ismail S, Nasrulhaq Boyce A (2016) Role of plant growth promoting rhizobacteria in agricultural sustainability—a review. Molecules 21(5):573. https://doi.org/10.3390/molecules21050573

Velten S, Leventon J, Jager N, Newig J (2015) What is sustainable agriculture? A systematic review. Sustainability 7(6):7833–7865. https://doi.org/10.3390/su7067833

Vergani L, Mapelli F, Suman J, Cajthaml T, Uhlik O, Borin S (2019) Novel PCB-degrading *Rhodococcus* strains able to promote plant growth for assisted rhizoremediation of historically polluted soils. PLoS One 14(8):e0221253. https://doi.org/10.1371/journal.pone.0221253

Vílchez JI, García-Fontana C, Román-Naranjo D, González-López J, Manzanera M (2016) Plant drought tolerance enhancement by trehalose production of desiccation-tolerant microorganisms. Front Microbiol 7. https://doi.org/10.3389/fmicb.2016.01577

Viscardi S, Ventorino V, Duran P, Maggio A, De Pascale S, Mora M, Pepe O (2016) Assessment of plant growth promoting activities and abiotic stress tolerance of *Azotobacter chroococcum* strains for a potential use in sustainable agriculture. J Soil Sci Plant Nutr 16(3):848–863. https://doi.org/10.4067/S0718-95162016005000060

Voigt CA (2006) Genetic parts to program bacteria. Curr Opin Biotechnol 17(5):548–557. https://doi.org/10.1016/j.copbio.2006.09.001

Volis S, Ormanbekova D, Yermekbayev K, Song M, Shulgina I (2015) Multi-approaches analysis reveals local adaptation in the emmer wheat (Triticum dicoccoides) at macro-but not micro-geographical scale. PLoS One 10(3):e0121153. https://doi.org/10.1371/journal.pone.0121153

Walters DR, Ratsep J, Havis ND (2013) Controlling crop diseases using induced resistance: challenges for the future. J Exp Bot 64(5):1263–1280. https://doi.org/10.1093/jxb/ert026

Wang X, Dong Z, Zhang J, Liu L (2004) Modern dust storms in China: an overview. J Arid Environ 58(4):559–574. https://doi.org/10.1016/j.jaridenv.2003.11.009

Wang Z, Gerstein M, Snyder M (2009) RNA-Seq: a revolutionary tool for transcriptomics. Nat Rev Genet 10(1):57–63. https://doi.org/10.1038/nrg2484

Wang J, Zhang Y, Li Y et al (2015) Endophytic microbes *Bacillus* sp. LZR216-regulated root development is dependent on polar auxin transport in *Arabidopsis* seedlings. Plant Cell Rep 34(6):1075–1087. https://doi.org/10.1007/s00299-015-1766-0

Wang B, Wang Q, Liu W et al (2017) Biosurfactant-producing microorganism *Pseudomonas* sp. SB assists the phytoremediation of DDT-contaminated soil by two grass species. Chemosphere 182:137–142. https://doi.org/10.1016/j.chemosphere.2017.04.123

Wang D-C, Jiang C-H, Zhang L-N, Chen L, Zhang X-Y, Guo J-H (2019a) Biofilms positively contribute to *Bacillus amyloliquefaciens* 54-induced drought tolerance in tomato plants. Int J Mol Sci 20(24):6271. https://doi.org/10.3390/ijms20246271

Wang J, Wang W, Wang H et al (2019b) Improvement of stress tolerance and riboflavin production of *Bacillus subtilis* by introduction of heat shock proteins from thermophilic bacillus strains. Appl Microbiol Biotechnol 103(11):4455–4465. https://doi.org/10.1007/s00253-019-09788-x

Wang F, Wei Y, Yan T et al (2022) *Sphingomonas* sp. Hbc-6 alters physiological metabolism and recruits beneficial rhizosphere bacteria to improve plant growth and drought tolerance. Front Plant Sci 13. https://doi.org/10.3389/fpls.2022.1002772

Wani PA, Khan MS (2010) *Bacillus* species enhance growth parameters of chickpea (*Cicer arietinum* L.) in chromium stressed soils. Food Chem Toxicol 48(11):3262–3267. https://doi.org/10.1016/j.fct.2010.08.035

Wetzler DE, Fuchs Wightman F, Bucci HA, Rinaldi J, Caramelo JJ, Iusem ND, Ricardi MM (2018) Conformational plasticity of the intrinsically disordered protein ASR1 modulates its function as a drought stress-responsive gene. PLoS One 13(8):e0202808. https://doi.org/10.1371/journal.pone.0202808

Win KT, Tanaka F, Okazaki K, Ohwaki Y (2018) The ACC deaminase expressing endophyte *Pseudomonas* spp. enhances NaCl stress tolerance by reducing stress-related ethylene production, resulting in improved growth, photosynthetic performance, and ionic balance in tomato plants. Plant Physiol Biochem 127:599–607. https://doi.org/10.1016/j.plaphy.2018.04.038

Wu Y, Ma L, Liu Q et al (2020) The plant-growth promoting bacteria promote cadmium uptake by inducing a hormonal crosstalk and lateral root formation in a hyperaccumulator plant Sedum alfredii. J Hazard Mater 395:122661. https://doi.org/10.1016/j.jhazmdat.2020.122661

Wungrampha S, Joshi R, Singla-Pareek S, Pareek A (2018) Photosynthesis and salinity: are these mutually exclusive? Photosynthetica 56:366–381. https://doi.org/10.1007/s11099-017-0763-7

Xiong L, Zhu J-K (2001) Abiotic stress signal transduction in plants: molecular and genetic perspectives. Physiol Plant 112(2):152–166. https://doi.org/10.1034/j.1399-3054.2001.1120202.x

Xu J, Li X-L, Luo L (2012) Effects of engineered *Sinorhizobium meliloti* on cytokinin synthesis and tolerance of Alfalfa to extreme drought stress. Appl Environ Microbiol 78(22):8056–8061. https://doi.org/10.1128/AEM.01276-12

Yadav SK (2010) Cold stress tolerance mechanisms in plants. a review. Agron Sustain Dev 30(3):515–527. https://doi.org/10.1051/agro/2009050

Yaghoubi Khanghahi M, Leoni B, Crecchio C (2021) Photosynthetic responses of durum wheat to chemical/microbiological fertilization management under salt and drought stresses. Acta Physiol Plant 43(8):123. https://doi.org/10.1007/s11738-021-03289-z

Yaghoubi Khanghahi M, AbdElgawad H, Verbruggen E, Korany SM, Alsherif EA, Beemster GTS, Crecchio C (2022) Biofertilisation with a consortium of growth-promoting bacterial strains improves the nutritional status of wheat grain under control, drought, and salinity stress conditions. Physiol Plant 174(6):e13800. https://doi.org/10.1111/ppl.13800

Yan Y, Yang J, Dou Y et al (2008) Nitrogen fixation island and rhizosphere competence traits in the genome of root-associated *Pseudomonas stutzeri* A1501. Proc Natl Acad Sci 105(21):7564–7569. https://doi.org/10.1073/pnas.0801093105

Yancheshmeh JB (2011) Evaluation of inoculation of plant growth-promoting rhizobacteria on cadmium and lead uptake by canola and barley. Afr J Microbiol Res 5(14):1747–1754. https://doi.org/10.5897/AJMR10.625

You C, Zhou F (1989) Non-nodular endorhizospheric nitrogen fixation in wetland rice. Can J Microbiol 35(3):403–408. https://doi.org/10.1139/m89-062

Zhang J-T, Mu C-S (2009) Effects of saline and alkaline stresses on the germination, growth, photosynthesis, ionic balance and anti-oxidant system in an alkali-tolerant leguminous forage *Lathyrus quinquenervius*. Soil Sci Plant Nutr 55(5):685–697. https://doi.org/10.1111/j.1747-0765.2009.00411.x

Zhang J, Jia W, Yang J, Ismail AM (2006) Role of ABA in integrating plant responses to drought and salt stresses. Field Crop Res 97(1):111–119. https://doi.org/10.1016/j.fcr.2005.08.018

Zhang DD, Lee HF, Wang C, Li B, Zhang J, Pei Q, Chen J (2011) Climate change and large-scale human population collapses in the pre-industrial era. Global Ecol Biogeogr 20(4):520–531. https://doi.org/10.1111/j.1466-8238.2010.00625.x

Zhang T, Yan Y, He S et al. (2012) Involvement of the ammonium transporter AmtB in nitrogenase regulation and ammonium excretion in Pseudomonas stutzeri A1501. Res Microbiol 163(5):332–339. https://doi.org/10.1016/j.resmic.2012.05.002

Zhang Y-Y, Wu W, Liu H (2019) Factors affecting variations of soil pH in different horizons in hilly regions. PLoS One 14(6):e0218563. https://doi.org/10.1371/journal.pone.0218563

Zhou C, Guo J, Zhu L et al (2016) *Paenibacillus polymyxa* BFKC01 enhances plant iron absorption via improved root systems and activated iron acquisition mechanisms. Plant Physiol Biochem 105:162–173. https://doi.org/10.1016/j.plaphy.2016.04.025

Zhou L, Zhou J, Xiong Y, Liu C, Wang J, Wang G, Cai Y (2018) Overexpression of a maize plasma membrane intrinsic protein ZmPIP1;1 confers drought and salt tolerance in *Arabidopsis*. PLoS One 13(6):e0198639. https://doi.org/10.1371/journal.pone.0198639

Climate Change and Soil Ecosystems: Impacts on Microbial Communities, Pathogen Dynamics, and Agricultural Sustainability

8

Abhishek Sharma, F. Seyma Gokdemir, Gudhal Patanjali, and Fusun Eyidogan

Abstract

Increasing global temperatures, altered precipitation patterns, and extreme weather events profoundly impact soil microbial communities and pathogen dynamics. The current chapter examines the intricate relationships between climate change and soil ecosystems, particularly on pathogen biodiversity and ecosystem functions. Research demonstrates that climate warming leads to decreased microbial diversity and altered soil biogeochemical conditions, significantly impacting ecosystem multifunctionality. Through analysis of temperature, precipitation, extreme weather events, CO_2 levels, and soil moisture variations, this work explores their influence on pathogen behavior, distribution, and virulence. Additionally, it addresses agricultural sustainability and food security implications, presenting adaptation strategies and management approaches for mitigating climate change impacts on soil ecosystems. The chapter synthesizes current research on soil-borne pathogen responses to climate change and provides

A. Sharma (✉)
Amity Food and Agriculture Foundation, Amity University Noida, Noida, Uttar Pradesh, India
e-mail: asharma5@amity.edu

F. S. Gokdemir
Institue of Food, Agriculture and Livestock Development, Başkent Universiy, Ankara, Turkey

G. Patanjali
Amity Institute of Horticulture Studies and Research, Amity University Noida, Noida, Uttar Pradesh, India

F. Eyidogan
Institue of Food, Agriculture and Livestock Development, Başkent Universiy, Ankara, Turkey

Department of Molecular Biology and Genetics, Faculty of Science and Letters, Başkent University, Ankara, Turkey

P. Pandey et al. (eds.), *Microorganisms Resilience to Climate Change*, Microorganisms for Sustainability 56, https://doi.org/10.1007/978-981-96-3748-5_8

insights into future challenges and opportunities for sustainable agriculture under changing climatic conditions.

Keywords

Climate change · Soil microbial diversity · Pathogen dynamics · Ecosystem functions · Agricultural adaptation

8.1 Introduction

Climate change significantly impacts soil ecosystems, primarily through alterations in microbial biodiversity, soil properties, and ecosystem functions. The ongoing rise in global temperatures has been shown to reduce microbial diversity in soil, which is critical for maintaining ecosystem health and functionality. Wu et al. (2022) provide evidence that long-term climate warming leads to a consistent decline in the taxonomic and phylogenetic diversity of soil microorganisms, including bacteria, fungi, and protists. This reduction in diversity is largely attributed to changes in biotic interactions and soil biogeochemical conditions, particularly soil moisture. This finding is further corroborated by Guo et al. (2019), who highlight that climate warming accelerates the temporal scaling of microbial biodiversity. The implications of these changes are profound, as soil biodiversity is closely linked to ecosystem multifunctionality, which includes nutrient cycling, organic matter decomposition, and plant growth (Wagg et al. 2014).

Moreover, the interactions between climate change and soil properties are complex and multifaceted. For instance, changes in precipitation patterns and increased aridity can lead to soil degradation, affecting nutrient availability and soil structure (Durán et al. 2018). The work of Ni and Vellend (2023) emphasizes that soil properties can constrain plant distributions, particularly under changing climatic conditions. This scenario initiates a feedback loop where shifts in plant communities further affect soil health and biodiversity, intensifying the impacts of climate change (Naylor et al. 2020). Additionally, the role of soil in carbon cycling is increasingly recognized, with the soil microbiome being pivotal in mediating soil organic carbon dynamics (Naylor et al. 2020; Eisenhauer 2024). As climate change alters soil moisture and temperature regimes, the capacity of soils to sequester carbon may be compromised, potentially leading to increased greenhouse gas emissions (Pi et al. 2021).

Indirect effects of climate change on soil ecosystems also deserve attention. Guerra et al. (2021) argue that alterations in soil communities are often driven indirectly by climate factors through changes in vegetation cover and soil pH. This underscores the necessity of examining direct and indirect pathways when assessing climate impacts on soil ecosystems. Additionally, soil ecosystem resilience is heavily influenced by its biodiversity. As noted by Eisenhauer (2024), diverse soil ecosystems are more adept at regulating nutrient cycling and maintaining fertility, crucial for ecosystem stability and productivity.

Understanding soil-borne pathogen responses to climate change is crucial due to the significant implications for agricultural productivity, ecosystem health, and public health. Climate change can lead to shifts in the geographic distribution of pathogens, allowing them to invade new areas and potentially increase disease prevalence in previously unaffected host populations. Pathogens are highly adaptable and may exhibit increased mutation rates under stress, resulting in more virulent strains (Caffarra et al. 2012). Insights from historical climate changes can inform predictions about future pathogen dynamics. In contrast, the interactions between pathogens and their hosts are influenced by various environmental factors, necessitating a comprehensive understanding of these relationships (Mordecai 2014). Additionally, changes in pathogen dynamics can disrupt ecosystem functionality, impacting nutrient cycling and overall ecological balance (Delgado-Baquerizo et al. 2017). The implications extend to public health, as climate change has been linked to the resurgence of infectious diseases, highlighting the need for ongoing research to develop effective management strategies and mitigate the impacts of climate change on pathogens and their associated ecosystems (Lafferty and Mordecai 2016).

This chapter examines the critical relationship between climate change and soil-borne pathogen dynamics, with particular emphasis on how various climate parameters influence pathogen behavior, distribution, and virulence. We will explore five key climate factors: temperature changes, precipitation patterns, extreme weather events, CO_2 levels, and soil moisture dynamics, analysing their individual and combined effects on pathogen ecology. Understanding these interactions is crucial for several reasons: first, it helps predict future disease patterns and potential threats to agricultural systems; second, it aids in developing effective management strategies for plant diseases under changing climatic conditions; and third, it contributes to our broader understanding of ecosystem resilience and adaptation. By synthesizing current research and identifying knowledge gaps, this chapter provides a comprehensive framework for understanding how climate change shapes the future of soil-borne plant diseases and their management.

8.2 Effects of Climate Change Variables on Soil Pathogen Dynamics

Climate change profoundly influencing soil pathogens through various parameters, including temperature changes, precipitation patterns, extreme weather events, CO_2 levels, and soil moisture dynamics. Each of these factors interacts with soil ecosystems, potentially altering the prevalence and virulence of soil-borne pathogens (Fig. 8.1).

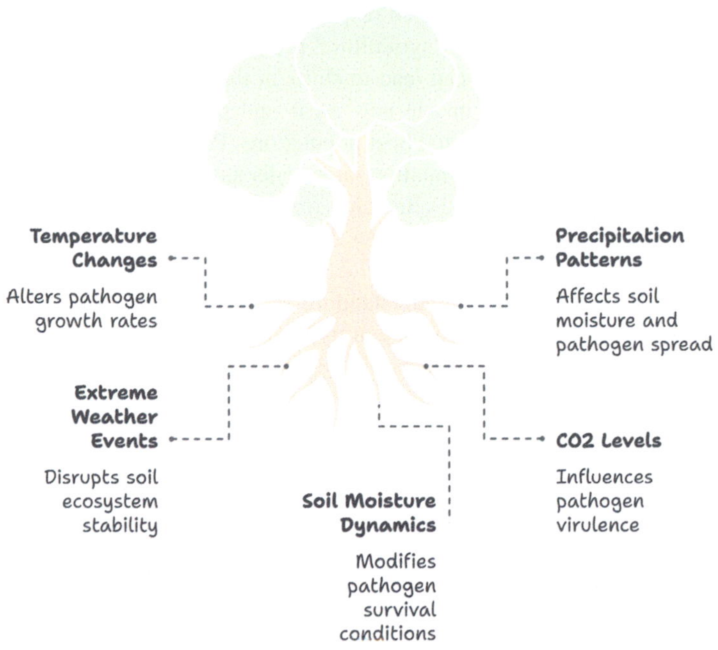

Fig. 8.1 Climate change impact on plant pathogens

8.2.1 Temperature Changes

Rising temperatures due to climate change intensify the adverse effects of soil pathogens on plant health. Liu and He (2021) observed that elevated temperatures amplify the negative feedback of soil pathogens on temperate tree species, suggesting increased pathogen virulence in warmer conditions. This aligns with Pritchard (2011), who noted that climate changes influence soil organisms, including pathogens, through direct warming effects and indirect changes in plant-mediated soil carbon inputs. Elevated temperatures can enhance pathogen development rates, contributing to more severe plant diseases as pathogens often thrive in warmer conditions.

Similarly, Marfenina and Danilogorskaya (2017) reported that a 3 °C rise in temperature led to significant shifts in fungal community structures, favoring the proliferation of pathogenic fungi such as *Fusarium* and *Rhizoctonia* while reducing the diversity of beneficial fungi like mycorrhizal species. The relative abundance of pathogenic fungi increased by approximately 25% under warmer conditions, posing heightened risks to plant health and ecosystem stability.

Temperature changes also influence pathogen survival and management strategies Williams et al. (2024). However, temperature is not the sole determinant of pathogen dynamics. Environmental variables such as soil moisture, precipitation, and plant health also play significant roles. Zhang (2023) emphasized how temperature shapes the diversity and community structure of soil fungal pathogens but stressed that plant-soil interactions and climatic factors like precipitation significantly influence pathogen aggregation and movement. This is supported by Tajul et al. (2011), who found that favorable conditions, such as optimal soil moisture, enhance the survival of *Ralstonia solanacearum*, a bacterial wilt pathogen, leading to increased disease outbreaks.

Overall, rising temperatures not only accelerate pathogen activity but also alter microbial community dynamics in ways that challenge plant health and agricultural productivity. Understanding these changes is crucial for developing adaptive management strategies to mitigate pathogen risks in a warming climate.

8.2.2 Precipitation Patterns

Shifts in precipitation patterns exert significant influence on the dynamics of soil pathogens, impacting their diversity, abundance, and virulence. Precipitation, as a key driver of soil moisture, directly affects the survival, growth, and activity of soil-borne pathogens. Delavaux et al. (2020) demonstrated that climatic factors, particularly precipitation, play a pivotal role in shaping the diversity and composition of root-associated plant pathogen communities in grasslands. Changes in rainfall can alter the microbial equilibrium, leading to shifts in pathogen community structures and their interactions with host plants. Similarly, Zhang (2023) emphasized that the interplay between precipitation and temperature is crucial for understanding the diversity and distribution of soil fungal pathogens, indicating that variations in rainfall patterns can profoundly influence pathogen ecology. Changes in rainfall can alter the microbial equilibrium, leading to shifts in pathogen community structures and their interactions with host plants. Similarly, Zhang (2023) emphasized that the interplay between precipitation and temperature is crucial for understanding the diversity and distribution of soil fungal pathogens, indicating that variations in rainfall patterns can profoundly influence pathogen ecology.

Precipitation fosters soil moisture levels essential for spore germination, pathogen survival, and infection processes. According to Zhang (2023), increased precipitation enhances plant biomass and regulates soil pH and nutrient availability, which are critical factors driving pathogen diversity. Chen et al. (2021) observed that heightened rainfall promotes fungal pathogen abundance by creating favorable conditions for their proliferation. Conversely, drought conditions, often associated with reduced precipitation, can enhance fungal diversity by altering competitive dynamics among soil pathogens, as noted by Hawkes et al. (2011). The effects of precipitation on pathogen virulence are, however, complex and highly context dependent. Conversely, Ochoa-Hueso et al. (2018) highlighted that altered precipitation regimes can yield varied impacts on soil microbial communities, as the outcomes often depend on pathogen taxa and specific environmental contexts. In addition to direct effects, precipitation indirectly influences pathogen dynamics through its impact on soil respiration and nutrient cycling. Miao et al. (2017) noted that soil respiration exhibits non-linear responses to precipitation changes, reflecting the complexity of microbial interactions under varying moisture conditions. For example, Thomey et al. (2010) observed that fluctuations in rainfall can influence soil organic matter availability, potentially altering pathogen populations and their ecological roles.

In conclusion, understanding the multifaceted effects of precipitation on soil pathogens is crucial for predicting and mitigating disease outbreaks, particularly in the face of climate change. The interplay between moisture, nutrient cycling, and pathogen diversity underscores the need for comprehensive studies to unravel these dynamics and develop sustainable agricultural management strategies under shifting environmental conditions.

8.2.3 Extreme Weather Events

Extreme weather events, including heavy rainfall and droughts, can disrupt soil ecosystems and significantly influence pathogen dynamics. The increased frequency of wetting and drying cycles, for instance, can enhance the resilience of certain pathogens, as observed by Orwin et al. (2015), who studied the effects of such cycles on soil microbial communities. These fluctuations create environments that favor the survival and proliferation of specific pathogens, potentially leading to more frequent disease outbreaks in plants. Furthermore, Wahdan et al. (2020) demonstrated that climate change could alter fungal plant pathogen dynamics during the decomposition of wheat litter, highlighting the cascading effects extreme weather events can have on pathogen behavior and interactions with other soil microorganisms.

Increased moisture levels are particularly influential in the spread of certain pathogens. Wu et al. (2016) found that heightened moisture conditions could lead to a higher incidence of *Phytophthora infestans*, the causative agent of potato late blight, as the pathogen thrives in wet environments. The study revealed that virulence factors of *P. infestans* are expressed more vigorously under moist conditions, resulting in more severe disease outbreaks. The relationship between heavy rainfall and pathogen virulence is intricate; Lindauer et al. (2020) noted that increased moisture levels can promote the growth of lethal amphibian fungal pathogens, which can then impact amphibian populations. This emphasizes how heavy rainfall can not only facilitate pathogen spread but also enhance their virulence, leading to more severe disease outbreaks among susceptible host populations. Additionally, Ashrafi et al. (2017) found that pathogens such as *Vibrio* species exhibited increased virulence at higher temperatures, which can be further exacerbated by heavy rainfall, stressing host organisms. These findings underscore the complex interactions between extreme weather events and pathogen dynamics, pointing to the potential for increased disease severity under shifting climatic conditions. Drought conditions also have a profound effect on soil pathogens, their virulence, and the biological interactions within ecosystems. Drought stress weakens plant defenses, making them more susceptible to infections. Sinha et al. (2019) demonstrated that simultaneous drought stress and pathogen infection in chickpeas resulted in more severe disease symptoms, highlighting the detrimental effects of combined stressors on plant health. Their study revealed a significant (40%) increase in disease symptoms caused by *Fusarium* and *Rhizoctonia* pathogens under drought conditions, showing that drought can exacerbate the impact of existing infections. Huot and Tamborindeguy (2017) found that drought stress increased the vulnerability of *Solanum lycopersicum* (tomato) to colonization by *Bactericera cockerelli*, a vector for the pathogen causing "zebra chip" disease. Their research showed a 30% increase in vector populations during drought conditions, which, in turn, enhanced pathogen spread, suggesting that drought not only impacts the plants directly but also alters the dynamics of vector-borne diseases, leading to increased pathogen transmission.

Drought can also create environments conducive to the proliferation of pests, which may further facilitate pathogen spread. Research has shown that during severe drought years, tree mortality rates can rise, creating favorable conditions for pathogen establishment and spread. For instance, Hart et al. (2014) reported that prolonged drought conditions significantly increased outbreaks of pine caterpillars (*Dendrolimus spp.*) in Shandong Province, China. Their study found that drought enhanced the nutritional quality of tree foliage, making it more attractive for caterpillar feeding, which in turn increased the likelihood of pathogen infections in stressed trees. Although the reference to Hart et al. (2017) was not directly relevant to these interactions, it is essential to understand that drought conditions can weaken tree defenses, making them more vulnerable to both insect infestations and pathogen infections, thus exacerbating forest decline.

In summary, extreme weather events such as heavy rainfall and drought play critical roles in influencing pathogen dynamics and virulence. These events often lead to increased disease severity and outbreaks, necessitating a deeper understanding of these interactions for effective management strategies in agricultural and forest ecosystems, especially in the context of climate change.

8.2.4 CO_2 Levels

Elevated CO_2 levels, a critical component of climate change, have substantial implications for soil pathogens. Cherian and Lambat (2015) found that increased CO_2 concentrations could accelerate the evolution of plant pathogens, thereby enhancing their virulence. This is largely due to altered canopy environments that favor the proliferation of pathogens. Supporting this, Hayden et al. (2012) showed that elevated CO_2, coupled with warming conditions, leads to significant shifts in soil microbial community structures, potentially disrupting the balance between beneficial organisms and harmful pathogens. The interaction of CO_2 with other environmental factors can create more favorable conditions for pathogens, consequently increasing their impact on plant health.

Atmospheric CO_2 levels play a significant role in influencing soil and plant pathogens, altering their virulence and overall biological dynamics. Elevated CO_2 concentrations can modify soil properties and microbial communities, enhancing pathogen virulence. Zhao et al. (2017) reported that naturally leaking CO_2 in the

Qinghai-Tibet Plateau resulted in changes to soil conditions and plant root systems, which, in turn, affected the diversity of soil-dwelling microorganisms, including pathogens. Their study underscored that higher CO_2 levels could increase the adaptability of soil microbes, potentially boosting their pathogenicity. Additionally, Lu et al. (2018) found that elevated CO_2 levels can maintain variability in virulence among *Fusarium* populations. Their research indicated that, under elevated CO_2 conditions, the virulence and reproduction rates of pathogen strains varied considerably, suggesting that atmospheric CO_2 can influence the evolutionary dynamics of plant pathogens. This variation in pathogenicity is crucial for understanding disease patterns and predicting future outbreaks.

Furthermore, Hay et al. (2021) demonstrated that elevated CO_2 levels increased the susceptibility of wheat to *Fusarium graminearum*, leading to more severe disease damage and higher contamination of deoxynivalenol (DON). Their findings revealed that disease severity under elevated CO_2 conditions varied depending on the host cultivar and pathogen strain, establishing a direct link between CO_2 levels and pathogen virulence. He et al. (2019) also observed that high soil CO_2 concentrations could significantly alter the pH of root tissues, impairing membrane function and osmoregulation in root cells. This physiological stress can reduce water absorption, increasing the susceptibility of plants to pathogens and further complicating plant-pathogen interactions. Finally, Bencze et al. (2017) highlighted that rising atmospheric CO_2 concentrations could heighten the risk of *Fusarium* mycotoxin contamination in wheat grains. While disease severity was not markedly affected by elevated CO_2 in most wheat varieties, they found that the spread of *Fusarium* increased in certain cultivars, leading to reduced grain weight and quality.

In conclusion, variations in atmospheric CO_2 levels significantly influence soil and plant pathogen dynamics, affecting virulence and disease outcomes. Understanding these complex interactions is essential for developing effective management strategies in agricultural systems facing the challenges of a changing climate.

8.2.5 Soil Moisture Dynamics

Soil moisture is a fundamental factor that governs pathogen dynamics, influencing microbial activity and plant health. Bell et al. (2010) underscored those fluctuations in soil moisture, resulting from changes in precipitation patterns and temperature, can profoundly impact plant biomass and, consequently, the nitrogen and carbon cycles within ecosystems. This interplay between soil moisture and ecosystem processes is crucial, as it also dictates the behavior of soil pathogens. Chang et al. (2014) further emphasized that soil respiration, a critical process in the carbon cycle, is highly responsive to soil moisture levels, thereby modulating pathogen activity. When soil moisture levels rise, it can create optimal conditions for pathogens to thrive, enhancing their survival and infectivity, particularly in regions with wetter climates. The dynamics of soil moisture are instrumental in regulating the survival, movement, growth, and invasiveness of soil and plant pathogens. Bunn et al. (2023) revealed that soil moisture and competition jointly shape the impact of soil biota on invasive species such as *Centaurea stoebe*. Their research showed that in arid soils, the presence of soil pathogens significantly suppressed the growth of native grasses, highlighting how moisture variations can shift competitive dynamics between invasive and native species, thereby influencing pathogen spread and virulence. In parallel, Wilschut and Kleunen (2021) reported that drought conditions modify plant-soil feedback mechanisms, affecting biomass allocation without markedly influencing overall plant performance. This underscores the role of soil moisture fluctuations in altering plant-pathogen interactions, potentially rendering plants more susceptible to diseases, particularly during water stress.

Ngo and Cavagnaro (2018) investigated the combined effects of compost and pre-planting soil moisture on plant biomass and nutrition. Their findings indicated that optimal soil moisture levels foster mycorrhizal colonization, which bolsters plant resistance to pathogens. In contrast, insufficient moisture resulted in reduced mycorrhizal development, thereby increasing susceptibility to soil-borne pathogens. Similarly, Fry et al. (2018) demonstrated that drought stress can neutralize plant-soil feedbacks that typically enhance plant growth and health. Their research showed that under drought conditions, pathogen accumulation in the soil increases, adversely affecting plant performance and emphasizing the significant impact of moisture dynamics on pathogen virulence and plant health. Devadoss et al. (2020) explored the relationship between soil moisture and the virulence of soil-borne pathogens, revealing that drought conditions amplify pathogen virulence. Their study reported a 30% increase in disease severity under drought, stressing how moisture fluctuations directly affect pathogen dynamics. Additionally, Gonsamo et al. (2019) examined the impact of soil moisture on root characteristics of herbaceous plants along riparian zones. Their findings indicated that moisture variability directly influences root growth and health, which in turn affects plant susceptibility to pathogens. Plants exposed to lower moisture levels showed a 25% higher incidence of root diseases, underscoring the intricate relationship between soil moisture and pathogen behavior.

In conclusion, soil moisture dynamics are crucial in shaping pathogen interactions, influencing their virulence and overall disease outcomes. The way moisture levels affect pathogen survival, growth, and movement is vital to understanding the broader ecological implications of soil health and plant disease management. Understanding these interactions is essential for developing targeted management strategies, especially in agricultural and forest ecosystems that face the challenges of changing climate conditions. The ability to predict and mitigate pathogen impacts concerning soil moisture will be key to safeguarding plant health and ecosystem stability in the future.

8.3 Climate Change and Pathogen Distribution

Shifts in temperature, humidity and precipitation significantly impact the distribution and range of pathogens. Increased temperatures hasten microbial decomposition activities, leading to faster Carbon dioxide emissions, which can transform soils from carbon sinks into Carbon dioxide sources. This warming effect also enhances the growth of Rhizosphere colonizing bacteria and plant growth promoting fungi, although it can reduce colonization by arbuscular mycorrhizal fungi due to drought conditions. Rise in temperature has led to an increase in soilborne fungal pathogens leading to the poleward movement of numerous pathogens and pests averaging a shift of approximately 3 kilometers annually since 1960 (Ibáñez et al. 2023). This shift is attributed to the changing climate which facilitates pathogens to expand into new geographic regions. Increased precipitation can lead to pathogens being washed into waterways, while rise in sea temperatures trigger mechanisms for the proliferation and spreading of the pathogen. Singh et al. (2023) reported that warming temperatures can lead to the development of new strains of pathogens which are adapted better and more virulent which was clear from the shift from *Fusarium culmorum* to the more aggressive *Fusarium graminearum* in wheat. It was also reported by Singh et al. (2023) that elevated temperatures result in the suppression of plant immunity and inflation of susceptibility to pathogens. For instance, higher temperatures have been linked to enhanced disease severity of Potato blight pathogen i.e. *Phytophthora infestans.* High humidity generally elevates the virulence of pathogens infecting aerial plant tissues. For example, increased humidity is related to higher infection rates of *Sclerotinia sclerotiorum* in lettuce and *Phytophthora sojae* in soybeans. On the contrary, lower moisture conditions can increase pathogen numbers and disease severity for pathogens like *Magnaporthe oryzae*, the causal agent of Rice blast.

The interaction between climatic changes and evolutionary processes can directly influence future pathogen outbreaks, as increases in the frequency and intensity of extreme climatic conditions can aid in spreading pathogens to new locations. This is further aggravated by the fact that pathogens tend to cause more damage in newly invaded regions and on new hosts than in their native regions and hosts. Pathogens can evolve which allows them to infect different plant species and become more virulent, thereby overcoming chemical controls and resistant cultivars.

Singh et al. (2023) reported that the hybrid species of *Phytophthora alni* which originated from the hybridization of *P. uniformis* and *P. multiformis*, has resulted in the severe decline in Ader populations across Europe. Another finding of invasion of pathogens into new areas was reported by Ibáñez et al. (2023) that in Europe, there has been a noticeable spread of various agricultural diseases moving towards northern regions. *Pyricularia oryzae* and *Bipolaris oryzae* which are found to be most detrimental rice pathogens are increasingly prevalent across all European rice districts, with Northern Italy experiencing the most critical and grave situation. A similar case of *Plasmopara viticola*, A pathogen on grapevines, is reported to rise by 5–20% throughout Europe, highlighting a shift in the distribution of these pathogens.

Climate change has a profound impact on life cycles, reproduction rates and virulence of plant pathogens. One significant effect is the change in diurnal temperature amplitude which results in increased pathogen reproduction due to shorter latency periods. For example, warmer winters can abbreviate the latency period, allowing more aggressive pathogens to emerge when conditions become favourable and conducive for infection (Chen et al. 2023). The research presented in (Khanal and Land 2023) suggests that climate change has a complex effect on Nematode species. While warmer temperatures may result in reduced reproduction rates, it can also increase nematode virulence, leading to greater crop damage. Higher temperatures of soil, which is the result of climate change, can directly impact nematode survival. (Khanal and Land 2023) found that the survival rate of *Rotylenchulus reniformis* increased as the exposure temperature increased. The study found that elevated soil temperatures might lead to decreased nematode reproduction. Despite the potential decrease in reproduction, the study found that the virulence of *R. reniformis* was higher at elevated soil temperatures. This suggests that while there might be fewer nematodes, they are more likely to cause greater damage to crops.

According to (Kumar et al. 2024), Stressed plants often have decreased turgor pressure and impaired cellular function, making it much easier for pathogens to invade. Heat stress can disrupt the expression of genes associated with plant defense, compromising their ability to fight off pathogens. Increased carbon dioxide levels, while potentially promoting growth can alter plant tissue composition, making them more appealing to certain pathogens. (Kumar et al. 2024) also explained in his study about the key changes in plant physiology and defense mechanisms under climate stress. Elevated levels of carbon dioxide can initially boost photosynthesis rates, but this effect might be offset by other stressors like heat and drought in the long run. Plants activate various mechanisms to cope with stress, such as accumulating osmoprotectants such as proline to maintain cell function under drought and producing heat shock proteins to protect cellular components from heat damage. Heat stress can slow down the hypersensitive response, a critical plant defense mechanism involving programmed cell death to limit pathogen spread, making plants more susceptible to infection. Hou et al. (2024) reported that climate stress leads to the buildup of harmful molecules known as Reactive oxygen species in plant cells. While ROS play a role in signaling and defense, enormous amounts damage plant tissues and compromise immunity.

8.4 Pathogen and Host Reactions: Adaptation, Virulence, and Defense Mechanisms

Climate change affects many systems, such as soil microbiomes and plant-pathogen interactions. Bacteria, fungi and many pathogens that are part of this system can be sensitive to changing climatic conditions and many environmental factors such as temperature, humidity and changing CO_2 (Jansson and Hofmockel 2020). Climate change is driving the evolution and adaptation of soilborne pathogens, altering their infection strategies and virulence. This change is leading to increased outbreaks of plant diseases. Plant disease outbreaks cause significant crop losses, global food security and biodiversity loss in many regions of the world. Annual crop losses due to pathogens that cause disease and limit host health and productivity are estimated at $220 billion (Rohr et al. 2019; Ristaino et al. 2021; FAO 2022; Singh et al. 2023). Changing climatic conditions regulate pathogen virulence, promoting pathogen evolution, host-pathogen interactions, and the emergence of new pathogenic strains (Singh et al. 2023; Lahlali et al. 2024). Therefore, it is crucial to understand the dynamics of climate-induced changes on pathogens and their behavior and to develop strategies to mitigate these negative effects.

The dynamic interactions between pathogens and their hosts are a complex process that needs to be addressed from many angles. This is because the changes occurring in the process of climate change affect a wide range of parameters. For example, the onset of global warming, characterized by changes in average temperatures, changes in annual precipitation patterns, and prolonged droughts, can disrupt plant growth and development and ultimately lead to crop losses (Ebi et al. 2016). Environmental stresses induced by climate change threaten plant health by reducing plant resistance to bacterial and fungal pathogens (Devendra 2012). Many of the factors that influence pathogen-host interactions are related to environmental factors (Table 8.1). For example, environmental carbon dioxide (ECO_2) affects plant-pathogen interactions, physiology, and pathogen resistance mechanisms. It can also disrupt agricultural systems. Increases in atmospheric CO_2 concentrations

Table 8.1 Plant-pathogen interactions related to climate change factors

Climate change factor	Plant-host	References
ECO_2	*Acer rubrum-Phyllosticta minima*	McElrone et al. (2005)
ECO_2	*Quercus robur-Erysiphe alphitoides*	Sanchez-Lucas et al. (2023)
ECO_2	*Zea mays-Heterorhabditis bacteriophora*	Hiltpold et al. (2020)
ECO_2	*Betula pendula-Pyrenopeziza betulicola*	Riikonen et al. (2008)
ECO_2	*Glycine max-Septoria glycines, Peronospora manshurica, Fusarium virguliforme*	Eastburn et al. (2010)
Temperature	Various plant-*Macrophomina phaseolina*	Shirai and Eulgem (2023)
Temperature	Various plant-*Fusarium oxysporum*	Gay et al. (2014)
Temperature	Various plant-*Puccinia triticina*	Ahmed et al. (2024)
Flood	*Citrus junos-Diversispora spurca*	Wu et al. (2013)
Flood	*Quercus robur-Tuber aestivum*	Thomas (2021)
Flood	*Pinus densiflora*-several pathogens	Cho et al. (2021)
Drought	*Arabidopsis thaliana-Pseudomonas syringae*	Melotto et al. (2006)
Drought	*Arabidopsis thaliana-Pseudomonas* and *Peronospora parasitica*	Mohr and Cahill (2003)
Drought	*Solanum lycopercicum-Oidium neolycopersici*	Sunarti et al. (2022)

as a result of anthropogenic activities can potentially enhance plant growth, leading to a phenomenon called " CO_2 fertilization". However, recent studies show that the effects of CO2 on plant-pathogen interactions are complex and contradictory. While experiments under controlled conditions help to determine the effects of CO_2 levels on disease severity, factors such as pathogen adaptation under natural conditions can alter these effects (Smith and Luna 2023).

Under the effect of ECO_2, *Arabidopsis thaliana* infected with the pathogen *Erysiphe cichoracearum* shows an increase in several traits such as stomatal density, guard cell number, and trichome number (Lake and Wade 2009). Another study shows that increasing environmental carbon dioxide levels due to climate change have significantly increased the risk of *Fusarium* head blight and *Septoria tritici* blotch diseases in an important cereal crop such as wheat, which may have important implications for food safety. Adaptation of both pathogens and wheat plants to the high carbon dioxide environment has increased disease severity even in disease-resistant wheat cultivars, resulting in yield losses (Váry et al. 2015).

Another important factor in climate change is flooding. Flooding is one of the most important environmental factors affecting the dynamics and distribution of soil-borne bacteria. Rhizosphere bacteria, root mycorrhizae and plant endophytes are most affected by flooding. Flooding fills the soil with water and leads to a gradual depletion of oxygen. Oxygen-depleted soils increase the number of facultative anaerobes and solid anaerobes (Li et al. 2023). The effect of flooding on the rhizosphere microbiota may be highly dependent on specific host characteristics. The bacterial composition of the rhizosphere may be determined by host species-specific characteristics. For example, the water tolerance of the host plant also affects the root rhizosphere. In addition, soil factors such as drainage conditions, pH, redox status, and C dynamics have the potential to alter the structure and composition of the rhizosphere microbiota (Liu et al. 2014; Bonito et al. 2014; Moche et al. 2015).

The inhibition of arbuscular mycorrhizae by flooding varies depending on the fungal species. However, the habitat of the mycorrhizal plants is also important. For example, mycorrhizal inhibition by flooding is lowest in coastal vegetation with a certain water level or in wetlands where crops such as rice grow. However, it is higher in regions with continental climates, polar regions, subtropical regions, far from coastal climates. In restoration efforts for wet forest ecosystems, the use of seedlings inoculated with flood-tolerant mycorrhizal species can be an important strategy for plant health and ecosystem restoration. By increasing plant vigor under pathogen pressure, mycorrhizal symbiosis can protect against diseases such as root rot, which is particularly common in wetlands. There is a decrease in microbial endophyte colonization associated with flooding (Li et al. 2010). This is because most endophytes in terrestrial ecosystems are obligatory aerobes. Tian et al. (2015) reported that anaerobic conditions negatively affected endophyte diversity in *Myricaria laxiflora* species exposed to periodic flooding. Vishwanathan et al. (2020) also reported a decrease in bacterial diversity in the roots of rice plants subjected to flooding.

As climate change accelerates, rising temperatures alter the dynamics of pathogen-host interactions. This change affects pathogen virulence, host defense mechanisms, and the balance of soil microbiota. For example, temperature-induced increases in the secretion of simple sugars and organic acids create an optimal microenvironment for pathogens such as *Phytophthora cinnamomi* Temperature-induced changes in root exudate profiles can increase pathogen colonization in species such as *P. cinnamomi*. (Shirai and Eulgem 2023). Similarly, drought stress is a critical factor that profoundly affects plant health and host-pathogen interactions. Drought also alters interactions between plants and soil-borne pathogens. The complex dynamics of pathogen adaptation, host defense mechanisms, and the soil microbiome are profoundly affected by prolonged water shortages and the presence of other stressors. For example, water-stressed plants become more susceptible to disease caused by *Neofusicoccum eucalyptorum*. This is an indication of the synergistic effects resulting from the combination of biotic and abiotic stresses (Barradas et al. 2018; Santos et al. 2024).

Pathogen-host responses to climate change are not only influenced by many environmental factors but are also evolutionary dynamic. Microorganisms, especially pathogens, struggle for survival. Soil pathogens develop various adaptation mechanisms to overcome host defenses and gain access to a particular habitat. This adaptation process can depend on changes in the virulence of the pathogen and on physiological and genetic characteristics of the host. Soil pathogens produce a variety of mechanisms for survival and reinforcement of adaptive strategies. Examples include host attachment, invasion, and immunosuppression. *Listeria monocytogenes* develops mechanisms such as actin polymerization to overcome host cellular defenses (Theriot et al. 1992). Other alternatives to this process include traits developed by pathogens, such as antibiotic resistance, or metabolic adaptations of pathogens that modify their metabolism to survive in low-oxygen or nutrient-limited environments. For example, the soil-dwelling bacterium *Ralstonia solanacearum*, which infects plants, produces a specialized enzyme called beta-lactamase. This enzyme breaks down and neutralizes many antibiotics used in agriculture. This allows the bacteria to survive in the presence of antibiotics and increases the risk of infecting plants. Especially under stressful conditions (drought, salinity, etc.), the ability of this bacterium to cause disease becomes stronger (Theriot et al. 1992; Larsson and Flach 2022).

8.5 Management Strategies, Practical Approaches and Evaluations

Climate change has been identified as an important factor in the adaptation, virulence and spread of soil-borne pathogens. In this context, it is critical to develop effective adaptation strategies to ensure sustainable agriculture and food security. We need different strategies to manage the interactions in this process. These strategies should include innovative strategies (Table 8.2) that support soil health, mitigate environmental factors such as heat and drought, reduce pathogen virulence due to climate change, include education and policy measures, and incorporate innovation.

Table 8.2 Key adaptation strategies to mitigate climate change impacts on soil pathogens

Strategies to support soil health	Strategies to reduce heat and drought impacts	Strategies to reduce climate change-induced pathogen virulence	Education and policy measures	Innovative approaches
Organic soil management	Heat-tolerant plant varieties	Biological control agents	Farmer education	Gene editing technologies
Protecting the soil microbiome	Covering and shading technique	Use of natural antimicrobials	Policies and incentives	Microbiota engineering
Drainage and soil aeration	Water management systems	Climate-sensitive pathogen management	Use of data and technology	Carbon management

As strategies to support soil health in key adaptation strategies to mitigate the effects of climate change on soil pathogens, organic soil management strategies and the use of compost, green manures and organic mulches to increase organic matter content support the proliferation of beneficial microorganisms and enhance pathogen suppression. Organic fertilizers and microbial inoculants can be used to maintain microbial diversity and increase beneficial microbiota. Water management, drainage and soil aeration: Improving water management and soil structure can reduce waterborne transport of pathogens and infections in the root zone. It is important to develop crop varieties that are resistant to heat and drought stress as well as pathogens. However, cover and shade techniques: It is recommended that agricultural shading techniques be used to reduce soil temperature and prevent water loss. Water management systems: Optimum water use should be ensured through techniques such as drip irrigation or sprinkling during drought conditions. The use of biological control agents is also very important. For example, the use of biocontrol agents such as *Trichoderma* spp. and *Bacillus subtilis* can suppress soil pathogens.

Similarly, to promote the use of natural antimicrobials such as garlic, turmeric, etc., the inhibitory effects of natural compounds on soil pathogens need to be evaluated. Dynamic decision support systems can be developed to monitor the response of pathogens to changes in temperature and humidity.

In addition, education and policy measures are crucial to address climatic processes and create global solutions. For example, farmers should receive regular training on soil health, climate change impacts and adaptation strategies. National and international policies should be established to promote climate-smart agricultural practices, and financial support should be provided to farmers for these practices. To expand the use of data and technology, rapid interventions can be made with the support of governments by predicting the spread of pathogens using precision agriculture technologies and data-based modeling. Gene editing technologies: Gene editing technologies such as CRISPR offer innovative approaches to develop innovative solutions that both increase plant resistance and reduce pathogen virulence. In addition, microbiota engineering can manipulate microbiomes to increase beneficial microbial communities in the plant root zone. Methods that increase the carbon storage capacity of soils both mitigate the effects of climate change and improve soil health.

Implementing all these strategies is critical to resolving the complex host-pathogen relationship and addressing climate change. Climate change has far-reaching implications for food security, agricultural/environmental sustainability, and socioeconomics. Approaches that analyze host-pathogen interactions on a global scale are useful for analyzing pathogen behavior and providing solution-oriented approaches to climate change. Effective implementation of these strategies will not only enhance the fight against the effects of climate change but will also lay the foundation for future food security and sustainable agricultural practices, a critical step in securing the future of our world.

8.6 Conclusion

The extensive review of climate change impacts on soil ecosystems reveals critical findings and significant implications for future research and management strategies. Climate change profoundly affects soil microbial communities through various pathways, leading to reduced biodiversity and altered ecosystem functions. These changes have far-reaching consequences for agricultural productivity and food security.

Rising temperatures and altered precipitation patterns not only disrupt microbial diversity but also influence the behavior and distribution of pathogens. These shifts can result in increased pathogen virulence and the emergence of new pathogenic strains in previously unaffected areas. Such developments highlight the complex interactions between climate variables and soil ecosystems, underscoring the need for integrated approaches to understanding and managing these challenges. The study emphasizes the importance of adaptive strategies, such as enhancing soil health management, developing climate-resilient crop varieties, and implementing innovative agricultural practices. These efforts should be supported by education and policy measures to ensure their effective implementation and sustainable outcomes. Policymakers and stakeholders must prioritize awareness programs and invest in resources to bridge the gap between research advancements and on-ground applications.

Additionally, the research underscores the urgent need for advanced modeling techniques and robust monitoring systems to predict and manage climate-induced changes in pathogen distribution and disease outbreaks. The development of such tools is crucial for informed agricultural planning and proactive risk management in the face of a changing climate. These systems should integrate data from diverse fields to enable accurate predictions and effective mitigation strategies.

Future research should focus on building resilient agricultural systems capable of withstanding multiple stressors while maintaining productivity. This includes exploring novel approaches to enhancing soil health, improving plant resistance to pathogens, and adopting sustainable management practices tailored to evolving environmental conditions. Efforts must also extend to understanding the feedback loops between soil ecosystems and broader climatic processes to design holistic interventions. The success of these endeavours depends on continued scientific research, international collaboration, and evidence-based policy implementation. Multidisciplinary partnerships between researchers, governments, and local communities are essential to develop and execute solutions that address the global nature of climate challenges. By fostering such comprehensive approaches, we can work towards maintaining food security and safeguarding ecosystem health in an increasingly uncertain climate future.

Acknowledgements First and third author acknowledge Dr. Ashok K Chauhan, Founder President, Amity University for their support and guidance.

References

Ahmed N, Alam M, Saeed M, Ullah H, Junaid M, Kanwal M, Ahmed S (2024) Role of plants in managing diseases. In: Ethnic knowledge and perspectives of medicinal plants. Apple Academic Press, pp 579–604. https://doi.org/10.1201/9781003415381-30

Ashrafi R, Bruneaux M, Sundberg L, Pulkkinen K, Valkonen J, Ketola T (2017) Thermal tolerance is linked with virulence in a fish pathogen. https://doi.org/10.1101/188185

Barradas C, Pinto G, Correia B, Castro BB, Phillips AJL, Alves A (2018) Drought × disease interaction in *Eucalyptus globulus* under *Neofusicoccum eucalyptorum* infection. Plant Pathol 67:87–96. https://doi.org/10.1111/ppa.12703

Bell J, Sherry R, Luo Y (2010) Changes in soil water dynamics due to variation in precipitation and temperature: an ecohydrological analysis in a tallgrass prairie. Water Resour Res 46:W03508. https://doi.org/10.1029/2009wr007908

Bencze S, Puskás K, Vida G, Karsai I, Balla K, Komáromi J, Veisz O (2017) Rising atmospheric CO_2 concentration may imply higher risk of *Fusarium* mycotoxin contamination of wheat grains. Mycotoxin Res 33:229–236. https://doi.org/10.1007/s12550-017-0281-2

Bonito G, Reynolds H, Robeson MS, Nelson J, Hodkinson BP, Tuskan G, Vilgalys R (2014) Plant host and soil origin influence fungal and bacterial assemblages in the roots of woody plants. Mol Ecol 23:3356–3370. https://doi.org/10.1111/mec.12821

Bunn R, Antunes P, Bullington L, Fahey C, Lekberg Y (2023) Soil moisture and competition determine soil biota effects on invasive *Centaurea stoebe*. Funct Ecol 37:1181–1196. https://doi.org/10.1111/1365-2435.14295

Caffarra A, Rinaldi M, Eccel E, Rossi, V, Pertot I (2012) Modelling the impact of climate change on the interaction between grapevine and its pests and pathogens: European grapevine moth and powdery mildew. Agric Ecosyst Environ 148:89–101. https://doi.org/10.1016/j.agee.2011.11.017

Chang C, Sabaté S, Sperlich D, Poblador S, Sabater F, Gracia C (2014) Does soil moisture overrule temperature dependence of soil respiration in Mediterranean riparian forests? Biogeosciences 11:6173–6185. https://doi.org/10.5194/bg-11-6173-2014

Chen Q, Hu H, Yan Z, Li C, Nguyen B, Zhu Y, He J (2021) Precipitation increases the abundance of fungal plant pathogens in *Eucalyptus* phyllosphere. Environ Microbiol 23:7688–7700. https://doi.org/10.1111/1462-2920.15728

Chen W, Modi D, Picot A (2023) Soil and phytomicrobiome for plant disease suppression and management under climate change: a review. Plants 12:Art. 1436. https://doi.org/10.3390/plants12142736

Cherian P, Lambat A (2015) Effect of changed CO_2 conditions on plant disease intensity in Nagpur region. IJRBAT. https://doi.org/10.29369/ijrbat.2015.03.ii.0007

Cho Y, Yoo S, Park MS, Kim JS, Kim CS, Lim YW (2021) Ectomycorrhizal fungi associated with *Pinus densiflora* seedlings under flooding stress. Sustain For 13:4367

Delavaux C, Schemanski J, House G, Tipton A, Sikes B, Bever J (2020) Root pathogen diversity and composition varies with climate in undisturbed grasslands. ISME J 15:304–317. https://doi.org/10.1038/s41396-020-00783-z

Delgado-Baquerizo M, Trivedi P, Trivedi C, Eldridge D, Reich P, Jeffries T, … & Singh B (2017) Microbial richness and composition independently drive soil multifunctionality. Funct Ecol 31(12):2330-2343. https://doi.org/10.1111/1365-2435.12924

Devadoss J, Falco N, Dafflon B, Wu Y, Franklin M, Hermes AL, Wainwright HM (2020) Remote sensing-informed zonation for understanding snow, plant and soil moisture dynamics within a mountain ecosystem. Remote Sens 12:2733. https://doi.org/10.3390/rs12172733

Devendra C (2012) Climate change threats and effects: challenges for agriculture and food security. Academy of Sciences Malaysia, Kuala Lumpur

Durán J et al (2018) Temperature and aridity regulate spatial variability of soil multifunctionality in drylands across the globe. Ecology 99:1184–1193. https://doi.org/10.1002/ecy.2199

Eastburn DM, Degennaro MM, Delucia EH, Dermody O, Mcelrone AJ (2010) Elevated atmospheric carbon dioxide and ozone alter soybean diseases at SoyFACE. Glob Change Biol 16:320–330. https://doi.org/10.1111/j.1365-2486.2009.01978.x

Ebi KL, Ziska LH, Yohe GW (2016) The shape of impacts to come: lessons and opportunities for adaptation from uneven increases in global and regional temperatures. Clim Chang 139:341–349. https://doi.org/10.1007/s10584-016-1816-9

Eisenhauer N (2024) A belowground perspective on the nexus between biodiversity change, climate change, and human well-being. J Sustain Agric Environ 3:12108. https://doi.org/10.1002/sae2.12108

FAO (2022) Climate change and plant pathogen interactions. FAO Reports

Fry E, Johnson G, Hall A, Pritchard W, Bullock J, Bardgett R (2018) Drought neutralises plant–soil feedback of two mesic grassland forbs. Oecologia 186:1113–1125. https://doi.org/10.1007/s00442-018-4082-x

Gay P, Ricauda Aimonino D, Tortia C, Gilardi G (2014) Time-temperature effect on soil-borne pathogens. In: VIII international symposium on chemical and non-chemical soil and substrate disinfestation. ISHS Acta Horticulturae, pp 231–236. https://doi.org/10.17660/ActaHortic.2014.1044.27

Gonsamo A, Ter-Mikaelian M, Chen J, Chen J (2019) Does earlier and increased spring plant growth lead to reduced summer soil moisture and plant growth on landscapes typical of tundra-taiga interface? Remote Sens 11:1989. https://doi.org/10.3390/rs11171989

Guerra C et al (2021) Global projections of the soil microbiome in the Anthropocene. Glob Ecol Biogeogr 30:987–999. https://doi.org/10.1111/geb.13273

Guo X et al (2019) Climate warming accelerates temporal scaling of grassland soil microbial biodiversity. Nat Ecol Evol 3:612–619. https://doi.org/10.1038/s41559-019-0848-8

Hart SJ, Veblen TT, Eisenhart KS, Jarvis D, Kulakowski D (2014) Drought induces spruce beetle (*Dendroctonus rufipennis*) outbreaks across northwestern Colorado. Ecology 95:930–939. https://doi.org/10.1890/13-0230.1

Hart SJ, Veblen TT, Schneider D, Molotch NP (2017) Summer and winter drought drive the initiation and spread of spruce beetle outbreak. Ecology 98:2698–2707. https://doi.org/10.1002/ecy.1963

Hawkes C, Kivlin SN, Rocca JD, Huguet V, Thomsen MA, Suttle KB (2011) Fungal community responses to precipitation. Glob Change Biol 17(5):1637–1645. https://doi.org/10.1111/j.1365-2486.2010.02327.x

Hay WT, McCormick SP, Vaughan M (2021) Effects of atmospheric CO_2 and temperature on wheat and corn susceptibility to *Fusarium graminearum* and deoxynivalenol contamination. Plan Theory 10:2582. https://doi.org/10.3390/plants10122582

Hayden H, Mele P, Bougoure D, Allan C, Norng S, Piceno Y (2012) Changes in the Microbial Community Structure of Bacteria, Archaea and Fungi in Response to Elevated CO_2 and Warming in an Australian Native Grassland Soil. Environmental Microbiology 14:3081–3096. https://doi.org/10.1111/j.1462-2920.2012.02855.x

He W, Yoo G, Moonis M, Kim Y, Chen X (2019) Impact assessment of high soil CO2 on plant growth and soil environment: a greenhouse study. PeerJ 7:e6311. https://doi.org/10.7717/peerj.6311

Hiltpold I, Moore BD, Johnson SN (2020) Elevated atmospheric carbon dioxide concentrations alter root morphology and reduce the effectiveness of entomopathogenic nematodes. Plant Soil 447:29–38

Hou S, Rodrigues O, Liu Z, Shan L, He P (2024) Small holes, big impact: stomata in plant–pathogen–climate epic trifecta. Mol Plant 17:26–49. https://doi.org/10.1016/j.molp.2023.11.011

Huot OB, Tamborindeguy C (2017) Drought stress affects *Solanum lycopersicum* susceptibility to *Bactericera cockerelli* colonization. Entomol Exp Appl 165:70–82. https://doi.org/10.1111/eea.12627

Ibáñez A, Garrido-Chamorro S, Barreiro C (2023) Microorganisms and climate change: a not so invisible effect. Microbiol Res 14:918–947. https://doi.org/10.3390/microbiolres14030064

Jansson JK, Hofmockel KS (2020) Soil microbiomes and climate change. Nat Rev Microbiol 18:35–46. https://doi.org/10.1038/s41579-019-0265-7

Khanal C, Land J (2023) Study on two nematode species suggests climate change will inflict greater crop damage. Sci Rep 13:41466. https://doi.org/10.1038/s41598-023-41466-x

Kumar S, Choudhary M, Reddy KJ, Vishwakarma VK, Kashyap VK, Sahoo S, Mukhopadhyay S (2024) A review on the impact of climate change on plant pathogen interactions. J Adv Microbiol 24:11–27. https://doi.org/10.9734/jamb/2024/v24i8843

Lahlali R, Taoussi M, Laasli SE, Gachara G, Ezzouggari R, Belabess Z, Barka EA (2024) Effects of climate change on plant pathogens and host-pathogen interactions. Crop Environ 3:159–170

Lafferty KD, Mordecai EA (2016) The rise and fall of infectious disease in a warmer world. F1000Research, 5, 2040. https://doi.org/10.12688/f1000research.8766.1

Lake JA, Wade RN (2009) Plant–pathogen interactions and elevated CO_2: morphological changes in favour of pathogens. J Exp Bot 60:3123–3131

Larsson DG, Flach CF (2022) Antibiotic resistance in the environment. Nat Rev Microbiol 20:257–269

Li HY, Zhao CA, Liu CJ, Xu XF (2010) Endophytic fungi diversity of aquatic/riparian plants and their antifungal activity in vitro. J Microbiol 48:1–6. https://doi.org/10.1007/s12275-009-0163-1

Li Z, Wang T, Yang X, Wen X, Chen W, He Y, Zhang C (2023) Microbial community function and methylmercury production in oxygen-limited paddy soil. Ecotoxicol Environ Saf 266:115585. https://doi.org/10.1016/j.ecoenv.2023.115585

Lindauer A, Maier P, Voyles J (2020) Daily fluctuating temperatures decrease growth and reproduction rate of a lethal amphibian fungal pathogen in culture. BMC Ecol 20:1–7. https://doi.org/10.1186/s12898-020-00286-7

Liu Y, He Y (2021) Warming intensifies soil pathogen negative feedback on a temperate tree. New Phytol 229:1–12. https://doi.org/10.1111/nph.17409

Liu Z, Cheng R, Xiao W, Guo Q, Wang N (2014) Effect of off-season flooding on growth, photosynthesis, carbohydrate partitioning, and nutrient uptake in *Distylium chinense*. PLoS One 9:e107636

Lu X, He M, Ding J, Siemann E (2018) Latitudinal variation in soil biota: testing the biotic interaction hypothesis with an invasive plant and a native congener. ISME J 12:2811–2822. https://doi.org/10.1038/s41396-018-0219-5

Marfenina O, Danilogorskaya A (2017) Effect of elevated temperatures on composition and diversity of microfungal communities in natural and urban boreal soils, with emphasis on potentially pathogenic species. Pedobiologia 60:11–19. https://doi.org/10.1016/j.pedobi.2016.11.002

Mcelrone AJ, Reid CD, Hoye KA, Hart E, Jackson RB (2005) Elevated CO_2 reduces disease incidence and severity of a red maple fungal pathogen via changes in host physiology and leaf chemistry. Glob Change Biol 11:1828–1836

Melotto M, Underwood W, Koczan J, Nomura K, He SY (2006) Plant stomata function in innate immunity against bacterial invasion. Cell 126:969–980

Miao Y, Han H, Du Y, Zhang Q, Jiang L, Hui D, Wan S (2017) Nonlinear responses of soil respiration to precipitation changes in a semiarid temperate steppe. Sci Rep 7:45782. https://doi.org/10.1038/srep45782

Moche M, Gutknecht J, Schulz E, Langer U, Rinklebe J (2015) Monthly dynamics of microbial community structure and their controlling factors in three floodplain soils. Soil Biol Biochem 90:169–178

Mohr PG, Cahill DM (2003) Abscisic acid influences the susceptibility of *Arabidopsis thaliana* to *Pseudomonas syringae* pv. *tomato* and *Peronospora parasitica*. Funct Plant Biol 30:461–469

Mordecai EA (2014) Pathogen impacts on plant diversity in variable environments. Oikos, 124(4):414–420. https://doi.org/10.1111/oik.01328

Naylor D et al (2020) Soil microbiomes under climate change and implications for carbon cycling. Annu Rev Environ Resour 45:29–59. https://doi.org/10.1146/annurev-environ-012320-082720

Ngo H, Cavagnaro T (2018) Interactive effects of compost and pre-planting soil moisture on plant biomass, nutrition and formation of mycorrhizas: a context dependent response. Sci Rep 8:1–9. https://doi.org/10.1038/s41598-017-18780-2

Ni M, Vellend M (2023) Soil properties constrain predicted poleward migration of plants under climate change. New Phytol 241:131–141. https://doi.org/10.1111/nph.19164

Ochoa-Hueso R, Collins S, Delgado-Baquerizo M, Hamonts K, Pockman W, Sinsabaugh R, Power S (2018) Drought consistently alters the composition of soil fungal and bacterial communities in grasslands from two continents. Glob Change Biol 24:2818–2827. https://doi.org/10.1111/gcb.14113

Orwin K, Buckland SM, Johnson D, Turner BL, Smart S, Oakley S, Bardgett RD (2015) Microbial structure and resilience in wetting-drying cycles. Funct Ecol 29:1504–1514. https://doi.org/10.1111/1365-2435.12443

Pi K, Wu L, He Y, Wang S, Liu S, Yu G (2021) The cold region critical zone in transition: responses to climate warming and land use change. Annu Rev Environ Resour 46:111–134. https://doi.org/10.1146/annurev-environ-012220-125703

Pritchard SG (2011) Soil organisms and global climate change. Plant Pathol 60(3):351–359. https://doi.org/10.1111/j.1365-3059.2010.02405.x

Riikonen J, Syrjälä L, Tulva I, Mänd P, Oksanen E, Poteri M, Vapaavuori E (2008) Stomatal characteristics and infection biology of *Pyrenopeziza betulicola* in *Betula pendula* trees grown under elevated CO2 and O3. Environ Pollut 156:536–543

Ristaino JB, Anderson PK, Bebber DP, Brauman KA, Cunniffe NJ, Fedoroff NV, Wei Q (2021) The persistent threat of emerging plant disease pandemics to global food security. Proc Natl Acad Sci USA 118:e2022239118

Rohr JR, Barrett CB, Civitello DJ, Craft ME, Delius B, DeLeo GA, Tilman D (2019) Emerging human infectious diseases and the links to global food production. Nat Sustain 2:445–456

Sanchez-Lucas R, Mayoral C, Raw M, Mousouraki MA, Luna E (2023) Elevated CO_2 alters photosynthesis, growth and susceptibility to powdery mildew of oak seedlings. Biochem J 480:1429–1443

Santos B, Pinto G, Berenguer H, Correia B, Amaral J, Gómez-Cadenas A, Alves A (2024) Physiological and hormonal responses of drought-stressed *Eucalyptus* seedlings infected with *Neofusicoccum kwambonambiense*. Plant Pathol 73:810–823

Shirai M, Eulgem T (2023) Molecular interactions between the soilborne pathogenic fungus *Macrophomina phaseolina* and its host plants. Front Plant Sci 14:1264569

Singh BK, Delgado-Baquerizo M, Egidi E, Guirado E, Leach JE, Liu H, Trivedi P (2023) Climate change impacts on plant pathogens, food security and paths forward. Nat Rev Microbiol 21:640–656

Sinha R, Irulappan V, Mohan-Raju B, Suganthi A, Senthil-Kumar M (2019) Impact of drought stress on simultaneously occurring pathogen infection in field-grown chickpea. Sci Rep 9:41463. https://doi.org/10.1038/s41598-019-41463-z

Smith F, Luna E (2023) Elevated atmospheric carbon dioxide and plant immunity to fungal pathogens: do the risks outweigh the benefits? Biochem J 480:1791–1804

Sunarti S, Kissoudis C, Van Der Hoek Y, Van Der Schoot H, Visser RG, Van Der Linden CG, Bai Y (2022) Drought stress interacts with powdery mildew infection in tomato. Front Plant Sci 13:845379

Tajul AR, Toyota K, Ohta H (2011) Effect of cold-water irrigation on bacterial wilt pathogen of tomato. Int J Pest Manag 57:341–348. https://doi.org/10.1080/09670874.2011.617134

Theriot JA, Mitchison TJ, Tilney LG, Portnoy DA (1992) The rate of actin-based motility of intracellular *Listeria monocytogenes* equals the rate of actin polymerization. Nature 357:257–260

Thomas PW (2021) Ectomycorrhiza resilience and recovery to extreme flood events in *Tuber aestivum* and *Quercus robur*. Mycorrhiza 31:511–517

Thomey M, Collins S, Vargas R, Johnson J, Brown R, Natvig D, Friggens M (2010) Effect of precipitation variability on net primary production and soil respiration in a Chihuahuan desert grassland. Glob Change Biol 17:1505–1515. https://doi.org/10.1111/j.1365-2486.2010.02363.x

Tian BY, Cao Y, Zhang KQ (2015) Metagenomic insights into communities, functions of endophytes and their associates with infection by root-knot nematode, *Meloidogyne incognita*, in tomato roots. Sci Rep 5:17087

Váry Z, Mullins E, McElwain JC, Doohan FM (2015) The severity of wheat diseases increases when plants and pathogens are acclimatized to elevated carbon dioxide. Glob Change Biol 21:2661–2669

Vishwanathan K, Zienkiewicz K, Liu Y, Janz D, Feussner I, Polle A, Haney CH (2020) Ectomycorrhizal fungi induce systemic resistance against insects on a nonmycorrhizal plant in a CERK1-dependent manner. New Phytol 228:728–740

Wahdan FMS, Hossen S, Tanunchai B, Schädler M, Buscot F, Purahong,W (2020). Future Climate Significantly Alters Fungal Plant Pathogen Dynamics during the Early Phase of Wheat Litter Decomposition. Microorganisms, 8(6):908. https://doi.org/10.3390/microorganisms8060908

Wagg C, Bender SF, Widmer F, van der Heijden MGA (2014) Soil biodiversity and soil community composition determine ecosystem multifunctionality. Proc Natl Acad Sci USA 111:5266–5270. https://doi.org/10.1073/pnas.1320054111

Wilschut R, Kleunen M (2021) Drought alters plant-soil feedback effects on biomass allocation but not on plant performance. Plant Soil 462:285–296. https://doi.org/10.1007/s11104-021-04861-9

Williams SL, Toda M, Chiller T, Brunkard JM, Litvintseva AP (2024) Effects of climate change on fungal infections. PLoS Pathog. 20(5): e1012219. https://doi.org/10.1371/journal.ppat.1012219

Wu QS, Zou YN, Huang YM (2013) The arbuscular mycorrhizal fungus *Diversispora spurca* ameliorates effects of waterlogging on growth, root system architecture and antioxidant enzyme activities of citrus seedlings. Fungal Ecol 6:37–43

Wu E, Yang L, Zhu W, Chen X, Shang L, Zhan J (2016) Diverse mechanisms shape the evolution of virulence factors in the potato late blight pathogen *Phytophthora infestans* sampled from China. Sci Rep 6:26182. https://doi.org/10.1038/srep26182

Wu L, Jing X, Zhang Q, Dobrovolskaya TG, Nawaz A, Shein EV, Tiedje JM, Yin H (2022) Reduction of microbial diversity in grassland soil is driven by long-term climate warming. Nat Microbiol 7:1054–1062. https://doi.org/10.1038/s41564-022-01147-3

Zhang Y (2023) The interaction of climate, plant, and soil factors drives putative soil fungal pathogen diversity and community structure in dry grasslands. Environ Microbiol Rep. https://doi.org/10.1111/1758-2229.13223

Zhao X, Deng H, Wang W, Feng H, Li C, Zhang H, Dai Z (2017) Impact of naturally leaking carbon dioxide on soil properties and ecosystems in the Qinghai-Tibet Plateau. Sci Rep 7:2500. https://doi.org/10.1038/s41598-017-02500-x

Part III

Microbial Solutions for Climate Resilience: Plant Interactions, Biofortification, and Extremophiles

Role of Soil Microbial Communities in Nutrient Homeostasis and Climate Resilience

9

Tanya Singh, Mohd. Mogees Ansari, and Puneet Singh Chauhan

Abstract

Soil health reflects its capacity to sustain crop growth, productivity, and long-term environmental quality. Central to this concept is beneficial microorganisms, particularly plant growth-promoting rhizobacteria (PGPR), which fulfil essential ecological functions in the rhizosphere, the area surrounding plant roots. PGPR enhance nutrient cycling and uptake suppresses harmful phytopathogens, boosts plant immune responses, and promotes growth through the production of phytohormones and other metabolites. They also play a role in bioremediation, helping to clean polluted soils. PGPR supports plant development by increasing the availability of crucial nutrients like nitrogen, phosphorus, and iron and regulating growth and stress-related hormones, such as reducing ethylene levels through ACC deaminase activity. These microbes also contribute indirectly by protecting plants against pathogens through competition for nutrients, producing biocontrol compounds, synthesizing enzymes that break down pathogen cell walls, and triggering systemic resistance in plants. Under nutrient-deprived conditions, PGPR enhances plant fitness, bolsters stress tolerance, and can assist in nutrient cycling for better bioavailability of nutrients. This chapter highlights the mechanisms of PGPR and their potential as eco-friendly biotechnological tools for sustainable agriculture, reducing dependence on agrochemicals and encouraging more resilient, climate-adapted farming practices.

T. Singh · M. M. Ansari · P. S. Chauhan (✉)
CSIR-National Botanical Research Institute (CSIR-NBRI), Rana Pratap Marg, Lucknow, India

Academy of Scientific and Innovative Research (AcSIR), Ghaziabad, India
e-mail: puneet@nbri.res.in

P. Pandey et al. (eds.), *Microorganisms Resilience to Climate Change*,
Microorganisms for Sustainability 56, https://doi.org/10.1007/978-981-96-3748-5_9

Keywords
Biofertilizers · Nitrogen fixation · Plant-microbe interactions · Soil microbiome

9.1 Introduction

Soil microbial communities are integral to maintaining ecosystem health, enhancing soil productivity, and promoting agricultural sustainability. These communities consist of diverse microorganisms, including bacteria, fungi, protozoa, archaea, and viruses, all of which perform critical functions within soil ecosystems. Their activities, such as organic matter decomposition, nutrient cycling, and facilitation of plant growth, are fundamental to sustaining soil health and supporting plant development. Moreover, microbial activity influences essential ecosystem services, including carbon sequestration, nutrient availability, and water regulation, underscoring their importance in both natural and managed ecosystems (Bertini and Azevedo 2022; Elbasiouny et al. 2022; Mason et al. 2023).

The diversity within soil microbial communities is remarkable, with a single gram of soil housing billions of organisms and thousands of microbial species (Torsvik and Øvreås 2002). This diversity encompasses key functional groups of microorganisms, each contributing uniquely to soil processes. For example, bacteria play pivotal roles in nutrient cycling, such as nitrogen fixation and phosphorus solubilization. Specific bacteria, such as *Rhizobium* (which forms nitrogen-fixing symbioses with legumes) and *Nitrosomonas* (which facilitates nitrification), are instrumental in enhancing nutrient availability for plants (Zahran 1999; Rodriguez et al. 2021). Similarly, fungi, particularly saprotrophic and mycorrhizal species, are essential for breaking down complex organic matter like lignin and cellulose, which are otherwise resistant to decomposition. Mycorrhizal fungi, in particular, form symbiotic associations with plant roots, improving nutrient uptake and plant growth (Wahab et al. 2023). Protozoa, another crucial group, regulate bacterial populations through predation, promoting nutrient cycling by releasing nutrients back into the soil for plant use. Archaea, although less studied, also play important roles, particularly in nitrogen cycling under anaerobic conditions (Taffner et al. 2018). Soil viruses, such as bacteriophages, further influence microbial community dynamics by modulating bacterial populations, which can affect overall soil functionality.

The diversity and complexity of soil microbial communities are critical for ecosystem resilience, productivity, and stability. High microbial diversity ensures that multiple soil functions are maintained, even under environmental stressors like drought, disease, and pollution (Chauhan et al. 2023). This resilience is particularly valuable in agricultural systems, where fostering microbial diversity can enhance crop productivity, reduce dependence on chemical fertilizers and pesticides, and sustain soil health over the long term. Moreover, diverse microbial communities contribute to suppressing soilborne pathogens, improving crop resilience to diseases and reducing the need for chemical interventions (Ansari et al. 2024a). Despite their significance, soil microbial communities are increasingly threatened by intensive agriculture, deforestation, and pollution, which can disrupt microbial diversity and impair soil functionality (Qiu et al. 2021; Dixit et al. 2024). Degraded microbial communities can reduce soil fertility and diminish ecosystem services, posing challenges for sustainable land use and food production (Dixit et al. 2024).

Given the essential roles that soil microbial communities play in ecosystem functioning and agricultural sustainability, there is a growing need to understand better their diversity, interactions, and contributions to soil health. Advancing our knowledge in this area will enable the development of more effective strategies for sustainable land management, climate resilience, and mitigating agriculture's environmental impacts. Preserving the diversity and functionality of soil microbial communities is imperative to ensuring the long-term productivity and sustainability of both natural and managed ecosystems.

9.2 Soil Microbes and Nutrient Cycling

Soil microbial communities are essential drivers of nutrient cycling, a complex series of biochemical processes through which nutrients are transformed and made available for plant uptake (Table 9.1).

Microbes in the soil play key roles in breaking down organic matter, transforming and mobilizing nutrients, and contributing to carbon storage. These processes underpin soil fertility and support plant growth, contributing to environmental resilience and climate regulation (Ansari et al. 2024b).

9.2.1 Role in Decomposition and Organic Matter Breakdown

Decomposition is a fundamental process in nutrient cycling where organic matter, such as dead plants, animals, and microbial cells, is broken down into simpler compounds. Soil microbes, including bacteria and fungi, are the primary agents of decomposition. These microorganisms secrete enzymes that break down complex organic molecules like cellulose, lignin, and proteins into smaller molecules (Datta et al. 2017; Datta 2024). As a result, essential nutrients such as nitrogen (N), phosphorus (P), and sulfur (S) are released into the soil in forms that plants can readily absorb.

Fungi are particularly effective in decomposing lignin, a complex polymer found in plant cell walls, which most bacteria cannot break down. Fungi, such as those in the genera *Trichoderma* and *Penicillium*, play a significant role in decomposing plant materials, contributing to the formation of humus—a stable, organic component of soil that enhances soil structure, water-holding capacity, and nutrient retention (Macías-Rodríguez et al. 2020; Corbu et al. 2023). Bacteria, on the other hand, are more effective at breaking down simpler organic molecules. Together, fungi and bacteria facilitate organic matter decomposition, supporting soil fertility and enabling the recycling of nutrients that would otherwise remain locked in dead material (Ahmed et al. 2023).

Table 9.1 Mechanism of PGPR for nutrient bioavailability

Nutrient	Bacterial Strain	Mechanism	References
Nitrogen	*Azospirillum brasilense*	Nitrogen fixation and ammonium release	Damasceno et al. (2024)
	Rhizobium leguminosarum	Symbiotic nitrogen fixation in root nodules	Lindström and Mousavi (2020)
	Gluconacetobacter diazotrophicus	Endophytic nitrogen fixation in sugarcane	Kumar et al. (2024)
Phosphate	*Pseudomonas fluorescens*	Phosphate solubilization via organic acid production	Ahmad et al. (2023)
	Bacillus megaterium	Phosphate mineralization and organic acid secretion	Rawat et al. (2021)
	Bacillus subtilis	Phosphate solubilization, organic acid secretion	Timofeeva et al. (2022)
Potassium	*Bacillus mucilaginosus*	Potassium solubilization from silicate minerals	Maysoon and Hassan (2024)
	Paenibacillus polymyxa	Organic acid production facilitating potassium release	Tian et al. (2024)
Micronutrients			
Iron (Fe)	*Serratia marcescens*	Siderophore secretion	Srivastava (2023)
	Azospirillum brasilens	Siderophore production	Gul et al. (2023)
	Pseudomonas putida	Siderophore production to chelate iron, enhance uptake	Lozano-González et al. (2023)
Zinc (Zn)	*Bacillus subtilis*	Zinc solubilization, siderophore activity	Masood et al. (2022)
	Burkholderia cepacia	Zinc mobilization through acid production	Bhakat et al. (2021)
Manganese (Mn)	*Rhizobium leguminosarum*	Chelation and solubilization	Khoshru et al. (2023)
Copper (Cu)	*Pseudomonas putida*	Organic acid secretion for solubilization	Kafi et al. (2021)
Molybdenum (Mo)	*Enterobacter cloacae*	Phosphate solubilization, enzyme activity	Yakasai et al. (2021)
Cobalt (Co)	*Bradyrhizobium japonicum*	Siderophore production, nutrient uptake	Li et al. (2024)
Nickel (Ni)	*Acinetobacter calcoaceticus*	Chelation and solubilization	Kour et al. (2024)

9.2.2 Nitrogen Cycling: Nitrogen Fixation, Nitrification, and Denitrification

Nitrogen is a critical nutrient for plants, as it is a fundamental component of amino acids, nucleic acids, and chlorophyll. However, atmospheric nitrogen (N_2) is inert and unusable by most plants (Ansari et al. 2024b). Soil microbes convert this nitrogen into biologically accessible forms through three key processes: nitrogen fixation, nitrification, and denitrification.

1. **Nitrogen Fixation**: Some bacteria, including free-living bacteria (e.g., *Azotobacter*) and symbiotic bacteria (e.g., *Rhizobium* associated with legumes), are capable of fixing atmospheric nitrogen by converting it into ammonia (NH_3), which plants can utilize (Fatima et al. 2019). This process is essential for replenishing nitrogen in soils, particularly in ecosystems where nitrogen is limited (Abd-Alla et al. 2023). Symbiotic nitrogen fixation occurs when specific bacteria infect plant roots, forming nodules in which the bacteria convert atmospheric nitrogen to ammonia in exchange for carbon from the plant. Notably, *Azospirillum* has improved nitrogen uptake and growth in wheat, increasing productivity without heavy reliance on synthetic fertilizers (Galindo et al. 2022).
2. **Nitrification**: This is a two-step process where ammonia is converted first into nitrite (NO_2^-) by bacteria such as *Nitrosomonas*, and then into nitrate (NO_3^-) by *Nitrobacter*. Nitrate is a preferred nitrogen source for most plants because it is highly mobile in soil and readily taken up by plant roots. Nitrification is crucial in agricultural soils, where nitrate availability is often linked to crop productivity. However, the process also has implications for environmental sustainability, as nitrates can leach into groundwater and contribute to water pollution.
3. **Denitrification**: In contrast to nitrification, denitrification is a process in which nitrate is reduced back to gaseous forms of nitrogen, including N_2 and nitrous oxide (N_2O), by denitrifying bacteria under anaerobic (oxygen-limited) conditions. Denitrification, carried out by bacteria such as *Pseudomonas* and *Clostridium*, helps regulate the amount of nitrogen in the soil and prevents the excessive buildup of nitrates (De Mandal et al. 2020; Salas et al. 2021). Although denitrification is essential for balancing the nitrogen cycle, it also releases nitrous oxide, a potent greenhouse gas, which can impact climate.

Together, these nitrogen cycling processes maintain a balance of nitrogen in the soil, replenishing nitrogen lost through plant uptake, leaching, or volatilization and ensuring a sustainable supply for plants and soil organisms (Stein and Klotz 2016).

9.2.3 Phosphorus Solubilization and Mobilization

Phosphorus is another essential nutrient for plants and is involved in energy transfer, photosynthesis, and cellular division. However, phosphorus in soil is often present in insoluble forms and unavailable to plants (Singh et al. 2024a). Soil microbes, particularly phosphate-solubilizing bacteria (PSB) and mycorrhizal fungi, play a key role in converting insoluble phosphorus compounds into bioavailable forms through a process known as phosphorus solubilization and mobilization.

Phosphate-solubilizing bacteria, such as species of *Pseudomonas* and *Bacillus*, produce organic acids that lower the soil pH, releasing phosphorus bound to minerals. These organic acids chelate metal ions, freeing phosphate ions for plant uptake (Elhaissoufi et al. 2022). Mycorrhizal fungi, particularly arbuscular mycorrhizal fungi (AMF), form symbiotic relationships with plant roots and help in phosphorus acquisition. The hyphal network of these fungi extends into the soil, reaching phosphorus resources beyond the root zone and transferring it to the plant in exchange for carbohydrates (Bonfante and Genre 2010). This association is especially beneficial in soils with low phosphorus availability, enhancing plant growth and productivity (Bonfante and Genre 2010; Etesami et al. 2021).

The microbial-mediated phosphorus mobilisation is vital for agricultural systems, where phosphorus is often a limiting nutrient. By promoting phosphorus solubilization, soil microbes reduce the need for chemical fertilizers and improve phosphorus use efficiency in crops, contributing to sustainable soil management practices (Ansari et al. 2024b). For example, *Bacillus megaterium* increases phosphorus uptake in maize, leading to a 20% increase in biomass (de Sousa et al. 2021). Additionally, inoculating wheat with PSB has been shown to boost phosphorus uptake and improve growth and yield (Wang et al. 2022). These findings highlight the role of PSB in enhancing phosphorus availability, particularly in calcareous soils.

9.2.4 Potassium Solubilization

Potassium is vital for plant growth, but its availability in soil is often limited as it primarily exists in forms that plants cannot directly use. Potassium-solubilizing bacteria (KSB), such as *Bacillus mucilaginosus* and *Pseudomonas* sp., help convert these insoluble minerals into absorbable forms through the release of organic acids and enzymes (Sindhu et al. 2016). These bacteria are especially valuable in potassium-deficient soils, enhancing uptake and improving plant resilience (Nawaz et al. 2023). KSB has been shown to boost potassium absorption in crops like rice and tomatoes, resulting in better growth and drought tolerance (Muthuraja et al. 2023). For example, rice inoculated with *Bacillus mucilaginosus* exhibited significant increases in potassium uptake and yield, underscoring the role of these bacteria in sustainable nutrient management (Vijayakumar et al. 2024).

9.2.5 Carbon Cycling: Microbial Contributions to Carbon Sequestration

Carbon cycling in soil is a dynamic process mediated mainly by microorganisms that break down organic matter and transform carbon-containing compounds. Microbes play dual roles in the carbon cycle: they release carbon dioxide (CO_2) into the atmosphere through respiration and contribute to carbon sequestration by forming stable organic matter that stores carbon in the soil (Falkowski et al. 2000).

Decomposer microbes, mainly fungi and bacteria, break down plant and animal residues, releasing CO_2 as a byproduct of respiration. However, not all organic carbon is lost as CO_2; some is converted into stable organic forms, such as humus, that are resistant to further decomposition. This stable organic matter, known as soil organic carbon (SOC), acts as a long-term carbon sink, reducing atmospheric CO_2 levels and mitigating climate change (Gao et al. 2024).

Mycorrhizal fungi, particularly ectomycorrhizal fungi associated with forest trees, contribute significantly to carbon storage in forest soils. By forming extensive networks around roots, these fungi stabilize carbon in soil aggregates and reduce the decomposition of soil organic matter (Husain et al. 2024). In agricultural soils, practices that promote microbial diversity and organic matter inputs, such as cover cropping, crop rotation, and reduced tillage, can enhance microbial activity and contribute to soil carbon sequestration (Esmaeilzadeh-Salestani et al. 2021; Zhu et al. 2024).

Carbon sequestration in soils is critical to climate mitigation, as it reduces atmospheric CO_2 levels while enhancing soil fertility. Microbial contributions to carbon cycling and sequestration are therefore essential for sustaining soil productivity and addressing climate change.

9.3 Mechanisms of Nutrient Homeostasis in Soil

Soil nutrient homeostasis, or the balance of essential nutrients within the soil, is crucial for sustainable plant growth, ecosystem stability, and agricultural productivity. Microbial communities play a central role in regulating nutrient availability, mediated through various mechanisms involving interactions with plant roots, symbiotic associations, microbial adaptations, and metabolite production (Shi et al. 2024). These microbial processes are fundamental for the soil ecosystem and highlight the dynamic interplay between soil biota and nutrient cycles.

9.3.1 Microbial Interactions with Plant Roots

The rhizosphere, the narrow region of soil influenced by root exudates and microbial activity, is a critical zone for nutrient cycling. Here, plant roots actively interact with diverse microbial communities, altering nutrient availability and cycling processes. Roots release various organic compounds, such as amino acids, sugars, organic acids, and secondary metabolites, which attract and stimulate the growth of specific microbial populations (Singh et al. 2024b). These root exudates serve as carbon sources for microbes and signals that can influence microbial metabolic pathways. Microbes in the rhizosphere play a significant role in nutrient solubilization, especially for relatively immobile nutrients like phosphorus, iron, and potassium (Fig. 9.1).

Certain bacteria and fungi produce organic acids and enzymes that break down complex nutrient compounds, making nutrients more accessible to plants (Ansari et al. 2024b). For example, phosphate-solubilizing bacteria convert insoluble forms of phosphate into bioavailable forms through organic acid production, which chelates metal ions and releases phosphate (Singh et al. 2024a).

Similarly, rhizosphere bacteria can enhance iron bioavailability by producing siderophores, molecules that bind tightly to iron ions and facilitate iron uptake by both microbes and plants. This mechanism improves iron uptake and decreases iron

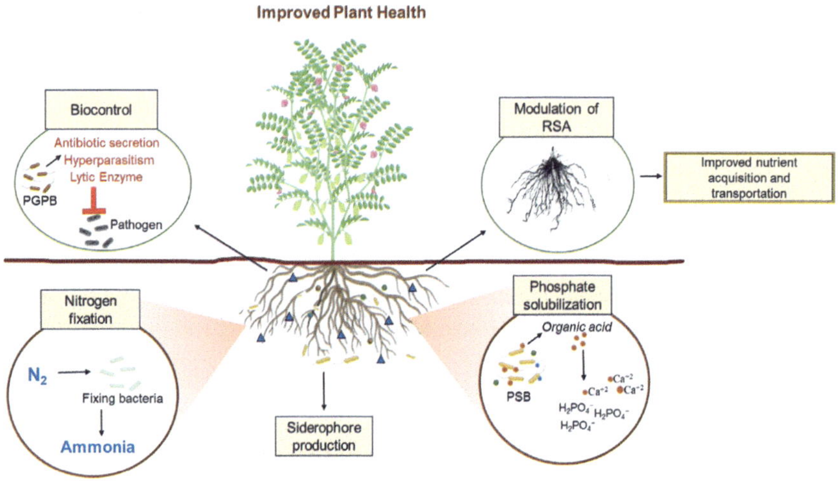

Fig. 9.1 Role of soil microorganisms in enhancing nutrient availability and providing biocontrol mechanisms for improved plant growth

availability to pathogens, providing indirect disease protection (Ansari et al. 2024a). Microbes like *Pseudomonas* and *Azotobacter vinelandii* are known for their siderophore production, supporting plant health in iron-deficient environments as a sustainable alternative to synthetic chelators (Singh et al. 2022a, 2022b). Additionally, other beneficial bacteria assist in mobilizing micronutrients like zinc, nickel and molybdenum. Zinc-solubilizing bacteria facilitate zinc availability, essential for enzyme function and protein synthesis, thereby enhancing plant health and improving yield and nutritional quality (Gorain et al. 2022). The active nutrient cycling in the rhizosphere establishes a balanced nutrient environment that supports plant growth while fostering microbial diversity.

9.3.2 Symbiotic Relationships

Symbiotic relationships between plants and microbes, particularly mycorrhizal associations and nitrogen-fixing bacteria, are vital for nutrient homeostasis in many ecosystems. Mycorrhizal fungi, which form associations with plant roots, expand the root's absorptive area, allowing for more effective nutrient uptake (Ansari et al. 2024b). Two major types of mycorrhizal fungi are AMF and ectomycorrhizal fungi, both of which aid in the uptake of phosphorus, nitrogen, and other nutrients. In exchange for nutrients, plants provide mycorrhizal fungi with a steady supply of carbon compounds. Through this mutualistic relationship, mycorrhizal fungi improve plant nutrient uptake, particularly under conditions where nutrients are scarce or immobilized within the soil matrix (Owiny and Dusengemungu 2024).

Rhizobia bacteria form another essential symbiotic association, particularly with legumes. Through the formation of root nodules, rhizobia fix atmospheric nitrogen into ammonia, a form that plants can readily use. This process is energy-intensive and requires a specialized enzyme complex, nitrogenase, which is effective only under low oxygen conditions. To support this, leguminous plants provide a protective nodule environment, maintaining low oxygen levels while supplying the bacteria with carbon resources. Rhizobial nitrogen fixation not only enhances nitrogen availability for the host plant but also contributes to the overall nitrogen pool in the soil, benefiting surrounding plants as well (Granada Agudelo et al. 2023).

9.3.3 Microbial Adaptations to Maintain Nutrient Equilibrium

Microbial communities in the soil are continuously adapting to maintain nutrient equilibrium in response to fluctuating nutrient availability and environmental conditions. Microbes employ various strategies, such as metabolic flexibility, dormancy, and competitive exclusion, to optimize nutrient use and storage (Pin et al. 2013; Chen et al. 2021). For instance, some bacteria and fungi are capable of shifting between different metabolic pathways depending on nutrient availability. When nitrogen is scarce, certain bacteria switch to processes such as denitrification or nitrogen fixation, while other microbes may rely on alternative nitrogen sources, such as amino acids or nitrates, to sustain their growth (Wang et al. 2016).

Dormancy is another adaptive strategy that allows microbes to withstand periods of nutrient scarcity. Certain microbial populations can form spores or enter a quiescent state, reducing their metabolic activity and conserving energy until conditions become favorable. This ability to transition between active and dormant states ensures that microbial communities can endure environmental stress while conserving essential nutrients within the soil (Wang et al. 2014). Microbial interactions, such as competition and cooperation, also contribute to nutrient homeostasis. In nutrient-limited soils, microbes compete for resources, which can lead to the exclusion of less competitive species (Hammarlund et al. 2019). However, in some cases, microbes exhibit cooperative behavior, such as cross-feeding, where one species produces metabolites that serve as nutrients for another species (Hammarlund et al. 2019). Such interactions promote a stable and resilient microbial community, which in turn maintains soil nutrient equilibrium.

9.3.4 Influence of Microbial Metabolites on Nutrient Availability

Microbial metabolites, including organic acids, enzymes, and signaling molecules, play a significant role in modifying soil nutrient dynamics. For instance, organic acids, produced by various rhizosphere microbes, can chelate metal ions, thus mobilizing nutrients like phosphorus, calcium, and magnesium (Ansari et al. 2024b). By altering the soil pH, organic acids create a favorable environment for nutrient solubilization, enhancing nutrient availability (Singh et al. 2024a).

Enzymes produced by microbes further facilitate nutrient cycling by breaking down complex organic compounds in the soil. Extracellular enzymes like phosphatases, cellulases, and proteases decompose organic matter into simpler forms, releasing nutrients such as nitrogen and phosphorus (Daunoras et al. 2024). The release of these enzymes is often regulated by nutrient availability; for example, microbes may upregulate phosphatase production in response to phosphorus deficiency, ensuring a steady supply of this nutrient. Microbial signaling molecules, such as phytohormones and other secondary metabolites, can also influence nutrient availability indirectly by affecting plant physiology and root architecture (Singh et al. 2024b). Certain bacteria, known as plant growth-promoting rhizobacteria (PGPR), produce phytohormones like auxins, cytokinins, and gibberellins, which enhance root growth and increase root surface area, improving the plant's ability to access soil nutrients (Singh et al. 2024b). These metabolites not only foster better nutrient uptake by plants but also influence the surrounding microbial community, promoting a self-sustaining nutrient cycle within the rhizosphere.

9.4 Soil Microbial Influence on Plant Health and Productivity

Soil microorganisms play crucial roles in enhancing plant health and productivity by facilitating nutrient acquisition, providing biocontrol, and promoting growth. Through symbiotic relationships with plants, these microorganisms improve nutrient availability, offer protection against pathogens, and enhance resilience under stress, all of which contribute to increased crop yield and soil health (Harman et al. 2021; Astapati and Nath 2023). The integration of microbial inoculants, particularly PGPRs and biocontrol agents, into agricultural practices can lead to more sustainable and productive cropping systems, ultimately supporting global food security in an environmentally friendly manner (Sammauria et al. 2020; Elnahal et al. 2022).

9.4.1 Microbial Enhancement of Plant Nutrient Uptake

Soil microorganisms are crucial for improving plant nutrient uptake, particularly in challenging soil conditions (Das et al. 2022). Their symbiotic relationships with plants enhance nutrient acquisition, improve soil structure, and bolster plant resilience (Astapati and Nath 2023). The primary microbial contributors to nutrient acquisition are various beneficial bacteria that work both directly and indirectly to enhance plants' access to essential nutrients such as phosphorus, nitrogen, and micronutrients under stressful conditions (Table 9.2) (Singh et al. 2022a, 2022b).

Nutrient availability in soil doesn't always ensure plant access, as root development and microbial competition can limit nutrient uptake. While soil may contain essential nutrients in sufficient quantities, plants rely on their root systems for uptake, and nutrient access depends heavily on root growth and functionality. Microorganisms in the soil, particularly PGPR, play an influential role in helping plants overcome these limitations and improve nutrient acquisition (Ansari et al. 2024b).

Table 9.2 Role of PGPR for enhanced yield

Host crop	Bacterial Strain	Mechanism	References
Wheat	*Azospirillum brasilense*	Nitrogen fixation, root growth stimulation	Santos et al. (2017)
Maize	*Pseudomonas fluorescens*	Phosphate solubilization, siderophore production	Timofeeva et al. (2023)
Rice	*Bacillus subtilis*	Induced systemic resistance, enzyme secretion	Velmurugan et al. (2023)
Tomato	*Rhizobium leguminosarum*	Auxin production, root development	Backer et al. (2018)
Soybean	*Bradyrhizobium japonicum*	Nitrogen fixation, plant hormone modulation	Chen et al. (2023)
Potato	*Azotobacter chroococcum*	Phytohormone production, nutrient uptake	Wani et al. (2016)
Pea	*Mesorhizobium ciceri*	Nitrogen fixation, siderophore production	Pandey et al. (2018)
Sugarcane	*Gluconacetobacter diazotrophicus*	Nitrogen fixation, biofilm formation	Schwister et al. (2022)
Sunflower	*Pseudomonas putida*	Phosphate solubilization, auxin production	Shahid et al. (2012)
Chickpea	*Enterobacter cloacae*	Growth hormone production, antifungal activity	Mukherjee et al. (2020)

Roots, especially root hairs, are the main sites for water and nutrient absorption (Singh et al. 2023). However, studies show that microorganisms can initially outcompete plant roots, especially for nutrients like nitrogen. According to Kuzyakov and Xu (2013), soil microbes often access nutrients more quickly than roots due to their rapid life cycles and decomposition abilities. Although microbes have an early advantage in nutrient uptake, expanding root systems gradually improve plants' ability to compete by providing increased surface area for nutrient transport, which becomes more effective as the season progresses. PGPR are particularly beneficial for enhancing root development and nutrient uptake (Table 9.3).

These bacteria produce important compounds like indole-3-acetic acid (IAA), a plant hormone that promotes the formation of lateral roots, which expands the root surface area. This increase in root surface enables plants to absorb water and nutrients more efficiently, supporting overall growth and resilience (Singh et al. 2023). By encouraging lateral and adventitious roots, IAA-producing PGPR effectively extends the plant's reach in the soil, maximizing its ability to access essential resources.

Another critical way PGPR aids root growth is through the production of ACC deaminase, an enzyme that helps plants under stress. Environmental stresses such as heavy metals, salinity, and drought can cause plants to produce excessive ethylene, a hormone that, at high levels, can hinder root growth. ACC deaminase breaks down ACC, the precursor to ethylene, thus lowering ethylene levels and allowing roots to grow even under adverse conditions. By maintaining root growth during stress, this

Table 9.3 PGPR influencing root traits under nutrient-deprived condition

Host plant	Bacterial strain	Root Traits	References
Wheat	*Azospirillum brasilense*	Inhibited root length while enhancing root hair formation	Silva et al. (2022)
Maize	*Azospirillum lipoferum*	Increased root biomass, root length, diameter, larger numbers of tips	Czarnes et al. (2020)
Rice	*Bacillus amyloliquefaciens*	Increased root surface area, improved root biomass	Cheng et al. (2021)
	Burkholderia pyrrocinia	Improved root length and diameter, expansion of cortex and aerenchyma	Rego et al. (2014)
Soybean	*Bradyrhizobium japonicum*	Improved root length, surface area, root volume, number of nodules per plant root	Egamberdieva et al. (2018)
Tomato	*Rhizobium leguminosarum*	Root branching, root length extension	Mayak et al. (2004)
Pea	*Bacillus simplex, Rhizobium leguminosarum*	Emergence of more lateral roots, highly branched vascular bundles, improved root dry weight	Grover et al. (2021)
Sunflower	*Pseudomonas putida*	Root hair formation, root system expansion	Shahid et al. (2012)
Chickpea	*Mesorhizobium ciceri*	Enhanced nodulation, root biomass increase	Zhang et al. (2024)
Sugarcane	*Gluconacetobacter diazotrophicus*	Root biomass increase, fine root development	dos Santos et al. (2019)
Barley	*Paenibacillus polymyxa*	Increased lateral root growth, root hair density	Xu et al. (2023)
Peanut	*Bacillus megaterium*	Improved root length, root dry weight	Yao et al. (2021)
Common bean	*Rhizobium phaseol*	Improved root length, root dry weight, nodule dry weight	Sofi et al. (2021)

enzyme helps create larger and healthier root systems that improve water and nutrient uptake capacity (Zahir et al. 2009; Etesami and Adl 2020).

9.4.2 Biocontrol Agents Against Pathogens

Soil microbes play a vital role in protecting plants from pathogenic infections, serving as biocontrol agents that suppress or eliminate disease-causing organisms. They employ various mechanisms, including competition, antagonism, and the induction of plant defense responses (Ansari et al. 2024a). The action of biocontrol agents (BCA) against pathogens relies on multiple mechanisms. Competitive exclusion is one such method, where BCAs outcompete pathogens for nutrients and space, reducing pathogen survival. The production of antimicrobial compounds is another effective method, with many BCAs secreting antibiotics, toxins, or enzymes that directly inhibit pathogen growth or degrade pathogen cell structures. Biofilm formation is another protective mechanism, where certain BCAs create a biofilm on plant surfaces, serving as a physical and chemical barrier against pathogen colonization. Additionally, some BCAs can stimulate plant defense mechanisms, such as ISR, wherein plants develop a heightened state of alertness that enables them to respond more effectively to pathogen attacks (Ansari et al. 2024a).

A prominent class of microbial BCAs are antagonistic bacteria, particularly species within the *Bacillus*, *Pseudomonas*, and *Streptomyces* genera (Lee et al. 2023). *Bacillus* species, are capable of producing a broad spectrum of antibiotics, such as iturins, surfactins, and fengycins, which directly inhibit the growth of fungal pathogens (Meena and Kanwar 2015). These bacteria also have robust environmental adaptability, making them suitable for different soil types and climate conditions. Similarly, *Pseudomonas* species are known for synthesizing secondary metabolites like phenazines and pyoluteorin, compounds that effectively hinder pathogen development (Liu et al. 2022). *Streptomyces* species, known for their prolific production of antibiotics, play an essential role in suppressing soilborne pathogens through their antimicrobial activity and growth-promoting properties (Dow et al. 2023).

Fungal BCAs, such as *Trichoderma* species, are also widely studied and used for pathogen control due to their versatility in different agricultural environments. *Trichoderma* fungi exhibit antagonistic activity against a wide range of fungal pathogens by producing hydrolytic enzymes like chitinases and glucanases, which degrade the cell walls of target pathogens (Tyśkiewicz et al. 2022). Additionally, *Trichoderma* can colonize plant roots, creating a protective barrier that limits pathogen access while also stimulating plant growth and resistance through ISR pathways. These fungi are especially effective against soilborne diseases, as they can persist in soil environments and provide long-term protection (Tyśkiewicz et al. 2022).

9.5 Impact of Human Activity on Soil Microbial Communities

Human activities, including intensive agricultural practices, land use changes, urbanization, and pollution, have profound impacts on soil microbial communities, altering their composition, diversity, and function.

9.5.1 Impact of Agricultural Practices

Tillage, the mechanical turning of soil, disrupts its structure and microbial habitats. While it controls weeds and prepares the ground for planting, intensive tillage exposes organic matter to oxygen, accelerating decomposition and nutrient depletion. This adversely affects soil microbial diversity, particularly beneficial fungi like mycorrhizae that aid in nutrient absorption (Mo et al. 2024). Additionally, tillage increases erosion, resulting in the loss of nutrient-rich topsoil and reduced soil productivity (Popoola et al. 2024). Monoculture, the repeated cultivation of the same crop, significantly reduces plant and soil microbial diversity. This practice encourages crop-specific pests and increases reliance on pesticides, ultimately leading to soil degradation (Mir et al. 2022). In contrast, polyculture systems that use crop rotation enhance microbial diversity and soil health by providing varied organic matter and root structures that improve nutrient cycling (Ramasamy et al. 2024).

Pesticides are used to control pests but can harm non-target soil microorganisms, particularly beneficial bacteria and fungi. Broad-spectrum pesticides decrease microbial diversity and disrupt key processes like nitrogen fixation and organic matter decomposition (Zhou et al. 2024). Some pesticides accumulate in the soil, leading to long-term changes in microbial communities. For instance, fungicides and herbicides have been linked to declines in mycorrhizal fungi, negatively affecting plant nutrient uptake and soil health (Gikas et al. 2022; Yu et al. 2023). Applying synthetic fertilizers alters soil pH and disrupts microbial balance by increasing nutrient levels, particularly nitrogen and phosphorus (Pahalvi et al. 2021). While this can boost short-term plant growth, prolonged use of fertilizers may create an imbalance in soil nutrients, often depleting essential trace minerals and diminishing microbial diversity. For instance, excess nitrogen can negatively affect nitrogen-fixing bacteria, interrupting the natural nitrogen cycle. Furthermore, fertilizers can contribute to soil acidification, creating unfavourable conditions for many beneficial microbes and ultimately leading to soil degradation over time (Tripathi et al. 2020; Dincă et al. 2022).

9.5.2 Influence of Land Use Changes and Urbanization

Land use changes and urbanization profoundly affect soil microbial communities, significantly altering their composition, diversity, and ecological functions. When natural landscapes are converted into urban areas or used for agriculture, the soil ecosystem experiences shift in both the abundance and diversity of microorganisms. Urbanization causes various physical and chemical disturbances, including soil compaction, a decrease in organic matter, and the introduction of pollutants, all of which impact soil microbial life. Research indicates that urban soils often harbor distinct microbial communities when compared to rural soils, influenced by changes in pH, organic carbon content, and moisture levels (Sapkota et al. 2021). For example, urban forest soils typically demonstrate a shift from a dominance of Acidobacteria in rural areas to a prevalence of Proteobacteria in urban environments, reflecting their responses to altered nutrient availability and increased disturbances (Bouchali 2022).

The conversion of natural landscapes into urban or agricultural areas reduces habitat complexity and limits niches for diverse microbial communities. Agricultural practices that emphasize monoculture can further suppress microbial diversity and promote certain groups over others. Additionally, urbanization fragments ecosystems, reducing overall biodiversity and favoring generalist species. These changes in microbial community structure can lead to decreased ecosystem functions, such as carbon sequestration and nutrient cycling (Gao et al. 2022).

9.5.3 Soil Pollution and Soil Degradation

Human activity has significantly contributed to soil pollution and degradation, primarily through practices associated with industrialization, agriculture, and urbanization. Soil pollution refers to the accumulation of harmful chemicals or biological contaminants in the soil, which often stems from unsustainable practices (Kumari et al. 2024). Common sources of such pollution include the excessive use of pesticides and fertilizers in agriculture, industrial discharges, mining activities, and improper waste disposal. These contaminants disrupt soil ecosystems, diminish soil fertility, and can cause long-term harm to both human health and the environment. For instance, pollutants such as heavy metals and toxic organic compounds can accumulate in the soil, threatening water quality and food safety as they move through ecosystems and enter the food chain (UNEP, 2021).

Soil degradation, closely linked to pollution, involves a decline in soil quality that reduces productivity and biodiversity (Lal 2015). Factors such as deforestation, overgrazing, andintensive agricultural practices, like monocultures and extensive tilling, exacerbate this issue by depleting soil organic matter, disrupting nutrient cycles, and causing erosion (Rhodes 2014). For instance, in Haiti, deforestation and erosion have decreased land fertility, severely impacting agriculture and food security. Similarly, the U.S. Dust Bowl illustrates how poor agricultural management can lead to significant economic and ecological consequences (Williams 2011; Lee and Gill 2015).

The environmental and economic consequences of soil pollution and degradation are considerable. Degraded soils often result in reduced crop yields and heightened food insecurity, particularly in vulnerable regions. Furthermore, the decline in ecosystem health contributes to global climate change, as soil carbon is released into the atmosphere during erosion. Addressing these challenges necessitates a combination of strategies aimed at restoring soil health and minimizing contaminant inputs. This includes adopting sustainable farming practices, enforcing environmental protections, and rehabilitating degraded lands through reforestation and organic soil amendments. These approaches are vital for preserving biodiversity, enhancing soil fertility, and fostering the overall resilience of ecosystems (Begum 2021; UNEP, 2021).

9.6 Strategies for Enhancing Soil Microbial Function for Sustainability

In sustainable agriculture, fostering a balanced and diverse microbial community is fundamental to achieving high soil quality and resilience against degradation. Sustainable practices that enhance microbial function in soil, such as organic farming, crop rotation, the use of biofertilizers, and soil management strategies, contribute to nutrient cycling, soil structure improvement, and carbon sequestration.

9.6.1 Organic Farming and Its Role in Promoting Beneficial Microbes

Organic farming is an agricultural practice that avoids synthetic fertilizers, pesticides, and genetically modified organisms, emphasizing natural soil fertility and ecological balance. By relying on organic inputs like compost, green manure, and animal waste, organic farming enriches the soil with organic matter, providing a conducive environment for beneficial microbial communities. Organic matter serves as a food source for soil microbes, enhancing microbial diversity and activity (Ghimire et al. 2024). This increased microbial activity contributes to the decomposition of organic residues and the release of nutrients, which are then made available for plant uptake.

The absence of synthetic chemicals in organic farming reduces the risk of microbial toxicity, allowing beneficial organisms such as nitrogen-fixing bacteria, phosphorus-solubilizing bacteria, and mycorrhizal fungi to thrive (Bisht and Chauhan 2020). Moreover, organic farming practices often lead to a balanced microbial community structure that can suppress soilborne pathogens through competition and production of antimicrobial compounds. The buildup of beneficial microbes in organically managed soils enhances soil health, and contributes to long-term sustainability (Lori et al. 2022).

9.6.2 Crop Rotation and Agroecological Practices

Crop rotation and agroecological practices play a vital role in sustaining microbial diversity and preventing soil degradation. Crop rotation involves the systematic planting of different crops in a sequence over multiple seasons, which breaks the cycle of pests and pathogens while diversifying root exudates and nutrient inputs. Different plant species release unique exudates into the soil, stimulating distinct microbial populations that can vary in their functional roles (Dhungana et al. 2023). Legume crops, for example, promote nitrogen-fixing bacteria, enhancing the nitrogen pool in the soil. When legumes are followed by nitrogen-demanding crops, the residual nitrogen from biological fixation supports plant growth without the need for additional synthetic fertilizers (Regassa et al. 2023).

Agroecological practices such as intercropping, cover cropping, and conservation tillage also promote soil microbial function. Intercropping, where multiple crops are grown simultaneously in the same field, fosters a complex belowground environment with diverse root architectures and exudates, supporting a wider array of microbial species. Cover cropping, the practice of planting non-harvested crops like clover or rye between growing seasons, enhances organic matter in the soil while preventing erosion and leaching. Cover crops also maintain microbial populations during fallow periods, ensuring a consistent nutrient cycling process (Scavo et al. 2022). Conservation tillage, which minimizes soil disturbance, preserves soil structure, and prevents the disruption of microbial communities, allowing microbes to function optimally in nutrient cycling and soil aggregation. Through these agroecological practices, the soil microbial ecosystem remains balanced, resilient, and capable of supporting sustainable crop productivity.

9.6.3 Use of Biofertilizers and Microbial Inoculants

Biofertilizers and microbial inoculants are biological products that introduce beneficial microbes into the soil to enhance nutrient availability, soil structure, and plant growth. Biofertilizers contain living microorganisms, such as bacteria, fungi, and algae, that fix atmospheric nitrogen, and produce growth-promoting compounds. *Rhizobium, Azotobacter,* and *Azospirillum* are common biofertilizers for nitrogen fixation, while phosphate-solubilizing bacteria (PSB) like *Pseudomonas* and *Bacillus* help release phosphorus from insoluble forms (Ansari et al. 2024b). For instance, *Azospirillum brasilense* has been shown to increase root growth in wheat, enhancing water and nutrient uptake and ultimately improving yield (Galindo et al. 2022).

Microbial inoculants are often applied as seed treatments or soil amendments, introducing specific microbes that confer growth benefits to plants. Mycorrhizal inoculants are particularly effective in phosphorus-limited soils, where the fungi can increase phosphorus availability to plants by extending hyphal networks far beyond the root zone (Bisht and Chauhan 2020). Using biofertilizers and inoculants as alternatives or supplements to chemical fertilizers promotes soil health by reducing the dependency on synthetic inputs that can harm microbial communities and lead to nutrient leaching and soil acidification (Bisht and Chauhan 2020). These products provide plants with an environmentally friendly source of nutrients, encourage sustainable nutrient cycling, and reduce the ecological footprint of agriculture. By enhancing soil microbial function and nutrient bioavailability, biofertilizers and microbial inoculants contribute to sustainable agricultural practices and soil preservation.

9.6.4 Sustainable Soil Management for Microbial Preservation

Sustainable soil management practices aim to preserve and enhance soil health while maintaining an environment conducive to microbial life. Key aspects of sustainable soil management include minimizing soil disturbance, preventing soil erosion, managing soil pH, and maintaining organic matter levels (Johannsen and Carter 2005). Minimal tillage or no-till practices reduce the physical disruption of soil structure, preserving the habitats of soil microbes and enhancing soil aggregation. Undisturbed soil allows microbes to form stable communities that contribute to nutrient cycling and soil organic matter retention (Johannsen and Carter 2005).

Organic matter management is essential for supporting soil microbial life, as organic matter serves as a primary energy source for microbes and helps retain soil moisture. Practices such as adding compost, mulching, and using crop residues return organic carbon to the soil, providing food for microbial communities and contributing to soil structure improvement (Gatkal et al. 2024). This organic input also aids in carbon sequestration, a process that mitigates climate change by capturing carbon in the soil.

Soil pH management is crucial, as extreme pH levels can negatively impact microbial diversity and enzyme activity. Liming acidic soils or adding sulfur to alkaline soils can help maintain an optimal pH range for microbial growth and activity (Hillel and Hatfield 2005). Additionally, soil moisture management, through proper irrigation practices, prevents both drought stress and waterlogging, which can harm microbial communities and impact crop productivity (Bogati and Walczak 2022). By preserving soil structure, optimizing organic matter and pH, and regulating moisture levels, sustainable soil management supports a healthy microbial ecosystem that sustains nutrient cycling, soil fertility, and crop productivity.

9.7 Conclusions

Soil microbial communities play a crucial role in nutrient cycling, soil structure maintenance, and overall ecosystem health, making them essential for sustainable agriculture. By fostering beneficial microbial interactions through practices like organic farming, crop rotation, biofertilizer application, and sustainable soil management, agricultural systems can improve soil fertility, reduce dependency on synthetic inputs, and enhance resilience against environmental stressors. These practices not only promote soil microbial diversity but also contribute to carbon sequestration, helping mitigate climate change.

Integrating microbial management into agricultural and climate policies is increasingly important as we face the dual challenges of feeding a growing population and reducing environmental impacts. Policies that encourage sustainable soil management practices and microbial-friendly approaches can support long-term soil health and productivity, aligning agricultural goals with climate action targets. Furthermore, recognizing the role of microbes in soil health fosters a holistic perspective on soil management, where microbial functions are valued alongside physical and chemical properties. Achieving these goals requires collaborative efforts in research and policy development. Scientists, policymakers, and farmers must work together to advance our understanding of soil microbial ecology and translate this knowledge into actionable frameworks. By supporting research that explores innovative microbial management practices and promoting policies that protect and enhance soil biodiversity, we can create resilient agricultural systems that are both productive and sustainable, ensuring a healthy environment for future generations.

References

Abd-Alla MH, Al-Amri SM, El-Enany AWE (2023) Enhancing rhizobium–legume symbiosis and reducing nitrogen fertilizer use are potential options for mitigating climate change. Agriculture 13(11):2092

Ahmad A, Moin SF, Liaqat I, Saleem S, Muhammad F, Mujahid T, Zafar U (2023) Isolation, solubilization of inorganic phosphate, and production of organic acids by individual and co-inoculated microorganisms. Geomicrobiol J 40(1):111–121

Ahmed T, Noman M, Qi Y, Shahid M, Hussain S, Masood HA et al (2023) Fertilization of microbial composts: a technology for improving stress resilience in plants. Plants 12(20):3550

Ansari MM, Bisht N, Singh T, Chauhan PS (2024a) Symphony of survival: insights into cross-talk mechanisms in plants, bacteria, and fungi for strengthening plant immune responses. Microbiol Res:127762

Ansari MM, Singh T, Bisht N, Agarwal R, Chauhan PS (2024b) Beneficial microorganisms for nutrient homeostasis in plants. In: Microbial biostimulants for plant growth and abiotic stress amelioration. Elsevier, pp 151–166

Astapati AD, Nath S (2023) The complex interplay between plant-microbe and virus interactions in sustainable agriculture: Harnessing phytomicrobiomes for enhanced soil health, designer plants, resource use efficiency, and food security. Crop Des 2(1):100028

Backer R, Rokem JS, Ilangumaran G, Lamont J, Praslickova D, Ricci E, Subramanian S, Smith D L (2018) Plant growth-promoting rhizobacteria: context, mechanisms of action, and roadmap to commercialization of biostimulants for sustainable agriculture. Frontiers in plant science 9:1473

Begum T (2021) Soil degradation: the problems and how to fix them. Natural History Museum, p 16

Bertini SCB, Azevedo LCB (2022) Soil microbe contributions in the regulation of the global carbon cycle. In: Microbiome under changing climate. Woodhead Publishing, pp 69–84

Bhakat K, Chakraborty A, Islam E (2021) Characterization of zinc solubilization potential of arsenic tolerant *Burkholderia* spp. isolated from rice rhizospheric soil. World J Microbiol Biotechnol 37:1–13

Bisht N, Chauhan PS (2020) Excessive and disproportionate use of chemicals cause soil contamination and nutritional stress. In: Soil contamination-threats and sustainable solutions, 2020, pp 1–10

Bogati K, Walczak M (2022) The impact of drought stress on soil microbial community, enzyme activities and plants. Agronomy 12(1):189

Bonfante P, Genre A (2010) Mechanisms underlying beneficial plant–fungus interactions in mycorrhizal symbiosis. Nat Commun 1(1):48

Bouchali R (2022) Impact of urbanization on the structuration of bacterial communities built from spillovers and coalescence processes leading to an emergence of synurbic strains (Doctoral dissertation, Université Claude Bernard-Lyon I)

Chauhan P, Sharma N, Tapwal A, Kumar A, Verma GS, Meena M, Seth CS, Swapnil P (2023) Soil microbiome: diversity, benefits and interactions with plants. Sustain For 15(19):14643

Chen W, Li J, Yuan H, You L, Wei Q, Feng R, Jiang S, Zhao X (2023) Plant growth regulators improve nitrogen metabolism, yield, and quality of soybean–rhizobia symbiosis. Ann Microbiol 73(1):15

Chen YJ, Leung PM, Wood JL, Bay SK, Hugenholtz P, Kessler AJ et al (2021) Metabolic flexibility allows bacterial habitat generalists to become dominant in a frequently disturbed ecosystem. ISME J 15(10):2986–3004

Cheng C, Wang R, Sun L, He L, Sheng X (2021) Cadmium-resistant and arginine decarboxylase-producing endophytic Sphingomonas sp. C40 decreases cadmium accumulation in host rice (*Oryza sativa* Cliangyou 513). Chemosphere 275:130109

Corbu VM, Gheorghe-Barbu I, Dumbravă AŞ, Vrâncianu CO, Şesan TE (2023) Current insights in fungal importance—a comprehensive review. Microorganisms 11(6):1384

Czarnes S, Mercier PE, Lemoine DG, Hamzaoui J, Legendre L (2020) Impact of soil water content on maize responses to the plant growth-promoting rhizobacterium *Azospirillum lipoferum* CRT1. J Agron Crop Sci 206(5):505–516

Damasceno LJ, Amaral VM, Mariano DDC, Maciel RP, Oliveira Neto CFD, Franco AAN, Viégas IDJM, Pedroso AJS, Simões PHO, Okumura RS (2024) Economic analysis of *Azospirillum brasilense* inoculation associated with enhanced-efficiency nitrogen fertilizers in corn production in the Brazilian Amazon. Nitrogen 5(3):544–552

Das PP, Singh KR, Nagpure G, Mansoori A, Singh RP, Ghazi IA, Kumar A, Singh J (2022) Plant-soil-microbes: a tripartite interaction for nutrient acquisition and better plant growth for sustainable agricultural practices. Environ Res 214:113821

Datta R (2024) Enzymatic degradation of cellulose in soil: a review. Heliyon 10:e24022
Datta R, Kelkar A, Baraniya D, Molaei A, Moulick A, Meena RS, Formanek P (2017) Enzymatic degradation of lignin in soil: a review. Sustain For 9(7):1163
Daunoras J, Kačergius A, Gudiukaitė R (2024) Role of soil microbiota enzymes in soil health and activity changes depending on climate change and the type of soil ecosystem. Biology 13(2):85
De Mandal S, Laskar F, Panda AK, Mishra R (2020) Microbial diversity and functional potential in wetland ecosystems. In: Recent advancements in microbial diversity. Academic Press, pp 289–314
de Sousa SM, de Oliveira CA, Andrade DL, de Carvalho CG, Ribeiro VP, Pastina MM, Marriel IE, de Paula Lana UG, Gomes EA (2021) Tropical Bacillus strains inoculation enhances maize root surface area, dry weight, nutrient uptake and grain yield. J Plant Growth Regul 40(2):867–877
Dhungana I, Kantar MB, Nguyen NH (2023) Root exudate composition from different plant species influences the growth of rhizosphere bacteria. Rhizosphere 25:100645
Dincă LC, Grenni P, Onet C, Onet A (2022) Fertilization and soil microbial community: a review. Appl Sci 12(3):1198
Dixit M, Ghoshal D, Meena AL, Ghasal PC, Rai A, Choudhary J, Dutta D (2024) Changes in soil microbial diversity under present land degradation scenario. Tot Environ Adv:200104
dos Santos SG, Chaves VA, da Silva Ribeiro F, Alves GC, Reis VM (2019) Rooting and growth of pre-germinated sugarcane seedlings inoculated with diazotrophic bacteria. Appl Soil Ecol 133:12–23
Dow L, Gallart M, Ramarajan M, Law SR, Thatcher LF (2023) Streptomyces and their specialised metabolites for phytopathogen control–comparative in vitro and in planta metabolic approaches. Front Plant Sci 14:1151912
Egamberdieva D, Jabborova D, Wirth SJ, Alam P, Alyemeni MN, Ahmad P (2018) Interactive effects of nutrients and Bradyrhizobium japonicum on the growth and root architecture of soybean (Glycine max L.). Frontiers in microbiology 9:1000
Elbasiouny H, El-Ramady H, Elbehiry F, Rajput VD, Minkina T, Mandzhieva S (2022) Plant nutrition under climate change and soil carbon sequestration. Sustain For 14(2):914
Elhaissoufi W, Ghoulam C, Barakat A, Zeroual Y, Bargaz A (2022) Phosphate bacterial solubilization: a key rhizosphere driving force enabling higher P use efficiency and crop productivity. J Adv Res 38:13–28
Elnahal AS, El-Saadony MT, Saad AM, Desoky ESM, El-Tahan AM, Rady MM, AbuQamar SF, El-Tarabily KA (2022) The use of microbial inoculants for biological control, plant growth promotion, and sustainable agriculture: A review. Eur J Plant Pathol 162(4):759–792
Esmaeilzadeh-Salestani K, Bahram M, Seraj RGM, Gohar D, Tohidfar M, Eremeev V et al (2021) Cropping systems with higher organic carbon promote soil microbial diversity. Agric Ecosyst Environ 319:107521
Etesami H, Adl SM (2020) Plant growth-promoting rhizobacteria (PGPR) and their action mechanisms in availability of nutrients to plants. In: Phyto-microbiome in stress regulation. Springer Singapore, pp 147–203
Etesami H, Jeong BR, Glick BR (2021) Contribution of arbuscular mycorrhizal fungi, phosphate–solubilizing bacteria, and silicon to P uptake by plant. Front Plant Sci 12:699618
Falkowski P, Scholes RJ, Boyle EEA, Canadell J, Canfield D, Elser J et al (2000) The global carbon cycle: a test of our knowledge of earth as a system. Science 290(5490):291–296
Fatima P, Mishra A, Om H, Saha B, Kumar P (2019) Free living nitrogen fixation and their response to agricultural crops. In: Biofertilizers and biopesticides in sustainable agriculture. Apple Academic Press, pp 173–200
Galindo FS, Pagliari PH, Fernandes GC, Rodrigues WL, Boleta EHM, Jalal A, Céu EGO, Lima BHD, Lavres J, Teixeira Filho MCM (2022) Improving sustainable field-grown wheat production with Azospirillum brasilense under tropical conditions: a potential tool for improving nitrogen management. Front Environ Sci 10:821628
Gao D, Zhang N, Liu S, Ning C, Wang X, Feng S (2022) Urbanization imprint on soil bacterial communities in forests and grasslands. Forests 14(1):38

Gao G, Li G, Liu M, Liu J, Ma S, Li D et al (2024) Microbial carbon metabolic activity and bacterial cross-profile network in paddy soils of different fertility. Appl Soil Ecol 195:105233

Gatkal NR, Nalawade SM, Sahni RK, Walunj AA, Kadam PB, Bhanage GB, Datta R (2024) Present trends, sustainable strategies and energy potentials of crop residue management in India: a review. Heliyon 10:e39815

Ghimire OP, Lazo A, Parajuli B, Nepal J (2024) Fostering microbial activity and diversity in agricultural systems. CSA News 69(6):43–47

Gikas GD, Parlakidis P, Mavropoulos T, Vryzas Z (2022) Particularities of fungicides and factors affecting their fate and removal efficacy: A review. Sustain For 14(7):4056

Gorain B, Paul S, Parihar M (2022) Role of soil microbes in micronutrient solubilization. In: New and future developments in microbial biotechnology and bioengineering. Elsevier, pp 131–150

Granada Agudelo M, Ruiz B, Capela D, Remigi P (2023) The role of microbial interactions on rhizobial fitness. Front Plant Sci 14:1277262

Grover M, Bodhankar S, Sharma A, Sharma P, Singh J, Nain L (2021) PGPR mediated alterations in root traits: way toward sustainable crop production. Front Sustain Food Syst 4:618230

Gul N, Wani IA, Mir RA, Nowshehri JA, Aslam S, Gupta R, Verma S, Aslam S (2023) Plant growth promoting microorganisms mediated abiotic stress tolerance in crop plants: a critical appraisal. Plant Growth Regul 100(1):7–24

Hammarlund SP, Chacón JM, Harcombe WR (2019) A shared limiting resource leads to competitive exclusion in a cross-feeding system. Environ Microbiol 21(2):759–771

Harman G, Khadka R, Doni F, Uphoff N (2021) Benefits to plant health and productivity from enhancing plant microbial symbionts. Front Plant Sci 11:610065

Hillel D, Hatfield JL (eds) (2005) Encyclopedia of soils in the environment, vol 3. Elsevier, Amsterdam

https://www.unep.org/resources/report/global-assessment-soil-pollution

Husain H, Keitel C, Dijkstra FA (2024) Fungi are more important than bacteria for soil carbon loss through priming effects and carbon protection through aggregation. Appl Soil Ecol 195:105245

Johannsen CJ, Carter PG (2005) Site-specific soil management. Elsevier

Kafi SA, Arabhosseini S, Karimi E, Koobaz P, Mohammadi A, Sadeghi A (2021) *Pseudomonas putida* P3-57 induces cucumber (*Cucumis sativus* L.) defense responses and improves fruit quality characteristics under commercial greenhouse conditions. Sci Hortic 280:109942

Khoshru B, Mitra D, Nosratabad AF, Reyhanitabar A, Mandal L, Farda B, Djebaili R, Pellegrini M, Guerra-Sierra BE, Senapati A, Panneerselvam P (2023) Enhancing manganese availability for plants through microbial potential: a sustainable approach for improving soil health and food security. Bacteria 2(3):129–141

Kour D, Khan SS, Kour H, Kaur T, Devi R, Rai AK, Yadav AN (2024) ACC deaminase producing phytomicrobiomes for amelioration of abiotic stresses in plants for agricultural sustainability. J Plant Growth Regul 43(4):963–985

Kumar A, Meena SK, Sinha SK, Singh AK, Singh SK (2024) Isolation and biochemical characterization of Endophytic bacterium from native Gluconacetobacter diazotrophocus sugarcane cultivar of Middle Gangetic Plains of India. Indian J Ecol 51(1):104–112

Kumari D, Thakur A, Tiwari AK, Singh R (2024) Types and sources of agricultural pollution: Identifying the different pollutants generated by farming practices and their origins. In: A comprehensive exploration of soil, water, and air pollution in agriculture, p 71

Kuzyakov Y, Xu X (2013) Competition between roots and microorganisms for nitrogen: mechanisms and ecological relevance. New Phytol 198(3):656–669

Lal R (2015) Restoring soil quality to mitigate soil degradation. Sustain For 7(5):5875–5895

Lee JA, Gill TE (2015) Multiple causes of wind erosion in the Dust Bowl. Aeolian Res 19:15–36

Lee J, Kim S, Jung H, Koo BK, Han JA, Lee HS (2023) Exploiting bacterial genera as biocontrol agents: Mechanisms, interactions and applications in sustainable agriculture. J Plant Biol 66(6):485–498

Li Y, Liu Q, Zhang DX, Zhang ZY, Xu A, Jiang YL, Chen ZC (2024) Metal nutrition and transport in the process of symbiotic nitrogen fixation. Plant Commun 5:100829

Lindström K, Mousavi SA (2020) Effectiveness of nitrogen fixation in rhizobia. Microb Biotechnol 13(5):1314–1335

Liu J, Wang X, Dai G, Zhang Y, Bian X (2022) Microbial chassis engineering drives heterologous production of complex secondary metabolites. Biotechnol Adv 59:107966

Lori M, Armengot L, Schneider M, Schneidewind U, Bodenhausen N, Mäder P, Krause HM (2022) Organic management enhances soil quality and drives microbial community diversity in cocoa production systems. Sci Total Environ 834:155223

Lozano-González JM, Valverde S, Montoya M, Martín M, Rivilla R, Lucena JJ, López-Rayo S (2023) Evaluation of siderophores generated by Pseudomonas bacteria and their possible application as Fe biofertilizers. Plan Theory 12(23):4054

Macías-Rodríguez L, Contreras-Cornejo HA, Adame-Garnica SG, Del-Val E, Larsen J (2020) The interactions of Trichoderma at multiple trophic levels: inter-kingdom communication. Microbiol Res 240:126552

Mason ARG, Salomon MJ, Lowe AJ, Cavagnaro TR (2023) Microbial solutions to soil carbon sequestration. J Clean Prod

Masood F, Ahmad S, Malik A (2022) Role of rhizobacterial bacilli in zinc solubilization. In: Microbial biofertilizers and micronutrient availability: the role of zinc in agriculture and human health. Springer International Publishing, pp 361–377

Mayak S, Tirosh T, Glick BR (2004) Plant growth-promoting bacteria confer resistance in tomato plants to salt stress. Plant physiology and Biochemistry 42(6):565–572

Maysoon AK, Hassan KU (2024) Synergistic effect of *Bacillus mucilaginosus* and *Pseudomonas fluorescens* on the availability of soil potassium, growth and yield of *Cucurbita pepo* L. Iraqi J Agric Sci 55(5):1650–1656

Meena KR, Kanwar SS (2015) Lipopeptides as the antifungal and antibacterial agents: applications in food safety and therapeutics. Biomed Res Int 2015(1):473050

Mir MS, Saxena A, Kanth RH, Raja W, Dar KA, Mahdi SS, Bhat TA, Naikoo NB, Nazir A, Amin Z, Mansoor T (2022) Role of intercropping in sustainable insect-pest management: a review. Int J Environ Clim Change 12(11):3390–3403

Mo Y, Bier R, Li X, Daniels M, Smith A, Yu L, Kan J (2024) Agricultural practices influence soil microbiome assembly and interactions at different depths identified by machine learning. Commun Biol 7(1):1349

Mukherjee A, Singh BK, Verma JP (2020) Harnessing chickpea (Cicer arietinum L.) seed endophytes for enhancing plant growth attributes and bio-controlling against Fusarium sp. Microbiol Res 237:126469

Muthuraja R, Muthukumar T, Natthapol C (2023) Drought tolerance of Aspergillus violaceofuscus and Bacillus licheniformis and their influence on tomato growth and potassium uptake in mica amended tropical soils under water-limiting conditions. Front Plant Sci 14:1114288

Nawaz A, Qamar ZU, Marghoob MU, Imtiaz M, Imran A, Mubeen F (2023) Contribution of potassium solubilizing bacteria in improved potassium assimilation and cytosolic K+/Na+ ratio in rice (Oryza sativa L.) under saline-sodic conditions. Front Microbiol 14:1196024

Owiny AA, Dusengemungu L (2024) Mycorrhizae in mine wasteland reclamation. Heliyon 10:e33141

Pahalvi HN, Rafiya L, Rashid S, Nisar B, Kamili AN (2021) Chemical fertilizers and their impact on soil health. In: Microbiota and biofertilizers, Vol 2: Ecofriendly tools for reclamation of degraded soil environs. Springer International Publishing, pp 1–20

Pandey RP, Srivastava AK, Gupta VK, O'Donovan A, Ramteke PW (2018) Enhanced yield of diverse varieties of chickpea (Cicer arietinum L.) by different isolates of Mesorhizobium ciceri. Environ Sustain 1(4):425–435

Pin C, Metris A, Baranyi J (2013) Next generation of predictive models. In: Advances in microbial food safety. Woodhead Publishing, pp 498–515

Popoola OJ, Ogundele OD, Ladapo EA, Senbore S (2024) The impact of gold mining on soil biogeochemistry and environmental health. In: Soil microbiome in green technology sustainability. Springer Nature Switzerland, Cham, pp 483–509

Qiu L, Zhang Q, Zhu H, Reich PB, Banerjee S, van der Heijden MG et al (2021) Erosion reduces soil microbial diversity, network complexity and multifunctionality. ISME J 15(8):2474–2489

Ramasamy M, Ghosh S, Yadav KK, Chitra M, Karthickraja A (2024) The role of organic farming in enhancing soil structure and crop performance: a comprehensive review. J Sci Res Rep 30(10):890–904

Rawat P, Das S, Shankhdhar D, Shankhdhar SC (2021) Phosphate-solubilizing microorganisms: mechanism and their role in phosphate solubilization and uptake. J Soil Sci Plant Nutr 21(1):49–68

Regassa H, Elias E, Tekalign M, Legese G (2023) The nitrogen fertilizer replacement values of incorporated legumes residue to wheat on vertisols of the Ethiopian highlands. Heliyon 9(11):e22119

Rego MCF, Ilkiu-Borges F, Filippi MCCD, Gonçalves LA, Silva GBD (2014) Morphoanatomical and biochemical changes in the roots of rice plants induced by plant growth-promoting microorganisms. J Bot 2014(1):818797

Rhodes CJ (2014) Soil erosion, climate change and global food security: challenges and strategies. Sci Prog 97(2):97–153

Rodriguez J, Chakrabarti S, Choi E, Shehadeh N, Sierra-Martinez S, Zhao J, Martens-Habbena W (2021) Nutrient-Limited enrichments of nitrifiers from soil yield consortia of Nitrosocosmicus-affiliated AOA and Nitrospira-affiliated NOB. Front Microbiol 12:671480

Salas A, Cabrera JJ, Jiménez-Leiva A, Mesa S, Bedmar EJ, Richardson DJ et al (2021) Bacterial nitric oxide metabolism: Recent insights in rhizobia. Adv Microb Physiol 78:259–315

Sammauria R, Kumawat S, Kumawat P, Singh J, Jatwa TK (2020) Microbial inoculants: potential tool for sustainability of agricultural production systems. Arch Microbiol 202(4):677–693

Santos KFDN, Moure VR, Hauer V, Santos ARS, Donatti L, Galvão CW, Pedrosa FO, Souza EM, Wassem R, Steffens MBR (2017) Wheat colonization by an Azospirillum brasilense ammonium-excreting strain reveals upregulation of nitrogenase and superior plant growth promotion. Plant Soil 415:245–255

Sapkota M, Young J, Slaughter L, Acosta-Martínez V, Coldren C (2021) Soil microbial biomass and composition from urban landscapes in a semiarid climate. Appl Soil Ecol 158:103810

Scavo A, Fontanazza S, Restuccia A, Pesce GR, Abbate C, Mauromicale G (2022) The role of cover crops in improving soil fertility and plant nutritional status in temperate climates. A review. Agron Sustain Dev 42(5):93

Schwister EM, Dietz BR, Knutson CM, Olszewski NE, Barney BM (2022) Gluconacetobacter diazotrophicus gene fitness during diazotrophic growth. Appl Environ Microbiol 88(23):e01241–e01222

Shahid M, Hameed S, Imran A, Ali S, van Elsas JD (2012) Root colonization and growth promotion of sunflower (Helianthus annuus L.) by phosphate solubilizing Enterobacter sp. Fs-11. World J Microbiol Biotechnol 28:2749–2758

Shi X, Zhao Y, Xu M, Ma L, Adams JM, Shi Y (2024) Insights into plant–microbe interactions in the rhizosphere to promote sustainable agriculture in the new crops era. New Crops 1:100004

Silva PST, Cassiolato AMR, Galindo FS, Jalal A, Nogueira TAR, Oliveira CEDS, Filho MCMT (2022) *Azospirillum brasilense* and zinc rates effect on fungal root colonization and yield of wheat-maize in tropical savannah conditions. Plan Theory 11(22):3154

Sindhu SS, Parmar P, Phour M, Sehrawat A (2016) Potassium-solubilizing microorganisms (KSMs) and its effect on plant growth improvement. In: Potassium solubilizing microorganisms for sustainable agriculture. Springer India, pp 171–185

Singh P, Chauhan PK, Upadhyay SK, Singh RK, Dwivedi P, Wang J, Jain D, Jiang M (2022a) Mechanistic insights and potential use of siderophores producing microbes in rhizosphere for mitigation of stress in plants grown in degraded land. Front Microbiol 13:898979

Singh SK, Wu X, Shao C, Zhang H (2022b) Microbial enhancement of plant nutrient acquisition. Stress Biol 2(1):3

Singh T, Ansari MM, Bisht N, Agarwal R, Chauhan PS (2024b) Role of bacteria in controlling root system behavior. In: Microbial biostimulants for plant growth and abiotic stress amelioration. Elsevier, pp 125–135

Singh T, Bisht N, Ansari MM, Chauhan PS (2023) The hidden harmony: Exploring ROS-phytohormone nexus for shaping plant root architecture in response to environmental cues. Plant Physiol Biochem:108273
Singh T, Bisht N, Ansari MM, Chauhan PS (2024a) Pseudomonas putida triggers phosphorus bioavailability and P-transporters under different phosphate regimes to enhance maize growth. Plant Physiol Biochem:109279
Sofi PA, Rehman K, Gull M, Kumari J, Djanaguiraman M, Prasad PVV (2021) Integrating root architecture and physiological approaches for improving drought tolerance in common bean (Phaseolus vulgaris L.). Plant Physiol Rep 26:4–22
Srivastava N (2023) Siderophore production in iron uptake and plant biofortification. In: Plant Microbiome for Plant Productivity and Sustainable Agriculture. Springer Nature Singapore, Singapore, pp 313–329
Stein LY, Klotz MG (2016) The nitrogen cycle. Curr Biol 26(3):R94–R98
Taffner J, Erlacher A, Bragina A, Berg C, Moissl-Eichinger C, Berg G (2018) What is the role of Archaea in plants? New insights from the vegetation of alpine bogs. MSphere 3(3):10–1128
Tian S, Xu Y, Zhong Y, Qiao Y, Wang D, Wu L, Yang X, Yang M, Wu Z (2024) Exploring the organic acid secretion pathway and potassium solubilization ability of *Pantoea vagans* ZHS-1 for enhanced rice growth. Plan Theory 13(14)
Timofeeva A, Galyamova M, Sedykh S (2022) Prospects for using phosphate-solubilizing microorganisms as natural fertilizers in agriculture. Plan Theory 11(16):2119
Timofeeva AM, Galyamova MR, Sedykh SE (2023) Plant growth-promoting soil bacteria: nitrogen fixation, phosphate solubilization, siderophore production, and other biological activities. Plan Theory 12(24):4074
Torsvik V, Øvreås L (2002) Microbial diversity and function in soil: from genes to ecosystems. Curr Opin Microbiol 5(3):240–245
Tripathi S, Srivastava P, Devi RS, Bhadouria R (2020) Influence of synthetic fertilizers and pesticides on soil health and soil microbiology. In: Agrochemicals detection, treatment and remediation. Butterworth-Heinemann, pp 25–54
Tyśkiewicz R, Nowak A, Ozimek E, Jaroszuk-Ściseł J (2022) Trichoderma: The current status of its application in agriculture for the biocontrol of fungal phytopathogens and stimulation of plant growth. Int J Mol Sci 23(4):2329
Velmurugan S, Ashajyothi M, Charishma K, Kumar S, Balamurugan A, Javed M, Karwa S, Prakash G, Subramanian S, Gogoi R, Eke P (2023) Enhancing defense against rice blast disease: unveiling the role of leaf endophytic firmicutes in antifungal antibiosis and induced systemic resistance. Microb Pathog 184:106326
Vijayakumar S, Gobinath R, Kannan P, Murugaiyan V (2024) Optimizing potassium mining in rice-wheat system: strategies for promoting sustainable soil health-A review. Farm Syst 2(3):100099
Wahab A, Muhammad M, Munir A, Abdi G, Zaman W, Ayaz A et al (2023) Role of arbuscular mycorrhizal fungi in regulating growth, enhancing productivity, and potentially influencing ecosystems under abiotic and biotic stresses. Plan Theory 12(17):3102
Wang G, Mayes MA, Gu L, Schadt CW (2014) Representation of dormant and active microbial dynamics for ecosystem modeling. PLoS One 9(2):e89252
Wang J, Yan D, Dixon R, Wang YP (2016) Deciphering the principles of bacterial nitrogen dietary preferences: a strategy for nutrient containment. MBio 7(4):10–1128
Wang Z, Zhang H, Liu L, Li S, Xie J, Xue X, Jiang Y (2022) Screening of phosphate-solubilizing bacteria and their abilities of phosphorus solubilization and wheat growth promotion. BMC Microbiol 22(1):296
Wani SA, Chand S, Wani MA, Ramzan M, Hakeem KR (2016) Azotobacter chroococcum–a potential biofertilizer in agriculture: an overview. In: Soil science: agricultural and environmental prospectives. Springer International Publishing, pp 333–348
Williams VJ (2011) A case study of desertification in Haiti. J Sustain Dev 4(3):20
Xu F, Liao H, Yang J, Zhang Y, Yu P, Cao Y, Fang J, Chen S, Li L, Sun L, Du C (2023) Auxin-producing bacteria promote barley rhizosheath formation. Nat Commun 14(1):5800

Yakasai HM, Rahman MF, Manogaran M, Yasid NA, Syed MA, Shamaan NA, Shukor MY (2021) Microbiological reduction of molybdenum to molybdenum blue as a sustainable remediation tool for molybdenum: A comprehensive review. Int J Environ Res Public Health 18(11):5731

Yao X, Chen P, Cheng T, Sun K, Megharaj M, He W (2021) Inoculation of Bacillus megaterium strain A14 alleviates cadmium accumulation in peanut: effects and underlying mechanisms. J Appl Microbiol 131(2):819–832

Yu Z, Lu T, Qian H (2023) Pesticide interference and additional effects on plant microbiomes. Sci Total Environ 888:164149

Zahir ZA, Ghani U, Naveed M, Nadeem SM, Asghar HN (2009) Comparative effectiveness of Pseudomonas and Serratia sp. containing ACC-deaminase for improving growth and yield of wheat (Triticum aestivum L.) under salt-stressed conditions. Arch Microbiol 191:415–424

Zahran HH (1999) Rhizobium-legume symbiosis and nitrogen fixation under severe conditions and in an arid climate. Microbiol Mol Biol Rev 63(4):968–989

Zhang J, Wang J, Zhu C, Singh RP, Chen W (2024) Chickpea: Its origin, distribution, nutrition, benefits, breeding, and symbiotic relationship with Mesorhizobium species. Plan Theory 13(3):429

Zhou W, Li M, Achal V (2024) A comprehensive review on environmental and human health impacts of chemical pesticide usage. Emerg Contamin 11:100410

Zhu S, Sainju UM, Zhang S, Tan G, Wen M, Dou Y et al (2024) Cover cropping promotes soil carbon sequestration by enhancing microaggregate-protected and mineral-associated carbon. Sci Total Environ 908:168330

Pseudomonas fluorescens: An Influencer for Food Crop Biofortification by Siderophore-Mediated Iron Acquisition

10

Shilpa Mishra, Chaitanya Kumar Jha, Dweipayan Goswami, and Meenu Saraf

Abstract

Iron is one of the essential micronutrients imperatives for normal growth and development of plants and humans. Iron (Fe) deficiency is one of the major causes of death and diseases, especially in developing countries. A report submitted by the World health Organization (WHO) reveals that approximately one-fourth of the world's population suffers from anemia because of Fe deficiency.

The process of enhancement of iron concentration in edible parts of the food crops by the means of agronomic intervention, plant breeding or genetic selection is known as biofortification. Biofortification is an effective approach to defy the deficiency of iron, also known as malnutrition or hidden hunger. Siderophores producing plant growth promoting rhizobacteria (PGPR) are known to enhance plant's ability to absorb iron from the soil. *Pseudomonas fluorescens* is one of the major bacteria that comprise PGPR.

This chapter outlines the importance and deficiency effects of iron, contribution of *Pseudomonas fluorescens* in biofortification of various crops with iron, biofortification mechanisms of *pseudomonas fluorescens*, overview of siderophores, efficiency of the process and future prospects to escalate the impact of the research.

S. Mishra · D. Goswami · M. Saraf (✉)
Department of Microbiology and Biotechnology, University School of Sciences, Gujarat University, Ahmedabad, Gujarat, India

C. K. Jha
Microbiology Department, Gujarat Arts and Science College, Ahmedabad, Gujarat, India

P. Pandey et al. (eds.), *Microorganisms Resilience to Climate Change*, Microorganisms for Sustainability 56, https://doi.org/10.1007/978-981-96-3748-5_10

Keywords

Biofortification · *Pseudomonas fluorescens* · Iron · Malnutrition · PGPR · Siderophore

10.1 Introduction

Humans are majorly dependent on plant-based foods for the primary diet. Balanced diet including micronutrients and macronutrients are essential for proper growth and development of plants and humans. The green revolution has led to agricultural benefits quantitatively but not qualitatively. The main concern of the green revolution was not to produce high quality food but to increase yields. This reduced the productivity of the land that was followed by production of less nutritious edible grains. Micronutrients are required in trace amounts on a daily basis. Lack of important micronutrients such as iron, zinc, selenium, copper, vitamins, in the food causes malnutrition. In general, the deficiency symptoms of malnutrition are not visual and thus it is also called hidden hunger. This is caused by lower consumption of a quality diet enriched with vitamins, minerals and proteins. The major impact of malnutrition is seen in developing countries and poor people. Resource poor people mainly consume cereal-based diets and cannot afford diverse foods or food supplements in their diets on a regular basis. Malnutrition is also prevalent in urban areas that lack the awareness of a balanced diet. Malnourishment can cause serious health problems that may lead to death (Khan et al. 2019). Around 7.3% of the global disease misfortune is due to lack of vitamins and minerals in the diet. Approximately 18% of Indian infants have birth weight less than the standard and 38% children under the age of five are underweight (Ghosh et al. 2019). According to Global Hunger Index 2021, out of 116 countries India got the 101rd rank and India's score is 27.5 (Index 2021). More than two billion people suffer from iron, zinc and iodine deficiencies. Among these, iron deficiency is pervasive that affects approximately 2.5 to 5 billion people around the world that predominantly includes pregnant women and pre-school children due to their heavy demand of iron for metabolic processes (Khalid et al. 2015). The recommended dietary allowances (RDA) in Indian context for adult men is 19 mg/day. The RDA (mg/day) for normal adult women, pregnant women, adolescent boys and girls is 29, 27, 22 and 30 respectively (Jaiwal et al. 2019). Anaemia is one of the major health problems caused due to deficiency of iron. An evaluation by the World Health Organization suggests that approximately one-fourth of the world's population suffers from anaemia (Khare et al. 2018). According to the global nutrition report 2021, globally one in every three women aged 15–49 are affected with anemia (Summary 2021). A survey conducted by comprehensive national nutritional survey (CNNS) under the Union Ministry of Health and Family Welfare, India in 2016–2018 suggests that about 30–32% of pre-school children and adolescent girls had iron deficiency. In the case of children aged 5–9 years, 11–15% were iron deficient. Besides anemia, deficiency of iron also causes defects in the functions of the immune system and also affects the nervous system in humans. Damage in neurocognitive development occurs if iron deprivation occurs in humans. Iron deficiency in plants results in chlorosis and reduction in crop yield (Khan et al. 2019).

There is a rich amount of iron present in the earth's crust but still it remains unavailable to the plants because it is present as insoluble complex forms. Deficiency of iron may occur because of lower concentration of available iron in soil or in the edible portion of crops. Another major factor contributing to lower iron content is postharvest processes such as milling and polishing. The presence of some antinutritional factors in the plants also lowers the micronutrient concentration. For example, phytates are anti nutritional factors present in the plants that hinders the iron absorption of plants (Verma et al. 2021). Phytic acids are negatively charged and thus have the ability to bind with positively charged micronutrients such as iron. This reduces the bioavailability of iron in humans as they do not have the enzyme phytase that is necessary to degrade phytic acid. Another serious problem that has aggravated malnutrition is the growing population. It is estimated that by 2050, the human population may reach up to 9.2 billion from the current 7.2 billion on the earth (Liu et al. 2021). Land area, soil fertility, water availability, and thus nutrient availability, are vulnerable to deterioration due to industrial activity and overcultivation, as the world's population is expected to grow over the next few years (Yadav et al. 2019). To feed this large population agricultural land area needed to be increased for increment in crop production. But this seems to be not feasible as increasing urbanisation and industrialisation has decreased the land availability for agricultural purposes. Thus, there is a dire need for crop improvement to produce crops with a balanced number of macronutrients and micronutrients over increasing crop quantity. The Sustainable Development Goals (SDG) or Global Goals adopted by the United Nations is a universal call to focus on ending poverty and malnutrition. The SDGs aim to cease all varieties of starvation and malnutrition by 2030, ensuring all people–mainly children have enough and nutritious meals all year. This entails selling sustainable agriculture.

10.2 Strategies for Biofortification

To overcome iron malnutrition, iron supplements can be provided for dietary diversification, and food fortification can be done. However, none of the approaches are sustainable and inexpensive (Patel et al. 2018). A more affordable and sustainable approach to defeat malnutrition is biofortification. Biofortification is the process of growing and providing nutrients in the food crops by relatively cheap, sustainable and long term means to supply more micronutrients. It is the practice of increasing concentration of bioavailable micronutrients in the edible portion of the crop. Ghosh et al. 2019 has suggested certain criteria for the crops to be qualified as biofortified. The biofortified crop should have higher yield and productivity. The micronutrient should have a positive influence on human health when it is biofortified. It should be stable so that sustainability is maintained. The quality of the biofortified crops should be superior to the normal crop. Over the last decade, some biofortified crops (together with rice, sweet potato, beans, legumes and cassava) had been evolved via the Harvest plus biofortification programme, utilizing traditional breeding approaches. There are several approaches for biofortification. The major approaches include agronomic intervention, plant breeding, beneficial microorganisms, fertilizers, transgenic approaches, genetic manipulation. Traditional breeding methods have been employed to form a number of biofortified crops in addition to iron supplementation. However, these measures have not achieved much success as desired due to a variety of reasons. This includes changing consumer demands and acceptance, unwanted sensorial changes in biofortified foods, socio-economic and physiological conditions (Yadav et al. 2019). Plants live in association with certain bacteria that makes insoluble nutrients such as iron present in the soil available to the plants. These bacteria are called plant growth promoting rhizobacteria (PGPR). Beneficial microorganisms include PGPR that are recently more favoured for biofortification due to their efficiency for iron acquisition. These bacteria also exert other beneficial effects that directly or indirectly promotes plant growth and biofortification. The zone of soil immediately surrounding the root of a plant is called rhizosphere that influences the biochemical properties of the soil. The edge is not defined but generally the rhizosphere is approximately 1 mm wide. The PGPR present in the rhizospheric zone affects the plant growth by several direct and indirect mechanisms and also influences the soil physicochemical characteristics and fertility. PGPR directly promotes plant growth by nitrogen fixation, phosphate solubilization, secreting hormones and enzymes like ACC deaminase. The indirect mechanisms by which PGPR facilitates plant growth includes production like siderophores, antibiosis, induced systemic resistance, etc. (Jha and Saraf 2015). Using the PGPR approach minimizes the use of fertilizers, reduces pollution and supports sustainable agriculture (Shaikh and Saraf 2017). Some of the prominent genera that comprise PGPR include *Pseudomonas, Enterobacter, Azospirillum, Bacillus, Serratia, Burkholderia, Variovorax and Klebsiella. Pseudomonas* are present abundantly in the rhizosphere and have gained consideration because of their phenomenal growth promoting attributes. Donate-Correa et al. 2005 reported that *Pseudomonas fluorescens* is the dominant strain present in the rhizosphere and can be used for plant growth in fields due to its excellent PGP properties. Some of the *Pseudomonas* can increase iron absorption by converting insoluble iron into soluble form that can be taken up by the plants.

One of the important attributes is production of metabolites such as siderophores. PGPR chelates insoluble iron by means of releasing siderophores. Siderophores are low molecular weight compounds produced by many species of rhizobacteria that can chelate iron present as a complex form in the soil and make it available to the plants. Siderophores have higher affiliation constant for iron and thus are capable of chelating iron from the complex insoluble hydroxides and oxyhydroxides form (Khare et al. 2018). In aerobic conditions, iron is present as a ferric Fe^{+3} compound that is insoluble at neutral or basic pH soils such as calcareous soils. Plants take up ferrous (Fe^{2+}) forms of iron that are soluble at neutral pH. Siderophores have affinity for ferric iron and they chelate iron irrespective of the pH conditions. Different microbes synthesize different sorts of siderophores.

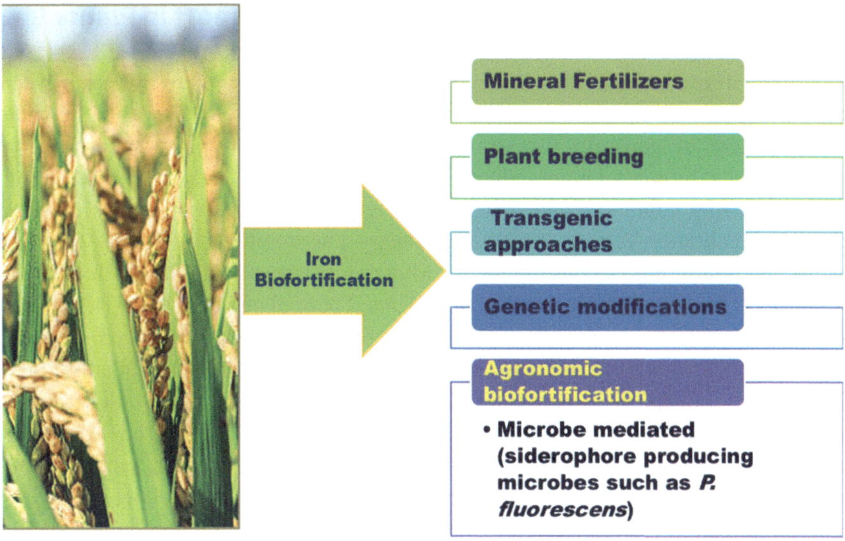

Fig. 10.1 Different strategies for iron biofortification

The major kinds of siderophores are: catecholates, hydroxamate, pyoverdine and carboxylates (Liu et al. 2021). *Pseudomonas fluorescens* have been reported for the production of hydroxamate and pyoverdine type siderophores in various literatures. Callanan et al. 1996 had reported the siderophore pseudobactin MI14 with its outer membrane receptor, PbuA, present in the *P. fluorescens*. This chapter provides a bird's eye view on the general mechanism for iron uptake and distribution in plants, the method by which siderophore produced by *P. fluorescens* helps in iron acquisition as well as present and future prospects in this research (Fig. 10.1).

10.2.1 Need for Iron in Humans and Plants

Iron is an essential element for every life form including human beings and microbes for their proper growth and development. Iron in haemoglobin and myoglobin is required for the transportation of gases in the body. Iron is an integral part of several enzymes that are vital for metabolic processes involved in growth and development of humans (Verma et al. 2021). It is a vital part of the electron transport carrier and helps in nucleotide synthesis, oxidative metabolism, cell division and growth, etc. Plants require very little amount of iron but it plays a critical role in their proper growth and development. Just like in humans, iron also acts as cofactors of enzymes involved in essential biochemical reactions. Iron is required for various metabolic processes in plants such as photosynthesis, respiration, synthesis of chlorophyll and some hormones, etc. (Verma et al. 2021).

10.3 Mechanism of Iron Acquisition and Transport in Plants

A plethora of research has been done to know about the mechanisms by which every plant transport iron from the soil to its roots and distribution of iron within the plant tissues. There are generally two strategies by which different plants mobilize iron into root cells. Strategy I plant reduce ferric iron at the root surface, acidify the soil to increase iron solubility, and transport the resultant ferrous iron across the root plasma membrane (Vert et al. 2002). Strategy I is employed in dicotyledonous plants and non-graminaceous monocots. Major studies exploring the mechanism of strategy I is done on Arabidopsis and rice plants. These plant cells have AHA pumps that release protons into the rhizosphere. This results in reduction of the mobilized iron into the ferrous form (Fe^{2+}) with the help of FRO2 gene product. In addition to this, strategy I plants also express ferric-chelate reductase (FRO) that reduces ferric iron to ferrous at the root surface (Nagata 2017). After this, the ferrous formed is imported into the roots by using a transporter called IRT1 present in the cell membrane (Vert et al. 2002). Natural resistance-associated macrophages proteins (Nramp) are another transporter that helps in Fe uptake by the plants (Nagata 2017). Apart from releasing protons for acidification, some strategy I plants also secrete certain compounds into the apoplastic region in order to solubilise the iron. This involves release of phenolics such as protocatechuic acid (PCA) and caffeic acid in rice plants as reported by Ishimaru et al. 2011. Jin et al. 2007 reported the release of flavonoids by red clover that reduces ferric form of iron and takes up Fe^{2+} form. Apart from this some coumarin derivatives have also been reported for iron absorption (Yadav et al. 2019). Plants tightly modulate absorption, use, and storage in response to environmental availability in order to acquire enough Fe while avoiding toxicity (Kobayashi and Nishizawa 2012). Strategy I is regulated at transcriptional level in the epidermis of the root cell by transcription factor (TF) FIT/FRU/FER (Ling et al. 2002; Yuan et al. 2005). This transcription factor has been known to interact with bHLH038 and bHLH039 to induce the expression of IRON-REGULATED TRANSPORTER 1 (IRT1) and ferric chelate reductase (FRO2) under iron limiting conditions (Yuan et al. 2008). Some of the many H^+-ATPase (HA) genes are activated in the presence of Fe shortage (Kobayashi and Nishizawa 2012). bHLH100 and bHLH101 are induced independently under iron deficiency and are independent of FIT. These transcription factors are also required for regulating the genes for iron remobilization and utilization within the plant and thus are necessary during iron deficiency response. Cailliatte et al. 2010 reported that Natural Resistance Associated Macrophage Protein 1 (NRAMP1) is another divalent metal transporter found in Arabidopsis thaliana that has the ability of iron uptake instead of IRT1 (Lanquar et al. 2005). Another network functional in the root pericycle that helps in iron acquisition is POPEYE (PYE)-PYE LIKE (PYEL)-BRUTUS (BTS). The iron deficiency response is positively regulated by POPEYE and PYEL (that are bHLH transcription factors). The PYEL includes bHLH104, bHLH115 and IAA-LEUCINE RESISTANT ILR3 (Long et al. 2010). BRUTUS degrades the PYEL proteins using 26S proteasomes and thus negatively regulates the iron deficiency response. There is a link between these two regulatory domains. bHLH104, bHLH115 and bHLH34 directly activate the bHLH subgroup 1b genes (Wang et al. 2013; Li et al. 2016; Liang et al. 2017; Yadav et al. 2019). Moreover, bHLH104 also activates PYE and subgroup 1b genes by interacting with PYEL ILR3. Thus, coordination is maintained in the iron deficiency response signalling cascade (Trapet et al. 2016).

In Graminaceous plants such as maize that employ strategy II for iron uptake are able to secrete phytosiderophores. Phytosiderophores are siderophores that are secreted by plants and have high affinity for iron. They can directly chelate iron from the soils (Römheld and Marschner 1986). The phytosiderophores are chemically derivatives of mugineic acids (MA) that are species-specific. S-adenosyl methionine (SAM) is synthesized by methionine cycle and subsequently trimerized to form a molecule of nicotianamine (NA) by the enzyme nicotianamine synthase (NAS). This NA is finally converted into different kinds of mugineic acids such as deoxy-MA in rice, epi-hydroxy MA in barley (Nakanishi et al. 2000). These phyosiderophores are secreted into the rhizosphere via the TRANSPORTER-OF-MAIZE (TOM1) transporter. MA are non-proteinic amino acids that are negatively charged and thus have the ability to find ferric ions bound to several soil compounds. They chelate the iron and make it available for the plants. The ferric-phytosiderophore (Fe^{3+}-phytosiderophore) complex is then internalized by proteins called YELLOW STRIPE 1 (YS1-LIKE (YS1/YSL) present in the epidermal membrane of roots (Curie et al. 2001). Like strategy I, this strategy is also regulated by a fine coordination between positive and negative regulation. This facilitates iron uptake under iron limiting conditions, distribution throughout the plant and prevention of iron overload. Positive regulation in strategy II plants is mediated by IRO2 protein, a transcriptional activator for mugineic acid synthesis genes (Ogo et al. 2007). Another transcription factor IDEF1 from the ABI3/VP1 family monitors the concentrations by binding to the ion. IDE1 subsequently binds to IDE1 cis-element in the promoter region of IRO2 as well as MA synthesis pathway genes and activates them (Kobayashi et al. 2007; Kobayashi and Nishizawa 2012). These activate both early and late response genes for iron deficiency. IDEF2 is another transcription activator belonging to the NAC (NAM, ATAF and CUC) family that has a role in including various iron deficiency responsive genes such as rice YSL2 (Ogo et al. 2008). The negative regulation is established by rice IRO3, that is, activated only under iron-deficient conditions and its targets are IRT1, NAS and YSL genes. This regulation is needed probably to check the excessive iron uptake. It also manages the energy-exhaustive MA synthesis (Fig. 10.2).

10.4 A Propitious Agent for Iron Biofortification: *Pseudomonas fluorescens*

Pseudomonas fluorescens is a gram-negative bacterium that is present as PGPR in the rhizosphere of many food crops. Inoculation with *P. fluorescens* ensured significant improvement in morphological, physiological as well as biochemical traits of spinach plant as iron uptake ability of the plant improved. *P. fluorescens* showed a

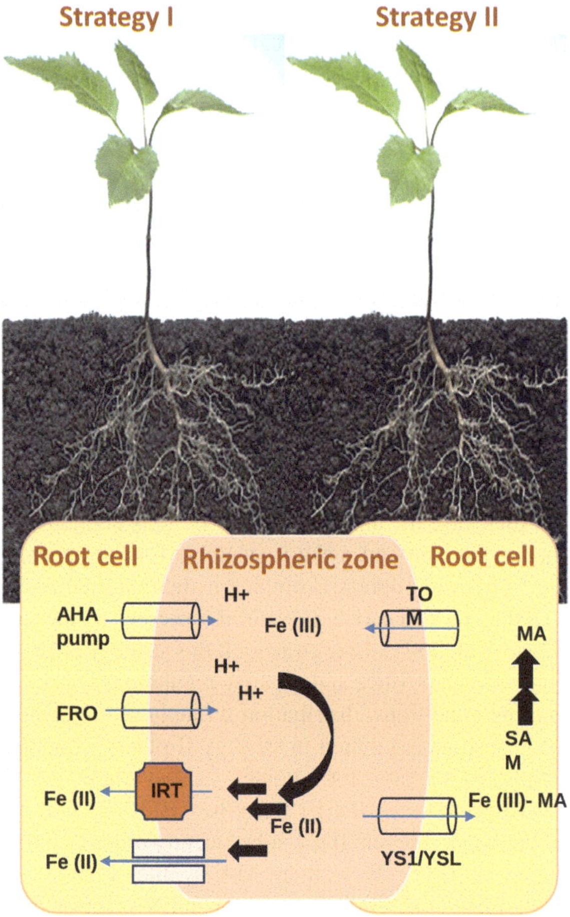

Fig. 10.2 Iron uptake in plants using strategy I and strategy II

positive effect on seed germination, root and shoot length. The seed germination was increased by 30% as compared to control when inoculated with the bacterium. The root length of uninoculated plants was 1.1 cm whereas the root length of plants inoculated with *P. fluorescens* ranged from 1.2 to 2.4 cm. This indicates that root length is doubled due to the influence of bacterium. The shoot length also increased by almost 70% over control when inoculated with the culture. It produced siderophores for the iron acquisition. The siderophore produced is of pyoverdine type (Khare et al. 2018). Sayyed et al. 2005 reported that *Pseudomonas spp*. Such as *P. fluorescens* and *P. putida* synthesize exceptionally high siderophores of hydroxamate type in modified succinic acid media. The hydroxamate siderophores elevated cowpea growth in soil with nickel contamination. The siderophores chelate iron from the soil by binding to iron and nickel. These increased the availability of iron to the plants besides safeguarding them from nickel toxicity. Baysse et al. 2001 studied the impact of mutations in pyoverdine producing gene on siderophore production by *Pseudomonas fluorescens*. The two most common methods for iron fertilizers are soil and foliar applications. In addition to siderophore production for iron chelation, *P. fluorescens* also have the ability to solubilize iron that makes iron available to plants (Liu et al. 2021). Sharma et al. 2013 had taken peroxidase activity as marker for Fe uptake. *P. fluorescens*, *A. lipoferum* and *B15* treated rice plants exhibited 2–3 times higher peroxidase activity as compared to control. These results were seen in all the genotypes of rice taken in study. The bacterial cultures increased the translocation efficiency of iron from the roots to grains that resulted in increased Fe uptake. Inoculation of different rice genotypes with *P. fluorescens* increased the iron content in vegetative parts like roots and shoots as well in addition to increase in grain iron content. Few plants may show decreased uptake of iron from siderophores. Rasouli-Sadaghiani et al. (2014) compared the Fe uptake by wheat leaves provided with various siderophores produced by different strains of fluorescent pseudomonads including *P. putida, P. fluorescens, P. aeruginosa* with standard Fe-siderophore desferrioxamine B (DFOB) complexes. The results were in favour of standard DFOB complexes in terms of iron acquisition when compared to microbial siderophores. The research concluded that *P. fluorescens* has enhanced Fe uptake as compared to *P. aeruginosa* irrespective of the wheat cultivars.

Iron is absorbed by plants generally by two strategies. Strategy I involve absorption of ferrous iron when its concentration is $>10^{-7.7}$ mol/L (Liu et al. 2021). This level of iron is maintained to avoid iron deficient conditions. The other strategy (strategy II) includes secretion of phytosiderophores. Phytosiderophores are secreted by some plants and act as iron carriers by forming iron-phytosiderophore complexes that can be taken up by plants through transporters present in the root cell membrane. The drawback of this strategy is that all plants do not have the ability to produce phytosiderophores and thus they require microbes for iron chelation. Marschner and Crowley (1998) reported that *Pseudomonas fluorescens Pf-5* was present in the rhizosphere of barley and had the ability to exploit phytosiderophores. Due to this reason, this microorganism was able to flourish in an iron deficient environment (Jaiwal et al. 2019). To determine the ability of *Pseudomonas fluorescens Pf-5* to use phytosiderophores for iron acquisition under Fe stress conditions, a microcosm experiment was conducted. The result showed that in iron deficient barley, the release of plant siderophores appears to provide bacterium with sufficient iron. *Pseudomonas fluorescens Pf-5* meets its iron requirement and reduces pyoverdine (siderophore of bacteria) production in the rhizosphere of this plant species. The efficiency of the bacterial siderophore is due to higher stability constant (K) than phytosiderophores (Meyer and Abdallah 1978; Yehuda et al. 1996). This experiment also determined the effect of iron condition and different iron sources on the production of siderophores. When barley plants were treated with foliar iron, they had higher chlorophyll amount compared to plants that were not treated with foliar. Jurkevitch et al. 1993 proposed that bacteria can utilize phytosiderophores by indirect mechanism of ligand exchange. Rachid and Ahmed (2005) also studied the effect of iron content on siderophore production by *P. fluorescens*. The study concluded that the presence of ferric iron and iron limiting conditions increase the growth yield by the bacterium. This suggests that *P. fluorescens* is an excellent choice for biofortification of iron in iron limiting conditions. The *Pseudomonas fluorescens* influence plants' other growth parameters that directly or indirectly enhance the growth of the plants in addition to increase in iron concentration. For example, a study conducted for determining the growth of tomato plants under the influence of the culture concluded that the bacterium increases the iron concentration in the roots and leaves of the plants. It also influenced the concentration of photosynthetic pigments, expression of proteins that is involved in the iron transport in plants like IRT1 (Iron regulated transporter), Nramp (Natural resistance-associated macrophages protein), FRO (ferric-chelate reductase, etc. The research concluded that the tomato plants grown in symbiosis with *P. fluorescens* ATCC 13525 had leaves with iron content higher than those of plants grown without the symbiosis. The concentration of iron in roots of tomato plants with symbiosis was found to be three times greater than in the roots of control after 4 days of incubation. *P. fluorescens* ATCC 13525 also doubled the content of photosynthetic pigments like chlorophyll, carotenoid, and anthocyanin in the plants (Nagata 2017). The same culture was used to determine the efficacy of pyoverdine synthesized by it on the supply of iron to the tomato plant. Iron deficiency signs were present in the tomato that had been fed with ferric iron. However, compared to tomatoes treated with ferrous iron, those exposed to pyoverdine showed no effects (Nagata et al. 2013). Table 10.1 enlists major crops whose iron content was increased by inoculating *P. fluorescens*.

10.5 Uptake of Iron by Siderophores Produced by *P. fluorescens*: Dramatis Personae

As siderophores, approximately 500 different substances have been identified. Siderophore architecture is significantly influenced by denticity. The majority of siderophores are hexadentate, which is the ideal denticity to bind to Fe′s six coordination sites (III) (Boukhalfa and Crumbliss 2002). Since *P. fluorescens* is a gram-negative bacterium, two strategies are mainly used in gram-negative bacteria to take up iron from the soil. These include either uptake of iron in the form of heme or uptake of iron-siderophore complexes (Cornelis 2010). Several reports suggest that *P. fluorescens* mostly secretes pyoverdine and hydroxamate, a kind of siderophore. However, several other sorts of siderophores secreted by *P. fluorescens* have also been reported. Pseudobactin M114 siderophore produced by *Pseudomonas fluorescens* M114 (Callanan et al. 1996). Youard et al. 2011 investigated the enantio-pyochelin produced by *P. fluorescens* as secondary siderophores. This siderophore is an enantiomer of pyochelin and remains physiologically active in their producer species only. A novel siderophore pseudomonine was reported to be produced by *P. fluorescens WCS374* as secondary siderophore as pseudobactin siderophore production was made defective in this bacterium (Mercado-Blanco et al. 2001). *P. fluorescens ATCC 17400* reported to produce the secondary siderophore, quinolobactin apart from the pyoverdine (Mossialos et al. 2002). Secondary siderophores produced by many species of *Pseudomonas* are of low affinity siderophores and are produced in lower concentration next to primary siderophores. Since the activation

Table 10.1 Details of siderophores produced by *P. fluorescens* and their effect on crops

Name of the crop	Name of the strain	Type of siderophore	Additional effects of siderophores on growth of plant	References
Apple rootstock	*P. fluorescens* SP3	Pyoverdine	Increase in the biomass, root development, decreased chlorosis	Gao et al. (2022)
Castor	*Pseudomonas fluorescens 21, 23, 34, 36*	Hydroxamate	Biocontrol against *Botryotinia ricini,* causing grey mold disease in castor	Basha et al. (2021)
Arabidopsis thaliana ecotype Columbia	*P. fluorescens* C7	Pyoverdine	Enhanced growth and chlorophyll content	Vansuyt et al. (2007)
Sorghum bicolor	*P. fluorescens* T17–24	Siderophore	Increased zinc, copper, manganese and dissolved organic carbon (DOC)	Abbaszadeh-Dahaji et al. (2019)
Spinacia olaracea	*P. fluorescens*	Pyoverdine	Increased seed germination, root and shoot length	Khare et al. (2018)
Arachis hypogaea	*P. fluorescens* PSF02	Hydroxamate	Increased root and shoot length, iron and oil content, gain in fresh weight and dry weight	Subramanium and Sundaram (2020)
Triticum aestivum	*P. fluorescens* NCIM 5096	Hydroxamate		Sayyed et al. (2005)
Arabidopsis thaliana	*P. fluorescens* C7R12	Pyoverdine	Modification in phytohormones production	Trapet et al. (2016)
Sugarcane	*P. fluorescens*	Mixed (Hydroxamate and catecholate)	Biocontrol agent	Tailor and Joshi (2012)
Triticum aestivum	*P. fluorescens* 2–79	Pyoverdine	Biocontrol agent against *Gaeumannomyces graminis var. tritici*	Hamdan et al. (1991)
Solanum lycopersicum	*P. fluorescens* ATCC 13525	Pyoverdine	Reduced symptoms of chlorosis under iron deficient conditions, increased chlorophyll and carotenoid concentration, higher ferric-chelate reductase (FRO) activity	Nagata (2017)

(continued)

Table 10.1 (continued)

Name of the crop	Name of the strain	Type of siderophore	Additional effects of siderophores on growth of plant	References
Solanum lycopersicum	*P. fluorescens* ATCC 13525	Pyoverdine	Decreased expression of SIFR01 and SIIRT1 genes in leaves and roots respectively	Nagata et al. (2013)
Bean	*P. fluorescens* Pf-5	Pyoverdine	Uptake of xenosiderophores resulting in increased iron uptake	Loper and Henkels (1997)
Pea	*P. fluorescens* C7R12	Pyoverdine	Modification of Pea ionome composition	Lurthy et al. (2020)
Triticum aestivum	*P. fluorescens* JY3	Siderophore	Biocontrol against *Fusarium oxysporum* and *Rhizoctonia solani*	Abo-Zaid et al. (2020)
Bermuda grass and St. Augustine grass	*P. fluorescens* Q6 and P28	Pseudobactin	Increased growth	Alexander and Zuberer (1991)
Sugarcane	*P. fluorescens*	Hydroxamate	Potential biofertilizer	Joshi et al. (2018)

and regulation of primary siderophores (mostly pyoverdine) biosynthesis genes are an energy exhaustive process, secondary siderophores are produced to conserve the energy. A large number of genes are involved in pyoverdine synthesis, thus when iron deficiency is not drastic secondary siderophores help in iron uptake without expenditure of much energy. Another advantage of releasing secondary siderophores is that most of them have antimicrobial properties. These siderophores provide plants with other benefits in addition to iron acquisition (Cornelis 2010). The *Pseudomonas* species *P. aeruginosa, P. chlororaphis, P. fluorescens, P. putida, and P. syringae* are included in the *Pseudomonas* RNA-homology group I, according to *Bergey's Manual of Systematic Bacteriology* based on the feature that they can produce pyoverdine (Meyer 2000). Rachid and Ahmed (2005) reported three strains of *P. fluorescens* (P1–14, P5–18 and P20–46) that produced pyoverdine siderophores that show few variations in their structures as they are produced from different biovars. This section will focus on the mechanism of iron uptake by siderophores along with the structure and regulation of some major siderophores and their production respectively. A multicomponent system is involved in the uptake of iron via siderophores. The major components are: an iron regulated outer membrane receptor that is dependent on TonB complex; TonB, an inner membrane receptor that forms TonB complex together with ExbB and ExbD; a periplasmic binding protein (PBP) present in the periplasm; and an ABC transporter in the inner membrane (Cornelis 2010). These components have the role in binding the Fe-siderophore complex present in the soil environment outside the cells and intake of this complex inside the cell. TonB dependent outer membrane receptor is also involved in uptake of iron via heme method.

There is lower concentration of bioavailable iron in the soil as compared to inside the cell and thus iron cannot be transported by simple diffusion process. Outer membrane receptor protein (OMRP) present on the plasma membrane of root which is a transmembrane protein bind to the Fe-siderophore (ferric siderophore) complexes (Boukhalfa and Crumbliss 2002; Cornelis 2010). Four outer membrane receptor proteins that transport ferric-siderophore complexes have had their crystal structures determined. Out of these, the OMRP responsible for the transport of ferric pyoverdine is FpvA (Cobessi et al. 2005; Storey 2005). All the types of OMRP show basic structural similarities. It is a gated porin made up of β-barrel that is composed of 22 anti-parallel β-strands. Its N-terminal conserved domain is a globular domain that is called TonB box or 'cork' or 'plug' that interacts with the TonB complex (Storey 2005; Cornelis 2010). The interaction of the TonB box of OMRP with the TonB complex allows the transmission of energy generated via proton motive force (pmf) from the TonB complex to the OMRP. The energy of pmf regulates the opening of the gate of the receptor and transport of ferric siderophore to the periplasmic region of the cell. Inside the periplasm, proteins called PBP binds this ferric complex and release it into the ABC transporter whose function is dependent on the ATP. When compared to PBPs that bind amino acids or sugars, these PBP proteins are often generated at a lower level. They also have a lesser affinity for ferric siderophore complexes than OMRPs do (Sprencel et al. 2000). Transferrin and Lactoferrin, two human iron-binding proteins, are likely to have developed from bacterial PBPs. The ABC transporter is an integral protein in the inner membrane of the cells and is generally composed of four subunits. Two are membrane spanning subunits while the other two are cytoplasmic subunits that bind to the nucleotides (Locher 2004; Storey 2005).

10.5.1 Fate of Iron and Siderophore inside the Cell

There are two ways iron can be liberated from the siderophore once transit into the cytoplasm has taken place. The first pathway is spontaneous separation of iron from the siderophores. This involves reduction of ferric (Fe^{3+}) iron present in the complex to ferrous (Fe^{2+}) iron by a non-destructive method. The reduction is done by free electron donors present in the cells known as siderophore reductases. The ferrous iron has low affinity for siderophores and thus gets released from it inside the cell. The siderophore or apo-siderophore structure remains intact and is reused for the iron uptake purpose. The second pathway for iron-siderophore separation results in the intracellular enzymatic destruction of the siderophore. Thus, the siderophore cannot be recycled for further use. The well-known cellular enzymes that breakdown the siderophores are ferric esterases studied in *E. coli*. From several studies, the first pathway of iron release remains controversial due to lack of compelling evidence (fig. 10.3).

10.5.2 Structure of Some Major Siderophores Produced by *P. fluorescens*

As mentioned before pyoverdines are well-characterized siderophores, a green-yellow fluorescent pigment produced by *Pseudomonas fluorescens* in the low iron environment (Meyer 2000). Meyer and Abdallah 1978 reported that Free iron in the growing medium inhibits *P. fluorescens*' ability to synthesise pyoverdine, and the pyoverdine under study had a very high affinity for iron Fe^{3+} ions. Now that numerous pyoverdines have been extracted from various *Pseudomonas* strains and species that are unique to them, however, they can use a variety of pyoverdines. Generally, they consist of three unique structural components: a yellow-green fluorescent chromophore group (present in all pyoverdines). a small acyl side-chain that is either dicarboxylic acid or amide. Its monoamide coupled via amide bond to the NH_2-group of the chromophore, and a strain-specific peptide chain of L- and D-amino acids (6–12 amino acids, depending on the generating strain) (Matthijs et al. 2009). The peptide chain is linked generally to the carboxylic group of the chromophore through an amide linkage (Visca et al. 2007). There are both common amino acids and uncommon amino acids present in the peptide chain, some of which have a D-configuration, such as δN-acyl (acetyl or formyl or β-hydroxy butyryl)-δN-hydroxyornithine, cyclized δN-hydroxy ornithine, β-hydroxyaspartic acid, β-hydroxyhistidine, and diaminobutyric acid (Meyer 2000). Trapet et al. 2016 studied the pyoverdine produced by *P. fluorescens* strain and reported that pyoverdines increased the growth of plants that had been kept in the iron-deficient medium. The pyoverdine mediated iron chelation and transport is strain specific, although some investigations suggest that different microbial pyoverdines can be recognised by a single outer membrane receptor. This phenomenon can be called as heterologous recognition or cross-uptake of pyoverdines. For example, Meyer 2000 reported that pyoverdines from *P. fluorescens* ATCC 17400 and *P. fluorescens* 1.3 that have the same peptide chain motif, D-Ala-D-Lys-Gly-Gly-OHAsp-D-Gln/DAB show heterologous recognition. Since pyoverdines have been shown to be effective taxonomic markers at the species level, "siderotyping" based on the separation of pyoverdines isoforms by isoelectric focusing and cross-uptake of Fe-pyoverdines has been developed. Hydroxamate siderophores are also among prominent iron chelating molecules produced by the concerned bacterium. The structure of hydroxamate includes a carboxyl group that is attached to the nitrogen. This is responsible for the iron chelation. Examples of hydroxamate siderophores are desferrioxamine-E and ferrichrome. In general, hydroxamates have more complex structural makeup and are thought to be more hydrophilic in nature (Storey 2005). Catecholate siderophores contain hydroxyl groups in their catechol rings. These hydroxyl groups bind to ferric iron and chelate them. The structure of catecholate is derived from 2,3-dihydroxy benzoic acid (DHBA). Some of the siderophores that are included in catecholate groups are enterobactin and chrysobactin.

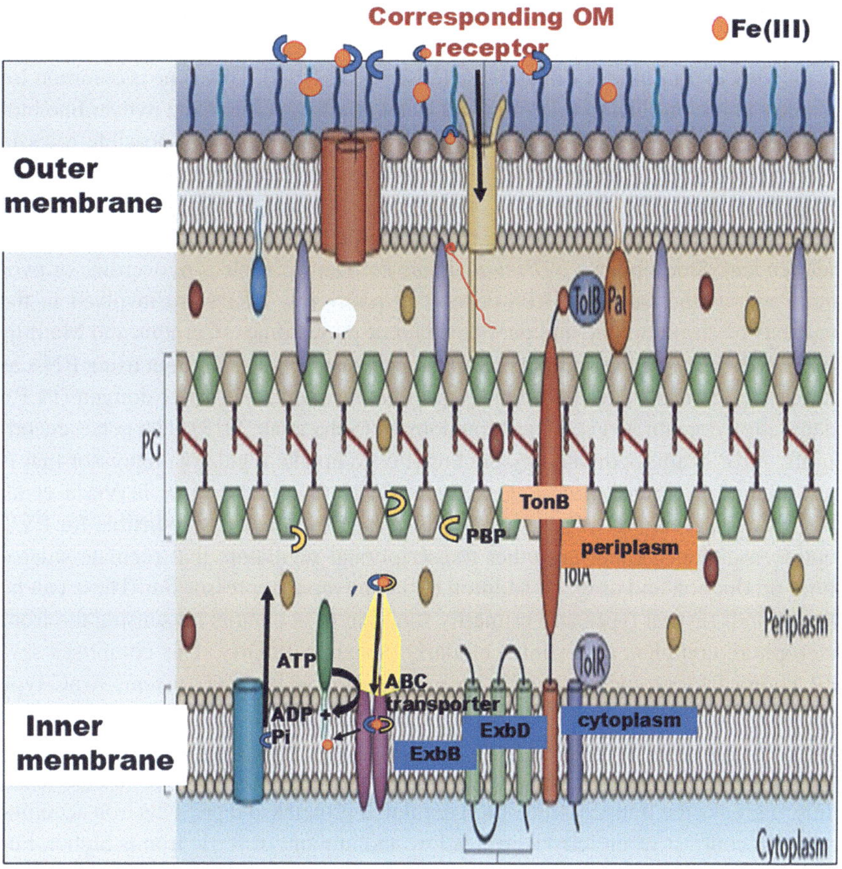

Fig. 10.3 Uptake of ferric siderophore in *Pseudomonas fluorescens*

10.6 Regulation of Iron Uptake and Utilization

As discussed earlier in this review, iron is an important micronutrient in the growth and development of living organisms. Iron can catalyse fenton reactions and the creation of reactive oxygen species (ROS), making an excess of it toxic. Therefore, iron uptake must be meticulously controlled to keep the metal's intracellular concentration within a desirable range. Pyoverdines have been demonstrated to be crucial, if not absolutely necessary, for the pathogenicity of both plant and animal pathogens as well as their ability to colonise host tissues. Studies on *P. fluorescens* siderophore genes reveals that all fluorescent pseudomonads have the same genomic arrangement and position except some variations. About 30 genes have been recognised in fluorescent *Pseudomonas* groups that are related to the function of siderophores. This also includes the genes for pyochelin and enterobactin siderophores.

Generally, three regions of chromosomes have been mapped. pvdL and fpvA are the genes encoding the biosynthesis and uptake of PVD genes. The gene PvdA encodes the enzyme L-ornithine N^5- oxygenase (Visca et al. 1994). This gene is common for all siderophore producing microbes and is not strain-specific in the pyoverdine biosynthetic pathway (Meyer 2000). 31 genes were identified with possible roles in PVD biosynthesis, transportation and regulation in *Pseudomonas fluorescens* by Moon et al. 2008. The synthesis of siderophores uses one of two pathways: one that is NRPS-dependent (Non ribosomal peptide synthetase), and the other that is NRPS-independent. Siderophores of *Pseudomonas fluorescens* such as pyoverdine or pyochelin are synthesised by NRPS-dependent pathways. NRPS are involved in the synthesis of chromophore and peptide chain of pyoverdines (Cornelis and Matthijs 2002). NRPS are enzyme complexes that synthesise peptides without using RNA as a template. It consists of three important domains: peptidyl carrier domain (PCP); adenylation domain and condensation domain (Saha et al. 2013). The genes encoding the NRPSs are modulated by the Fur (Ferric uptake regulator) repressor that is well studied in *Pseudomonas aeruginosa* PA01 (detailed review in Visca et al. 2007). Several reports suggest that NRPS in *P. fluorescens* is responsible for PVD peptide backbone. There are other transcriptional regulators that regulate siderophore production and usage in addition to the universal repressor Fur. These can be divided into several types and primarily function as activators by sensing the iron-siderophore complex, either intracellularly or extracellularly. This comprises several sigma factors such as PvdS, two-component signalling system, AraC-type regulators and many more. Fur repressor binds to ferrous iron and gets activated when the iron concentration is high. In its activated form it interacts with the Fur box present in the promoter region of various genes responsible for iron regulation. Thus, it blocks the transcription of ion regulated proteins and prevents iron accumulation. In contrast when ferrous iron is low and amount of ferric iron is higher, Fur repressor cannot interact with ferric iron and remains inactivated. Fur also transcriptionally regulates extracytoplasmic sigma factors (ECF) such as PvdS that are required for transcription of genes of pyoverdine biosynthesis. 27 genes for TonB dependent receptor OMRP have been identified in *P. fluorescens* (Cornelis and Matthijs 2002). Moon et al. 2008 investigated the homologous genes of *P. Aeruginosa* PAO1 for iron regulation in *P. fluorescens* SBW25. The SBW25 PVD gene is distributed at seven different loci in the genome. The PVD genes studied by Paulsen et al. 2005 in *P. fluorescens* Pf-5 were present within three loci. Two of these genes were for siderophores such as pyoverdine and pyochelin. Three NRPS are synthesised in this bacterium encoded by a 42-kb gene. The Fur box containing PVD genes in their promoter regions identified are pvdS and fpvI. The pvdS and fpvI genes encoded ECF sigma factors while their cognate inner membrane spanning anti-sigma factors were encoded by fpvR gene. fpvR gene contain Iron-starvation (IS) box motifs instead of Fur box and an alternative sigma factor (PvdS) with RNA polymerase binds to this motif. The fpvR gene is positively regulated by fpvI gene in *P. fluorescens*. When ferric siderophore binds to the outer membrane receptor, the ECF sigma factors are released and interact with core RNA polymerase. The expression of PvdS is regulated by a transmembrane signalling system. This system

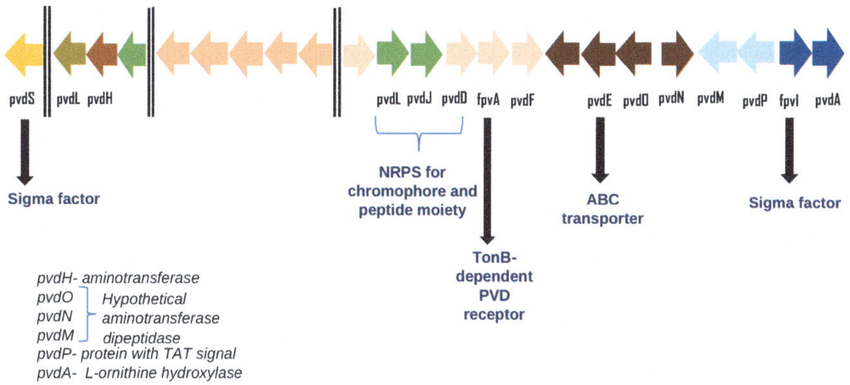

Fig. 10.4 Genetic arrangement of PVD genes and products encoded by some prominent genes. Intervening DNA are represented by double vertical lines along with their length shown above in kb. (Figure is adapted and manipulated from Ravel and Cornelis 2003)

comprises FpvR and FpvA, an outer membrane PVD receptor. The unique feature found in *P. fluorescens* SBW25 genome was that the position of fpvR gene was away from all other genes that were arranged in clusters. Out of all pyoverdine biosynthesis genes (PVD) identified in *P. fluorescens* SBW25, homologues of PA2411 and pvdJ and pvdD of *P. Aeruginosa* PAO1 were absent. AraC regulators are present for the synthesis and uptake of thioquinolobactin siderophore in *P. fluorescens*. The enterobactin uptake and regulation involves two-component system (Moon et al. 2008) (Fig. 10.4).

10.7 Conclusion

PGPR-mediated biofortification reaches maximum people, especially in developing countries, which do not have access to commercial food supplements, fortified food or diet diversification due to economic reasons. PGPR-mediated biofortification is a promising approach for agricultural sustainability also. *Pseudomonas fluorescens* has proved to be one of the prominent rhizospheric bacteria in biofortification. The siderophores released by *P. fluorescens* and its ability to utilize xeno siderophores gives it competitive advantage over other PGPR in iron biofortification. The bacteria have the ability to synthesize almost all sorts of siderophores including secondary siderophores that increases its efficiency in iron uptake and utilization. In addition to increase in iron content in food crops, *P. fluorescens* also exhibits important direct or indirect plant growth promoting attributes such as biocontrol against pathogenic microbes, increase in root and shoot growth and modification of ionome composition. The regulation of siderophore synthesis and uptake in pseudomonads is intricate and does not solely rely on the cell's iron state. The dynamic interaction of the bacteria and rhizosphere must receive more attention in order to accomplish integrated management of the rhizosphere population, the impact of environmental variables on *P. fluorescens*, and the sustainability of the approach under challenging circumstances. *P. fluorescens* alone or in consortia has the potential to defy iron malnutrition and to promote health of human beings.

Competing Interest The authors have declared that no competing interest exists.

References

Abbaszadeh-dahaji P, Masalehi F, Akhgar A (2019) Improved growth and nutrition of Sorghum (Sorghum bicolor) plants in a lowFertility calcareous soil treated with plant growth–promoting rhizobacteria and FeEDTA. J Soil Sci Plant Nutr 20(1):1–12. https://doi.org/10.1007/S42729-019-00098-9.Pdf

Abo-Zaid GA, Soliman NAM, Abdullah AS et al (2020) Maximization of Siderophores production from exponential fed-batch fermentation. Processes 8(455):1–20. https://doi.org/10.3390/pr8040455

Alexander B, Zuberer DA (1991) Use of chrome azurol S reagents to evaluate siderophore production by rhizosphere bacteria. Biol Fertil Soils 12:39–45

Basha SA, Ramya V, Begum AS et al (2021) Evaluation of *Pseudomonas fluorescens* strains, fungicides and non-conventional chemicals against *Botyrotinia ricini* causing Grey Mold disease in Castor. Int J Bio-Resource Stress Manag 12(5):339–347. https://doi.org/10.23910/1.2021.2293

Baysse C, Matthijs S, Pattery T et al (2001) Impact of mutations in *hem*A and *hem*H genes on pyoverdine production by *Pseudomonas fluorescens* ATCC17400. FEMS Microbiol Lett 205(1):57–63. https://doi.org/10.1016/S0378-1097(01)00441-4

Boukhalfa H, Crumbliss AL (2002) Chemical aspects of siderophore mediated iron transport. Biometals 15(4):325–339. https://doi.org/10.1023/A:1020218608266

Cailliatte R, Schikora A, Briat JF et al (2010) High-affinity manganese uptake by the metal transporter nramp1 is essential for Arabidopsis growth in low manganese conditions. Plant Cell 22(3):904–917. https://doi.org/10.1105/tpc.109.073023

Callanan M, Sexton R, Dowling DN et al (1996) Regulation of the iron uptake genes in *Pseudomonas fluorescens* M114 by pseudobactin M114: the pbrA sigma factor gene does not mediate the siderophore regulatory response. FEMS Microbiol Lett 144(1):61–66. https://doi.org/10.1016/0378-1097(96)00339-4

Cobessi D, Celia H, Folschweiller N et al (2005) The crystal structure of the pyoverdine outer membrane receptor FpvA from *Pseudomonas aeruginosa* at 3.6 Åresolution. J Mol Biol 347(1):121–134

Cornelis P (2010) Iron uptake and metabolism in pseudomonads. Appl Microbiol Biotechnol 86(6):1637–1645. https://doi.org/10.1007/s00253-010-2550-2

Cornelis P, Matthijs S (2002) Diversity of siderophore-mediated iron uptake systems in fluorescent pseudomonads: not only pyoverdines. Environ Microbiol 4(12):787–798. https://doi.org/10.1046/j.1462-2920.2002.00369.x

Curie C, Panaviene Z, Loulergue C et al (2001) Maize yellow stripe1 encodes a membrane protein directly involved in Fe (III) uptake. Nature 409(6818):346–349

Donate-Correa J, León-Barrios M, Pérez-Galdona R (2005) Screening for plant growth-promoting rhizobacteria in *Chamaecytisus proliferus* (tagasaste), a forage tree-shrub legume endemic to the Canary Islands. Plant Soil 266(1–2):261–272. https://doi.org/10.1007/s11104-005-0754-5

Gao B, Chai X, Huang Y et al (2022) Siderophore production in pseudomonas SP. strain SP3 enhances iron acquisition in apple rootstock. J Appl Microbiol. https://doi.org/10.1111/jam.155913

Ghosh A, Hasim Reja M, Nalia A et al (2019) Micronutrient biofortification in pulses: an agricultural approach. Curr J Appl Sci Technol 35:1–12. https://doi.org/10.9734/cjast/2019/v35i430194

Hamdan H, Weller DM, Thomashow LS (1991) Relative importance of fluorescent siderophores and other factors in biological control of Gaeumannomyces graminis var. tritici by *Pseudomonas fluorescens* 2-79 and M4-80R. Appl Environ Microbiol 57(11):3270–3277

Index GH (2021) Global hunger index 2021: India. October, 2019–2021

Ishimaru Y, Kakei Y, Shimo H et al (2011) A rice phenolic efflux transporter is essential for solubilizing precipitated apoplasmic iron in the plant stele. J Biol Chem 286(28):24649–24655

Jaiwal PK, Chhillar AK, Chaudhary D et al (2019) Nutritional quality improvement in plants. https://doi.org/10.1007/978-3-319-95354-0

Jha CK, Saraf M (2015) Plant growth promoting rhizobacteria (PGPR): a review. J Agric Res Dev 5(2):108–119

Jin CW, You GY, He YF et al (2007) Iron deficiency-induced secretion of phenolics facilitates the reutilization of root apoplastic iron in red clover. Plant Physiol 144(1):278–285

Joshi RS, Shaikh SH, Joshi SS (2018) Optimization and partial characterization of siderophore produced by *pseudomonas* species isolated from agricultural soil. J Glob Biosci 7(1):5342–5349

Jurkevitch E, Hadar Y, Chen Y et al (1993) Indirect utilization of the phytosiderophore mugineic acid as an iron source to rhizosphere fluorescent *Pseudomonas*. Biometals 6(2):119–123

Khalid S, Asghar HN, Akhtar MJ et al (2015) Biofortification of iron in chickpea by plant growth promoting rhizobacteria. Pak J Bot 47(3):1191–1194

Khan A, Singh J, Upadhayay VK et al (2019) Microbial biofortification: a green technology through plant growth promoting microorganisms. In: Sustainable green technologies for environmental management, pp 255–269. https://doi.org/10.1007/978-981-13-2772-8_13

Khare A, More H, Madliya S et al (2018) Biofortfication of Spinach (*Spinacia olaracea*) with plant growth promoting rhizobacteria – *Pseudomonas fluorescens*. JETIR 5(8):669–675

Kobayashi T, Nishizawa NK (2012) Iron uptake, translocation, and regulation in higher plants. Annu Rev Plant Biol 63(1):131–152

Kobayashi T, Ogo Y, Itai RN et al (2007) The transcription factor IDEF1 regulates the response to and tolerance of iron deficiency in plants. Proc Natl Acad Sci 104(48):19150–19155

Lanquar V, Lelièvre F, Bolte S et al (2005) Mobilization of vacuolar iron by AtNRAMP3 and AtNRAMP4 is essential for seed germination on low iron. EMBO J 24(23):4041–4051

Li Z, Zu C, Wang C, Yang J et al (2016) Different responses of rhizosphere and non-rhizosphere soil microbial communities to consecutive *Piper nigrum* L. monoculture. Sci Rep 6(1):1–8

Liang G, Zhang H, Li X et al (2017) bHLH transcription factor bHLH115 regulates iron homeostasis in *Arabidopsis thaliana*. J Exp Bot 68(7):1743–1755

Ling HQ, Bauer P, Bereczky Z et al (2002) The tomato fer gene encoding a bHLH protein controls iron-uptake responses in roots. Proc Natl Acad Sci 99(21):13938–13943

Liu C, Ye Y, Liu J et al (2021) Iron biofortification of crop food by symbiosis with beneficial microorganisms. J Plant Nutr 44(18):2793–2810. https://doi.org/10.1080/01904167.2021.1927089

Locher KP (2004) Structure and mechanism of ABC transporters. Curr Opin Struct Biol 14(4):426–431

Long TA, Tsukagoshi H, Busch W et al (2010) The bHLH transcription factor POPEYE regulates response to iron deficiency in Arabidopsis roots. Plant Cell 22(7):2219–2236

Loper JE, Henkels MD (1997) Availability of iron to *Pseudomonas fluorescens* in rhizosphere and bulk soil evaluated with an ice nucleation reporter gene. Appl Environ Microbiol 63(1):99–105

Lurthy T, Cantat C, Jeudy C et al (2020) Impact of bacterial siderophores on iron status and ionome in pea. Front Plant Sci 11:730

Marschner P, Crowley DE (1998) Phytosiderophores decrease iron stress and pyoverdine production of *Pseudomonas fluorescens* Pf-5 (pvd-inaZ). Soil Biol Biochem 30(10–11):1275–1280

Matthijs S, Laus G, Meyer JM, Abbaspour-Tehrani K, Schäfer M, Budzikiewicz H, Cornelis P (2009) Siderophore-mediated iron acquisition in the entomopathogenic bacterium *Pseudomonas entomophila* L48 and its close relative *Pseudomonas putida* KT2440. Biometals 22(6):951–964

Mercado-Blanco J, van der Drift KM, Olsson PE, Thomas-Oates JE, van Loon LC, Bakker PA (2001) Analysis of the pmsCEAB gene cluster involved in biosynthesis of salicylic acid and the siderophore pseudomonine in the biocontrol strain *Pseudomonas fluorescens* WCS374. J Bacteriol 183(6):1909–1920

Meyer JM (2000) Pyoverdines: pigments, siderophores and potential taxonomic markers of fluorescent *Pseudomonas* species. Arch Microbiol 174(3):135–142

Meyer JA, Abdallah MA (1978) The fluorescent pigment of *Pseudomonas fluorescens*: biosynthesis, purification and physicochemical properties. Microbiology 107(2):319–328

Moon CD, Zhang XX, Matthijs S, Schäfer M, Budzikiewicz H, Rainey PB (2008) Genomic, genetic and structural analysis of pyoverdine-mediated iron acquisition in the plant growth-promoting bacterium *Pseudomonas fluorescens* SBW25. BMC Microbiol 8(1):1–13

Mossialos D, Ochsner U, Baysse C, Chablain P, Pirnay JP, Koedam N, Cornelis P (2002) Identification of new, conserved, non-ribosomal peptide synthetases from fluorescent pseudomonads involved in the biosynthesis of the siderophore pyoverdine. Mol Microbiol 45(6):1673–1685

Nagata T (2017) Effect of *Pseudomonas fluorescens* inoculation on the improvement of iron deficiency in tomato. Plant Root 11:1–9

Nagata T, Oobo T, Aozasa O (2013) Efficacy of a bacterial siderophore, pyoverdine, to supply iron to *Solanum lycopersicum* plants. J Biosci Bioeng 115(6):686–690

Nakanishi H, Yamaguchi H, Sasakuma T, Nishizawa NK, Mori S (2000) Two dioxygenase genes, Ids3 and Ids2, from *Hordeum vulgare* are involved in the biosynthesis of mugineic acid family phytosiderophores. Plant Mol Biol 44(2):199–207

Ogo Y, Nakanishi Itai R, Nakanishi H, Kobayashi T, Takahashi M, Mori S, Nishizawa NK (2007) The rice bHLH protein OsIRO2 is an essential regulator of the genes involved in Fe uptake under Fe-deficient conditions. Plant J 51(3):366–377

Ogo Y, Kobayashi T, Itai RN, Nakanishi H, Kakei Y, Takahashi M, Nishizawa NK (2008) A novel NAC transcription factor, IDEF2, that recognizes the iron deficiency-responsive element 2 regulates the genes involved in iron homeostasis in plants. J Biol Chem 283(19):13407–13417

Patel P, Trivedi G, Saraf M (2018) Iron biofortification in mungbean using siderophore producing plant growth promoting bacteria. Environ Sustain 1(4):357–365. https://doi.org/10.1007/s42398-018-00031-3

Paulsen IT, Press CM, Ravel J, Kobayashi DY, Myers GS, Mavrodi DV, Loper JE (2005) Complete genome sequence of the plant commensal *Pseudomonas fluorescens* Pf-5. Nat Biotechnol 23(7):873–878

Rachid D, Ahmed B (2005) Effect of iron and growth inhibitors on siderophores production by *Pseudomonas fluorescens*. Afr J Biotechnol 4(7):697–702

Rasouli-Sadaghiani M, Malakouti MJ, Khavazi K, Miransari M (2014) Siderophore efficacy of fluorescent pseudomonades affecting labeled iron (59 Fe) uptake by wheat (*Triticum aestivum* L.) genotypes differing in Fe efficiency. In: Use of microbes for the alleviation of soil stresses. Springer, New York, pp 121–132

Ravel J, Cornelis P (2003) Genomics of pyoverdine-mediated iron uptake in pseudomonads. Trends Microbiol 11(5):195–200

Römheld V, Marschner H (1986) Evidence for a specific uptake system for iron phytosiderophores in roots of grasses. Plant Physiol 80(1):175–180

Saha R, Saha N, Donofrio RS, Bestervelt LL (2013) Microbial siderophores: a mini review. J Basic Microbiol 53(4):303–317

Sayyed RZ, Badgujar MD, Sonawane HM, Mhaske MM, Chincholkar SB (2005) Production of microbial iron chelators (siderophores) by fluorescent Pseudomonads. Indian J Biotechnol 4(4):484–490

Shaikh S, Saraf M (2017) Biofortification of *Triticum aestivum* through the inoculation of zinc solubilizing plant growth promoting rhizobacteria in field experiment. Biocatal Agric Biotechnol 9(October 2016):120–126. https://doi.org/10.1016/j.bcab.2016.12.008

Sharma A, Shankhdhar D, Shankhdhar SC (2013) Enhancing grain iron content of rice by the application of plant growth promoting rhizobacteria. Plant Soil Environ 59(2):89–94

Sprencel C, Cao Z, Qi Z, Scott DC, Montague MA, Ivanoff N, Klebba PE (2000) Binding of ferric enterobactin by the *Escherichia coli* periplasmic protein FepB. J Bacteriol 182(19):5359–5364

Storey EP (2005) *Isolation, purification, and chemical characterization of the dihydroxamate-type siderophore, "Schizokinen," produced by Rhizobium leguminosarum IARI 917* (Doctoral dissertation, East Tennessee State University)

Subramanium N, Sundaram L (2020) Siderophore producing *Pseudomonas* spp. isolated from rhizospheric soil and enhancing iron content in Arachis hypogaea L. plant. J Agric Technol 16:429–442

Summary E (2021) Global nutrition report: the state of global nutrition. Executive Summary. Global Nutrition Report. https://globalnutritionreport.org/reports/2021-global-nutrition-report/

Tailor AJ, Joshi BH (2012) Characterization and optimization of siderophore production from *Pseudomonas fluorescens* strain isolated from sugarcane rhizosphere. J Environ Res Dev 6(3A):688–694

Trapet P, Avoscan L, Klinguer A, Pateyron S, Citerne S, Chervin C, Besson-Bard A (2016) The *Pseudomonas fluorescens* siderophore pyoverdine weakens *Arabidopsis thaliana* defense in favor of growth in iron-deficient conditions. Plant Physiol 171(1):675–693

Vansuyt G, Robin A, Briat JF, Curie C, Lemanceau P (2007) Iron acquisition from Fe-pyoverdine by *Arabidopsis thaliana*. Mol Plant-Microbe Interact 20(4):441–447

Verma S, Chakdar H, Kumar M, Varma A, Saxena AK (2021) Microorganisms as a sustainable alternative to traditional biofortification of iron and zinc: status and Prospect to combat hidden hunger. J Soil Sci Plant Nutr 21(2):1700–1717. https://doi.org/10.1007/s42729-021-00473-5

Vert G, Grotz N, Dédaldéchamp F, Gaymard F, Guerinot ML, Briat JF, Curie C (2002) IRT1, an Arabidopsis transporter essential for iron uptake from the soil and for plant growth. Plant Cell 14(6):1223–1233

Visca P, Alessandra C, Nicola O (1994) Cloning and nucleotide sequence of the pvdA gene encoding the pyoverdin biosynthetic enzyme L-ornithine N5-oxygenase in *Pseudomonas aeruginosa*. J Bacteriol 176(4):1128–1140

Visca P, Imperi F, Lamont IL (2007) Pyoverdine synthesis and its regulation in fluorescent pseudomonads. In: Microbial siderophores. Springer, Berlin/Heidelberg, pp 135–163

Wang N, Cui Y, Liu Y, Fan H, Du J, Huang Z, Ling HQ (2013) Requirement and functional redundancy of Ib subgroup bHLH proteins for iron deficiency responses and uptake in *Arabidopsis thaliana*. Mol Plant 6(2):503–513

Yadav K, Patel P, Ganapathi TR (2019) All roads leading to: iron phytofortification. In: Nutritional quality improvement in plants, pp 25–54

Yehuda Z, Shenker M, Romheld V, Marschner H, Hadar Y, Chen Y (1996) The role of ligand exchange in the uptake of iron from microbial siderophores by gramineous plants. Plant Physiol 112(3):1273–1280

Youard ZA, Wenner N, Reimmann C (2011) Iron acquisition with the natural siderophore enantiomers pyochelin and enantio-pyochelin in *Pseudomonas* species. Biometals 24(3):513–522. https://doi.org/10.1007/s10534-010-9399-9

Yuan YX, Zhang J, Wang DW, Ling HQ (2005) AtbHLH29 of *Arabidopsis thaliana* is a functional ortholog of tomato FER involved in controlling iron acquisition in strategy I plants. Cell Res 15(8):613–621

Yuan Y, Wu H, Wang N, Li J, Zhao W, Du J, Wang D, Ling HQ (2008) FIT interacts with AtbHLH38 and AtbHLH39 in regulating iron uptake gene expression for iron homeostasis in Arabidopsis. Cell Res 18(3):385–397

Extremophiles in Climate Change Mitigation: Harnessing Resilient Microbes for a Sustainable Future

11

Shubham Pandey, Bhavna Parmar, Simran Yadav, A. S. Arthananair, Alvin Huang, and Ram Karan

Abstract

In the face of escalating climate change challenges such as rising global temperatures, unpredictable weather patterns, and the degradation of ecosystems, extremophiles present a promising avenue for sustainable solutions. These resilient organisms are capable of thriving in extreme conditions and can play a pivotal role in mitigating the effects of climate change by enhancing carbon sequestration, facilitating the bioremediation of polluted environments, and promoting sustainable agricultural practices. This chapter delves into the unique abilities of extremophiles to transform and detoxify pollutants into less harmful substances, harness renewable energy sources, and contribute to the bioeconomy, thereby offering innovative strategies to address and adapt to the complex challenges posed by climate change. Their integration into bio-based technologies exemplifies a strategic approach in enhancing ecosystem resilience and advancing global sustainability efforts.

Keywords

Climate adaptation · Extremophiles · Environmental resilience · Bioremediation · Carbon sequestration · Sustainable agriculture · Biofuel production · Bioeconomy

Authors Shubham Pandey and Bhavna Parmar have equally contributed to this chapter.

S. Pandey · B. Parmar · S. Yadav · A. S. Arthananair · R. Karan (✉)
Department of Microbiology, University of Delhi, New Delhi, India
e-mail: ramkaran@south.du.ac.in

A. Huang
Case Western Reserve University, Cleveland, OH, USA

P. Pandey et al. (eds.), *Microorganisms Resilience to Climate Change*, Microorganisms for Sustainability 56, https://doi.org/10.1007/978-981-96-3748-5_11

11.1 Introduction

Human civilization has profoundly impacted our planet's climate by altering the environment's abiotic and biotic components. Rapid industrialization, escalating greenhouse gas emissions, and unsustainable exploitation of natural resources have significantly contributed to climate change. Such relentless consumption and alteration of natural resources have compromised the viability of ecosystems, leading to soil depletion, loss of biodiversity, and a decrease in land fertility due to the accumulation of heavy metals and microplastics (Sharma et al. 2023).

Amid these global challenges, extremophiles—microorganisms adept at surviving under extreme conditions—emerge as innovative allies. These organisms offer potential solutions for mitigating the adverse effects of climate change and promoting environmental sustainability (Kumari et al. 2023). Extremophiles are classified based on their preferred extreme environments, such as thermophiles for extreme heat, halophiles for high salinity, acidophiles for acidic conditions, alkaliphiles for alkaline environments, psychrophiles for cold temperatures, and xerophiles for dry conditions. Each group exhibits unique adaptations that allow them to thrive under harsh conditions and make them valuable in biotechnological applications aimed at sustainability (Chatterjee et al. 2023). For example, thermophiles utilize long-chain fatty acids and have upregulated glycolysis proteins, while halophiles employ potassium uptake strategies and osmoprotectants like ectoine to manage high salinity environments (Karan et al. 2020).

Extremophiles, with their remarkable ability to thrive in the most inhospitable environments on Earth, are poised to make significant contributions to the United Nations Sustainable Development Goals (SDGs). These organisms offer innovative solutions to challenges in health, energy sustainability, and environmental protection. By harnessing their unique biological processes, extremophiles can help advance our goals of reducing pollution, enhancing health outcomes through novel medical applications, and creating sustainable energy solutions (Cowan et al. 2024). For instance, alkaliphilic bacteria such as *Citricoccus alkalitolerans* CSB1 and halophiles like *Parapedobacter sp.* ISTM3 have shown effectiveness in chromium reduction and heavy metal bioremediation, respectively (Alavi et al. 2020; Tyagi et al. 2020). Furthermore, extremophiles such as *Haloarcula japonica* and *Halococcus morrhuae* demonstrate significant resistance to cadmium, showcasing their potential for advancing research in bioremediation and the development of biotechnological applications to mitigate heavy metal pollution (Tavoosi et al. 2023).

Extremophiles also hold promise in developing sustainable technologies for bio-based products and bioenergy. Their unique enzymes and metabolic pathways could revolutionize industries by reducing reliance on fossil fuels and enhancing biotechnological applications. These organisms not only help degrade non-biodegradable polymers such as polyethylene and nylon but also contribute to novel bio-catalytic processes under harsh environmental conditions (Mahdi et al. 2016; Dang et al. 2018; Yan et al. 2021a).

This chapter explores the role of extremophiles in climate change mitigation, emphasizing their application across various domains, such as greenhouse gas reduction, pollution control, sustainable agriculture, biofuel production, and the broader bioeconomy. By thoroughly examining their adaptive mechanisms and biotechnological applications, we highlight how extremophiles can contribute to a sustainable future, addressing ecological and economic challenges.

11.2 Extremophiles and Greenhouse Gas Mitigation

Global warming, primarily driven by human activities, has significantly increased global temperatures and contributed to the depletion of non-repairable resources (Abbass et al. 2022). The rising levels of greenhouse gases, including CO_2, CH_4, and N_2O, respectively, about 40%, 150%, and 20% higher than pre-industrial levels, exemplify climate change's acute challenges (Danovaro et al. 2017). This issue intersects ecological, environmental, sociopolitical, and economic domains, demanding comprehensive mitigation strategies (Neil Adger et al. 2005; Leal Filho et al. 2021; Feliciano et al. 2022).

Recent focus areas in research include optimizing green and sustainable industrial processes and developing biotechnological systems for environmental monitoring and remediation (Gallo et al. 2021). Extremophiles, capable of thriving under extreme conditions, are pivotal in such strategies, particularly in carbon sequestration and methane oxidation—key processes for reducing greenhouse gas emissions and managing global climate (Gallo et al. 2021).

The accumulation of CO_2 in the atmosphere, a significant concern through the twentieth century, highlights the urgency to address its impact as a potent greenhouse gas (Lemaire et al. 2020). Extremophiles facilitate carbon sequestration through various processes, including using CO_2 as a carbon source for growth via photosynthesis or chemosynthesis, contributing significantly to its removal from the atmosphere (Enrich-Prast et al. 2014). Additionally, some extremophiles catalyse the synthesis of carbonate minerals such as calcium carbonate ($CaCO_3$), effectively solidifying CO_2 through microbial-induced calcite precipitation (MICP) processes (Bai et al. 2021; Robles-Fernández et al. 2022; Ni et al. 2023). Furthermore, extremophiles often form biofilms that serve as carbon sinks, sequestering organic carbon and facilitating its conversion to stable forms (Flemming and Wuertz 2019; Jeong and Choi 2020).

Gamma carbonic anhydrase (γ-CA), identified from the extreme environment of the Red Sea brine pool Discovery Deep, exemplifies the potential for innovative applications in greenhouse gas mitigation. As described in a recent study, this enzyme's ability to convert CO_2 into bicarbonate under high-salt conditions, supported by its structural resilience and enhanced activity, underscores its integration into carbon capture technologies (Vogler et al. 2020). Using γ-CA in such applications could offer an environmentally friendly solution in reducing atmospheric CO_2, aligning with broader efforts to address climate change impacts. This enzyme's unique properties and adaptability make it a valuable tool in developing practical and sustainable greenhouse gas reduction strategies.

Methane, another significant greenhouse gas, is addressed by methanotrophic extremophiles that oxidize methane into the less potent carbon dioxide, reducing its impact on climate change. These extremophiles are crucial in environments like landfills, deep-sea hydrothermal vents, and wetlands where methane is prevalent (Rani et al. 2024; Gęsicka et al. 2021). Anaerobic methanotrophs, for example, oxidize methane in oxygen-free environments using alternative electron acceptors such as sulfate or nitrate, reducing methane emissions from anoxic zones (Guerrero-Cruz et al. 2021; Fenibo et al. 2023).

These capabilities underscore the essential role of extremophiles in biogeochemical cycles, particularly in extreme environments that are significant sources of greenhouse gases. For instance, microbial communities in hot springs and acid mine drainage (AMD) sites exhibit remarkable resilience and functionality in the biogeochemical cycling of sulfur and iron, impacting overall environmental chemistry and climate gas regulation (Acharya and Kharel 2020; Massello et al. 2020).

Acidophilic bacteria such as *Acidithiobacillus ferrooxidans* play a significant role in these settings, transforming iron and sulfur compounds and thus influencing the sulfur cycle crucial for environmental remediation strategies (Huang et al. 2021; Johnson et al. 2014).

In summary, extremophiles not only adapt and thrive under harsh conditions but also are pivotal in the mitigation of greenhouse gas emissions. Their unique metabolic adaptations provide invaluable insights into sustainable environmental management and offer promising avenues for developing effective strategies against global warming.

11.3 Bioremediation and Pollution Control

Bioremediation utilizes biological agents, especially microbes and enzymes, to transform hazardous chemicals into less toxic forms, essential for preventing the accumulation of harmful pollutants in ecosystems. This process, which includes degradation, detoxification, mineralization, and transformation of pollutants, is critical for environmental cleanup. Extremophiles, organisms that thrive under extreme conditions, are particularly valuable in bioremediation because they can function in environments where typical organisms capable of bioremediation would not survive, thus expanding the possibilities and efficacy of pollution control (Karan et al. 2020).

For the bioremediation of heavy metals, dyes, and oil degradation, extremophiles are invaluable, especially in hostile environments like acid mine drainages and saline waters (Table 11.1). Acidophilic microorganisms such as *Acidithiobacillus ferrooxidans* play a crucial role in bioleaching heavy metals from contaminated soils and waters, thus reducing environmental toxicity and facilitating the recovery of valuable metals (Romero-González et al. 2016; Zhang et al. 2018). These microbes flourish in low pH conditions, mobilizing metals through bio-oxidation and bioleaching, stabilizing and rendering them less harmful. Halophilic microorganisms, such as species of *Halomonas*, are critical for cleaning up saline

Table 11.1 Extremophilic bacteria utilized for the bioremediation

Trait	Bioremediation	Pollutant	Extremophilic bacteria	References
Thermophile	Heavy metal	Cr (VI)	*Anoxybacillus flavithermus* ABF1	Yang et al. (2022)
		U(VI)	*Anoxybacillus flavithermus* ST15	Acer et al. (2022)
		Cu (II)	*Acidianus manzaensis* YN25	Li et al. (2020)
		Ni (II)	*Sulfobacillus thermosulfidooxidans*	Huang et al. (2020)
		Eu	*T. scotoductus* SA-01	Maleke et al. (2019)
		Cd	*Geobacillus stearothemophilus*, *G. thermocatenulatus*	Hetzer et al. (2006)
	Dye	Red E azo dyes	*Geobacillus thermoleovorans* KNG 112	Rajashekarappa et al. (2022)
		Organic dyes	*Geobacillus yumthangensis*	Sharma and Leung (2021)
		Acid Orange 7	*Caldanaerobacter*	Yan et al. (2021b)
		Direct Black G	*Anoxybacillus* sp. PDR2	Chen et al. (2021)
	Organic compound	Hexadecane	*Candidatus Cerberiarchaeum oleivorans*	Benito Merino et al. (2024)
		Polychlorinated biphenyls	*Burkholderia* sp., *Sphingomonas* sp., *Bacillus* sp., *Micrococcus* sp., *Corynebacterium* sp., *Thioclava* sp. and *Paenibacillus* sp.	Calogero et al. (2024)
		Aromatic hydrocarbon	*Hydrogenibacillus* sp.	Qiu et al. (2022)
			Geobacillus stearothermophilus strain A-2	Zhou et al. (2016)
		Crude oil	*Geobacillus* sp., *Anoxybacillus* sp.	Yusoff et al. (2020)
			Geobacillus pallidus	Zheng et al. (2011)
		Phenol, n-alkanes	*Nocardia otitidiscaviarum* TSH1	Zeinali et al. (2007)
		Naphthalene	*B. thermoleovorans* Hamburg 2	Annweiler et al. (2000)
	Radionuclides	U(VI), Cr(VI), Tc(VII)	*Deinococcus geothermalis*	Brim et al. (2000)

(continued)

Table 11.1 (continued)

Trait	Bioremediation	Pollutant	Extremophilic bacteria	References
Psychrophile	Heavy metal	Pb (II)	*Pseudomonas* sp. 4-2	Yao et al. (2022)
		Cu (II)	*Dietzia psychralcaliphila* JI1D	Ausuri et al. (2022)
	Dye	Methyl red	*Zhihengliuella* sp.	Takkar et al. (2022)
		Azo dye	*Bacillus* sp	AYDIN et al. (2021)
	Oil	Crude oil	*Sphingomonas* sp. AR-OL41 strain	Semenova et al. (2023)
		Crude oil	*Chryseobacterium, Bacillus* and *Pseudomonas*	Wang et al. (2016)
		Aromatic hydrocarbon	*Psychrobacter* sp. ANT206	Wang et al. (2022)
			Pseudomonas sp.	Hassan and Aly (2018)
			Sphingomonas sp. strain Ant 17	Baraniecki et al. (2002)
		Petroleum hydrocarbon	*Halorientalis hydrocarbonoclasticus*	Zhao et al. (2017)
			Haloferax sp.	Bonfá et al. (2011)
			Pseudomonas sp. ST41	Stallwood et al. (2005)
	Radionuclides	U(VI), Co-60, I-125	*Deinococcus radiodurans*	Appukuttan et al. (2006), Gogada et al. (2015), Choi et al. (2017)

(continued)

Table 11.1 (continued)

Trait	Bioremediation	Pollutant	Extremophilic bacteria	References
Halophile	Heavy metal	Pb (II)	*Halomonas* WQL9	Abdel-Razik et al. (2020)
		Cr (VI)	*Microbacterium* sp. M5	Kumar and Saini (2019)
		Mn (II)	*Arthrobacter* sp. NI-2	Nakayama et al. (2019)
	Dye	Acid Brilliant Scarlet GR	*Halomonas* sp.	Tian et al. (2019)
	Oil	Aromatic hydrocarbon	*Halomonas* sp. RM	Govarthanan et al. (2020)
			Haloferax mucsum	Temerk (2019)
			Kocuria turfanesis, Halomonas alkaliphila and *Pseudomonas balearica*	Ahmadi et al. (2017)
			Enterobacter cloacae (MU-1)	Hua et al. (2010)
			Bacterial strain TM-1	Hao and Lu (2009)
		Petroleum hydrocarbon	*Alcanivorax borkumensis*	(Kadri et al. (2022))
			Natrialba sp. C21	Khemili-Talbi et al. (2015)
		Crude oil	*Halomonas titanicae, Marinobacter lutaoensis*	Tourova et al. (2022)
			Acinetobacter	Chen et al. (2020b)
	Radionuclides	U(VI)	*Shewanella putrefaciens*	Wildung Gorby (2000)
Alkaliphile	Heavy metal	Cr (VI)	*Paenibacillus pabuli*	Rangasamy et al. (2021)
		Mn (II)	*Bacillus cereus* SO-16	Ozdemir et al. (2023)
	Dye	Azo dyes	*Pseudomonas mendocina*	Sridharan et al. (2021)
			Bacillus subtilis	Krithika et al. (2021)
	Oil	n-Alkanes	*Pseudomonas toyotomiensis*	Hirota et al. (2011)
		Aromatic hydrocarbon	*Rhodococcus erythropolis, Arthrobacter, Actinobacteria, Hydrogenophaga, Pseudomonas*	Fahy et al. (2008)

(continued)

Table 11.1 (continued)

Trait	Bioremediation	Pollutant	Extremophilic bacteria	References
Acidophile	Heavy metal	Sr	*Acidithiobacillus thiooxidans* PTCC 1717	Hosseini et al. (2022)
		Zn (II)	*Acidithiobacillus ferrooxidans*	Kremser et al. (2020)
		Cu (II)	*Acidomyces acidothermus*	Inoue et al. (1991)
	Dye	Mixed dye	*Microbacterium* sp. CIAB417	Purohit and Yadav (2020)
	Oil	Aromatic hydrocarbon	*Acidocella* sp. SRS 1, *Acidiphilium facilis* SRS 2	Stapleton et al. (1998)

environments affected by oil spills (Gutierrez et al. 2013). They degrade hydrocarbons in salt-rich waters, converting these pollutants into benign substances vital for restoring marine ecosystems where traditional microbial treatments are ineffective.

Extremophiles also provide unique solutions for the microbial treatment of radioactive waste by leveraging their intrinsic resistance to extreme conditions. The radio-resistant *Deinococcus radiodurans*, for instance, has shown potential in reducing and precipitating uranium, which helps in transforming radioactive material into forms that are less soluble and mobile, thus facilitating safer disposal (Jin et al. 2019). Extremophiles such as *Shewanella* and *Geobacter* species can also reduce other radionuclides like plutonium and thorium, demonstrating their utility in managing radioactive waste through biomineralization and biosorption (Jeong and Choi 2020).

In wastewater remediation, extremophiles are essential for breaking down pollutants under challenging conditions. The thermophilic bacterium *Geobacillus thermantarcticus* effectively removes heavy metals from industrial effluents through robust biosorption capabilities (Özdemir et al. 2013). Halophilic bacteria like *Halomonas* sp. treat saline industrial wastewater, degrading organic contaminants and aiding in detoxifying toxic substances (Neifar et al. 2019).

Biodegradation by extremophiles enables the breakdown of complex pollutants under harsh environmental conditions. Psychrotolerant bacteria, such as *Pseudomonas* sp., have been noted for their ability to degrade caffeine efficiently at low temperatures, an essential capability for bioremediation in cooler climates (Thathola et al. 2022). Extremophilic fungi, including *Talaromyces emersonii* and *Thermomucorindica eseudaticae*, are used for the bioremediation of contaminated waters and dye decolorization, showcasing their efficiency in high temperatures and acidic environments (Singh et al. 2016).

Moreover, extremophiles contribute to the development of biosensors for pollution detection. While accurate, traditional analytical techniques like gas chromatography and mass spectrometry are often costly and labor-intensive. In contrast, biosensors employing extremozymes from organisms such as *Thermobacullum terrenum* and *Haloferax volcanii* provide a cost-effective and rapid method for environmental monitoring. These biosensors are especially valuable in extreme conditions, offering enhanced stability and sensitivity for real-time pollution assessment (Uthandi et al. 2010; Brander et al. 2015; Purcarea et al. 2024).

In summary, integrating extremophiles in bioremediation and pollution control broadens the scope of treatable pollutants and enhances the efficiency and sustainability of environmental restoration efforts, by leveraging the unique capabilities of extremophiles, complex pollution challenges can be addressed more effectively, promoting cleaner ecosystems and a healthier environment.

11.4 Sustainable Agricultural Practices

By 2030, the global population is expected to reach 8.5 billion, presenting significant challenges for the agricultural sector, including resource depletion, evolving diseases, and rising consumer expectations. After the Green Revolution, chemical fertilizers surged by 233%, leading to a substantial increase in atmospheric nitrous oxide, a potent greenhouse gas (Sabalpara and Mahatma 2019). Innovations in nitrogen utilization and nitrogen inhibitors are crucial for reducing N_2O emissions, contributing to environmental sustainability (Kumar et al. 2023).

Biofertilizers represent a cost-effective and environmentally safe alternative to chemical fertilizers, playing a vital role in enhancing soil fertility and reducing the ecological footprint of traditional farming practices. The adverse effects of pesticides, including risks of hormonal imbalances, cancer, and reproductive issues, particularly in humans and beneficial soil organisms, underscore the urgent need for safer agricultural practices (Santoyo et al. 2021; Noor-Hassim et al. 2023). Due to their unique adaptations, extremophiles offer potential solutions by promoting plant growth and improving stress resistance in harsh agricultural environments (Masmoudi et al. 2023).

Specifically, halophilic and thermophilic bacteria help plants adapt to high salinity and temperature. Acidophilic extremophiles enhance nutrient availability in acidic soils, thus supporting robust plant growth in otherwise inhospitable conditions. Furthermore, xerophilic bacteria such as *Xanthomonas campestris* and *Deinococcus radiodurans* produce exopolysaccharides that help retain soil moisture, aiding plant survival during dry conditions. Additionally, these extremophiles secrete growth-promoting hormones like auxins and cytokinins, which enhance root development and nutrient absorption (Rao et al. 2022b). By applying these unique biological traits, researchers are developing innovative biotechnological strategies to increase agricultural output and resilience in the face of climate challenges (Chia et al. 2024).

In sustainable agriculture, extremophiles are used as bioinoculants to enhance plant growth, provide stress tolerance, and increase productivity. They also help in providing resistance against various plant pathogens (Table 11.2). These organisms,

Table 11.2 Significant extremophilic bacteria for agricultural application

Traits	Agricultural application	Bacteria	References
Halophile	Stress tolerance	*Aphanothecea halophytica*	Daoud and Ali (2020)
		Bacillus halodenitrificans	Ramadoss et al. (2013)
		Bacillus sp.	El Arbi et al. (2016)
	Nitrogen fixer	*Halorhodospira halophila, Halomonas maura*	George et al. (2020)
		Virgibacillus halodenitrificans ASH15	Sharma et al. (2023)
		Salinibacter ruber	Oren (2016)
	P solubilizer	*Natrinema* sp. strain IARI-WRAB2, *Halococcus hamelinensis* strain IARI-SNS2	Yadav et al. (2015)
	Biofertilizer	*Halomonas salina*	Mukhtar et al. (2019)
	Biopesticide	*Salinicoccus* strains	Santhaseelan et al. (2022)
		Halomonas	Chia et al., (2024)
	PGPR	*Metabacillus dongyingensis* BY2G20	Kumar et al. (2021)
	Biocontrol	*Trichoderma asperellum*	Hamed et al. (2015)
Psychrophile	P solubilizer	*Pseudomonas* sp.,	Adhikari et al. (2021)
	Stress tolerance	*Pseudomonas*, *Serratia,* and *Staphylococcus*	Araya et al. (2020)
Xerophile	Stress tolerance	*Xanthomonas campestris* and *Deinococcus radiodurans*	Rao et al. (2022a, b)
Thermophiles	PGPR	*Bacillus licheniformis, Bacillus subtilis*	Verma et al. (2018)

part of the phyto-microbiome, form symbiotic relationships with plants, often in extreme conditions such as saline soils and arid environments (Maitra et al. 2021). For example, halophilic plant growth-promoting rhizobacteria (PGPR) thrive in high-salinity conditions, improving plant growth and productivity in saline-affected areas (Santoyo et al. 2021).

Nitrogen-fixing extremophiles convert atmospheric nitrogen into ammonia, a more accessible form for plants. *Halomonas maura*, typically found in hypersaline environments, is known for its nitrogen-fixing capabilities, which are crucial in nutrient-poor settings (Argandoña et al. 2003; George et al. 2020). Although not traditionally known as nitrogen fixers, other halophiles like *Salinibacter ruber* also contribute to the nitrogen cycle in their respective habitats (Oren 2016).

Halophiles such as species of *Halomonas* and *Halobacterium* play a significant role in solubilizing phosphate, making this essential nutrient more available to plants. This ability is precious in saline agricultural settings where phosphate availability is often limited (Daoud and Ali 2020).

Extremophiles such as *Pseudomonas*, *Serratia*, and *Bacillus* enhance plant growth by solubilizing nutrients, controlling pathogens, and producing growth-promoting hormones. These adaptations support plant health in adverse conditions and align with sustainable farming practices, reducing reliance on synthetic fertilizers (Kulkova et al. 2024; Kelbessa et al. 2023).

Extremophiles enhance soil fertility and plant health by improving nutrient cycling and resisting environmental stresses. Their role as biofertilizers and biopesticides is increasingly recognized as essential for sustainable agriculture, particularly in challenging environments (Moopantakath et al. 2023).

These sustainable practices underscore the importance of integrating extremophiles into modern agricultural systems to enhance food security and environmental sustainability.

11.5 Biofuel Production and Biomass Utilization

As global resource consumption continues at an alarming rate, the search for sustainable alternatives, particularly in biofuel production, becomes increasingly critical. *Saccharomyces cerevisiae* is notably effective in utilizing food waste and paper pulp industry sludge for biofuel production, showcasing its potential for large-scale application (Roukas and Kotzekidou 2022). Furthermore, microbes such as *Clostridium thermosaccharolyticum*, *Zymomonas mobilis*, *C. thermohydrosulfuricum*, *Thermoanaerobacter brockii*, *T. mathranii*, and *T. ethanolicus* have demonstrated diverse metabolic capabilities conducive to biofuel production via fermentation, particularly using substrates like lignocellulosic biomass, starch, and sugars (Arora et al. 2015). While the production of first-generation biofuels primarily involves mesophilic microorganisms, second-generation biofuels, which utilize mainly lignocellulosic biomass, present a more significant challenge due to the complex nature of the substrate (Padder et al. 2024). Extremophiles offer a promising solution, given their ability to produce robust enzymes that break down lignocellulosic biomass—a critical step in converting this abundant but recalcitrant material into usable fuel (Karić et al. 2022).

Caldicellulosiruptor bescii, a thermophilic bacterium, is particularly adept at degrading lignocellulosic biomass, producing a range of enzymes, including cellulases, hemicellulases, and xylanases. Its ability to withstand high temperatures and various ionic liquids makes it an ideal candidate for biofuel production processes, which often require such conditions (Thapa et al. 2020). Additionally, extremophiles like *Thermotoga neapolitana* and *Pyrococcus furiosus* reduce the need for extensive biomass preprocessing, significantly lowering production costs (Lipscomb et al. 2023).

Biohydrogen production also highlights the capabilities of extremophilic microorganisms. For example, *Caloranaerobacter azorensis*, found in deep-sea hydrothermal vents, efficiently produces hydrogen at substantial rates, demonstrating the potential of extremophiles in renewable energy sectors (Jiang et al. 2014). Acidophilic microorganisms like *Acidithiobacillus ferrooxidans* further contribute to bioenergy through their ability to oxidize ferrous iron, releasing energy crucial for various bioenergy applications, including microbial fuel cells (Wang and Zhou 2012).

Remarkably, extremophiles are not only pivotal in producing traditional biofuels but are also involved in generating novel bioenergy sources (Table 11.3). *Arthrobacter livingstonensis* and *Pseudoalteromonas arctica*, isolated from Antarctic soil, produce undecane, a potential hydrocarbon for aviation fuel, showcasing the versatility of extremophiles in biofuel sectors (Passarini et al. 2020).

Microbial fuel cells represent another innovative application, utilizing bacteria to convert organic matter into electrical energy. This technology offers sustainable energy solutions and addresses water treatment challenges, especially in regions with limited freshwater resources (Pugazhendi et al. 2022). Halophilic bacteria operating in high-salt and anoxic environments are particularly suited for generating electricity in microbial fuel cells while treating saline wastewater (Sathiyanarayanan et al. 2021).

The potential of extremophiles in biofuel production is further enhanced by advances in genetic engineering, such as CRISPR-Cas technologies, which allow for the tailored modification of these organisms to optimize biofuel production

Table 11.3 Biofuel production by extremophilic bacteria

Biofuel	Trait	Bacteria	Biomass	References
Bioethanol	Thermophile	*Sulfolobus solfataricus*, *Sulfolobus acidocaldarius*	Wood, straw, grass, lignocellulose	Quehenberger et al. (2017)
		Clostridium thermohydrosulfuricum	Starch, xylose	Wagner and Wiegel (2008)
	Psychrophile	*Rhodobacter ovatus*	Starch crops and sugars	Srinivas et al. (2008)
		Pseudomonas antarctica [KX186936.1]	Cheese whey powder, wheat straw hydrolysate, sugarcane molasses	Alvarez-Guzmán et al. (2020)
	Halophile	*Nesterenkonia* sp.	Starch crops and sugars	Amiri et al. (2016)
		Bacillus sp. and *Halomonas* spp.	Bread and rice waste	Al-Mahrouqi et al. (2021)
		Aquisalibacillus elongatus	Starch crops and sugars	Rezaei et al. (2017)
	Alkaliphile	*Butyribacterium methylotrophicum*	Lignocellulosic agricultural waste	Kumari and Singh (2018)

(continued)

Table 11.3 (continued)

Biofuel	Trait	Bacteria	Biomass	References
Biohydrogen	Thermophile	*Pyrococcus furiosus*	Sugars, starch crops	Basen et al. (2014)
		Thermococcus eurythermalis A501	Lipid-extracted *Dunaliella* residue	Chen et al. (2020a, b)
		Thermotoga maritima	Starch and xylan polymers	Auria et al. (2016)
	Psychrophile	*Sejongia marina*	Starch crops and sugars	Zhang et al. (2008)
		Brevumdimonas sp.	Lignocellulosic agricultural waste	Bao et al. (2012)
		Trichococcus collinsii	Starch crops and sugars	Bottos et al. (2014)
	Halophile	*Kocuria varians*	Starch crops and sugars	Taroepratjeka et al. (2019)
		Halanaerobium fermentans and *Halanaerobacter lacunarum*	Lignocellulosic waste	Taroepratjeka et al. (2021)
		Halolamina pelagica	Lignocellulosic agricultural waste	Gaba et al. (2017)
		Flammeovirga pacifica	Lignocellulosic agricultural waste	Cai et al. (2018)
	Alkaliphile	*Halanaerobium hydrogeniformans*	Lignocellulosic agricultural waste	Begemann et al. (2012)
Biodiesel	Psychrophile	*Pseudomonas fluorescens*	Lignocellulosic agricultural waste and seeds	Luo et al. (2010)
	Halophile	*Bacillus atrophaeus*	Lignocellulosic agricultural waste and seeds	Amiri et al. (2016)

processes (Garg et al. 2023). Extremophiles, with their unique environmental adaptations and robust metabolic pathways, are thus crucial to developing sustainable bioenergy solutions that align with global environmental and energy sustainability goals.

11.6 Contribution of Extremophiles to the Bioeconomy and Their Role in the Circular Economy

As the consumption of non-renewable resources continues unabated, the need to develop sustainable methods for coexisting with nature becomes increasingly pressing. The bioeconomy encompasses producing and transforming renewable biological resources into food, feed, bio-based products, and bioenergy in environmentally friendly ways. This paradigm shift promotes economic growth while reducing dependency on non-renewable resources (Mallick et al. 2024).

As a result, extremophiles, with their unique resilience and capability to withstand harsh environmental conditions, play a pivotal role in further advancing the bioeconomy. They are harnessed across various scientific fields, including biotechnology, environmental science, and astrobiology, primarily for their efficiency in bioremediation and resource recovery. Extremophiles effectively remove heavy metals using diverse mechanisms such as metal exclusion, extracellular and intracellular sequestration, and enzymatic detoxification, showcasing their capability to operate in environments where traditional organisms cannot (Jeong and Choi 2020).

For example, *Anoxybacillus flavithermus* ABF1 has demonstrated the ability to remove up to 93.71% of Cr(VI) within 72 h in conditions ranging from 50 °C to 70 °C, supporting its use in thermally extreme environments (Yang et al. 2022). Similarly, *Acidithiobacillus* species are utilized in the sulfidogenesis process to precipitate copper from zinc-copper solutions at a pH of 2.6, illustrating their potential in bioremediation of polluted waters (Chia et al. 2024).

A classic example of extremophiles in biotechnology is the discovery and application of Taq polymerase from *Thermus aquaticus*. This enzyme has revolutionized molecular biology by enabling the polymerase chain reaction (PCR), crucial for various applications, including DNA cloning, gene manipulation, and medical diagnostics. The robustness of Taq polymerase at high temperatures makes it invaluable in PCR processes, highlighting the significant contributions of extremophiles to modern biotechnology (Saiki 1989).

In the circular economy, extremophiles offer innovative waste management and resource recovery solutions, which are crucial for sustainable practices. These organisms are adept at degrading, detoxifying, and transforming various pollutants and waste materials into reusable products, thus being valuable assets in closing the resource utilization loop and reducing environmental impacts. Their abilities are critical in bioremediating contaminated sites and converting organic waste into valuable bioproducts, including bioleaching of metals.

Moreover, extremophiles contribute to the production of polyhydroxyalkanoates (PHAs), addressing the growing demand for sustainable bioplastics. Species such as *Halomonas hydrothermalis* and *Halomonas neptunia* use crude waste oil to synthesize PHA copolymers, showcasing significant yields and emphasizing the potential of extremophiles in sustainable industrial applications (Satoh et al. 2011; Pernicova et al. 2019).

Extremophiles also play a vital role in medical applications. For instance, peptides derived from extremophilic archaea are explored for their immune-supporting and anti-aging properties, and extremophiles like *Deinococcus radiodurans* offer antioxidant properties that contribute to metal detoxification and improved cardiovascular health (Ranawat and Rawat 2017; Chandra and Banik 2021). Extremophiles are a valuable source of nanoparticles that play a crucial role in targeted drug delivery (DasSarma et al. 2013). The gas vesicle nanoparticles from haloarchaea are important in vaccine development as carriers for Salmonella antigens (DasSarma et al. 2014, 2015; Karan et al. 2023). Furthermore, research has shown that they also enhance cell growth (Alshehri et al. 2022).

Gas vesicle nanoparticles (GVNPs) from haloarchaea have gained attention due to their utility in vaccine development and cell growth enhancement (DasSarma et al. 2013). Used as carriers for antigens, GVNPs have shown promise in immunological applications (DasSarma et al. 2014, 2015; Karan et al. 2023) and have been found to promote cellular proliferation (Alshehri et al. 2022). These buoyant, protein-based structures are inherently modular and biocompatible, making them suitable for a range of biotechnological applications. Innovations in haloarchaeal GVNPs involve their production using *Haloferax volcanii*, which offers a system free from antibiotics and efficient protein purification due to the organism's fragile cell walls.

Significant advancements include the bioengineering of the GVNPs' gvp genes, enhancing their utility by integrating functional proteins. This bioengineering effort opens new avenues for using GVNPs in medical and industrial applications, underscoring their potential to significantly impact biotechnology and health sciences. In summary, extremophiles significantly contribute to the bioeconomy by enabling more sustainable environmental and industrial processes. Their unique abilities not only help mitigate the impacts of human activity on the environment but also open new avenues for future technological and medical advancements, playing a crucial role in advancing the principles of the circular economy.

11.7 Conclusion

Climate change represents a formidable challenge to humanity in the twenty-first century. Extremophiles, capable of thriving in some of the Earth's most forbidding environments, are pivotal in the battle against global climate change and driving sustainable practices. These resilient microorganisms significantly contribute to the cycling of crucial elements like carbon, nitrogen, and phosphorus, processes that inherently sequester considerable amounts of greenhouse gases.

The remarkable ability of extremophiles to prosper in extreme conditions uniquely positions them as invaluable allies in mitigating the impacts of climate change. Their adaptation to life in severe environments, from scalding thermal vents to frozen tundras, allows them to perform biochemical processes that few other organisms can, making them essential in biotechnological applications for environmental restoration and sustainability.

However, harnessing the full potential of extremophiles presents notable challenges. Large-scale cultivation remains a significant hurdle, primarily due to the specialized conditions required for their growth. Advances in microfluidic technologies have begun to address this challenge, enabling precise control of microenvironments for the cultivation of extremophiles. Continued innovations are essential to scale these processes effectively.

Another concern is the ecological risk of introducing extremophiles into non-native environments, particularly during in-situ bioremediation efforts. The potential for unintended environmental impacts necessitates rigorous research to thoroughly assess the benefits and the possible adverse effects of such introductions.

Despite these obstacles, the distinctive capabilities of extremophiles offer substantial promise for addressing the pervasive effects of climate change. As research advances, our understanding of extremophiles will expand, enhancing our ability to leverage their unique traits for ecological benefit. The ongoing exploration of these remarkable organisms broadens our scientific horizons and fortifies our toolkit for protecting and sustaining our planet in the face of environmental challenges.

References

Abbass K, Qasim MZ, Song H et al (2022) A review of the global climate change impacts, adaptation, and sustainable mitigation measures. Environ Sci Pollut Res 29:42539–42559. https://doi.org/10.1007/s11356-022-19718-6

Abdel-Razik MA, Azmy AF, Khairalla AS, AbdelGhani S (2020) Metal bioremediation potential of the halophilic bacterium, *Halomonas* sp. strain WQL9 isolated from Lake Qarun, Egypt. Egypt J Aquat Res 46:19–25. https://doi.org/10.1016/j.ejar.2019.11.009

Acer Ö, Kılınç E, Özdemir S (2022) Bioaccumulation, resistance, removal of U(VI) and Th(IV) and their effects on antioxidant enzymes on thermophilic *Anoxybacillus flavithermus* ST15. Geomicrobiol J 39:54–65. https://doi.org/10.1080/01490451.2021.2005187

Acharya BS, Kharel G (2020) Acid mine drainage from coal mining in the United States—an overview. J Hydrol 588:125061. https://doi.org/10.1016/j.jhydrol.2020.125061

Adhikari P, Jain R, Sharma A, Pandey A (2021) Plant growth promotion at low temperature by phosphate-solubilizing *Pseudomonas* Spp. isolated from high-altitude Himalayan soil. Microb Ecol 82:677–687. https://doi.org/10.1007/s00248-021-01702-1

Ahmadi M, Jorfi S, Kujlu R et al (2017) A novel salt-tolerant bacterial consortium for biodegradation of saline and recalcitrant petrochemical wastewater. J Environ Manag 191:198–208. https://doi.org/10.1016/j.jenvman.2017.01.010

Alavi S, Rafieyan S, Yavari-Bafghi M, Amoozegar MA (2020) Extremophiles: a powerful choice for bioremediation of toxic oxyanions. In: Microbial bioremediation & biodegradation. Springer, Singapore, pp 203–249

Al-Mahrouqi N, Muthukrishnan T, Nallusamy S et al (2021) Isolation of extremophilic bacteria from microbial mats and the applicability of their amylases for bioethanol production from food waste. Energy Sources Part A: Recovery Util Environ Eff:1–12. https://doi.org/10.1080/15567036.2021.1910384

Alshehri S, Karan R, Ghalayini S et al (2022) Air-loaded gas vesicle nanoparticles promote cell growth in three-dimensional bioprinted tissue constructs. Int J Bioprint 8:489. https://doi.org/10.18063/ijb.v8i3.489

Alvarez-Guzmán CL, Balderas-Hernández VE, De Leon-Rodriguez A (2020) Coproduction of hydrogen, ethanol and 2,3-butanediol from agro-industrial residues by the Antarctic psychrophilic GA0F bacterium. Int J Hydrog Energy 45:26179–26187. https://doi.org/10.1016/j.ijhydene.2020.02.105

Amiri H, Azarbaijani R, Parsa Yeganeh L et al (2016) *Nesterenkonia* sp. strain F, a halophilic bacterium producing acetone, butanol and ethanol under aerobic conditions. Sci Rep 6:18408. https://doi.org/10.1038/srep18408

Annweiler E, Richnow HH, Antranikian G et al (2000) Naphthalene degradation and incorporation of naphthalene-derived carbon into biomass by the thermophile *Bacillus thermoleovorans*. Appl Environ Microbiol 66:518–523. https://doi.org/10.1128/AEM.66.2.518-523.2000

Appukuttan D, Rao AS, Apte SK (2006) Engineering of *Deinococcus radiodurans* R1 for bioprecipitation of uranium from dilute nuclear waste. Appl Environ Microbiol 72:7873–7878. https://doi.org/10.1128/AEM.01362-06

Araya MA, Valenzuela T, Inostroza NG et al (2020) Isolation and characterization of cold-tolerant hyper-ACC-degrading bacteria from the Rhizosphere, Endosphere, and Phyllosphere of Antarctic vascular plants. Microorganisms 8:1788. https://doi.org/10.3390/microorganisms8111788

Argandoña M, Martínez-Checa F, Llamas I et al (2003) Megaplasmids in Gram-negative, moderately halophilic bacteria. FEMS Microbiol Lett 227:81–86. https://doi.org/10.1016/S0378-1097(03)00651-7

Arora R, Behera S, Kumar S (2015) Bioprospecting thermophilic/thermotolerant microbes for production of lignocellulosic ethanol: a future perspective. Renew Sust Energ Rev 51:699–717. https://doi.org/10.1016/j.rser.2015.06.050

Auria R, Boileau C, Davidson S et al (2016) Hydrogen production by the hyperthermophilic bacterium *Thermotoga maritima* Part II: modeling and experimental approaches for hydrogen production. Biotechnol Biofuels 9:268. https://doi.org/10.1186/s13068-016-0681-0

Ausuri J, Dell'Anno F, Vitale GA et al (2022) Bioremediation of Multiple Heavy Metals Mediated by Antarctic Marine Isolated *Dietzia psychralcaliphila* JI1D. J Mar Sci Eng 10:1669. https://doi.org/10.3390/jmse10111669

Aydin M, Ağaoğlu A, Bariş Ö (2021) Azo dye decolorization by using four different psychrotolerant *Bacillus* species. Nat Eng Sci 6:19–29. https://doi.org/10.28978/nesciences.868061

Bai H, Liu D, Zheng W et al (2021) Microbially-induced calcium carbonate precipitation by a halophilic ureolytic bacterium and its potential for remediation of heavy metal-contaminated saline environments. Int Biodeterior Biodegradation 165:105311. https://doi.org/10.1016/j.ibiod.2021.105311

Bao M, Su H, Tan T (2012) Biohydrogen production by dark fermentation of starch using mixed bacterial cultures of *Bacillus* sp and *Brevumdimonas* sp. Energy Fuel 26:5872–5878. https://doi.org/10.1021/ef300666m

Baraniecki CA, Aislabie J, Foght JM (2002) Characterization of Sphingomonas sp. Ant 17, an aromatic hydrocarbon-degrading bacterium isolated from antarctic soil. Microb Ecol 43:44–54. https://doi.org/10.1007/s00248-001-1019-3

Basen M, Schut GJ, Nguyen DM et al (2014) Single gene insertion drives bioalcohol production by a thermophilic archaeon. Proc Natl Acad Sci 111(49):17618–17623. https://doi.org/10.1073/pnas.1413789111

Begemann MB, Mormile MR, Sitton OC et al (2012) A streamlined strategy for biohydrogen production with *Halanaerobium hydrogeniformans*, an Alkaliphilic bacterium. Front Microbiol 3:93. https://doi.org/10.3389/fmicb.2012.00093

Benito Merino D, Lipp JS, Borrel G et al (2024) Anaerobic hexadecane degradation by a thermophilic Hadarchaeon from Guaymas Basin. ISME J 18. https://doi.org/10.1093/ismejo/wrad004

Bonfá MRL, Grossman MJ, Mellado E, Durrant LR (2011) Biodegradation of aromatic hydrocarbons by Haloarchaea and their use for the reduction of the chemical oxygen demand of hypersaline petroleum produced water. Chemosphere 84:1671–1676. https://doi.org/10.1016/j.chemosphere.2011.05.005

Bottos EM, Scarrow JW, Archer SDJ et al (2014) Bacterial community structures of Antarctic soils. In: Antarctic Terrestrial microbiology. Springer, Berlin/Heidelberg, pp 9–33

Brander S, Mikkelsen JD, Kepp KP (2015) TtMCO: a highly thermostable laccase-like multicopper oxidase from the thermophilic *Thermobaculum terrenum*. J Mol Catal B Enzym 112:59–65. https://doi.org/10.1016/j.molcatb.2014.12.002

Brim H, McFarlan SC, Fredrickson JK et al (2000) Engineering *Deinococcus radiodurans* for metal remediation in radioactive mixed waste environments. Nat Biotechnol 18:85–90. https://doi.org/10.1038/71986

Cai ZW, Ge HH, Yi ZW et al (2018) Characterization of a novel psychrophilic and halophilic β-1, 3-xylanase from deep-sea bacterium, *Flammeovirga pacifica* strain WPAGA1. Int J Biol Macromol 118:2176–2184. https://doi.org/10.1016/j.ijbiomac.2018.07.090

Calogero R, Arcadi E, Fabiano F et al (2024) PCB bioremediation potential of thermophilic strains from shallow hydrothermal vent (Vulcano Island). J Water Process Eng 61:105330. https://doi.org/10.1016/j.jwpe.2024.105330

Chandra R, Banik A (2021) Detoxification and bioconversion of arsenic and chromium. In: Nanobiotechnology. Elsevier, pp 253–270

Chatterjee A, Puri S, Sharma PK et al (2023) Nature-inspired Enzyme engineering and sustainable catalysis: biochemical clues from the world of plants and extremophiles. Front Bioeng Biotechnol 11:1229300. https://doi.org/10.3389/fbioe.2023.1229300

Chen S, Qu D, Xiao X, Miao X (2020a) Biohydrogen production with lipid-extracted *Dunaliella* biomass and a new strain of hyper-thermophilic archaeon Thermococcus eurythermalis A501. Int J Hydrog Energy 45:12721–12730. https://doi.org/10.1016/j.ijhydene.2020.03.010

Chen W, Kong Y, Li J et al (2020b) Enhanced biodegradation of crude oil by constructed bacterial consortium comprising salt-tolerant petroleum degraders and biosurfactant producers. Int Biodeterior Biodegradation 154:105047. https://doi.org/10.1016/j.ibiod.2020.105047

Chen G, An X, Li H et al (2021) Detoxification of azo dye Direct Black G by thermophilic *Anoxybacillus* sp. PDR2 and its application potential in bioremediation. Ecotoxicol Environ Saf 214:112084. https://doi.org/10.1016/j.ecoenv.2021.112084

Chia XK, Hadibarata T, Kristanti RA et al (2024) The function of microbial enzymes in breaking down soil contaminated with pesticides: a review. Bioprocess Biosyst Eng 47:597–620. https://doi.org/10.1007/s00449-024-02978-6

Choi MH, Jeong S-W, Shim HE et al (2017) Efficient bioremediation of radioactive iodine using biogenic gold nanomaterial-containing radiation-resistant bacterium, *Deinococcus radiodurans* R1. Chem Commun 53:3937–3940. https://doi.org/10.1039/C7CC00720E

Cowan DA, Albers SV, Antranikian G et al (2024) Extremophiles in a changing world. Extremophiles 28:26. https://doi.org/10.1007/s00792-024-01341-7

Dang TCH, Nguyen DT, Thai H et al (2018) Plastic degradation by thermophilic *Bacillus* sp. BCBT21 isolated from composting agricultural residual in Vietnam. Adv Nat Sci Nanosci Nanotechnol 9(1):015014. https://doi.org/10.1088/2043-6254/aaabaf

Danovaro R, Corinaldesi C, Dell'Anno A, Rastelli E (2017) Potential impact of global climate change on benthic deep-sea microbes. FEMS Microbiol Lett 364(23). https://doi.org/10.1093/femsle/fnx214

Daoud L, Ben Ali M (2020) Halophilic microorganisms: Interesting group of extremophiles with important applications in biotechnology and environment. In: Physiological and biotechnological aspects of extremophiles. Elsevier, pp 51–64

DasSarma S, Karan R, DasSarma P et al (2013) An improved genetic system for bioengineering buoyant gas vesicle nanoparticles from Haloarchaea. BMC Biotechnol 13:112. https://doi.org/10.1186/1472-6750-13-112

DasSarma P, Negi VD, Balakrishnan A et al (2014) Haloarchaeal gas vesicle nanoparticles displaying *Salmonella* SopB antigen reduce bacterial burden when administered with live attenuated bacteria. Vaccine 32:4543–4549. https://doi.org/10.1016/j.vaccine.2014.06.021

DasSarma P, Negi VD, Balakrishnan A et al (2015) Haloarchaeal gas vesicle nanoparticles displaying *Salmonella* antigens as a novel approach to vaccine development. Procedia Vaccinol 9:16–23. https://doi.org/10.1016/j.provac.2015.05.003

El Arbi A, Rochex A, Chataigné G et al (2016) The Tunisian oasis ecosystem is a source of antagonistic *Bacillus* spp. producing diverse antifungal lipopeptides. Res Microbiol 167:46–57. https://doi.org/10.1016/j.resmic.2015.09.003

Enrich-Prast A, Bastviken D, Crill P et al (2014) Chemosynthesis. In: Reference module in earth systems and environmental sciences, vol 1. Elsevier, pp 1–19

Fahy A, Ball AS, Lethbridge G et al (2008) Isolation of alkali-tolerant benzene-degrading bacteria from a contaminated aquifer. Lett Appl Microbiol 47:60–66. https://doi.org/10.1111/j.1472-765X.2008.02386.x

Feliciano D, Recha J, Ambaw G et al (2022) Assessment of agricultural emissions, climate change mitigation and adaptation practices in Ethiopia. Clim Pol 22:427–444. https://doi.org/10.1080/14693062.2022.2028597

Fenibo EO, Selvarajan R, Wang H et al (2023) Untapped talents: insight into the ecological significance of methanotrophs and its prospects. Sci Total Environ 903:166145. https://doi.org/10.1016/j.scitotenv.2023.166145

Flemming H-C, Wuertz S (2019) Bacteria and archaea on Earth and their abundance in biofilms. Nat Rev Microbiol 17:247–260. https://doi.org/10.1038/s41579-019-0158-9

Gaba S, Singh RN, Abrol S et al (2017) Draft Genome Sequence of *Halolamina pelagica* CDK2 Isolated from Natural Salterns from Rann of Kutch, Gujarat, India. Genome Announc 5(6):10–128. https://doi.org/10.1128/genomeA.01593-16

Gallo G, Puopolo R, Carbonaro M et al (2021) Extremophiles, a nifty tool to face environmental pollution: from exploitation of metabolism to genome engineering. Int J Environ Res Public Health 18:5228. https://doi.org/10.3390/ijerph18105228

Garg D, Samota MK, Kontis N et al (2023) Revolutionizing biofuel generation: unleashing the power of CRISPR-Cas mediated gene editing of extremophiles. Microbiol Res 274:127443. https://doi.org/10.1016/j.micres.2023.127443

George DM, Vincent AS, Mackey HR (2020) An overview of anoxygenic phototrophic bacteria and their applications in environmental biotechnology for sustainable resource recovery. Biotechnol Rep 28:e00563. https://doi.org/10.1016/j.btre.2020.e00563

Gęsicka A, Oleskowicz-Popiel P, Łężyk M (2021) Recent trends in methane to bioproduct conversion by methanotrophs. Biotechnol Adv 53:107861. https://doi.org/10.1016/j.biotechadv.2021.107861

Gogada R, Singh SS, Lunavat SK et al (2015) Engineered *Deinococcus radiodurans* R1 with NiCoT genes for bioremoval of trace cobalt from spent decontamination solutions of nuclear power reactors. Appl Microbiol Biotechnol 99:9203–9213. https://doi.org/10.1007/s00253-015-6761-4

Govarthanan M, Khalifa AYZ, Kamala-Kannan S et al (2020) Significance of allochthonous brackish water *Halomonas* sp. on biodegradation of low and high molecular weight polycyclic aromatic hydrocarbons. Chemosphere 243:125389. https://doi.org/10.1016/j.chemosphere.2019.125389

Guerrero-Cruz S, Vaksmaa A, Horn MA et al (2021) Methanotrophs: discoveries, environmental relevance, and a perspective on current and future applications. Front Microbiol 12:678057. https://doi.org/10.3389/fmicb.2021.678057

Gutierrez T, Berry D, Yang T et al (2013) Role of bacterial exopolysaccharides (EPS) in the fate of the oil released during the Deepwater horizon oil spill. PLoS One 8:e67717. https://doi.org/10.1371/journal.pone.0067717

Hamed ER, Awad HM, Ghazi EA et al (2015) *Trichoderma asperellum* isolated from salinity soil using rice straw waste as biocontrol agent for cowpea plant pathogens. J Appl Pharm Sci 2:91–98. https://doi.org/10.7324/JAPS.2015.58.S15

Hao R, Lu A (2009) Biodegradation of heavy oils by halophilic bacterium. Prog Nat Sci 19:997–1001. https://doi.org/10.1016/j.pnsc.2008.11.010

Hassan HA, Aly AA (2018) Isolation and characterization of three novel catechol 2,3-dioxygenase from three novel haloalkaliphilic BTEX-degrading *Pseudomonas* strains. Int J Biol Macromol 106:1107–1114. https://doi.org/10.1016/j.ijbiomac.2017.08.113

Hetzer A, Daughney CJ, Morgan HW (2006) Cadmium Ion Biosorption by the Thermophilic Bacteria *Geobacillus stearothermophilus* and *G. thermocatenulatus*. Appl Environ Microbiol 72:4020–4027. https://doi.org/10.1128/AEM.00295-06

Hirota K, Yamahira K, Nakajima K et al (2011) *Pseudomonas toyotomiensis* sp. nov., a psychrotolerant facultative alkaliphile that utilizes hydrocarbons. Int J Syst Evol Microbiol 61:1842–1848. https://doi.org/10.1099/ijs.0.024612-0

Hosseini SM, Vakilchap F, Baniasadi M et al (2022) Green recovery of cerium and strontium from gold mine tailings using an adapted acidophilic bacterium in one-step bioleaching approach. J Taiwan Inst Chem Eng 138:104482. https://doi.org/10.1016/j.jtice.2022.104482

Hua X, Wang J, Wu Z et al (2010) A salt tolerant *Enterobacter cloacae* mutant for bioaugmentation of petroleum- and salt-contaminated soil. Biochem Eng J 49:201–206. https://doi.org/10.1016/j.bej.2009.12.014

Huang Y, Li M, Yang Y et al (2020) *Sulfobacillus thermosulfidooxidans*: an acidophile isolated from acid hot spring for the biosorption of heavy metal ions. Int J Environ Sci Technol 17:2655–2666. https://doi.org/10.1007/s13762-020-02669-1

Huang Y, Li X-T, Jiang Z et al (2021) Key factors governing microbial community in extremely acidic mine drainage (pH<3). Front Microbiol 12:761579. https://doi.org/10.3389/fmicb.2021.761579

Inoue C, Sugawara K, Kusano T (1991) The merR regulatory gene in *Thiobacillus ferrooxidans* is spaced apart from the mer structural genes. Mol Microbiol 5:2707–2718. https://doi.org/10.1111/j.1365-2958.1991.tb01979.x

Jeong S-W, Choi YJ (2020) Extremophilic microorganisms for the treatment of toxic pollutants in the environment. Molecules 25:4916. https://doi.org/10.3390/molecules25214916

Jiang L, Long C, Wu X et al (2014) Optimization of thermophilic fermentative hydrogen production by the newly isolated *Caloranaerobacter azorensis* H53214 from deep-sea hydrothermal vent environment. Int J Hydrog Energy 39:14154–14160. https://doi.org/10.1016/j.ijhydene.2014.05.025

Jin M, Xiao A, Zhu L et al (2019) The diversity and commonalities of the radiation-resistance mechanisms of *Deinococcus* and its up-to-date applications. AMB Express 9:138. https://doi.org/10.1186/s13568-019-0862-x

Johnson DB, Hallberg KB, Hedrich S (2014) Uncovering a microbial enigma: isolation and characterization of the streamer-generating, iron-oxidizing, acidophilic bacterium "*Ferrovum myxofaciens*". Appl Environ Microbiol 80:672–680. https://doi.org/10.1128/AEM.03230-13

Kadri T, Miri S, Robert T et al (2022) Pilot-scale production and in-situ application of petroleum-degrading enzyme cocktail from *Alcanivorax borkumensis*. Chemosphere 295:133840. https://doi.org/10.1016/j.chemosphere.2022.133840

Karan R, Mathew S, Muhammad R et al (2020) Understanding high-salt and cold adaptation of a polyextremophilic enzyme. Microorganisms 8:1594. https://doi.org/10.3390/microorganisms8101594

Karan R, Renn D, Nozue S et al (2023) Bioengineering of air-filled protein nanoparticles by genetic and chemical functionalization. J Nanobiotechnol 21(1):108. https://doi.org/10.1186/s12951-023-01866-7

Karić N, Maia AS, Teodorović A et al (2022) Bio-waste valorisation: agricultural wastes as biosorbents for removal of (in)organic pollutants in wastewater treatment. Chem Eng J Adv 9:100239. https://doi.org/10.1016/j.ceja.2021.100239

Kelbessa BG, Dubey M, Catara V et al (2023) Potential of plant growth-promoting rhizobacteria to improve crop productivity and adaptation to a changing climate. CABI Rev. https://doi.org/10.1079/cabireviews.2023.0001

Khemili-Talbi S, Kebbouche-Gana S, Akmoussi-Toumi S et al (2015) Isolation of an extremely halophilic arhaeon *Natrialba* sp. C21 able to degrade aromatic compounds and to produce stable biosurfactant at high salinity. Extremophiles 19:1109–1120. https://doi.org/10.1007/s00792-015-0783-9

Kremser K, Thallner S, Schoen H et al (2020) Stirred-tank and heap-bioleaching of shredder-light-fractions (SLF) by acidophilic bacteria. Hydrometallurgy 193:105315. https://doi.org/10.1016/j.hydromet.2020.105315

Krithika A, Gayathri KV, Kumar DT, Doss CGP (2021) Mixed azo dyes degradation by an intracellular azoreductase enzyme from alkaliphilic *Bacillus subtilis*: a molecular docking study. Arch Microbiol 203:3033–3044. https://doi.org/10.1007/s00203-021-02299-2

Kulkova I, Wróbel B, Dobrzyński J (2024) *Serratia* spp. as plant growth-promoting bacteria alleviating salinity, drought, and nutrient imbalance stresses. Front Microbiol 15:1342331. https://doi.org/10.3389/fmicb.2024.1342331

Kumar M, Saini HS (2019) Reduction of hexavalent chromium (VI) by indigenous alkaliphilic and halotolerant *Microbacterium* sp. M5: comparative studies under growth and nongrowth conditions. J Appl Microbiol 127:1057–1068. https://doi.org/10.1111/jam.14366

Kumar A, Singh S, Mukherjee A et al (2021) Salt-tolerant plant growth-promoting *Bacillus pumilus* strain JPVS11 to enhance plant growth attributes of rice and improve soil health under salinity stress. Microbiol Res 242:126616. https://doi.org/10.1016/j.micres.2020.126616
Kumar R, Kumar A, Dhaka RK et al (2023) Climate change and agriculture: impact assessment and sustainable alleviation approach using rhizomicrobiome. In: Bioinoculants: biological option for mitigating global climate change. Springer, Singapore, pp 87–114
Kumari D, Singh R (2018) Pretreatment of lignocellulosic wastes for biofuel production: a critical review. Renew Sust Energ Rev 90:877–891. https://doi.org/10.1016/j.rser.2018.03.111
Kumari A, Dash M, Singh SK et al (2023) Soil microbes: a natural solution for mitigating the impact of climate change. Environ Monit Assess 195:1436. https://doi.org/10.1007/s10661-023-11988-y
Leal Filho W, Azeiteiro UM, Balogun A-L et al (2021) The influence of ecosystems services depletion to climate change adaptation efforts in Africa. Sci Total Environ 779:146414. https://doi.org/10.1016/j.scitotenv.2021.146414
Lemaire ON, Jespersen M, Wagner T (2020) CO_2-fixation strategies in energy extremophiles: what can we learn from acetogens? Front Microbiol 11:486. https://doi.org/10.3389/fmicb.2020.00486
Li M, Huang Y, Yang Y et al (2020) Heavy metal ions removed from imitating acid mine drainages with a thermoacidophilic archaea: *Acidianus manzaensis* YN25. Ecotoxicol Environ Saf 190:110084. https://doi.org/10.1016/j.ecoenv.2019.110084
Lipscomb GL, Crowley AT, Nguyen DMN et al (2023) Manipulating fermentation pathways in the hyperthermophilic archaeon *Pyrococcus furiosus* for ethanol production up to 95°C driven by carbon monoxide oxidation. Appl Environ Microbiol 89(6):e00012–e00023. https://doi.org/10.1128/aem.00012-23
Luo G, Xie L, Zou Z et al (2010) Evaluation of pretreatment methods on mixed inoculum for both batch and continuous thermophilic biohydrogen production from cassava stillage. Bioresour Technol 101:959–964. https://doi.org/10.1016/j.biortech.2009.08.090
Mahdi MS, Ameen RS, Ibrahim HK (2016) Study on Degradation of Nylon 6 by thermophilic bacteria *Anoxybacillus rupiensis* Ir3 (JQ912241). Int J Adv Res Biol Sci (IJARBS) 3:200–209. https://doi.org/10.22192/ijarbs.2016.03.09.027
Maitra S, Brestic M, Bhadra P et al (2021) Bioinoculants—natural biological resources for sustainable plant production. Microorganisms 10:51. https://doi.org/10.3390/microorganisms10010051
Maleke M, Valverde A, Vermeulen J-G et al (2019) Biomineralization and bioaccumulation of europium by a thermophilic metal resistant bacterium. Front Microbiol 10:81. https://doi.org/10.3389/fmicb.2019.00081
Mallick R, Dwivedi K, Rai S (2024) Bioeconomy: a sustainable approach for biomass waste management. In: Srivastav AL, Bhardwaj AK, Kumar M (eds) Valorization of biomass wastes for environmental sustainability. Springer, Cham, pp 239–250
Masmoudi F, Alsafran M, Jabri HAL et al (2023) Halobacteria-based biofertilizers: a promising alternative for enhancing soil fertility and crop productivity under biotic and abiotic stresses—a review. Microorganisms 11:1248. https://doi.org/10.3390/microorganisms11051248
Massello FL, Chan CS, Chan K-G et al (2020) Meta-analysis of microbial communities in hot springs: recurrent taxa and complex shaping factors beyond PH and temperature. Microorganisms 8:906. https://doi.org/10.3390/microorganisms8060906
Moopantakath J, Imchen M, Anju VT et al (2023) Bioactive molecules from haloarchaea: scope and prospects for industrial and therapeutic applications. Front Microbiol 14:1113540. https://doi.org/10.3389/fmicb.2023.1113540
Mukhtar S, Mehnaz S, Malik KA (2019) Microbial diversity in the rhizosphere of plants growing under extreme environments and its impact on crop improvement. Environ Sustain 2:329–338. https://doi.org/10.1007/s42398-019-00061-5
Nakayama H, Shin Y, Sumita T et al (2019) Characterization of manganese oxide-biomineralization by the psychrophilic marine bacterium, *Arthrobacter* sp. strain NI-2 and its spontaneous mutant strain NI-2′. Environ Nat Resour J 17:68–77. https://doi.org/10.32526/ennrj.17.4.2019.32

Neifar M, Chouchane H, Najjari A et al (2019) Genome analysis provides insights into crude oil degradation and biosurfactant production by extremely halotolerant *Halomonas desertis* G11 isolated from Chott El-Djerid salt-lake in Tunisian desert. Genomics 111:1802–1814. https://doi.org/10.1016/j.ygeno.2018.12.003

Neil Adger W, Arnell NW, Tompkins EL (2005) Successful adaptation to climate change across scales. Glob Environ Chang 15:77–86. https://doi.org/10.1016/j.gloenvcha.2004.12.005

Ni S, Lv W, Ji Z, Wang K (2023) CO_2 mineralized sequestration and assistance by microorganisms in reservoirs: development and outlook. Energies 16:7571. https://doi.org/10.3390/en16227571

Noor-Hassim M-FB, Chuen LNG, Teo HM et al (2023) The utilization of microbes for sustainable food production. Biotechnologia 104:209–216. https://doi.org/10.5114/bta.2023.127209

Oren A (2016) Probing Saltern brines with an oxygen electrode: what can we learn about the community metabolism in hypersaline systems? Life 6:23. https://doi.org/10.3390/life6020023

Özdemir S, Kılınç E, Poli A, Nicolaus B (2013) Biosorption of heavy metals (Cd^{2+}, Cu^{2+}, Co^{2+}, and Mn^{2+}) by Thermophilic Bacteria, *Geobacillus thermantarcticus* and *Anoxybacillus amylolyticus*: equilibrium and kinetic studies. Biorem J 17:86–96. https://doi.org/10.1080/10889868.2012.751961

Ozdemir S, Turkan Z, Kilinc E et al (2023) The removal of heavy metal pollution from wastewaters using thermophilic *B. cereus* SO-16 bacteria. Chemosphere 311:136986. https://doi.org/10.1016/j.chemosphere.2022.136986

Padder SA, Khan R, Rather RA (2024) Biofuel generations: new insights into challenges and opportunities in their microbe-derived industrial production. Biomass Bioenergy 185:107220. https://doi.org/10.1016/j.biombioe.2024.107220

Passarini MRZ, Silva TR, Bernal SPF et al (2020) Undecane production by cold-adapted bacteria from Antarctica. Extremophiles 24:863–873. https://doi.org/10.1007/s00792-020-01200-1

Pernicova I, Kucera D, Nebesarova J et al (2019) Production of polyhydroxyalkanoates on waste frying oil employing selected *Halomonas* strains. Bioresour Technol 292:122028. https://doi.org/10.1016/j.biortech.2019.122028

Pugazhendi A, Jamal MT, Al-Mur BA, Jeyakumar RB (2022) Bioaugmentation of electrogenic halophiles in the treatment of pharmaceutical industrial wastewater and energy production in microbial fuel cell under saline condition. Chemosphere 288:132515. https://doi.org/10.1016/j.chemosphere.2021.132515

Purcarea C, Ruginescu R, Banciu RM, Vasilescu A (2024) Extremozyme-based biosensors for environmental pollution monitoring: recent developments. Biosensors 14(3):143. https://doi.org/10.3390/bios14030143

Purohit A, Yadav SK (2020) Characterization of a thermotolerant and acidophilic mannanase producing *Microbacterium* sp. CIAB417 for mannooligosachharide production from agro-residues and dye decolorization. Int J Biol Macromol 163:1154–1161. https://doi.org/10.1016/j.ijbiomac.2020.07.114

Qiu X, Wang W, Zhang L et al (2022) A thermophile *Hydrogenibacillus* sp. strain efficiently degrades environmental pollutants polycyclic aromatic hydrocarbons. Environ Microbiol 24:436–450. https://doi.org/10.1111/1462-2920.15869

Quehenberger J, Shen L, Albers S-V et al (2017) *Sulfolobus* – a potential key organism in future biotechnology. Front Microbiol 8:2474. https://doi.org/10.3389/fmicb.2017.02474

Rajashekarappa KK, Mahadevan GD, Neelagund SE et al (2022) Decolorization of amaranth RI and fast red E azo dyes by thermophilic *Geobacillus thermoleovorans* KNG 112. J Chem Technol Biotechnol 97:482–489. https://doi.org/10.1002/jctb.6834

Ramadoss D, Lakkineni VK, Bose P et al (2013) Mitigation of salt stress in wheat seedlings by halotolerant bacteria isolated from saline habitats. Springerplus 2:1–7. https://doi.org/10.1186/2193-1801-2-6

Ranawat P, Rawat S (2017) Radiation resistance in thermophiles: mechanisms and applications. World J Microbiol Biotechnol 33:112. https://doi.org/10.1007/s11274-017-2279-5

Rangasamy A, Gandhi S, Tamilchelvan V (2021) Biosorption of Hexavalent Chromium by *Paenibacillus pabuli* and *Bacillus cereus* Isolated from Alkaline Industrial Contaminated Soil in Puducherry, India. NEPT 20(2):729–735. https://doi.org/10.46488/NEPT.2021.v20i02.032

Rani A, Pundir A, Verma M et al (2024) Methanotrophy: A Biological Method to Mitigate Global Methane Emission. Microbiol Res 15:634–654. https://doi.org/10.3390/microbiolres15020042

Rao AS, Nair A, More VS et al (2022a) Extremophiles for sustainable agriculture. In: New and future developments in microbial biotechnology and bioengineering. Elsevier, pp 243–264

Rao AS, Nair A, Nivetha K et al (2022b) Molecular adaptations in proteins and enzymes produced by extremophilic microorganisms. In: Extremozymes and their industrial applications. Elsevier, pp 205–230

Rezaei S, Shahverdi AR, Faramarzi MA (2017) Isolation, one-step affinity purification, and characterization of a polyextremotolerant laccase from the halophilic bacterium *Aquisalibacillus elongatus* and its application in the delignification of sugar beet pulp. Bioresour Technol 230:67–75. https://doi.org/10.1016/j.biortech.2017.01.036

Robles-Fernández A, Areias C, Daffonchio D et al (2022) The role of microorganisms in the nucleation of carbonates, environmental implications and applications. Minerals 12:1562. https://doi.org/10.3390/min12121562

Romero-González M, Nwaobi BC, Hufton JM, Gilmour DJ (2016) Ex-situ bioremediation of U(VI) from contaminated mine water using *Acidithiobacillus ferrooxidans* strains. Front Environ Sci 4:39. https://doi.org/10.3389/fenvs.2016.00039

Roukas T, Kotzekidou P (2022) From food industry wastes to second generation bioethanol: a review. Rev Environ Sci Biotechnol 21:299–329. https://doi.org/10.1007/s11157-021-09606-9

Sabalpara AN, Mahatma L (2019) Role of microbes in sustainable agriculture. In: Natural resource management: ecological perspectives, pp 147–163

Saiki RK (1989) The design and optimization of the PCR. In: PCR technology: principles and applications for DNA amplification. Palgrave Macmillan UK, pp 7–16

Santhaseelan H, Dinakaran VT, Dahms H-U et al (2022) Recent antimicrobial responses of halophilic microbes in clinical pathogens. Microorganisms 10:417. https://doi.org/10.3390/microorganisms10020417

Santoyo G, Urtis-Flores CA, Loeza-Lara PD et al (2021) Rhizosphere colonization determinants by plant growth-promoting rhizobacteria (PGPR). Biology 10:475. https://doi.org/10.3390/biology10060475

Sathiyanarayanan G, Chabert N, Tulumello J, Achouak W (2021) Building efficient biocathodes with *Acidithiobacillus ferrooxidans* for the high current generation. J Power Sources 514:230586. https://doi.org/10.1016/j.jpowsour.2021.230586

Satoh Y, Tajima K, Nakamoto S et al (2011) Isolation of a thermotolerant bacterium producing medium-chain-length polyhydroxyalkanoate. J Appl Microbiol 111:811–817. https://doi.org/10.1111/j.1365-2672.2011.05093.x

Semenova EM, Tourova TP, Babich TL et al (2023) Crude Oil degradation in temperatures below the freezing point by bacteria from hydrocarbon-contaminated arctic soils and the genome analysis of *Sphingomonas* sp. AR_OL41. Microorganisms 12:79. https://doi.org/10.3390/microorganisms12010079

Sharma N, Leung IKH (2021) Novel Thermophilic bacterial Laccase for the degradation of aromatic organic pollutants. Front Chem 9:711345. https://doi.org/10.3389/fchem.2021.711345

Sharma AK, Sharma M, Sharma AK et al (2023) Mapping the impact of environmental pollutants on human health and environment: a systematic review and meta-analysis. J Geochem Explor 255:107325. https://doi.org/10.1016/j.gexplo.2023.107325

Singh B, Poças-Fonseca MJ, Johri BN, Satyanarayana T (2016) Thermophilic molds: biology and applications. Crit Rev Microbiol 42:985–1006. https://doi.org/10.3109/1040841X.2015.1122572

Sridharan R, Krishnaswamy VG, Archana KM et al (2021) Integrated approach on azo dyes degradation using laccase enzyme and CuI nanoparticle. SN Appl Sci 3:370. https://doi.org/10.1007/s42452-021-04164-9

Srinivas TNR, Anil Kumar P, Sasikala C et al (2008) *Rhodobacter ovatus* sp. nov., a phototrophic alphaproteobacterium isolated from a polluted pond. Int J Syst Evol Microbiol 58:1379–1383. https://doi.org/10.1099/ijs.0.65619-0

Stallwood B, Shears J, Williams PA, Hughes KA (2005) Low temperature bioremediation of oil-contaminated soil using biostimulation and bioaugmentation with a *Pseudomonas* sp. from maritime Antarctica. J Appl Microbiol 99:794–802. https://doi.org/10.1111/j.1365-2672.2005.02678.x

Stapleton RD, Savage DC, Sayler GS, Stacey G (1998) Biodegradation of aromatic hydrocarbons in an extremely acidic environment. Appl Environ Microbiol 64:4180–4184. https://doi.org/10.1128/AEM.64.11.4180-4184.1998

Takkar S, Tyagi B, Kumar N et al (2022) Biodegradation of methyl red dye by a novel actinobacterium *Zhihengliuella* sp. ISTPL4: kinetic studies, isotherm and biodegradation pathway. Environ Technol Innov 26:102348. https://doi.org/10.1016/j.eti.2022.102348

Taroepratjeka DAH, Imai T, Chairattanamanokorn P, Reungsang A (2019) Investigation of hydrogen-producing ability of extremely halotolerant bacteria from a salt pan and salt-damaged soil in Thailand. Int J Hydrog Energy 44:3407–3413. https://doi.org/10.1016/j.ijhydene.2018.06.010

Taroepratjeka DAH, Imai T, Chairattanamanokorn P, Reungsang A (2021) Extremely halophilic biohydrogen producing microbial communities from high-salinity soil and salt evaporation pond. Fuels 2:241–252. https://doi.org/10.3390/fuels2020014

Tavoosi N, Akhavan Sepahi A, Amoozegar MA, Kiarostami V (2023) Toxic heavy metal/oxyanion tolerance in haloarchaea from some saline and hypersaline ecosystems. J Basic Microbiol 63:558–569. https://doi.org/10.1002/jobm.202200465

Temerk H (2019) Isolation and characterization of *Haloferax mucsum* from Red sea saltern as an halobacterial Archea capable of growing on Naphtaline as an aromatic compounds. Egypt J Exp Biol (Botany) 14(2):392. https://doi.org/10.5455/egyjebb.20181126081938

Thapa S, Mishra J, Arora N et al (2020) Microbial cellulolytic enzymes: diversity and biotechnology with reference to lignocellulosic biomass degradation. Rev Environ Sci Biotechnol 19:621–648. https://doi.org/10.1007/s11157-020-09536-y

Thathola P, Agnihotri V, Pandey A, Upadhyay SK (2022) Biodegradation of bisphenol A using psychrotolerant bacterial strain *Pseudomonas palleroniana* GBPI_508. Arch Microbiol 204:272. https://doi.org/10.1007/s00203-022-02885-y

Tian F, Guo G, Zhang C et al (2019) Isolation, cloning and characterization of an azoreductase and the effect of salinity on its expression in a halophilic bacterium. Int J Biol Macromol 123:1062–1069. https://doi.org/10.1016/j.ijbiomac.2018.11.175

Tourova TP, Sokolova DS, Semenova EM et al (2022) Genomic and physiological characterization of Halophilic bacteria of the genera *Halomonas* and *Marinobacter* from petroleum reservoirs. Microbiology 91(3):235–248. https://doi.org/10.1134/S0026261722300038

Tyagi B, Gupta B, Thakur IS (2020) Biosorption of Cr (VI) from aqueous solution by extracellular polymeric substances (EPS) produced by *Parapedobacte*r sp. ISTM3 strain isolated from Mawsmai cave, Meghalaya, India. Environ Res 191:110064. https://doi.org/10.1016/j.envres.2020.110064

Uthandi S, Saad B, Humbard MA, Maupin-Furlow JA (2010) LccA, an Archaeal Laccase secreted as a highly stable glycoprotein into the extracellular medium by *Haloferax volcanii*. Appl Environ Microbiol 76:733–743. https://doi.org/10.1128/AEM.01757-09

Verma JP, Jaiswal DK, Krishna R et al (2018) Characterization and screening of Thermophilic *Bacillus* strains for developing plant growth promoting consortium from hot spring of Leh and Ladakh Region of India. Front Microbiol 9:1293. https://doi.org/10.3389/fmicb.2018.01293

Vogler M, Karan R, Renn D et al (2020) Crystal structure and active site engineering of a halophilic γ-carbonic anhydrase. Front Microbiol 11:742

Wagner ID, Wiegel J (2008) Diversity of thermophilic anaerobes. Ann N Y Acad Sci 1125:1–43. https://doi.org/10.1196/annals.1419.029

Wang M, Zhou L (2012) Simultaneous oxidation and precipitation of iron using jarosite immobilized *Acidithiobacillus ferrooxidans* and its relevance to acid mine drainage. Hydrometallurgy 125–126:152–156. https://doi.org/10.1016/j.hydromet.2012.06.003

Wang J, Wang J, Zhang Z et al (2016) Cold-adapted bacteria for bioremediation of crude oil-contaminated soil. J Chem Technol Biotechnol 91:2286–2297. https://doi.org/10.1002/jctb.4814

Wang Y, Hou Y, Wang Y et al (2022) Immobilization of psychrophile *Psychrobacter* sp. ANT206 onto novel reusable magnetic nanoparticles and its application for nitro-aromatic compounds biodegradation under low temperature. Biodegradation 33:223–237. https://doi.org/10.1007/s10532-022-09978-x

Yadav AN, Sharma D, Gulati S et al (2015) Haloarchaea endowed with phosphorus solubilization attribute implicated in phosphorus cycle. Sci Rep 5:12293. https://doi.org/10.1038/srep12293

Yan F, Wei R, Cui Q et al (2021a) Thermophilic whole-cell degradation of polyethylene terephthalate using engineered *Clostridium thermocellum*. Microb Biotechnol 14:374–385. https://doi.org/10.1111/1751-7915.13580

Yan Y, Geng Z-Q, Dai K et al (2021b) Decoupling mechanism of Acid Orange 7 decolorization and sulfate reduction by a *Caldanaerobacter* dominated extreme-thermophilic consortium. J Hazard Mater 419:126498. https://doi.org/10.1016/j.jhazmat.2021.126498

Yang W, Hong W, Huang Y et al (2022) Exploration on the Cr(VI) resistance mechanism of a novel thermophilic Cr(VI)-reducing bacteria *Anoxybacillus flavithermus* ABF1 isolated from Tengchong geothermal region, China. Environ Microbiol Rep 14(5):795–803. https://doi.org/10.1111/1758-2229.13070

Yao J, Zhang M, Han F et al (2022) Thermophilic adsorption of Pb2+ onto bacterium from deep-sea hydrothermal vent. J Environ Chem Eng 10:107364. https://doi.org/10.1016/j.jece.2022.107364

Yusoff DF, Raja Abd Rahman RNZ, Masomian M et al (2020) Newly isolated alkane hydroxylase and lipase producing *Geobacillus* and *Anoxybacillus* species involved in crude oil degradation. Catalysts 10:851. https://doi.org/10.3390/catal10080851

Zeinali M, Vossoughi M, Ardestani SK (2007) Characterization of a moderate thermophilic *Nocardia* species able to grow on polycyclic aromatic hydrocarbons. Lett Appl Microbiol 45:622–628. https://doi.org/10.1111/j.1472-765X.2007.02241.x

Zhang G, Jiang N, Liu X, Dong X (2008) Methanogenesis from methanol at low temperatures by a novel psychrophilic methanogen, "*Methanolobus psychrophilus*" sp. nov., prevalent in Zoige wetland of the Tibetan plateau. Appl Environ Microbiol 74:6114–6120. https://doi.org/10.1128/AEM.01146-08

Zhang S, Yan L, Xing W et al (2018) *Acidithiobacillus ferrooxidans* and its potential application. Extremophiles 22:563–579. https://doi.org/10.1007/s00792-018-1024-9

Zhao D, Kumar S, Zhou J et al (2017) Isolation and complete genome sequence of *Halorientalis hydrocarbonoclasticus* sp. nov., a hydrocarbon-degrading haloarchaeon. Extremophiles 21:1081–1090. https://doi.org/10.1007/s00792-017-0968-5

Zheng C, He J, Wang Y et al (2011) Hydrocarbon degradation and bioemulsifier production by thermophilic *Geobacillus pallidus* strains. Bioresour Technol 102:9155–9161. https://doi.org/10.1016/j.biortech.2011.06.074

Zhou J, Li G, Xie J et al (2016) A novel bioemulsifier from *Geobacillus stearothermophilus* A-2 and its potential application in microbial enhanced oil recovery. RSC Adv 6:96347–96354. https://doi.org/10.1039/C6RA15626F

Part IV

Microbial Responses to Climate Change: Carbon Sequestration, Cryosphere Ecology, and Antimicrobial Resistance

Carbon Sequestration and Rhizoremediation: Strategies for Managing Xenobiotic Compounds and Restoring Ecological Balance to Mitigate Climate Change

12

Nandita Das and Piyush Pandey

Abstract

Anthropogenic activities, such as the combustion of fossil fuels and industrial processes, release substantial amounts of CO_2 into the atmosphere, exacerbating concerns about global warming and climate change. Conventional CO_2 capture technologies often prove inadequate and can be hazardous, contributing to secondary environmental pollution. In contrast, biological systems present a promising alternative for CO_2 conversion due to their high selectivity and adaptability. Numerous bacteria have the capability to utilize CO_2 as their sole carbon source, converting it into valuable products. Additionally, bio-based remediation and environmentally friendly mitigation strategies have been employed to protect and restore environments contaminated by crude oil. Rhizoremediation has gained prominence in bioremediation due to its effective in-situ application. Rhizoremediation approaches for the mitigation of crude oil contamination and delineates strategies for CO_2 conversion aimed at reducing greenhouse gas emissions. Furthermore, recent advancements in the engineering of bacteria to efficiently utilize CO_2 and other carbon compounds as substrates. Modern sequencing technologies, including metagenomics, transcriptomics, and proteomics, are discussed for their role in identifying and characterizing microbial communities and metabolic pathways involved in these processes. These sequencing approaches provide deep insights into microbial ecology and function, facilitating the development of more efficient bio-based CO_2 conversion and bioremediation strategies.

N. Das · P. Pandey (✉)
Department of Microbiology, Soil and Environmental Microbiology Laboratory, Assam University, Silchar, India

P. Pandey et al. (eds.), *Microorganisms Resilience to Climate Change*, Microorganisms for Sustainability 56, https://doi.org/10.1007/978-981-96-3748-5_12

Keywords

Climate change · Rhizoremediation · Carbon capture · NGS technologies · Metabolic engineering

12.1 Introduction

Fossil fuels, such as coal, oil, and gas, are the main contributors to global climate change, accounting for about 75% of worldwide greenhouse gas emissions and almost 90% of all CO_2 emissions (Perera 2018). Human activities release over 30 billion tons of CO_2 annually, with a 40% increase in CO_2 levels concentration from the eighteenth century to 2020, from 280 ppm to 414 ppm (Tan et al. 2021). In the last half-century, there has been a substantial rise in climate change-induced catastrophes. Nevertheless, as a result of progress in warning systems and disaster management, the fatality count has been reduced to two million, while economic losses have been contained at US$ 3.64 trillion (Kalogiannidis et al. 2022). According to the U.S. Energy Information Administration report in 2019, the United States emitted a staggering 5130 million metric tons of CO_2 from energy-related sources. By contrast, the global emissions of energy-related CO_2 were totalled approximately 33,621.5 million metric tons. China, the USA, and India are the leading nations in terms of CO_2 emissions globally, collectively responsible for around 50% of total CO_2 emissions worldwide. Additionally, the top 20 emitters, including these three countries, account for 80% of global CO_2 emissions (Mehmood et al. 2024).

In addition, the utilisation of fossil fuels has a detrimental impact on the environment as a result of spills and leaks that negatively impact both terrestrial and aquatic ecosystems (George et al. 2024). In 2010, Enbridge's oil leak in Michigan caused over one million gallons of crude oil to be released into nearby waters, causing contamination of land and water (Kharey 2024). The Niger Delta in Nigeria is one of the world's most oil-contaminated areas, with 13 million tonnes of fossil fuel causing significant damage. Around 2000 locations have been affected by oil spills (Das et al. 2024).

The frequency and intensity of natural disasters have seen a significant increase, accompanied by a rise in sea level (Etherton et al. 2024), and the loss of biodiversity. According to Dutta et al. (2024), the models predicting global yield loss for wheat, maize, and rice show that for every degree °C rise in temperature, there is a corresponding increase in yield losses ranging from 10% to 25%. According to Brás et al. (2021), the combination of heatwave and drought has caused agricultural losses to increase by almost three times over the last 50 years. The losses went from −2.2% during the period of 1964–1990 to −7.3% during the period of 1991–2015. The cumulative impact of climate-induced abiotic pressures on crop output is estimated to surpass US$ 170 billion year, posing a significant risk to global food security (Arias et al. 2024; Razzaq et al. 2021). Crude oil pollution can negatively impact agriculture by reducing crop yield, affecting seed germination, and reducing plant life lifespan. The accumulation of hydrocarbons and contaminants in plant components can lead to a decrease in seed germination index (Das et al. 2024). Moreover, the flagship Living Planet Report 2022 report, shows that since 1970, the average reduction in species populations has been 69% (Holden et al. 2024).

In light of this anticipated climatic crisis, several nations, particularly, The United States and the European Union have pledged to achieve net zero carbon emissions by 2050, while China aims for 2060 and India sets a target for 2070 (Schenuit et al. 2024). India also aims to reduce its reliance on fossil fuels by 40% by 2030. Over 100 nations aim to decrease global methane emissions by 30% by 2030 and end deforestation (Bhatt 2024).

The process of eliminating and storing atmospheric CO_2 is referred to as carbon sequestration. To mitigate the effects of global climate change, this strategy aims to minimize the amount of CO_2 in the atmosphere (Joel and Oguanobi 2024). Geologic and biologic sequestration are two important sources of carbon sequestration that are now being evaluated by the USGS (Moore et al. 2024). Geologic carbon sequestration involves the underground storage of CO_2 in geologic formations, such as porous rock formations, where it is compressed until it becomes liquid. This method is used during the later stages of an active oil well's lifespan and is sometimes incorporated into enhanced oil recovery (Chu et al. 2024). Biological carbon sequestration captures and stores carbon dioxide in natural components like plants, soil, wood, and water bodies, with advocates aiming to reduce atmospheric CO_2 levels by encouraging plant development, particularly larger ones like trees (Singh et al. 2024).

Approximately 25% of all known biodiversity on Earth resides in soil, mostly in the form of microbes, which are among the planet's most varied and abundant species. Many soil microorganisms play a crucial role in biogeochemical cycling and broader ecosystem function, including the fluxes of important greenhouse gases like carbon dioxide (CO_2), methane (CH_4), and nitrous oxide (N_2O). Plants distribute about 40% of the carbon that is fixed by photosynthesis to the soil via rhizodeposition (Jones et al. 2009). The organic substances released by roots function as nutrients for the soil organisms, leading to increased microbial activity in the rhizosphere (Bisht et al. 2015). Further, soil microorganisms are essential to plant and soil health by mineralizing and degrading crude oil contaminants (Paul 2014; Wardle et al. 2004; Whalen et al. 2022; Singh et al. 2008). Considering the importance of microbial populations in maintaining soil and plant health, as well as their involvement in biogeochemical cycles, the control and utilisation of plant microbe-pair can play a significant role in improving production systems and reducing atmospheric CO_2 levels (Buragohain et al. 2019; Ahmed et al. 2019; Vishwakarma et al. 2016; King 2011).

Advancements in genetic engineering and molecular biology have now enabled the use of microbes, which have long been involved in carbon sequestration and fixation in the carbon cycle, to overcome these cost obstacles (Correa et al. 2023). The process of microbial CO_2 sequestration and fixation is facilitated by the enzymes ribulose 1,5-bisphosphate carboxylase/oxygenase (RuBisCO) and Carbonic Anhydrase (CA), which are found in both archaeal and bacterial domains (Liu et al.

2024). These advancements in genetic engineering allow for the effective conversion of CO_2 into biomass and valuable molecules such CO, CH_4, CH_3OH, DME, olefins, and higher hydrocarbons (Kouhi et al. 2024).

This chapter discusses the profound impact of fossil fuel consumption on global climate change, highlighting the substantial contributions to greenhouse gas emissions and the subsequent rise. Further the detrimental effects of fossil fuel spills on ecosystems and agriculture, emphasizing the importance of effective disaster management and mitigation strategies has been explored. Additionally, it delves into the promising advancements in carbon sequestration technologies, particularly through biological methods, and the innovative use of genetic engineering and microbial processes to enhance carbon capture and conversion. The crucial role of soil microorganisms in biogeochemical cycles and their potential in degrading contaminants and improving plant health is also highlighted. Specifically, the chapter examines the role of rhizoremediation, a process involving the synergistic action of plant roots and associated soil microorganisms to degrade, remove, or stabilize contaminants in the soil. This method enhances soil health and promotes sustainable agricultural practices. Moreover, the chapter discusses the application of sequencing approaches, including genomics, proteomics and metabolomics, to better understand and optimize microbial processes for carbon sequestration. In addition, we suggest exploring the research on the possible use of microbes in the production of energy and the remediation of polluted environments.

12.2 Effect of Xenobiotic Pollution on Bacterial Soil Carbon Sequestration

The substantial quantities of hazardous chemicals discharged into the environment lead to extensive pollution of ecosystems (Rajak et al. 2024). Soil contains significant amounts of compounds such as polycyclic aromatic hydrocarbons (PAHs), heavy metal ions, pesticides, fertilisers, and oil derivatives (Borah and Deka 2024). Xenobiotics, including dioxins, 1,1,1-trichloro-2,2-bis (4-chlorophenyl) ethane (DDT), polychlorinated biphenyls (PCBs), chlordane, PAHs, and nitroaromatics, causes the higher risk to soil ecosystems in industrialised regions (Ebsa et al. 2024). Nevertheless, there have been studies indicating that some other contaminants, including benzene, nitrobenzene, toluene, xylene, aniline, ethylbenzene, trinitrotoluene/dibenzofurans, and chlorinated solvents, may potentially act as xenobiotics, particularly within the soil ecosystem (Malik and Garg 2024). Moreover, the bioavailability and dispersion of xenobiotics in soil are influenced by their chemical properties and site circumstances, with soil organic matter (SOM) playing a crucial role (Aggarwal et al. 2024).

Oil-contaminated soils show a 60% decrease in chlorophyll content, potentially hindering plant growth (Vocciante et al. 2024). Additionally, crude oil adherence to soil particles reduces porosity, increases dry density and shifts soil pH, causing damage to intercellular components (Zhang et al. 2024a, b). Additionally, it disrupts the formation of metabolites by microorganisms, hence affecting the biogeochemistry of the soil. Xenobiotic contaminants may also alter the water retention and infiltration rates as well as increase bulk density, erosion risk, and salinization in soil (Kumari et al. 2024). Xenobiotics undergo reactions with the soil, as well as the indigenous soil microflora, resulting in changes to both biological and physical structures. Consequently, these changes have an impact on enzymatic activity and biochemical reactions of soil microbial populations (Priya et al. 2024). Xenobiotic compounds have the ability to disturb microbial populations in soils, which are crucial for the breakdown of organic matter and the cycling of nutrients (Patel and Kulwant 2024). Petroleum hydrocarbons have the potential to harm the cell membrane of microbes in a non-specific manner. This can result in alterations in the fluidity, integrity, and functionality of the cell membrane, ultimately leading to the inhibition of certain metabolic processes in microbes (Parmar et al. 2024). It has been observed that only a subset of the microorganisms present in a community are capable of adapting to environments contaminated with petroleum hydrocarbons (Das et al. 2024). The temperature increase leads to a shift in microbial communities and accelerates metabolic activities like respiration, fermentation and methanogenesis (Abatenh et al. 2018).

The presence of TPH-degrading bacteria, such as *Arthrobacter, Pseudomonas, Mycobacterium, Bacillus, Rhodococcus, Streptomyces, Polaromonas, Comamonas,* and *Sphingomonas* in the oil-contaminated soil suggests that they may be used as markers to evaluate the impact of TPH stress on the soil ecosystem (Singha and Pandey 2021). These particular microorganisms exhibit rapid growth and increased activity when exposed to petroleum hydrocarbons (Kotoky and Pandey 2020). This disturbance may diminish the capacity of soil to function as a carbon sink, resulting in elevated amounts of CO_2 in the atmosphere.

12.3 Role of Microbes in Mitigating Anthropogenic Contaminants

Bioremediation is an environment friendly, cost-effective, and secure approach to pollution treatment that utilises organic substances such as plants and bacteria (Singha and Pandey 2020). The degradation efficacy is contingent upon the characteristics, geographical position, and degree of contamination. Microbes are preferred due to their ease of cultivation, rapid growth rate, and straightforward manipulability (Atasoy et al. 2024). Enhancing the use of microorganisms in bioremediation is essential for advancing a sustainable environment and mitigating pollution (Bala et al. 2022). Microorganisms have the ability to transform harmful substances into water, carbon dioxide and other substances that are less harmful (Bhuyan et al. 2020). These less harmful substances are then broken down by other microorganisms in a process known as mineralization (Fan et al. 2023). The bacteria include *Achromobacter, Arthrobacter, Alcaligenes, Bacillus, Corynebacterium, Pseudomonas, Flavobacterium, Mycobacterium, Nitrosomonas, Xanthobacter*, and other more (Shah and Ahmed 2023). These microorganisms have been shown to break down pesticides, hydrocarbons, alkanes and polyaromatic chemicals (Kumari and Das 2023). For instance, bacteria such as *Enterobacter asburiae* and *Bacillus cereus* have been shown to have a role in the process of immobilising heavy metals that contaminate the environment (Kelany et al. 2023).

However, there are only a limited number of microbial enzymes that play a role in detoxifying xenobiotics. These enzymes include cytochrome P450s, laccases, cellulase, phytase, proteases, and lipases (Piotrowska-Długosz 2017; Sharma et al. 2018). These enzymes have the ability to break down aromatic hydrocarbons, dyes and halogenated compounds using different mechanisms. Aliphatic hydrocarbons can undergo degradation through two different mechanisms. One involves the action of monooxygenases, which introduce a single oxygen atom to the terminal methyl functional group (Li et al. 2024a, b). The other mechanism involves dioxygenase, which adds two oxygen atoms, leading to the formation of peroxide that is then converted into a fatty acid (Ahmed et al. 2024). In the process of oxidation, the fatty acid molecule undergoes a transformation into TCA cycle intermediates, which then proceed to metabolise into CO_2 and H_2O (Parmar et al. 2024). In the initial stages, these compounds undergo conversion into cis-dihydrodiols and are subsequently broken down by dioxygenase enzymes, following either ortho- or meta-cleavage pathways. Next, the aromatic rings undergo fission, with the hydroxyl groups in ortho-cleavage and those adjacent to hydroxyl groups in meta-cleavage pathways being involved (Singha et al. 2018). This ultimately results in the formation of intermediates found in central pathways (Kotoky et al. 2018).

The microorganisms are capable of degrading petroleum hydrocarbons through aerobic metabolism, resulting in the production of CO_2 and H_2O. The emission of significant quantities of CO_2 will further exacerbate the "Greenhouse effect" in the atmosphere (Abatenh et al. 2018). Therefore, to mitigate such consequences rhizoremediation technique utilising beneficial plant-microbe pair is a highly effective method for the decontamination of environmental pollution and the reduction of greenhouse gas emissions at a same time (Singha and Pandey 2021).

12.4 Synergistic Action of Rhizoremediation in Carbon Sequestration and Contaminant Removal

According to the 2007 Energy Independence and Security Act, the USGS is legally obligated to carry out a comprehensive evaluation of carbon storage and movement, as well as the movement of other greenhouse gases, in ecosystems (Sissine 2008). In the soil system, carbon (C) can be divided into two primary pools: labile, which has a fast turnover rate and a short lifespan, and recalcitrant, which is stable and long-lasting (Cotrufo and Lavallee 2022). The labile C primarily comprises of coarse fraction material (>0.4 mm), referred to as particulate organic matter (POM). Small portion (<0.4 mm) of the resistant soil C pools is long-lasting stabilized C (Ballayan 2021). This includes the comparatively low molecular weight components of mineral-association organic matter (MAOM), such as lipids, polysaccharides, and other extensively degraded plant and microbial products (Reiter 2015). Soil microbes play a crucial role in the decomposition of organic matter. They consume plant carbon inputs for their growth and release carbon in the form of extracellular products and CO_2 (Horwath 2007). Firstly, they assimilate organic carbon into their cellular components and substances, which can later be stabilised through interactions with minerals. Secondly, they produce enzymes that facilitate the breakdown and alteration of plant and soil carbon (Cuadros 2017).

Rhizoremediation is a specialised technology that utilises specific plants and microorganisms to effectively reduce and break down hazardous substances (Singha et al. 2018). Plants employ a variety of techniques to carry out the process of phytoremediation on various contaminants. These methods include phytoextraction, rhizofiltration, phytostabilization, and phytovolatilization (Das et al. 2024). In order to expedite the process of Rhizoremediation, which involves the breakdown of contaminants in soil through the growth and multiplication of microorganisms with the ability to degrade these contaminants (Das et al. 2022). Plants have the capability to absorb and remediate a diverse array of pollutants, including heavy metals, organic pollutants, metalloids, radioactive elements, pesticides, antibiotics, and inorganic substances (Pant et al. 2016). Using plant-microbe combinations in contaminated areas provides additional advantages for the growth and development of both organisms (Bisht et al. 2015). Plant species play a crucial role in reducing atmospheric CO_2 levels by absorbing CO_2 from the environment. Additionally, they also help in mitigating the CO_2 emissions produced by microorganisms during the breakdown of crude oil pollutants through aerobic catabolism (Das et al. 2024). During different stages of plant growth and in different environmental conditions, the gene transcripts of rhizosphere bacteria experience alterations (Newman et al. 2016). Furthermore, the gene transcripts associated with SOM decomposition and carbohydrate depolymerization exhibit distinct variations between the rhizosphere and bulk soil as the plant matures (Sieradzki et al. 2020). The CO_2 is assimilated into the plant biomass through the process of photosynthesis. Later on, the plant releases this carbon through its rhizospheric regions, which helps in the growth of organisms in the rhizosphere. The process of efficient CO_2 sequestration can be achieved by plants utilising their extensive leaf and root systems (Grover et al. 2015).

12.4.1 Role of Root Exudates

Approximately 30–60% of carbon ingested by plants is transported to the roots. Additionally, up to 50% of this carbon is released into the rhizosphere in various forms (Mueller et al. 2024). A significant number of interactions between roots and the surrounding microbial community occur through chemical communication facilitated by root exudates (Jin et al. 2024). These interactions play a crucial role in plant defence, nutrient uptake and the control of soil bacterial community composition (Mishra et al. 2024). Studies have demonstrated that the higher levels of easily decomposable soil carbon in the vicinity of plant roots can have both positive and negative effects on the breakdown of soil organic carbon (Zhang et al. 2024). Some studies indicate that exudates have the same potential for persisting in soil as root tissue carbon. One particular challenge arises from the intricate composition of root exudate compounds, which differ based on plant genotype, root maturity and environmental stimuli (Li et al. 2024a, b). One challenge is accurately characterizing the chemical composition of exudate due to the significant background signal from soil and microbial components (Chenu et al. 2024). Increased levels of CO_2 can enhance photosynthesis rates and lead to greater distribution of carbon to roots and different carbon storage areas in the soil (Madhu and Hatfield 2013). These include higher rates of photosynthesis, increased production of belowground biomass and the deposition of labile C in the soil. Additionally, eCO_2 has been observed to lower transpiration rates and potentially increase soil water content by reducing stomatal conductance (De Kauwe et al. 2021). A study found that elevated levels of CO_2 had a significant impact on the fungal populations in the rhizosphere and the activity of carbon decomposition enzymes, led to a decrease in soil carbon content (Duan et al. 2015). Moreover, the presence of a substantial amount of dissolved matter in root exudates within forests can result in the formation of extensive organism clusters and the sequestration of carbon in the lower layers of soil (Ma et al. 2022). Root exudates that have simple compositions can play a role in the creation of MAOC, while those with more complex compositions may stimulate the formation of POC (Villarino et al. 2021). It is fascinating to observe that the composition of root exudates can have diverse impacts on the stability of soil organic carbon (SOC) in various regions. In addition, the amount and makeup of root exudates can also impact the carbon turnover in the soil (Lei et al. 2023).

The addition of root exudates, such as high concentrations of organic acids, to soil has been found to result in a reduced absorption of organic pollutants (Hayat et al. 2017). A study conducted by Matiasek and Hernes (2019), reported that the artificial application of root exudate substances, such as organic acids, glucose and serine can significantly enhance the degradation of PAHs. The presence of glucose in the root exudates stimulates the production of the dehydrogenase enzyme, which is responsible for breaking down pyrene and promoting the growth of *Mycobacterium* sp. (Liao et al. 2021). The process of horizontal gene transfer is employed by microorganisms as a means of adapting to a diverse environment (Burmeister and Smith 2016). However, legume symbiosis is also facilitated by horizontal gene transfer of symbiotic plasmids among legume symbionts. The secretion of root exudates involves the release of various chemical compounds, including flavonoids, which serve as chemical signals (Singla and Garg 2017). These signals play a role in promoting horizontal gene transfer between different microbial species and initiating a symbiotic process (Skorupska et al. 2010). Microorganisms have evolved chemotaxis signals in response to root exudates in order to obtain nutrients and energy for their survival and reproduction. The highest concentration of exudates can be found in root tips and lateral branches near the roots (Rohrbacher and St-Arnaud 2016). The concentration of root exudates may decrease as a result of their consumption by microorganisms, chemical degradation, or diffusion (Ma et al. 2022). The root exudes substances primarily through two transport mechanisms: active transport, which involves the secretion of compounds, and passive transport, which includes simple diffusion, ion channels, and vesicles (Ahlawat et al. 2024). Membrane transport proteins aid in the excretion of secondary metabolites, such as polysaccharides and diverse proteins, from the root cells. Four primary membrane transporters have been discovered by Gani et al. (2021). These encompass ABC (ATP-binding cassette) transporters, which have the role of excreting chemicals. MATEs, also known as multidrug and toxic compound extrusion transporters, facilitate the active transportation of a wide range of chemicals including antibiotics, alkaloids, phenolic substances, ions and citrate across the cell membrane (Nuruzzaman et al. 2014). This process relies on the utilisation of the electrochemical gradient of ions. The remaining two categories consist of MFS (major facilitator superfamily) and ALMT (aluminum-activated malate transporter) (Gani et al. 2021). Badri et al. (2012), found that there was a notable increase in the expression of 25 distinct genes associated with ABC transporters in the roots of Arabidopsis thaliana. These genes played a crucial role in the release of flavonoid chemicals. Bacterial ABC transporters function as both importers and exporters of substances across the plasma membrane. ABC transporters and MATEs facilitate the transportation of flavonoids to the vacuole (Nogia and Pati 2021). A study conducted by Badri et al. (2012), found that the eight distinct subspecies of *Arabidopsis* release specific exudates that selectively affect different microbiomes in the rhizosphere. The phenolic substance possesses the capability to alter the microbial composition in comparison to other compounds, such as sugar and other carbohydrates (Nihorimbere et al. 2011). The exudates from cucumber roots, specifically p-coumaric acid and vanillic acid, have been found to induce changes in the rhizosphere microbiome and increase its abundance. These alterations are attributed to the presence of phenolic compounds (Zhou and Wu 2012). Certain substances released by plant roots, known as root exudates, have the ability to imitate the signalling molecules called N-acyl homoserine lactones (AHLs) that are involved in bacterial quorum sensing. These compounds play a role in regulating the connection between plants and microbes (Sekar and Kandavel 2010). Additionally, the gene responsible for host infection, biofilm and biosurfactant synthesis, nitrogen fixation and enhanced production of degrading enzymes can be induced to modulate bacterial functions in the host plant (Molinatto 2017). Plant species such as *Coronilla varia*, *Pisum sativum*, and *Oryza sativa* generate compounds that are comparable to the AHLs (N-acyl homoserine lactones) produced by specific microorganisms. These compounds help regulate the population of these microbes in the rhizosphere, while also inhibiting the growth of other microbes (Massalha et al. 2017). Root exudates in polluted soil enhance the growth of hydrocarbon-degrading microorganisms, hence accelerating the process of soil restoration. The level of root exudates decreases as the concentration of contamination increases (Singha and Pandey 2021). In other words, as the contamination gradient increases, the lowest concentration of exudates is found at the surface of the root (Phillips et al. 2012). The rhizosphere of *Lolium perenne* L, perennial ryegrass, exhibited the greatest microbial activity and contamination level (Tannenbaum 2021). The predominant phenanthrene degraders were identified as *Pseudoxanthomonas* spp. and *Microbacterium* spp. However, when ryegrass exudates were introduced, there was a noticeable shift in the microbial population towards *Arthrobacter* spp. and Pseudomonas *stutzeri* (Afridi et al. 2022). Several species of microorganisms have the ability to utilise both root exudates and hydrocarbons as their source of carbon. Enzymes that exude, such as peroxidase, laccases, and phenol oxidases, have a role in directly oxidising hydrocarbon compounds into intermediate derivatives (Gramss 2012). Plant roots release flavonoid chemicals that can trigger the co-metabolism of polyaromatic hydrocarbons. This is because flavonoids have a similar structure to aromatic compounds, which enhances the breakdown and conversion of these hydrocarbons by microbes (Bisht et al. 2015). Gram-positive microbes are capable of degrading non-aromatic plant compounds, such as linoleic acid-induced pyrene and benzo[a]pyrene. Multiple studies have documented that the presence of low molecular weight organic acids, such as malic acid, citric acid, succinic acid, tartaric acid, and oxalic acid, released by plant roots, enhances the bioavailability of xenobiotic substances (Molina and Segura 2021).

12.4.2 Role of Bacterial Communities

Bacteria play a crucial role in enhancing plant nutrient uptake by providing readily available nutrients, acting as natural biofertilizers. It has been observed that bacteria have the ability to enhance plant growth and resilience by producing phytohormones (Kumar et al. 2021). Additionally, they can act as biological control agents, protecting plants from harmful phytopathogens. Microbes provide essential nutrients and growth hormones to plants, protect them against a range of plant diseases, and facilitate the breakdown of pollutants (Meena et al. 2017). In addition, bacteria encourage root growth and regulate the generation of stress hormones (Debnath et al. 2023). These findings highlight the potential of bacteria in promoting plant health. There have been numerous PGPB that have been discovered and made available for commercial use (Orozco-Mosqueda et al. 2021). According to research, belowground carbon inputs have been found to be more effective in forming MAOC compared to aboveground carbon inputs. This is mainly due to the fact that rhizosphere microbial communities are more efficient at forming stable soil carbon compared to bulk soil communities (Sokol and Bradford 2019). In contrast to the microbial materials generated by root and bud litter, the microbial available materials produced by rhizodeposition are converted into MAOC more efficiently (Brewer et al. 2023). This occurs due to the combination of organic matter with minerals, which results in a decrease in the interaction between soil microbes and enzymes. Some of the most notable ones include various species of *Bacillus, Rhizobia, Bradyrhizobium*, and *Pseudomonas*. Many different taxa have been identified and used in various studies (Glick 2015), which includes *Azospirillum sp., Streptomyces* sp., *Agrobacterium radiobacter, Azotobacter chroococcum, Burkholderia cepacia, Delfitia acidovorans, Paenobacillus macerans, Pantoea agglomerans*, and *Serratia entomophilia* (Velázquez et al. 2017).

Similarly, the bacterial community related to degradation of phenanthrene in the rhizosphere of ryegrass exhibits was significantly higher rate compared to the non-rhizospheric soil. The study observed a substantial decrease in TPH compounds from an initial concentration of 4330 mg/kg to below 120 mg/kg within a 22-week period. This reduction was achieved in the presence of rye grass (Azaizeh et al. 2010). An average removal of crude oil ranging from 41.6% to 63.2% was observed in contaminated soil treated with *Mirabilis jalapa*, while soils without plants showed a removal rate of 19.8–37.9% (Shahsavari et al. 2013). However, the diversity of carbon-fixing bacteria in soil significantly depending on the different soil types. For instance, the dominant microorganisms responsible for carbon fixation in the soil of Napahai plateau wetland are mostly *Archaea, Chloroeclobacter*, and *Proteobacteria* (Jiang et al. 2022). The other microorganisms associated with both crude oil degradation and carbon fixation in soil include *Acidobacteria, Actinomycetes, Archaea, Bacteroidetes, Chlorobacteria, Blastomonas, Firmicutes, Nitrocellulose, Proteobacteria,* and *Pumice* (Jiang et al. 2022).

12.4.3 Factors Effecting Carbon Sequestration and Rhizoremediation

Rhizoremediation is a biological process that is influenced by multiple elements. The elements influencing the environment are soil type, pollutants, pH levels, organic matter content, temperature, nutrition availability, oxygen availability, solubility of contaminants, microbial flora and the metabolic capacities of microorganisms (Masciandaro et al. 2013). The stability of soil organic carbon (SOC) is affected by a range of factors, including both biological ones such as the efficacy of various organic compounds utilised and generated by soil bacteria and abiotic factors such as the adsorption between soil minerals and organic matter derived from plants and soil microbes (Ramesh et al. 2019). The degradation process is affected by factors such as the plant's age and type, as well as the concentration of soluble nutrients in the rhizosphere (Mimmo et al. 2017). Root exudates supply carbon and nitrogen, which foster the decomposition of organic pollutants by microbes. The chemical composition of root exudates influences the microbial community in the rhizosphere. Interconnected elements such as pollution history, composition, porosity, permeability, soil density, pollutants distribution, metabolic functioning, and diversity can contribute to variations in rhizoremediation (Nwachukwu 2023).

Huang et al. (2023), revealed that petroleum contamination drastically modified the relationships among genes involved in carbon cycling and other soil activities. Decreased water levels in rhizosphere soils lead to observable alterations in bacterial community composition and functional gene profiles. However, the specific function of genes related to carbon fixation potential in controlling the carbon cycle during the breakdown of organic matter in soil is yet unknown. Therefore, it is essential to employ the abundant array of essential functional genes to evaluate the ability of microorganisms to capture carbon dioxide in soil contaminated with petroleum hydrocarbons (He et al. 2020). This approach can facilitate the investigation of the relationship between the prevalence of crucial functional genes and the eradication of petroleum hydrocarbons (Gougoulias et al. 2014). Figure 12.1 showing the role of plant microbe pair in carbon sequestration and xenobiotic degrartion.

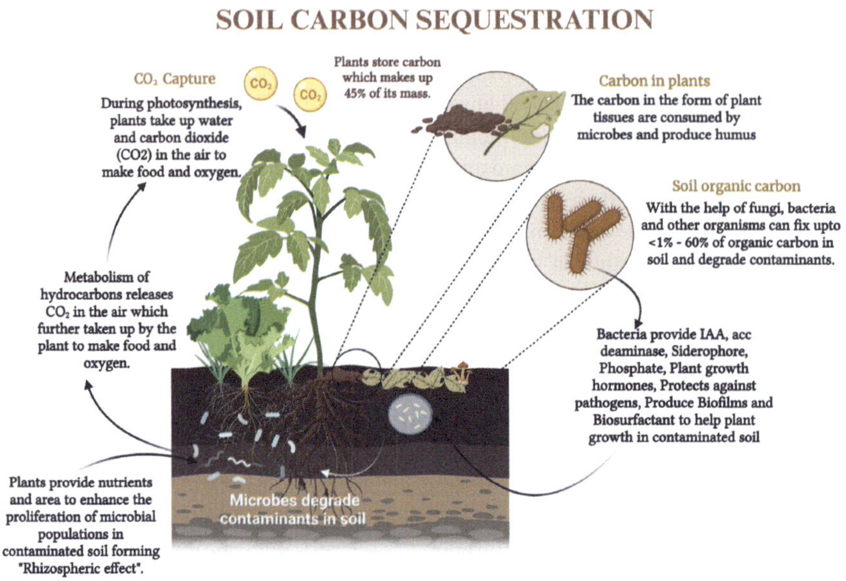

Fig. 12.1 Soil carbon sequestration and xenobiotic degradation with the help of plant-microbe interaction

12.5 Different Pathways Involved in Carbon Sequestration

Bacterial functional diversity in carbon sequestration pathways refers to the variety of metabolic processes that bacteria employ to capture and store carbon (Trivedi et al. 2013). Bacteria are essential in the global carbon cycle since they participate in several routes that transform inorganic carbon into organic forms that can be stored in biomass or sediments (Jiao et al. 2024). Globally soil microorganisms are predicted to fix from 0.5% to 4.1% of the atmospheric CO_2 annually. Corresponding to an annual carbon fixation quantity ranging from 0.3 to 3.7 Pg (Feng et al. 2022). Microbial carbon use efficiency (CUE) refers to the equilibrium between various metabolic pathways, and it directly impacts the retention of carbon within the system (Geyer et al. 2016). After microorganisms die, their remains, such as necromass, can undergo two processes. Several investigations propose that approximately 80% of soil organic carbon may be attributed to carbon found in microbial corpses (Aide and Braden 2022). The organic materials can be decomposed and utilized as a substrate for the generation of new microorganisms. Alternatively, they can be transformed into stable carbon reservoirs. The transformation can take place via micro-aggregation, leading to the creation of aggregated carbon (AggC), or via absorption onto mineral surfaces in the soil, resulting in the formation of mineral-associated organic matter (MAOM) (Gougoulias et al. 2014). In a recent study by Bhattacharyya et al. (2022), the relationship among the soil microbial communities

(SMC) and soil organic matter was examined, showing impact of SMC in the storage of soil carbon. The study found several important taxa that were involved, such *Rhizobium* sp., and *Actinobacteria*.

Soil microbes may transform organic materials present in the soil into gas, which is released into the atmosphere. Microbes residues contribute 10–80% to the creation of SOC and there is an advantageous relationship between SOC concentration and MBC content. Microorganisms have the ability to selectively utilize carbon fixation pathways in order to increase the efficiency of carbon fixation when the concentration of carbon dioxide increases in the external or soil environment (Wu et al. 2023).

12.5.1 Calvin–Benson–Bassham Cycle

In the 1940s and 1950s, Melvin Calvin, Andrew Benson, and James Bassham made a significant discovery in the field of photosynthesis. They identified the photosynthetic and pentose phosphate reducing pathway, also known as the Calvin-Benson-Bassham (CBB) pathway (Correa et al. 2023). The pathway is present in a wide range of organisms, such as algae, plants, cyanobacteria, *Proteobacteria, Firmicutes*, as well as photoautotrophic and chemoautotrophic bacteria. The carbon concentrating mechanism (CCM) is essential for the assimilation of carbon dioxide in microorganisms (Martin et al. 2018). This process involves the utilization of specific enzymes such as ribulose-1,5-bisphosphate carboxylase/oxygenase (RuBisCO) and phosphoribulokinase (PRK) (Meloni et al. 2023). Glyceraldehyde-3-phosphate dehydrogenase (GAPDH) is an essential enzyme involved in the CBB cycle. In a dark environment, the enzymes GAPDH and PRK interact with the regulatory protein CP12 to form an inactive complex, serves as a light/dark regulator (López-Calcagno et al. 2014). The disassembly of this complex occurs when it is exposed to light, thereby enabling the reactivation of GAPDH and PRK enzymes. Oxygenic photosynthetic organisms utilize the CBB cycle for CO_2 fixation and possess the CP12 protein. In contrast, nonoxygenic phototrophic bacteria do not have this protein (Jensen et al. 2017). The enzyme RuBisCO catalyzes the conversion of three molecules of CO_2 into one molecule of glyceraldehyde 3-phosphate. This process plays a crucial role in the transformation of atmospheric CO_2 into carbohydrates (Correa et al. 2023).

12.5.2 Reductive Tricarboxylic Acid Cycle (rTCA)

In 1966, Evans and collaborators provided a description of the second carbon-fixation pathway. Researchers discovered this pathway in *Chlorobium limicola*, green sulphur bacterium (Correa et al. 2023). Within the rTCA cycle, two cofactors play a crucial role in facilitating carbon fixation. One of these cofactors is responsible for converting acetyl-CoA to pyruvate, while the other helps produce ketoglutarate from succinyl-CoA. This process operates in a manner that is contrary to the oxidative citric acid cycle (oTCA), which is a stage of cellular respiration in heterotrophs. Unlike oTCA, this cycle occurs in the absence of O_2, resulting the production of acetyl-CoA using two CO_2 molecules (Correa et al. 2023).

12.5.3 Reductive Acetyl-CoA Pathway

The third method of carbon fixation, known as the reductive acetyl-*CoA* or WL pathway, was documented by Ljungdahl and colleagues in 1986. The noncyclic route occurs in anaerobic microbes and involves the conversion of CO_2 into acetyl-CoA through reduction (Schultz et al. 2024). The process involves the simultaneous transformation of two molecules of CO_2 into acetyl-*CoA* through two distinct branches: the methyl branch (also referred to as the eastern branch) and the carbonyl branch (also known as the western branch) (Saini et al. 2011). In the eastern branch, one CO_2 molecule is reduced by six electrons, leading to the formation of a methyl group. Conversely, the western branch entails the conversion of another CO_2 molecule into CO by a reduction process. Subsequently, this carbon monoxide (CO) is joined with the methyl group and CoA to generate acetyl-*CoA* (Correa et al. 2023). The molecule Acetyl-*CoA* is subsequently integrated into the carbon structure of the cell or transformed into acetylphosphate. The phosphoryl group obtained from acetylphosphate is subsequently transferred to adenosine diphosphate (ADP) to produce ATP and acetate, produced by acetogenic bacteria (Lamminjoki 2024). The eastern branch of the cellular cycle involves the enzymatic reduction of CO_2 into formate, which is facilitated by the formate dehydrogenase enzyme (Davies 2017). The process described involves the condensation of formate with tetrahydrofolate, which is driven by ATP. This condensation reaction leads to the production of 10-formyl-H4folate. Subsequently, 10-formyl-H4folate is converted into 5,10-methylenyl-H4folate and 5,10-methylene-H4folate (Pala et al. 2018; Correa et al. 2023). The methyltetrahydrofolate enzyme facilitates the transfer of a methyl group from N5 of (6S)–CH3-H4folate to the cobalt centre of the corrinoid iron-sulfur protein (CfeSP) (Correa et al. 2023). On the other hand, the western branch of the process only necessitates the presence of a single enzyme, known as CO-dehydrogenase/acetyl-*CoA* synthase, to convert CO_2 into CO (Montarnal 2023). Further, the organisms undergo a metabolic process in which CO_2 is converted into formate and then incorporated into the carbonyl group of acetyl-CoA (Karishma et al. 2024). The reductive acetyl-*CoA* pathway is observed in prokaryotes that flourish in thermodynamic conditions characterized by extremes, such as acetogenic and methanogenic Archaea (Schultz et al. 2024). The organisms utilize this particular pathway for the purpose of converting CO_2 into acetyl-CoA. Additionally, this pathway serves the function of generating an electrochemical gradient, which is essential for the conservation of energy (Liu et al. 2024).

12.5.4 3-Hydroxypropionate Bicycle Pathway

The 3-hydroxypropionate bicycle, also known as the Fuchs-Holo route, was discovered in 1989 in the green sulphur thermophilic bacteria *Chloroflexus aurantiacus* (Correa et al. 2023). This metabolic pathway uses HCO_3 molecules to produce glyoxylate through two cycles. The first cycle involves carboxylation of acetyl-*CoA*, resulting in malonyl-*CoA*, which is reduced to proprionyl-*CoA* through hydroxypropionate (Correa et al. 2023). This is then converted into succinyl-*CoA*, which is transformed into (S)-malyl-*CoA*, restoring acetyl-*CoA* and releasing glyoxylate. The second cycle involves incorporating glyoxylate, transforming it into methylmalyl-*CoA*, which is then transformed into citramailyl-*CoA*, which is converted into pyruvate and acetyl-CoA, replenishing the precursors needed for the cycle (Correa et al. 2023).

12.5.5 4-Hydroxybutyrate Pathway

The metabolism of acetyl-*CoA* involves two primary cycles: the HP/HB cycle and the DC/HB cycle. The HP/HB cycle is a biochemical process that entails the addition of CO_2 to acetyl-CoA, leading to the formation of succinyl-CoA (Gupta et al. 2024). The DC/HB cycle utilizes ferredoxin/NAD(P)H reduction cofactors, whereas the HP/HB cycle employs NAD(P)H as the reduction cofactor (Kurt et al. 2023). The enzymes involved in the HP/HB cycle demonstrate the ability to tolerate oxygen, whereas the 4-hydroxybutyryl-*CoA* dehydratase enzyme in the DC/HB cycle loses its activity upon exposure to oxygen (Correa et al. 2023). The primary distinction between the two cycles lies in their association with carbon metabolism. In the DC/HB process, the conversion of acetyl-*CoA* to pyruvate is facilitated by pyruvate synthase. On the other hand, the HP/HB pathway requires an additional half cycle to generate succinyl-CoA, which is then oxidized to produce pyruvate through the succinate pathway (Correa et al. 2023). The 4-hydroxybutyrate cycle (HP/HB) was initially identified in *Metallosphaera sedula*, a thermophilic Archaea organism classified under the *Sulfolobales* order. The dicarboxylate/4-hydroxybutyrate (DC/HB) has been discovered in *Ignicoccus hospitalis*, which is an Archaea species belonging to the *Desulfurococcales* group.

12.5.6 Microbial Conversion of CO_2 into CH_4

Microorganisms have the ability to convert CO_2 into CH_4, which is expected to be a highly efficient process (Zheng et al. 2024). To investigate the potential of CO_2 biofixation and bioconversion into CH4 in oil reservoirs, as well as to enhance our knowledge of the microorganisms involved and the impacts of prolonged CO_2 exposure on them, it is essential to carry out research that specifically examines functional genes (Liu et al. 2015). CO_2 may serve as a carbon source for chemolithoautotrophic bacteria, including methanogens. The first investigation focused on methanol-utilizing bacteria in soil using DNA-SIP (Friedrich 2010). Subsequent studies have expanded upon this research by employing DNA-SIP to establish a connection between function and identification in relation to methane cycle and the biodegradation of organic contaminants (Cheng et al. 2013). The genes *cbbL* and *cbbM* are responsible for encoding the enzymes ribulose 1,5-bisphosphate carboxylase/oxygenase (RubisCO) form I and II, respectively. These enzymes play a crucial role in the CBB cycle, which is the primary pathway for fixing CO_2 (Lin et al. 2014; Paoli et al. 1995). The homoacetogenic bacteria present in both thermophilic and mesophilic anaerobic sludge use the gene *fthfs*, which encodes for formyltetrahydrofolate synthetase. This enzyme plays a vital role in the reductive acetyl-*CoA* pathway. Methyl-Coenzyme M reductase (*mcr*) plays a crucial role in the production of CH_4 (Chen et al. 2020a, b). The α-subunit of MCR (*mcrA*) is often used for the purpose of identifying certain methanogenic populations (Juottonen et al. 2006). Moreover, the presence of H_2 is essential in the process of converting CO_2 into CH_4. H_2 may be produced by Prokaryotes that belong to diverse evolutionary lineages and have the ability to manufacture H_2 (Piché-Choquette and Constant 2019). The gene encoding [Fe-Fe]-hydrogenases, which play a catalytic role in fermentative microbes, is well recognized as a dependable indicator for identifying anaerobic organisms that generate H_2 (Greening et al. 2016).

12.6 Functional Diversity of Microorganisms for Soil Carbon Sequestration

Culture-based techniques are often used in the study of microbial population dynamics. Nevertheless, these methods have a tendency to underestimate the actual range of diversity within the soil microbial community. Indeed, a significant portion of the soil microbial population has yet to be fully understood or described (Blagodatskaya and Kuzyakov 2013). Recent sequencing methods have significantly enhanced our comprehension of the taxonomic classification of soil microbes. Regarding the stability of soil carbon, it is known that microbial communities consist of certain keystone taxa that demonstrate functional redundancy. As a result, these taxa have significant positive relationships throughout the cycling of carbon (Li et al. 2024a, b).

12.6.1 Metagenomics

Metagenomic techniques have been successfully employed to investigate the functionality of C-cycling genes in the context of various agricultural practices, including the application of organic matter and fertilizers, as well as different tillage methods (Liao et al. 2023). The following approaches can be utilized to predict the amount of carbon that is lost or stored in agricultural soils. The effects of traditional tillage and crop rotation management strategies on soil microbial populations have been extensively studied (Dignac et al. 2017). These activities have been linked to a rise in the prevalence of gene fragments relevant to glucose metabolism (Bosco et al. 2011). Carbon cycling enzymes, including glycosyltransferases, glycoside hydrolases, carbohydrate esterases, carbohydrate-binding modules, and polysaccharide lyases, play a crucial role in the biogeochemical processes that take place in soil (Arnosti et al. 2021). These enzymes are responsible for the degradation of biopolymers. Research has indicated variations in the relative abundance of these genes among various soil types. The detection of representative gene fragments in multiple soil samples from a fertilized tall grass prairie indicates that the microbial capacity to facilitate carbon cycling in soils is extensive (Souza et al. 2015).

12.6.2 Meta-transcriptomics

The field of soil meta-transcriptomics aims to capture the complete RNA content in soil. RNA expression and transcription are commonly used as indicators of the presence and activity of live members within a microbial community (Mukherjee and Reddy 2020). The utilization of soil meta-transcriptomics tools for investigating carbon stabilization in soil results in concurrent information about the population and functional dynamics of the soil community, providing insights into the ongoing processes in soil (Urich et al. 2008). A significant study was done to investigate the effects of integrating wood ash additives into agricultural soil. Wood ash, when used as an amendment, may affect many physicochemical properties of the soil (Kanu et al. 2021). Additionally, it may impact the makeup and operational manifestation of the soil microbial population, specifically in regards to soil chemical characteristics such as heightened pH, electrical conductivity, dissolved organic carbon, and phosphate levels (Bungau et al. 2021). The inclusion of wood ash amendments resulted in an elevation in the expression of genes linked to microbial subpopulations (rRNA). Overall, there was a greater degree of elevated levels of mRNA when compared to downregulation (Marchal et al. 2020). The genes can be primarily categorized into four functional groups: "post-translation alterations, turnover of proteins and chaperones"; "gene transcription"; "replication, recombination, and maintenance"; and "carbohydrate transportation and metabolic processes" (Overy et al. 2021). The genes under consideration play a crucial role in the carbon cycling process, which includes various metabolic and cellular growth processes. The findings indicate that the taxonomic makeup of microbiomes in paddy soil is affected by the current environmental circumstances, including the availability of oxygen (Li et al. 2019). The differences in microbiome composition and succession between the two oxygen zones can potentially be elucidated by the unique expression of transcripts derived from highly conserved and widely distributed genes that are crucial for metabolic activity in each respective zone (Sun et al. 2024). The symbiotic relationship between fungi and Xanthomonadales has been observed in the process of decomposing plant materials within the oxic zone (Kim and Liesack 2015). In contrast, the majority of the dominant groups in the anoxic zone worked together to transform organic matter (Jessen et al. 2017). A separate investigation conducted a shotgun sequencing analysis on the total extracted RNA from rice paddy soils. The results indicated that the predominant carbon transformations in this environment were anaerobic biodegradation of soil organic matter (SOM) and methanogenesis (Masuda et al. 2018). The results demonstrated the significant involvement of *Anaeromyxobacter* in the carbon and nitrogen cycles, as well as in iron reduction within the paddy soil ecosystem (Masuda et al. 2018).

12.6.3 Metaproteomics

Soil metaproteomics tools are employed to analyse the entirety of protein content in a soil sample. The aforementioned tools offer significant insights into individual proteins, such as enzymes, that are linked to the functionality of microbial communities (Abiraami et al. 2020). The proteins in question exhibit long-term persistence in the soil, in contrast to RNA which is known to be transient. The application of soil metaproteomics has proven to be effective in evaluating and describing the composition of metabolically-active bacteria in rhizosphere soils (Overy et al. 2020). The rhizosphere is the soil environment in close proximity to the root hairs of actively developing root tips. In this region, carbon inputs known as plant exudates are exchanged with microbial communities. This interaction often entails inorganic nutrients that are vital for the development of plants (Zancarini et al. 2013). Wang et al. (2011), detected a diverse array of proteins in the soil around the rice, sugarcane and tobacco crops. The 23% proteins were discovered to possess potential roles in many crucial biological processes, such as energy metabolism, protein degradation, amino acid production, secondary metabolism, nucleotide metabolism, signal transmission, and defence mechanisms in the rhizosphere. Furthermore, it was noted that *Proteobacteria* were the predominant microbes in both the metaproteomics (44%) and T-RFLP (48%) studies (Singh et al. 2021). Mattarozzi et al. (2017), the technique of soil metaproteomics was effectively employed to analyse rhizosphere soils collected from plants exhibiting metal tolerance and metal hyper-accumulation. The analysis was conducted using liquid chromatography (LC)-high-resolution MS (HRMS) techniques led to the identification of a total of 294 distinct proteins produced by bacteria.

12.6.4 Metabolomics

Soil metabolomics is a method used to analyse the types of metabolites (ranging from 100 to 2000 Da) present in the soil. These metabolites are the by-products, intermediates, and end products of biological activities (Parray et al. 2019). By studying soil metabolomics, we may get direct information on the transformation and stability of carbon, as well as the components of soil organic matter (SOM) (Swenson et al. 2015). Metabolomics has been widely utilized in various experimental settings to analyse the molecular composition of SOM (soil organic matter) (Swenson et al. 2015). Clemente et al. (2013), a comparative examination was conducted to examine the composition of soil organic matter (SOM) following incubation with various components of the maize plant. After a 36-week period of incubation, the findings showed that soil containing maize roots had the highest contribution of microbial SOM, compared to soil containing stems or leaves. This discovery aligns with other studies that have shown the influence of the quality of plant inputs on the diversity of the soil microbial community. In the study by Rochfort et al. (2015), showed the metabolite profiles of soils with different land use were examined and compared using nuclear magnetic resonance (NMR) spectroscopy-based soil metabolomics. The analysis revealed that the agricultural soil exhibited a significant abundance of metabolites related to lipids and sugars. On the other hand, the remnant site soil was primarily characterized by the presence of terpene metabolites. The presence of readily degradable carbon substrates in agricultural soils could potentially be associated with the higher levels of organic carbon being introduced. Pisani et al. (2016), conducted an evaluation of the composition of soil SOM across a broad temperature range, while considering variations in land usage. The utilization of NMR allowed researchers to determine that soil cultivation resulted in elevated microbial contributions at elevated temperatures, decreased plant contributions, and improved degradation of suberin and lignin. The researchers have determined that alterations in land use, specifically due to agricultural activities, have the potential to contribute to the emission of CO_2 into the atmosphere. Jenkins et al. (2017), measured 96 soil metabolites using LC-MS and GC-MS. Their results showed that 25 chemicals were unevenly distributed. They created two soil-based media with 23 and 46 metabolites, respectively. These media were created to cultivate new microorganisms. Ward and Cory (2015), showed the soil dissolved organic matter (DOM) may be used to assess bacterial growth and several factors related to microbial respiration, production, and growth efficiency. Warren (2014), suggested a technique using LC-MS to identify fatty acids as markers of microbial activity, in order to deduce the particular microbial involvement in the creation of SOM. In a similar manner, Mitchell et al. (2015), used untargeted NMR in combination with a targeted study of phospholipid fatty acids (PLFAs) to establish a connection between the effects of biochar amendment and its influence on the soil microbial population and the water-extractable organic matter (WEOM) component of SOM. The method employed untargeted NMR analysis of the DOM to investigate changes in the microbial population. The results demonstrated a transition from a predominantly fungal population to one that was increasingly dominated by bacteria. Specifically, there was a shift from Gram-negative bacteria to Gram-positive bacteria over a period of 16 weeks following the introduction of biochar. The study conducted by WEOM also observed elevated concentrations of short-chain carboxylic acids, while detecting reduced levels of carbohydrates and peptides. The researchers have reached a conclusion that the introduction of biochar into the soil has resulted in a notable increase in microbial activity, which in turn has led to elevated levels of CO_2 emissions.

12.7 Genetic Approaches to Enhance CO_2 Sequestration

Recent technological breakthroughs in comprehending microbial metabolic pathways, decoding genetic makeup, and other associated domains have revolutionised our methodology in unravelling the basic principles of biology (Zhou et al. 2022). These advancements have enabled us to make alterations that were previously considered inconceivable. Thus, the alteration of microbial metabolic pathways can be directed towards simpler approaches that employ established techniques in metabolic and bioprocess engineering to tackle clearly defined and widely acknowledged issues (Volk et al. 2022). Autotrophic bacteria possess inherent mechanisms to metabolise CO_2 as their exclusive carbon source, either by utilising light energy or inorganic compounds (Nybo et al. 2015). Methylotrophic bacteria may use formate, methanol, and other methylated substances as their only carbon and energy sources (Kremp and Müller 2021). It is noteworthy that these microorganisms have the potential to undergo genetic modifications in order to improve their adaptability for bioproduction processes, including those carried out in industrial environments (Benedetti et al. 2018). Subsequently, a number of inorganic-biological hybrid systems were developed for the purpose of capturing and utilising CO_2 (Xiao et al. 2021). The research shown that the metabolic flexibility of the bacteria was improved by using light-harvesting inorganic materials, which initiated the process of converting CO_2 into bio-based products (Onyeaka and Ekwebelem 2023). Synthetic biology focuses on altering and relocating natural pathways for CO_2 fixation, enhancing the supply of CO_2 by modifying CO_2 fixation pathways, and enhancing the efficiency and durability of CO_2 fixation enzymes to facilitate efficient fixation (Gong et al. 2016).

A recent study has shown that bacterial isolates (*Bacillus altitudinis*) identified from mangrove sediments in India, which have positive carbonic anhydrase (CA) activity, had a remarkable capacity to trap and sequester carbon dioxide (CO_2), leading to a significant decrease of 97% in CO_2 levels (Nathan and Ammini 2019). d Mattozzi et al. (2013), classified the 16 stages of the 3-hydroxypropionate cycle from *Chloroflexus aurantiacus* into four distinct sub-pathways. Subsequently, they effectively articulated each individual sub-pathway inside *Escherichia coli*. Further, similar study investigates genetic alterations in *Ralstonia eutropha*, a gram-negative bacterium, to improve its ability to produce isobutanol and 3-methyl-1-butanol using only CO_2 as carbon and electricity as energy (Lu et al. 2012). In a study conducted by Chen et al. (2020a, b), the researchers investigated the capacity of genetically engineered *E. coli* to thrive and proliferate by using methanol as a sustainable nutrition supply. The study explores the potential of formate and methanol as environmentally friendly bioproduction techniques for *E. coli*, which utilizes the reductive glycine pathway and relies on carbon dioxide and renewable energy sources (Kim et al. 2020). In a study conducted by Gleizer (2019), investigated the ability of *E. coli* to produce all of its biomass carbon by utilising atmospheric CO_2. In addition, the use of genetically manipulable microorganisms is a highly beneficial technology because it allows for the introduction of new pathways (Gleizer et al. 2020).

For instance, *E. coli* generates H_2 more efficiently than formate when it uses it as an energy source, making it ideal for bacterial growth and multiplication (Benoit et al. 2020). Antonovsky et al. (2016), have successfully achieved the conversion of CO_2 into biomass in *E. coli*. The strain was engineered with the ability to assimilate CO_2, although it only constituted a small fraction of the organism's overall carbon uptake. Most of the carbon originated from a specific chemical compound known as pyruvate, which was provided by a non-indigenous Calvin-Benson-Bassham (CBB) cycle in genetically modified *E. coli*. A recent study employed metabolic rewiring and directed evolution techniques to successfully engineer *E. coli* strains that can utilise CO_2 as their main carbon source. The strains utilise formate oxidation as a means to fulfil their requirements for both reducing power and energy through the utilisation of a non-native Calvin-Benson-Bassham cycle (Zhuang et al. 2022). Keller et al. (2013), successfully transferred a segment of the 3-hydroxypropionate/4-hydroxybutyrate cycle, derived from the archaea *Metallosphaera sedula* which thrives at an optimal growth temperature of 73 °C, into archaea species *Pyrococcus furiosus* (able to thrive with an optimal growth temperature of 100 °C). The engineered strain has the capability to efficiently synthesise 3-hydroxypropionic acid by utilising CO_2 and hydrogen as the energy source, in such high temperature range. A CO_2-fixing bypass was constructed in the central metabolic pathways of *E. coli* and *Saccharomyces cerevisiae*, resulting in observed CO_2 recycling and increased ethanol yield (Zhuang and Li 2013; Guadalupe-Medina et al. 2013).

According to Herz et al. (2017), it has been demonstrated that a non-native CBB cycle can support robust development with only five mutations, as long as all the necessary metabolic components are provided from sugar. The mutations in enzymes, specifically *prs*, serA, and *pgi*, have the potential to impact the efflux of intermediates from the autocatalytic CO_2 fixation cycle to biomass. In the study, it was found that mutations in crucial regulators, namely *crp* and *ppsR*, have the potential to interfere with carbon metabolism. Further research has revealed an impressive demonstration of carbon metabolic adaptability in carbon-fixing bacteria. In the cyanobacteria *Synechococcus elongatus* PCC 7942, modifications were made to the CBB cycle and Embden-Meyerhof pathway to increase the flow of carbon towards CO_2 fixation. Additionally, a specific portion of the operator gene (*cp12*) in the CBB cycle was removed to improve the conversion of ribulose-5-phosphate to ribose-1,5-bisphosphate and regulate the carbon metabolism of the cyanobacteria. As a result of the absence of light, there was an augmented production of 2,3-butanediol. (Santos-Merino et al. 2023). Lehtinen et al. (2018), conducted a study demonstrating the successful integration of *Acetobacterium woodii*, an acetogen, with *Acinetobacter baylyi* ADP1, a non-native alkane producer, in order to generate alkanes within a biosystem. The development and implementation of nine synthetic two-step alkane biosynthesis pathways in *A. baylyi*. The achievement was accomplished through the utilisation of a combination of enzymes responsible for aldehyde and alkane production. In addition, they showed their ability to produce drop-in liquid fuels from CO_2. Nevertheless, the productivity of alkane in the modular system was observed to be inadequate, thereby posing a noteworthy research concern for future investigations. Further, the bacterium *Moorella thermoacetica* was stimulated to self-photosensitize using cadmium sulfide nanoparticles, enabling photosynthesis to convert CO_2 into acetic acid (Rosenbaum and Müller 2023). These nanoparticles function as light harvesters, supporting cellular metabolism. The enhanced biological system generates acetic acid with high efficiency, demonstrating a self-replicating pathway for reducing CO_2 to manufacture chemicals using sunlight (Rosenbaum and Müller 2023).

Additionally, the organism *Ralstonia eutropha* H16 underwent genetic modification in order to produce isobutanol and 3-methyl-1-butanol. The objective was accomplished through the utilization of an electro-bioreactor, in which CO_2 served as the sole carbon source and electricity was the exclusive energy source. The experiment involved the utilization of a plasmid that harboured a *lacZ* gene under the control of a promoter derived from either the *katG, sodC*, or *norA* gene. The objective was to induce the production of b-galactosidase in the *Ralstonia* strain LH74D. The exposure of *Ralstonia* cells to oxygen and nitric oxide triggers a stress response, which inhibits their growth and development. The reactive chemicals of the anode may become depleted before reaching the cells that are located outside. This allows for the growth and production of over 140 mg/l of biofuels using only electricity and CO_2 as the carbon source. The presented technique is independent of biological light reactions and can be operated using photovoltaic cells, wind turbines, or off-peak grid power sources. The process described herein can be utilized to transform CO_2 into liquid fuels (Vamsi Krishna et al. 2022). A reverse microbial fuel cell converts electricity into energy forms compatible with microbes, reducing nitrite to ammonia. This ammonia can promote cell growth and facilitate CO_2 fixation in *Nitrosomonas europaea*, a chemoautotroph species of bacteria. The electricity is used to convert nitrite into ammonia, a viable energy source (Khunjar et al. 2012).

A study aimed to create an *E. coli* strain that efficiently metabolizes methanol as its carbon source. The artificial methylotroph was created by removing the gene *frmA*, which detoxifies formaldehyde, and redirecting it towards the RuMP pathways to make E. *coli* to synthetic methylotroph. The study found that conserving formaldehyde is crucial for sustaining biosynthetic pathways for methylotrophic growth. To maintain balance across metabolic pathways, two key metabolic genes, *Pfk* and *Gapdh*, were downregulated, essential for generating NADH and connecting glycolysis, the RuMP cycle, and the pentose phosphate pathway (Keller et al. 2022) (Fig. 12.2).

12.8 The Significance of Soil Carbon in Various Ecosystems

Soil carbon performs crucial functions in several ecosystems. Some of them are mentioned below:

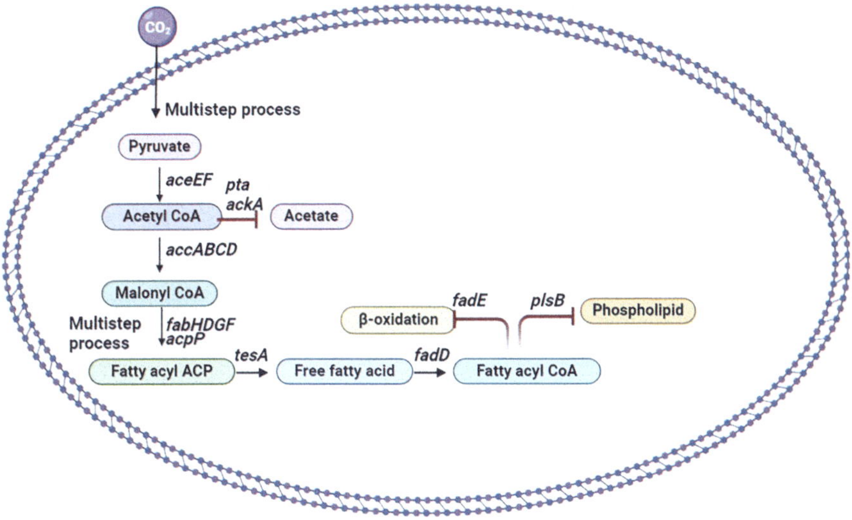

Fig. 12.2 Microorganisms utilize a diverse range of enzymes encoded by various genes in the synthesis of fatty acids. Gene *fadE* is responsible for the β-oxidation pathway, *tesA* encodes thioesterase, *fadD* encodes acyl-*CoA* synthetase, and *accABCD* plays a role in acetyl-CoA carboxylase. Gene *aceEF* is involved in the coding of enzymes related to pyruvate metabolism. In addition, the gene *plsB* glycerol-3-phosphate O-acyltransferase. Acetyl-*CoA* plays a significant role in the synthesis of fatty acids, with fatty acid synthase being the key enzyme involved in this process. It demonstrates the process of developing metabolic pathways in microorganisms for the purpose of genetically modifying them to produce fatty acids (Liu et al. 2022) (Table 12.1)

12.8.1 Climate Change Mitigation

Approximately 8.7 gigatons of carbon per year being released into the atmosphere, only 3.8 gigatons of carbon per year is left from human activities. The unexplained discrepancy of 4.9 gigatons of carbon per year is thought to be stored in various terrestrial bodies such as seas, forests, and soils, referred as 'missing sink' (Dymond 2010). This understanding has sparked interest in the capacity of the terrestrial sector, particularly soil, to capture and store carbon in long-lasting reservoirs, therefore decreasing the quantity of carbon in the atmosphere (Battle et al. 2000).

12.8.2 Land Management Practices

Soil carbon sequestration not only reduces greenhouse gas (GHG) levels in the atmosphere but also contributes to efforts to improve land productivity, whether it is for forests or agricultural purposes (Lal 2003). The rationale behind this is that any solutions that involve carbon storage in soil also have the added benefit of improving soil quality and increasing land productivity through the addition of organic matter to the soil (Mason et al. 2023). The addition of organic matter to soil has

Table 12.1 Genetic engineering of bacterial isolates to enhance byproduct synthesis from fixed carbon

Bacterial strain	Features	References
Moorella thermoacetica	Produce acetic acid from CO_2	Sakimoto et al. (2016)
Ralstonia eutropha H16	Bioconversion of CO_2 into isobutanol and 3-methyl-1-butanol	Li et al. (2012)
Ralstonia eutropha	Engineered to synthesize biomass, fuels, or chemicals from CO_2 concentrations	Liu et al. (2013)
E. coli BW25113	Use methanol as the sole carbon source	Chen et al. (2020a, b)
Sporomusa ovata	Solar-driven CO_2 fixation with high conversion to acetate	Su et al. (2020)
E. coli	Produce sugars from CO_2	Antonovsky et al. (2016)
Rhodopseudomonas palustris	Light-driven CO_2 reduction to methane	Fixen et al. (2016)

several beneficial effects. Firstly, it enhances the structural integrity of the soil, making it more stable and resistant to erosion. Secondly, organic matter increases the soil's capacity to retain water, which is important for plant growth and reduces the risk of drought (Jangir et al. 2019). Thirdly, it improves the availability of nutrients in the soil, making them more accessible to plants. Lastly, the presence of organic matter creates a favourable environment for soil organisms, promoting their growth and activity (Brown et al. 2000). Participating in carbon sequestration programs offers an opportunity to recover lost productivity, especially in agricultural systems. Reports indicate that managed ecosystems, such as agriculture, have seen a significant decline of 30–55% in their initial supply of soil organic carbon since the conversion process (van Noordwijk et al. 2015). Carbon sequestration methods provide an opportunity to recover 50–60% of the original carbon content in unproductive agricultural and degraded lands (Kumar et al. 2022a, b).

12.8.3 Secondary Advantages

It enhances water retention and infiltration, providing a substrate for soil organisms. Additionally, it acts as a source and storage of essential plant nutrients and improves the integrity of soil structure. These advantages have been documented in various studies (Sung et al. 2017). According to the findings, soil carbon sequestration offers environmental advantages in agroforestry that are 40–70% greater than the benefits in terms of the manufacturing process (Meena et al. 2020).

12.8.4 Quantification of Carbon Stocks

Parties to the United Nations Framework Convention on Climate Change (UNFCCC) are obligated to submit their individual nationally determined contributions (INDCs), which requires a comprehensive evaluation and assessment of all carbon sinks and sources in terrestrial and other sectors. The measurement and quantification of carbon in the Land Use, Land-Use Change, and Forestry (LULUCF) sector may be insufficient if the contribution of soil carbon is not taken into account, considering its substantial proportion (36–46%). The carbon inventory refers to the methodical assessment of variations in the amounts of carbon stored in soil and biomass over a specific period. This assessment encompasses both carbon emissions and removals. The procedure is performed on a regular basis for various purposes (Pacala and Socolow 2004).

12.9 Future Prospects and Conclusion

A significant challenge faced by researchers in addressing the problem of elevated CO_2 emissions, which contribute to the rise in greenhouse gases, is the task of enhancing the conversion of CO_2 through the application of conventional natural processes that have historically been employed to capture inorganic carbon sources. Thus, we suggest that there is a significant lack of commercialization in the area of rhizoremediation for the purpose of treating soil contaminated with crude oil. This approach could offer a sustainable option for restoring ecosystems. Rhizoremediation not only purifies soil contaminated with crude oil, but also improves soil fertility by reinstating its natural structure. The conventional bioremediation approach utilises microorganisms exclusively in contaminated areas or encourages the existing indigenous microorganisms in such polluted places. This procedure leads to the emission of CO_2 as a result of their catabolic respiration process, which can also contribute to the release of greenhouse gases. Therefore, by employing suitable plant-microbe combinations in polluted areas, this issue could be effectively resolved. Through this mechanism, the CO_2 produced during the degradation process can be absorbed by the corresponding plants and, through photosynthesis, integrated into the plant biomass. The plant biomass has the potential to be converted into green biofuel, which offers a sustainable alternative to fossil fuels. This conversion process not only reduces greenhouse gas emissions but also enhances energy security. In addition, the study has provided detailed explanations on the bioeconomy of the process, the viability of commercialization, the market share for oil-hydrocarbon biodegradation as an industry, and the strength of the knowledge network. These explanations serve to strengthen the suggested concept. The current economic drawback lies in cultivating techniques, which could be mitigated with the implementation of genetic engineering procedures that yield high productivity rates. Metabolic engineering presents viable approaches to enhance CO_2 fixation by altering central metabolisms. The strategies encompass: (i) dispersing the catalytic role of RuBisCO over numerous enzymes, (ii) substituting the CBB cycle with other routes, and (iii) replacing Rubisco with alternative carboxylation processes. These methods have the capacity to greatly improve the process of CO_2 fixation in the future. The field of biotechnological engineering is currently focused on the challenging issue of improving the efficiency of producing fuel, commodities, and food from CO_2.

The absence of adequate manufacturing and storage facilities for hydrogen produced from water, along with safety concerns and substantial capital needs, limits the utilisation of CO_2 in fuel production. This methodology has the potential to be an intriguing method for implementing and progressing new plant-bacterial systems that can gather and convert CO_2.

Acknowledgments ND and PP are grateful to Department of Biotechnology, Ministry of Science and Technology, Govt of India, for financial support.

Conflict of Interest Author(s) declares no conflict of interest.

References

Ahmed AAQ, Odelade KA, Babalola OO (2019) Microbial inoculants for improving carbon sequestration in agroecosystems to mitigate climate change. Handbook of climate change resilience 1–21.

Abatenh E, Gizaw B, Tsegaye Z, Tefera G (2018) Microbial function on climate change–a review. Open J Environ Biol 3(1):001–007

Abiraami TV, Singh S, Nain L (2020) Soil metaproteomics as a tool for monitoring functional microbial communities: promises and challenges. Rev Environ Sci Biotechnol 19(1):73–102

Afridi MS, Javed MA, Ali S, Medeiros D, Flavio HV, Ali B, Salam A, Sumaira, Marc RA, Alkhalifah DHM, Selim S, Santoyo G (2022) New opportunities in plant microbiome engineering for increasing agricultural sustainability under stressful conditions. Front Plant Sci 13:899464

Aggarwal H, Sharma N, Iqbal K, Varma A, Mishra A (2024) Fate and PPCPs transport to agricultural soils receiving reused water. In: Development in wastewater treatment research and processes. Elsevier, pp 33–54

Ahlawat OP, Yadav D, Walia N, Kashyap PL, Sharma P, Tiwari R (2024) Root exudates and their significance in abiotic stress amelioration in plants: a review. J Plant Growth Regul 43:1–26

Ahmed S, Kumari K, Singh D (2024) Different strategies and bio-removal mechanisms of petroleum hydrocarbons from contaminated sites. Arab Gulf J Sci Res 42(2):342–358

Aide MT, Braden I (eds) (2022) Soil science: emerging technologies, global perspectives and applications

Antonovsky N, Gleizer S, Noor E, Zohar Y, Herz E, Barenholz U, Zelcbuch L, Amram S, Wides A, Tepper N, Davidi D, On YB, Bareia T, Wernick DG, Shani I, Malitsky S, Jona G, Even AB, Milo R (2016) Sugar synthesis from CO_2 in Escherichia coli. Cell 166(1):115–125

Arias PA, Rivera JA, Sörensson AA, Zachariah M, Barnes C, Philip S, Kew S, Vautard R, Koren G, Pinto I, Vahlberg M, Singh R, Raju E, Li S, Yang W, Vecchi G, Otto FE (2024) Interplay between climate change and climate variability: the 2022 drought in Central South America. Clim Chang 177(1):6

Arnosti C, Wietz M, Brinkhoff T, Hehemann JH, Probandt D, Zeugner L, Amann R (2021) The biogeochemistry of marine polysaccharides: sources, inventories, and bacterial drivers of the carbohydrate cycle. Annu Rev Mar Sci 13(1):81–108

Atasoy M, Bartkova S, Çetecioğlu-Gürol Z, Mira N, O'Byrne C, Pérez-Rodríguez F, Possas A, Scheler O, Sedláková-Kaduková J, Sinčák M, Steiger M, Ziv C, Lund PA (2024) Methods for studying microbial acid stress responses: from molecules to populations. FEMS Microbiol Rev 48:fuae015

Azaizeh H, Castro PM, Kidd P (2010) Biodegradation of organic xenobiotic pollutants in the rhizosphere. In Organic Xenobiotics and Plants: From Mode of Action to Ecophysiology (pp. 191–215). Dordrecht: Springer Netherlands

Badri DV, Chaparro JM, Manter DK, Martinoia E, Vivanco JM (2012) Influence of ATP-binding cassette transporters in root exudation of phytoalexins, signals, and in disease resistance. Front Plant Sci 3:24907

Bala S, Garg D, Thirumalesh BV, Sharma M, Sridhar K, Inbaraj BS, Tripathi M (2022) Recent strategies for bioremediation of emerging pollutants: a review for a green and sustainable environment. Toxics 10(8):484

Ballayan MN (2021) Effect of land use types and soil macrofauna on soil aggregate stability and carbon storage in the Central Highlands of Kenya (Doctoral dissertation, Uon)

Battle M, Bender ML, Tans PP, White JWC, Ellis JT, Conway T, Francey RJ (2000) Global carbon sinks and their variability inferred from atmospheric O2 and δ13C. Science 287(5462):2467–2470

Benedetti M, Vecchi V, Barera S, Dall'Osto L (2018) Biomass from microalgae: the potential of domestication towards sustainable biofactories. Microb Cell Factories 17:1–18

Benoit SL, Maier RJ, Sawers RG, Greening C (2020) Molecular hydrogen metabolism: a widespread trait of pathogenic bacteria and protists. Microbiol Mol Biol Rev 84(1):10–1128

Bhatt RP (2024) Restoration of global forests and biodiversity through green energy

Bhattacharyya SS, Ros GH, Furtak K, Iqbal HM, Parra-Saldívar R (2022) Soil carbon sequestration–an interplay between soil microbial community and soil organic matter dynamics. Sci Total Environ 815:152928

Bhuyan B, Debnath S, Pandey P (2020) The rhizosphere microbiome and its role in plant growth in stressed conditions. In: Rhizosphere microbes: soil and plant functions, pp 503–529

Bisht S, Pandey P, Bhargava B, Sharma S, Kumar V, Sharma KD (2015) Bioremediation of polyaromatic hydrocarbons (PAHs) using rhizosphere technology. Braz J Microbiol 46:7–21

Blagodatskaya E, Kuzyakov Y (2013) Active microorganisms in soil: critical review of estimation criteria and approaches. Soil Biol Biochem 67:192–211

Borah P, Deka H (2024) Polycyclic aromatic hydrocarbon (PAH) accumulation in selected medicinal plants: a mini review. Environ Sci Pollut Res 31:1–19

Bosco D, Fava A, Plastino M, Montalcini T, Pujia A (2011) Possible implications of insulin resistance and glucose metabolism in Alzheimer's disease pathogenesis. J Cell Mol Med 15(9):1807–1821

Brás TA, Seixas J, Carvalhais N, Jägermeyr J (2021) Severity of drought and heatwave crop losses tripled over the last five decades in Europe. Environmental Research Letters 16(6):065012.

Brewer KM, Muñoz-Araya M, Martinez I, Marshall KN, Gaudin AC (2023) Long-term integrated crop-livestock grazing stimulates soil ecosystem carbon flux, increasing subsoil carbon storage in California perennial agroecosystems. Geoderma 438:116598

Brown GG, Barois I, Lavelle P (2000) Regulation of soil organic matter dynamics and microbial activityin the drilosphere and the role of interactionswith other edaphic functional domains. European journal of soil biology 36(3-4):177–198

Bungau S, Behl T, Aleya L, Bourgeade P, Aloui-Sossé B, Purza AL, Abid A, Samuel AD (2021) Expatiating the impact of anthropogenic aspects and climatic factors on long-term soil monitoring and management. Environ Sci Pollut Res 28(24):30528–30550

Buragohain P (2019) Role of microbes on carbon sequestration. Int J Microbiol Res. ISSN 0975–5276

Burmeister AR, Smith JJ (2016) Evolution across the curriculum: microbiology. J Microbiol Biol Educ 17(2):252–260

Chen H, Gan Q, Fan C (2020a) Methyl-coenzyme M reductase and its post-translational modifications. Front Microbiol 11:578356

Chen FYH, Jung HW, Tsuei CY, Liao JC (2020b) Converting Escherichia coli to a synthetic methylotroph growing solely on methanol. Cell 182(4):933–946

Cheng L, Ding C, Li Q, He Q, Dai LR, Zhang H (2013) DNA-SIP reveals that Syntrophaceae play an important role in methanogenic hexadecane degradation. PLoS One 8(7):e66784

Chenu C, Rumpel C, Védère C, Barré P (2024) Methods for studying soil organic matter: nature, dynamics, spatial accessibility, and interactions with minerals. In: Soil microbiology, ecology and biochemistry. Elsevier, pp 369–406

Chu H, Huang Z, Zhang Z, Yan X, Qiu B, Xu N (2024) Integration of carbon emission reduction policies and technologies: research progress on carbon capture, utilization and storage technologies. Sep Purif Technol 127153:127153

Clemente JS, Simpson MJ, Simpson AJ, Yanni SF, Whalen JK (2013) Comparison of soil organic matter composition after incubation with maize leaves, roots, and stems. Geoderma 192:86–96

Correa SS, Schultz J, Lauersen KJ, Rosado AS (2023) Natural carbon fixation and advances in synthetic engineering for redesigning and creating new fixation pathways. J Adv Res 47:75–92

Cotrufo MF, Lavallee JM (2022) Soil organic matter formation, persistence, and functioning: a synthesis of current understanding to inform its conservation and regeneration. Adv Agron 172:1–66

Cuadros J (2017) Clay minerals interaction with microorganisms: a review. Clay Minerals:52(2):235–261

d Mattozzi M, Ziesack M, Voges MJ, Silver PA, Way JC (2013) Expression of the sub-pathways of the Chloroflexus aurantiacus 3-hydroxypropionate carbon fixation bicycle in E. coli: toward horizontal transfer of autotrophic growth. Metab Eng 16:130–139

Das N, Bhuyan B, Pandey P (2022) Correlation of soil microbiome with crude oil contamination drives detection of hydrocarbon degrading genes which are independent to quantity and type of contaminants. Environmental Research 215:114185

Das N, Maheshwari DK, Pandey P (2024) Energy crop-based rhizoremediation and lignocellulosic biomass production as sustainable bioeconomy-driven solution for biofuel generation and waste mitigation. Prog Energy Combust Sci 103:101161

Davies JA (2017) Characterisation of the reversible formate dehydrogenases of Shewanella (Doctoral dissertation, University of East Anglia)

De Kauwe MG, Medlyn BE, Tissue DT (2021) To what extent can rising [CO_2] ameliorate plant drought stress? New Phytol 231(6):2118–2124

Debnath S, Chakraborty S, Langthasa M, Choure K, Agnihotri V, Srivastava A, Rai PK, Tilwari A, Maheshwari DK, Pandey P (2023) Non-rhizobial nodule endophytes improve nodulation, change root exudation pattern and promote the growth of lentil, for prospective application in fallow soil. Front Plant Sci 14:1152875

Dignac MF, Derrien D, Barré P, Barot S, Cécillon L, Chenu C, Chevallier T, Freschet GT, Garnier P, Guenet B, Hedde M, Klumpp K, Lashermes G, Moron PA, Nunan N, Roumet C, Basile-Doelsch I (2017) Increasing soil carbon storage: mechanisms, effects of agricultural practices and proxies. A review. Agron Sustain Dev 37:1–27

Duan B, Zhang Y, Xu G, Chen J, Paquette A, Peng S (2015) Long-term responses of plant growth, soil microbial communities and soil enzyme activities to elevated CO_2 and neighbouring plants. Agric For Meteorol 213:91–101

Dutta G, Paramanik B, Bhabai B, Layek J, Choudhury A, Dutta S, Bhattacharjee S, Rahman FH (2024) Climate change impacts and adaptation strategies for agronomic crops. In: Climate change impacts on soil-plant-atmosphere continuum. Springer Nature Singapore, Singapore, pp 383–404

Dymond JR (2010) Soil erosion in New Zealand is a net sink of CO2. Earth Surface Processes and Landforms 35(15):1763–1772

Ebsa G, Gizaw B, Admassie M, Degu T, Alemu T (2024) The role and mechanisms of microbes in dichlorodiphenyltrichloroethane (DDT) and its residues bioremediation. Biotechnol Rep:e00835

Etherton BA, Choudhury RA, Alcalá Briseño RI, Mouafo-Tchinda RA, Plex Sulá AI, Choudhury M, Adhikari A, Lei SL, Kraisitudomsook N, Buritica JR, Cerbaro VA, Ogero K, Cox CM,

Walsh SP, Andrade-Piedra JL, Omondi BA, Navarrete I, McEwan M, Garrett KA (2024) Disaster plant pathology: smart solutions for threats to global plant health from natural and human-driven disasters. Phytopathology 114(5):855–868
Fan Q, Fan L, Quach WM, Zhang R, Duan J, Sand W (2023) Application of microbial mineralization technology for marine concrete crack repair: a review. J Build Eng 69:106299
Feng H, Wang Z, Jia P, Gai J, Chen B, Wang S (2022) Diversity and distribution of CO_2-fixing microbial community along elevation gradients in meadow soils on the Tibetan Plateau. Sci Rep 12(1):9621
Fixen KR, Zheng Y, Harris DF, Shaw S, Yang ZY, Dean DR, Seefeldt LC, Harwood CS (2016) Light-driven carbon dioxide reduction to methane by nitrogenase in a photosynthetic bacterium. Proc Natl Acad Sci 113(36):10163–10167
Friedrich MW (2010) Trophic interactions in microbial communities and food webs traced by stable isotope probing of nucleic acids. In: Stable isotope probing and related technologies. ASM Press, pp 203–232
Gani U, Vishwakarma RA, Misra P (2021) Membrane transporters: the key drivers of transport of secondary metabolites in plants. Plant Cell Rep 40:1–18
George II, Nawawi MGM, Mohd ZJ, Farah BS (2024) Environmental effects from petroleum product transportation spillage in Nigeria: a critical review. Environ Sci Pollut Res 31(2):1719–1747
Geyer KM, Emily Kyker-Snowman A, Grandy S, Frey SD (2016) Microbial carbon use efficiency: accounting for population, community, and ecosystem-scale controls over the fate of metabolized organic matter. Biogeochemistry 127:173–188
Gleizer S (2019) Metabolic engineering and directed evolution of Escherichia coli towards autotrophic growth (Doctoral dissertation, The Weizmann Institute of Science (Israel)
Gleizer S, Bar-On YM, Ben-Nissan R, Milo R (2020) Engineering microbes to produce fuel, commodities, and food from CO_2. Cell Reports Physical Science 1(10).
Glick BR (2015) Beneficial plant-bacterial interactions, vol 243. Springer, Heidelberg
Gong F, Cai Z, Li Y (2016) Synthetic biology for CO_2 fixation. Sci China Life Sci 59:1106–1114
Gougoulias C, Clark JM, Shaw LJ (2014) The role of soil microbes in the global carbon cycle: tracking the below-ground microbial processing of plant-derived carbon for manipulating carbon dynamics in agricultural systems. J Sci Food Agric 94(12):2362–2371
Gramss G (2012) Potential contributions of oxidoreductases from alfalfa plants to soil enzymology and biotechnology: a review. J Nat Sci Sustain Technol 6(3):169
Greening C, Biswas A, Carere CR, Jackson CJ, Taylor MC, Stott MB, Cook GM, Morales SE (2016) Genomic and metagenomic surveys of hydrogenase distribution indicate H_2 is a widely utilised energy source for microbial growth and survival. ISME J 10(3):761–777
Grover M, Maheswari M, Desai S, Gopinath KA, Venkateswarlu B (2015) Elevated CO_2: plant associated microorganisms and carbon sequestration. Appl Soil Ecol 95:73–85
Guadalupe-Medina V, Wisselink HW, Luttik MA, de Hulster E, Daran JM, Pronk JT, van Maris AJ (2013) Carbon dioxide fixation by Calvin-Cycle enzymes improves ethanol yield in yeast. Biotechnology for biofuels 6:1–12.
Gupta R, Mishra A, Thirupathaiah Y, Chandel AK (2024) Biochemical conversion of CO_2 in fuels and chemicals: status, innovation, and industrial aspects. Biomass Convers Biorefinery 14(3):3007–3030
Hayat S, Faraz A, Faizan M (2017) Root exudates: composition and impact on plant–microbe interaction. In: Biofilms in plant and soil health. Wiley, pp 179–193
He Z, Deng Y, Xu M, Li J, Liang J, Xiong J, Yu H, Wu B, Wu L, Xue K, Shi S, Carrillo Y, Nostrand JDV, Hobbie SE, Reich PB, Schadt CW, Kent AD, Pendell E, Wallenstein M, Luo Y, Zhou J (2020) Microbial functional genes commonly respond to elevated carbon dioxide. Environ Int 144:106068
Herz E, Niv A, Yinon B, Dan D, Shmuel G, Noam P, Lianet N (2017) "The genetic basis for the adaptation of E. coli to sugar synthesis from CO_2." Nature communications 8, no. 1 (2017):1705.
Holden L, Lee RG, Orsini L, Eastwood N, Zhou J, Čavoški A (2024) Biodiversity management challenges: A policy brief. Environmental Law Review 26(2):141–150.

Horwath W (2007) Carbon cycling and formation of soil organic matter. In: Soil microbiology, ecology and biochemistry. Academic, pp 303–339

Huang Z, He X, Zhang C, Zhang M, Wang J, Hou Y, Wang D, Yao S, Yu Q, Ji K (2023) Microbial communities and functions changed in rhizosphere soil of Pinus massoniana provenances with different carbon storage. Front Microbiol 14:1264670

Jangir CK, Kumar S, Meena RS (2019) Significance of soil organic matter to soil quality and evaluation of sustainability. In: Sustainable agriculture. Scientific Publisher, Jodhpur, pp 357–381

Jenkins S, Swenson TL, Lau R, Rocha A, Aaring A, Hazen TC, Chakraborty R, Northen T (2017) Construction of soil defined media using quantitative exometabolomic analysis of soil metabolites. bioRxiv:151282

Jensen E, Clément R, Maberly SC, Gontero B (2017) Regulation of the Calvin–Benson–Bassham cycle in the enigmatic diatoms: biochemical and evolutionary variations on an original theme. Philos Trans R Soc B: Biol Sci 372(1728):20160401

Jessen GL, Lichtschlag A, Ramette A, Pantoja S, Rossel PE, Schubert CJ, Boetius A (2017) Hypoxia causes preservation of labile organic matter and changes seafloor microbial community composition (Black Sea). Science advances 3(2):e1601897

Jiang P, Xiao LQ, Wan X, Yu T, Liu YF, Liu MX (2022) Research progress on microbial carbon sequestration in soil: A review. Eurasian Soil Sci 55(10):1395–1404

Jiao N, Luo T, Chen Q, Zhao Z, Xiao X, Liu J, Jian Z, Xie S, Thomas H, Herndl GJ, Benner R, Gonsior M, Chen F, Cai WJ, Robinson C (2024) The microbial carbon pump and climate change. Nat Rev Microbiol 22:1–12

Jin X, Jia H, Ran L, Wu F, Liu J, Schlaeppi K, Dini-Andreote F, Wei Z, Zhou X (2024) Fusaric acid mediates the assembly of disease-suppressive rhizosphere microbiota via induced shifts in plant root exudates. Nat Commun 15(1):5125

Joel OT, Oguanobi VU (2024) Geological survey techniques and carbon storage: optimizing renewable energy site selection and carbon sequestration. Open Access Res J Sci Technol 11(1):039–051

Jones DL, Nguyen C, Finlay RD (2009) Carbon flow in the rhizosphere: carbon trading at the soil–root interface. Plant Soil 321:5–33

Juottonen H, Galand PE, Yrjälä K (2006) Detection of methanogenic Archaea in peat: comparison of PCR primers targeting the mcrA gene. Res Microbiol 157(10):914–921

Kalogiannidis S, Toska E, Chatzitheodoridis F (2022) Contribution of civil protection to the urban economy: evidence from a small-sized Greek City. Sustain For 14(2):981

Kanu MO, Joseph GW, George I (2021) Measurement of physicochemical properties, electrical and thermal conductivity of wood ash for effective soil amendment. Indonesian J Appl Phys 11(2):176–188

Karishma S, Kamalesh R, Saravanan A, Deivayanai VC, Yaashikaa PR, Vickram AS (2024) A review on recent advancements in biochemical fixation and transformation of CO_2 into constructive products. Biochem Eng J 208:109366

Kelany MS, El-Sawy MA, El-Gendy AR, Beltagy EA (2023) Bioremediation of industrial wastewater heavy metals using solo and consortium Enterobacter spp. Environ Monit Assess 195(11):1357

Keller MW, Schut GJ, Lipscomb GL, Menon AL, Iwuchukwu IJ, Leuko TT, Thorgersen MP, Nixon WJ, Hawkins AS, Kelly RM, Adams MW (2013) Exploiting microbial hyperthermophilicity to produce an industrial chemical, using hydrogen and carbon dioxide. Proc Natl Acad Sci 110(15):5840–5845

Keller P, Reiter MA, Kiefer P, Gassler T, Hemmerle L, Christen P, Noor E, Vorholt JA (2022) Generation of an Escherichia coli strain growing on methanol via the ribulose monophosphate cycle. Nat Commun 13(1):5243

Kharey G (2024) An exploration of controlled experimental crude oil and dilbit spills in a pristine boreal freshwater lake, and the response of the indigenous bacterial community

Khunjar WO, Sahin A, West AC, Chandran K, Banta S (2012) Biomass production from electricity using ammonia as an electron carrier in a reverse microbial fuel cell. PLoS One 7:e44846

Kim Y, Liesack W (2015) Differential assemblage of functional units in paddy soil microbiomes. PLoS One 10(4):e0122221

Kim S, Lindner SN, Aslan S, Yishai O, Wenk S, Schann K, Bar-Even A (2020) Growth of E. coli on formate and methanol via the reductive glycine pathway. Nat Chem Biol 16(5):538–545

King GM (2011) Enhancing soil carbon storage for carbon remediation: potential contributions and constraints by microbes. Trends in Microbiology 19(2):75–84

Kotoky R, Pandey P (2020) Difference in the rhizosphere microbiome of Melia azedarach during removal of benzo (a) pyrene from cadmium co-contaminated soil. Chemosphere 258:127175

Kotoky R, Rajkumari J, Pandey P (2018) The rhizosphere microbiome: significance in rhizoremediation of polyaromatic hydrocarbon contaminated soil. J Environ Manag 217:858–870

Kouhi RM, Moghaddam MMJ, Ardejani FD, Mirheydari A, Maghsoudy S, Gholizadeh F, Ghobadipour B (2024) Carbon utilization technologies & methods. In: *Carbon capture, utilization, and storage technologies*, p 1

Kremp F, Müller V (2021) Methanol and methyl group conversion in acetogenic bacteria: biochemistry, physiology and application. FEMS Microbiol Rev 45(2):fuaa040

Kumar S, Singh RK, Kumar SC, Banjare U, Patel AK, Singh P (2021) Estimation of plant growth hormone (IAA) production ability by bradyrhizobium strains isolated from root nodules of pigeon pea (Cajanus cajan). Plant Archives 21(1):1567–1570

Kumar S, Sindhu SS, Kumar R (2022a) Biofertilizers: an ecofriendly technology for nutrient recycling and environmental sustainability. Curr Res Microb Sci 3:100094

Kumar R, Singh A, Datta A, Yadav RP, Dinesh D, Verma K (2022b) Carbon sequestration in degraded lands: current prospects, practices, and future strategies. In: Plans and policies for soil organic carbon management in agriculture. Springer Nature Singapore, Singapore, pp 221–255

Kumari S, Das S (2023) Bacterial enzymatic degradation of recalcitrant organic pollutants: catabolic pathways and genetic regulations. Environ Sci Pollut Res 30(33):79676–79705

Kumari S, Devi M, Thakur K, Minhas B, Bhatt AK, Kaushik N (2024) Impact of anthropogenic activities on microbial diversity and soil health. In: Advancements in microbial biotechnology for soil health. Springer Nature Singapore, Singapore, pp 227–248

Kurt E, Qin J, Williams A, Zhao Y, Xie D (2023) Perspectives for using CO_2 as a feedstock for biomanufacturing of fuels and chemicals. Bioengineering 10(12):1357

Lal R (2003) Global potential of soil carbon sequestration to mitigate the greenhouse effect. Crit Rev Plant Sci 22(2):151–184

Lamminjoki L (2024) Functional expression of the carbon monoxide methylating acetyl-CoA synthase enzyme complexes and their β subunits from alkanotrophic archaea in Methanosarcina acetivorans and in Escherichia coli

Lehtinen T, Virtanen H, Santala S, Santala V (2018) Production of alkanes from CO_2 by engineered bacteria. Biotechnol Biofuels 11:1–11

Lei X, Shen Y, Zhao J, Huang J, Wang H, Yu Y, Xiao C (2023) Root exudates mediate the processes of soil organic carbon input and efflux. Plan Theory 12(3):630

Li H, Opgenorth PH, Wernick DG, Rogers S, Wu TY, Higashide W, Malati P, Huo YX, Cho KM, Liao JC (2012) Integrated electromicrobial conversion of CO_2 to higher alcohols. Science 335(6076):1596–1596

Li H, Su JQ, Yang XR, Zhu YG (2019) Distinct rhizosphere effect on active and total bacterial communities in paddy soils. Sci Total Environ 649:422–430

Li Z, Guo X, Ma Y, Hu B, Yang Y, Tian H, Liu X, Meng N, Zhu J, Yan D, Song H, Bao B, Li X, Dai X, Zheng Y, Jin Y, Zheng H (2024a) The hidden risk: changes in functional potentials of microbial keystone taxa under global climate change jeopardizing soil carbon storage in alpine grasslands. Environ Int 185:108516

Li Y, Yan Q, Wang J, Shao MA, Li Z, Jia H (2024b) Biodegradable plastics fragments induce positive effects on the decomposition of soil organic matter. J Hazard Mater 468:133820

Liao Q, Liu H, Lu C, Liu J, Waigi MG, Ling W (2021) Root exudates enhance the PAH degradation and degrading gene abundance in soils. Sci Total Environ 764:144436

Liao X, Zhao J, Yi Q, Li J, Li Z, Wu S, Zhang W, Wang K (2023) Metagenomic insights into the effects of organic and inorganic agricultural managements on soil phosphorus cycling. Agric Ecosyst Environ 343:108281
Lin MT, Occhialini A, Andralojc PJ, Parry MA, Hanson MR (2014) A faster Rubisco with potential to increase photosynthesis in crops. Nature 513(7519):547–550
Liu R, Liang L, Wu M, Chen K, Jiang M, Ma J, Wei P, Ouyang P (2013) CO_2 fixation for succinic acid production by engineered Escherichia coli co-expressing pyruvate carboxylase and nicotinic acid phosphoribosyltransferase. Biochem Eng J 79:77–83
Liu JF, Sun XB, Yang GC, Mbadinga SM, Gu JD, Mu BZ (2015) Analysis of microbial communities in the oil reservoir subjected to CO_2-flooding by using functional genes as molecular biomarkers for microbial CO_2 sequestration. Front Microbiol 6:236
Liu J, Wang X, Dai G, Zhang Y, Bian X (2022) Microbial chassis engineering drives heterologous production of complex secondary metabolites. Biotechnology Advances 59:107966
Liu X, Li L, Zhao G, Xiong P (2024) Optimization strategies for CO_2 biological fixation. Biotechnol Adv 73:108364
López-Calcagno PE, Howard TP, Raines CA (2014) The CP12 protein family: a thioredoxin-mediated metabolic switch? Front Plant Sci 5:9
Lu J, Brigham CJ, Gai CS, Sinskey AJ (2012) Studies on the production of branched-chain alcohols in engineered Ralstonia eutropha. Appl Microbiol Biotechnol 96:283–297
Ma W, Tang S, Dengzeng Z, Zhang D, Zhang T, Ma X (2022) Root exudates contribute to below-ground ecosystem hotspots: A review. Front Microbiol 13:937940
Madhu M, Hatfield JL (2013) Dynamics of plant root growth under increased atmospheric carbon dioxide. Agron J 105(3):657–669
Malik A, Garg VK (eds) (2024) Bioremediation for sustainable environmental cleanup. CRC Press
Marchal JLM, Dersjant-Li Y, Bello A, Sobotik EB, Archer GS (2020) Total inorganic phosphorus free and more economical broiler diets supplemented with a novel consensus bacterial 6-phytase variant maintained normal growth characteristics in all phase. In: Virtual 2020 PSA annual meeting abstract book, p 78
Martin WF, Bryant DA, Beatty JT (2018) A physiological perspective on the origin and evolution of photosynthesis. FEMS Microbiol Rev 42(2):205–231
Masciandaro G, Macci C, Peruzzi E, Ceccanti B, Doni S (2013) Organic matter–microorganism–plant in soil bioremediation: a synergic approach. Rev Environ Sci Biotechnol 12:399–419
Mason ARG, Salomon MJ, Lowe AJ, Cavagnaro TR (2023) Microbial solutions to soil carbon sequestration. J Clean Prod 412:137993
Massalha H, Korenblum E, Tholl D, Aharoni A (2017) Small molecules below-ground: the role of specialized metabolites in the rhizosphere. Plant J 90(4):788–807
Masuda Y, Itoh H, Shiratori Y, Senoo K (2018) Metatranscriptomic insights into microbial consortia driving methane metabolism in paddy soils. Soil Sci Plant Nutr 64(4):455–464
Matiasek SJ, Hernes PJ (2019) The chemical fingerprint of solubilized organic matter from eroded soils and sediments. Geochim Cosmochim Acta 267:92–112
Mattarozzi M, Manfredi M, Montanini B, Gosetti F, Sanangelantoni AM, Marengo E, Careri M, Visioli G (2017) A metaproteomic approach dissecting major bacterial functions in the rhizosphere of plants living in serpentine soil. Anal Bioanal Chem 409:2327–2339
Meena M, Swapnil P, Zehra A, Aamir M, Dubey MK, Goutam J, Upadhyay RS (2017) Beneficial microbes for disease suppression and plant growth promotion. In: Plant-microbe interactions in agro-ecological perspectives: volume 2: microbial interactions and agro-ecological impacts, pp 395–432
Meena RS, Kumar S, Yadav GS (2020) Soil carbon sequestration in crop production. In: Nutrient dynamics for sustainable crop production. Springer Singapore, pp 1–39
Mehmood K, Hassan ST, Qiu X, Ali S (2024) Comparative analysis of CO_2 emissions and economic performance in the United States and China: navigating sustainable development in the climate change era. Geosci Front 15(5):101843
Meloni M, Gurrieri L, Fermani S, Velie L, Sparla F, Crozet P, Henri J, Zaffagnini M (2023) Ribulose-1, 5-bisphosphate regeneration in the Calvin-Benson-Bassham cycle: focus on

the last three enzymatic steps that allow the formation of Rubisco substrate. Front Plant Sci 14:1130430
Mimmo T, Pii Y, Valentinuzzi F, Astolfi S, Lehto N, Robinson B, Brunetto G, Terzano R, Cesco S (2017) Nutrient availability in the rhizosphere: a review. In: VIII international symposium on mineral nutrition of fruit crops 1217, pp 13–28
Mishra S, Zhang X, Yang X (2024) Plant communication with rhizosphere microbes can be revealed by understanding microbial functional gene composition. Microbiol Res 284:127726
Mitchell PJ, Simpson AJ, Soong R, Simpson MJ (2015) Shifts in microbial community and water-extractable organic matter composition with biochar amendment in a temperate forest soil. Soil Biol Biochem 81:244–254
Molina L, Segura A (2021) Biochemical and metabolic plant responses toward polycyclic aromatic hydrocarbons and heavy metals present in atmospheric pollution. Plan Theory 10(11):2305
Molinatto G (2017) Study of the genome of Bacillus amyloliquefaciens subsp. plantarum and role of its plasmid (pS499) in rhizosphere interactions
Montarnal A (2023) Study of glyceraldehyde-3-phosphate oxidoreductases from archaea and bacteria (Doctoral dissertation, University of Nottingham)
Moore EJ, Karplus VJ, Morgan MG (2024) Expert elicitation of the timing and uncertainty to establish a geologic sequestration well for CO_2 in the United States. Proc Natl Acad Sci 121(1):e2307984120
Mueller CW, Baumert V, Carminati A, Germon A, Holz M, Kögel-Knabner I, Peth S, Schlüter S, Uteau D, Vetterlein D, Teixeira P, Vidal A (2024) From rhizosphere to detritusphere–soil structure formation driven by plant roots and the interactions with soil biota. Soil Biol Biochem 193:109396
Mukherjee A, Reddy MS (2020) Metatranscriptomics: an approach for retrieving novel eukaryotic genes from polluted and related environments. 3 Biotech 10(2):71
Nathan VK, Ammini P (2019) Carbon dioxide sequestering ability of bacterial carbonic anhydrase in a mangrove soil microcosm and its bio-mineralization properties. Water Air Soil Pollut 230:1–12
Newman MM, Lorenz N, Hoilett N, Lee NR, Dick RP, Liles MR, Ramsier C, Kloepper JW (2016) Changes in rhizosphere bacterial gene expression following glyphosate treatment. Sci Total Environ 553:32–41
Nihorimbere V, Ongena M, Smargiassi M, Thonart P (2011) Beneficial effect of the rhizosphere microbial community for plant growth and health. Biotechnol Agron Soc Environ 15(2):327
Nogia P, Pati PK (2021) Plant secondary metabolite transporters: diversity, functionality, and their modulation. Front Plant Sci 12:758202
Nuruzzaman M, Zhang R, Cao HZ, Luo ZY (2014) Plant pleiotropic drug resistance transporters: transport mechanism, gene expression, and function. J Integr Plant Biol 56(8):729–740
Nwachukwu BC (2023) Microbial diversity, community structure and functional characteristics of sunflower rhizospheric soil (Doctoral dissertation, North-West University (South Africa))
Nybo SE, Khan NE, Woolston BM, Curtis WR (2015) Metabolic engineering in chemolithoautotrophic hosts for the production of fuels and chemicals. Metab Eng 30:105–120
Onyeaka H, Ekwebelem OC (2023) A review of recent advances in engineering bacteria for enhanced CO_2 capture and utilization. Int J Environ Sci Technol 20(4):4635–4648
Orozco-Mosqueda MDC, Flores A, Rojas-Sánchez B, Urtis-Flores CA, Morales-Cedeño LR, Valencia-Marin MF, Chávez-Avila S, Rojas-Solis D, Santoyo G (2021) Plant growth-promoting bacteria as bioinoculants: attributes and challenges for sustainable crop improvement. Agronomy 11(6):1167
Overy DP, Bell MA, Habtewold J, Helgason B, Gregorich G, Overy D (2020) APPENDIX 2: linking 'omics' technologies to study soil carbon stabilization: a review. In: Developing statistical and analytical methods for untargeted analysis of complex environmental matrices, p 333
Overy DP, Bell MA, Habtewold J, Helgason BL, Gregorich EG (2021) "Omics" technologies for the study of soil carbon stabilization: a review. Front Environ Sci 9:617952
Pacala S, Socolow R (2004) Stabilization wedges: solving the climate problem for the next 50 years with current technologies. Science 305(5686):968–972

Pala U, Yelmazer B, Çorbacıoğlu M, Ruupunen J, Valjakka J, Turunen O, Binay B (2018) Functional effects of active site mutations in NAD+-dependent formate dehydrogenases on transformation of hydrogen carbonate to formate. Protein Eng Des Sel 31(9):327–335

Pant R, Pandey P, Kotoky R (2016) Rhizosphere mediated biodegradation of 1, 4-dichlorobenzene by plant growth promoting rhizobacteria of Jatropha curcas. Ecol Eng 94:50–56

Paoli GC, Morgan NS, Tabita FR, Shively JM (1995) Expression of the cbbLcbbS and cbbM genes and distinct organization of the cbb Calvin cycle structural genes of Rhodobacter capsulatus. Arch Microbiol 164:396–405

Parmar P, Dhurandhar R, Naik S (2024) Environmental fate and microbial reactions to petroleum hydrocarbon contamination in terrestrial ecosystems. In: Impact of petroleum waste on environmental pollution and its sustainable management through circular economy. Springer Nature Switzerland, Cham, pp 139–158

Parray JA, Yaseen Mir M, Shameem N, Parray JA, Yaseen Mir M, Shameem N (2019) Plant metabolomics: sustainable approach towards crop productivity. In: Sustainable agriculture: biotechniques in plant biology, pp 51–70

Patel D, Kulwant M (2024) Bioremediation of xenobiotic contamination in soil. In: Bioremediation of emerging contaminants from soils. Elsevier, pp 203–225

Paul EA (2014) Soil microbiology, ecology, and biochemistry: an exciting present and great future built on basic knowledge and unifying concepts. Soil microbiology, ecology, and biochemistry 1–13.

Perera F (2018) Pollution from fossil-fuel combustion is the leading environmental threat to global pediatric health and equity: solutions exist. Int J Environ Res Public Health 15(1):16

Phillips LA, Greer CW, Farrell RE, Germida JJ (2012) Plant root exudates impact the hydrocarbon degradation potential of a weathered-hydrocarbon contaminated soil. Appl Soil Ecol 52:56–64

Piché-Choquette S, Constant P (2019) Molecular hydrogen, a neglected key driver of soil biogeochemical processes. Appl Environ Microbiol 85(6):e02418–e02418

Piotrowska-Długosz A (2017) The use of enzymes in bioremediation of soil xenobiotics. In: Xenobiotics in the soil environment: monitoring, toxicity and management, pp 243–265

Pisani O, Haddix ML, Conant RT, Paul EA, Simpson MJ (2016) Molecular composition of soil organic matter with land-use change along a bi-continental mean annual temperature gradient. Sci Total Environ 573:470–480

Priya AK, Muruganandam M, Kumar A, Senthilkumar N, Shkir M, Pandit B, Imran M, Prakash C, Ubaidullah M (2024) Recent advances in microbial-assisted degradation and remediation of xenobiotic contaminants; challenges and future prospects. J Water Process Eng 60:105106

Rajak P, Ganguly A, Nanda S, Mandi M, Ghanty S, Das K, Biswas G, Sarkar S (2024) Toxic contaminants and their impacts on aquatic ecology and habitats. In: Spatial modeling of environmental pollution and ecological risk. Woodhead Publishing, pp 255–273

Ramesh T, Bolan NS, Kirkham MB, Wijesekara H, Kanchikerimath M, Rao CS, Sandeep S, Rinklebe J, Ok YS, Choudhury BU, Wang H, Tang C, Wang X, Song Z, Freeman OW II (2019) Soil organic carbon dynamics: impact of land use changes and management practices: A review. Adv Agron 156:1–107

Razzaq A, Wani SH, Saleem F, Yu M, Zhou M, Shabala S (2021) Rewilding crops for climate resilience: economic analysis and de novo domestication strategies. J Exp Bot 72(18):6123–6139

Reiter L (2015) Effect of crop residue incorporation on soil organic carbon dynamics: changes in carbon stocks and carbon fractions in an Italian long-term field experiment. Second cycle, A2E. Department of Ecology, Uppsala

Rochfort S, Ezernieks V, Mele P, Kitching M (2015) NMR metabolomics for soil analysis provide complementary, orthogonal data to MIR and traditional soil chemistry approaches–a land use study. Magn Reson Chem 53(9):719–725

Rohrbacher F, St-Arnaud M (2016) Root exudation: the ecological driver of hydrocarbon rhizoremediation. Agronomy 6(1):19

Rosenbaum FP, Müller V (2023) Moorella thermoacetica: a promising cytochrome-and quinone-containing acetogenic bacterium as platform for a CO_2-based bioeconomy. Green Carbon 1:2–13

Saini R, Kapoor R, Kumar R, Siddiqi TO, Kumar A (2011) CO_2 utilizing microbes—a comprehensive review. Biotechnol Adv 29(6):949–960
Sakimoto KK, Wong AB, Yang P (2016) Self-photosensitization of nonphotosynthetic bacteria for solar-to-chemical production. Science 351(6268):74–77
Santos-Merino M, Gargantilla-Becerra Á, de la Cruz F, Nogales J (2023) Highlighting the potential of Synechococcus elongatus PCC 7942 as platform to produce α-linolenic acid through an updated genome-scale metabolic modeling. Front Microbiol 14:1126030
Schenuit F, Brutschin E, Geden O, Guo F, Mohan A, Oliveira Fiorini AC, Saluja S, Schaeffer R, Riahi K (2024) Taking stock of carbon dioxide removal policy in emerging economies: developments in Brazil, China, and India. Clim Pol:1–20
Schultz J, Correa SS, dos Santos A, Rosado AS (2024) Disentangling the autotrophic thermophiles: concepts, diversity, and emerging trends. In: Microbial diversity in the genomic era. Elsevier, pp 281–300
Sekar S, Kandavel D (2010) Interaction of plant growth promoting rhizobacteria (PGPR) and endophytes with medicinal plants–new avenues for phytochemicals. J Phytology 2(7):91–100
Shah M, Ahmed S (2023) Bioremediation potential of rhizosphere microbes—current perspectives. Rhizobiome:81–94. https://doi.org/10.1016/b978-0-443-16030-1.00019-5
Shahsavari E, Adetutu EM, Anderson PA, Ball AS (2013) Tolerance of selected plant species to petrogenic hydrocarbons and effect of plant rhizosphere on the microbial removal of hydrocarbons in contaminated soil. Water Air Soil Pollut 224:1–14
Sharma B, Dangi AK, Shukla P (2018) Contemporary enzyme based technologies for bioremediation: a review. J Environ Manag 210:10–22
Sieradzki ET, Nuccio EE, Pett-Ridge J, Firestone MK (2020) Transcription of macromolecular N degrading genes provides a window into organic nitrogen decomposition in soil. bioRxiv. https://doi.org/10.1101/2020.12.14.422732
Singh A, Müller B, Schnürer A (2021) Profiling temporal dynamics of acetogenic communities in anaerobic digesters using next-generation sequencing and T-RFLP. Sci Rep 11(1):13298
Singh S, Kiran BR, Mohan SV (2024) Carbon farming: a circular framework to augment CO_2 sinks and to combat climate change. Environm Sci Adv 3(4):522–542
Singh P, Cameotra SS, (2008) Bioremediation of oil sludge using crude biosurfactants. International Biodeterioration & Biodegradation 62(3):274–280.
Singha LP, Pandey P (2020) Rhizobacterial community of Jatropha curcas associated with pyrene biodegradation by consortium of PAH-degrading bacteria. Appl Soil Ecol 155:103685
Singha LP, Pandey P (2021) Rhizosphere assisted bioengineering approaches for the mitigation of petroleum hydrocarbons contamination in soil. Crit Rev Biotechnol 41(5):749–766
Singha LP, Sinha N, Pandey P (2018) Rhizoremediation prospects of polyaromatic hydrocarbon degrading rhizobacteria, that facilitate glutathione and glutathione-S-transferase mediated stress response, and enhance growth of rice plants in pyrene contaminated soil. Ecotoxicol Environ Saf 164:579–588
Singla P, Garg N (2017) Plant flavonoids: key players in signaling, establishment, and regulation of rhizobial and mycorrhizal endosymbioses. In: Mycorrhiza-function, diversity, state of the art. Springer International Publishing, pp 133–176
Sissine FJ (2008) Energy Independence and security act of 2007: a summary of major provisions. Congressional Research Service
Skorupska A, Wielbo J, Kidaj D, Marek-Kozaczuk M (2010) Enhancing rhizobium–legume symbiosis using signaling factors. Springer Vienna, pp 27–54
Sokol NW, Bradford MA (2019) Microbial formation of stable soil carbon is more efficient from belowground than aboveground input. Nat Geosci 12(1):46–53
Souza RC, Hungria M, Cantão ME, Vasconcelos ATR, Nogueira MA, Vicente VA (2015) Metagenomic analysis reveals microbial functional redundancies and specificities in a soil under different tillage and crop-management regimes. Appl Soil Ecol 86:106–112
Su Y, Cestellos-Blanco S, Kim JM, Shen YX, Kong Q, Lu D, Liu C, Zhang H, Cao Y, Yang P (2020) Close-packed nanowire-bacteria hybrids for efficient solar-driven CO_2 fixation. Joule 4(4):800–811

Sun Y, Du P, Li H, Zhou K, Shou L, Chen J, Li M (2024) Prokaryotic community assembly patterns and nitrogen metabolic potential in oxygen minimum zone of Yangtze estuary water column. Environ Res 252:119011

Sung CTB, Ishak CF, Abdullah R, Othman R, Panhwar QA, Aziz MMA (2017) Soil properties (physical, chemical, biological, mechanical). In: Soils of Malaysia, pp 103–154

Swenson TL, Jenkins S, Bowen BP, Northen TR (2015) Untargeted soil metabolomics methods for analysis of extractable organic matter. Soil Biol Biochem 80:189–198

Tannenbaum I (2021) The assessment of intergenerational microbiomes of Lolium perenne seed (Doctoral dissertation, La Trobe)

Tan X, Tu T, Gu B, Zeng Y (2021) Scenario simulation of CO_2 emissions from light-duty passenger vehicles under land use-transport planning: A case of Shenzhen International Low Carbon City. Sustainable Cities and Society 75:103266.

Trivedi P, Anderson IC, Singh BK (2013) Microbial modulators of soil carbon storage: integrating genomic and metabolic knowledge for global prediction. Trends Microbiol 21(12):641–651

Urich T, Lanzén A, Qi J, Huson DH, Schleper C, Schuster SC (2008) Simultaneous assessment of soil microbial community structure and function through analysis of the meta-transcriptome. PLoS One 3(6):e2527

Vamsi Krishna K, Bharathi N, George Shiju S, Alagesan Paari K, Malaviya A (2022) An updated review on advancement in fermentative production strategies for biobutanol using Clostridium spp. Environ Sci Pollut Res 29(32):47988–48019

van Noordwijk M, Goverse T, Ballabio C, Banwart SA, Bhattacharyya T, Goldhaber M, Nikolaidis N, Noellemeyer E, Zhao YongCun ZY (2015) Soil carbon transition curves: reversal of land degradation through management of soil organic matter for multiple benefits. In: Soil carbon: science, management and policy for multiple benefits. CABI, Wallingford, pp 26–46

Velázquez E, Carro L, Flores-Félix JD, Martínez-Hidalgo P, Menéndez E, Ramírez-Bahena MH, Mulas R, González-Andrés F, Martínez-Molina E, Peix A (2017) The legume nodule microbiome: a source of plant growth-promoting bacteria. In: Probiotics and plant health. Springer Singapore, pp 41–70

Villarino SH, Pinto P, Jackson RB, Piñeiro G (2021) Plant rhizodeposition: A key factor for soil organic matter formation in stable fractions. Sci Adv 7(16):eabd3176

Vishwakarma K, Sharma S, Kumar N, Upadhyay N, Devi S, Tiwari A (2016) Contribution of microbial inoculants to soil carbon sequestration and sustainable agriculture. Microbial Inoculants in Sustainable Agricultural Productivity: Vol. 2: Functional Applications 101–113.

Vocciante M, Franchi E, Fusini D, Pedron F, Barbafieri M, Petruzzelli G, Reverberi AP (2024) Sustainable recovery of an agricultural area impacted by an oil spill using enhanced phytoremediation. Appl Sci 14(2):582

Volk MJ, Tran VG, Tan SI, Mishra S, Fatma Z, Boob A, Li H, Xue P, Martin TA, Zhao H (2022) Metabolic engineering: methodologies and applications. Chem Rev 123(9):5521–5570

Wang HB, Zhang ZX, Li H, He HB, Fang CX, Zhang AJ, Li QS, Chen RS, Guo XK, Lin HF, Wu LK, Lin S, Chen T, Lin RY, Peng XX, Lin WX (2011) Characterization of metaproteomics in crop rhizospheric soil. J Proteome Res 10(3):932–940

Ward CP, Cory RM (2015) Chemical composition of dissolved organic matter draining permafrost soils. Geochim Cosmochim Acta 167:63–79

Wardle DA, Bardgett RD, Klironomos JN, Setälä H, Van Der Putten WH, Wall DH (2004) Ecological linkages between aboveground and belowground biota. science 304(5677):1629–1633

Warren CR (2014) Response of osmolytes in soil to drying and rewetting. Soil Biol Biochem 70:22–32

Whalen ED, Grandy AS, Sokol NW, Keiluweit M, Ernakovich J, Smith RG, Frey SD (2022) Clarifying the evidence for microbial-and plant-derived soil organic matter, and the path toward a more quantitative understanding. Global Change Biology 28(24):7167–7185

Wu H, Cui H, Fu C, Li R, Qi F, Liu Z, Yang G, Xiao K, Qiao M (2023) Unveiling the crucial role of soil microorganisms in carbon cycling: A review. Sci Total Environ 909:168627

Xiao S, Fu Q, Li Z, Li J, Zhang L, Zhu X, Liao Q (2021) Solar-driven biological inorganic hybrid systems for the production of solar fuels and chemicals from carbon dioxide. Renew Sust Energ Rev 150:111375

Zancarini A, Lépinay C, Burstin J, Duc G, Lemanceau P, Moreau D, Munier-Jolain N, Pivato B, Rigaud T, Salon C, Mougel C (2013) Combining molecular microbial ecology with ecophysiology and plant genetics for a better understanding of plant–microbial communities' interactions in the rhizosphere. In: Molecular microbial ecology of the rhizosphere, vol 1. Wiley, pp 69–86

Zhang F, Wang Q, Zhang Y, Yao S, Wang Q, Ndzana G, Hamer U, Kuzyakov Y, Zhang B (2024) Soil organic carbon increase via microbial assimilation or soil protection against the priming effect is mediated by the availability of soil N relative to input C. Geoderma 444, p.116861

Zhang M, Chen Q, Gong Z (2024a) Microbial remediation of petroleum-contaminated soil focused on the mechanism and microbial response: a review. Environ Sci Pollut Res 31:1–22

Zhang F, Wang Q, Zhang Y, Yao S, Wang Q, Ndzana G, Hamer U, Kuzyakov Y, Zhang B (2024b) Soil organic carbon increase via microbial assimilation or soil protection against the priming effect is mediated by the availability of soil N relative to input C. Geoderma 444:116861

Zheng Y, Wang H, Liu Y, Liu P, Zhu B, Zheng Y, Li J, Chistoserdova L, Ren ZJ, Zhao F (2024) Electrochemically coupled CH_4 and CO_2 consumption driven by microbial processes. Nat Commun 15(1):3097

Zhou Z, Tran PQ, Breister AM, Liu Y, Kieft K, Cowley ES, Karaoz U, Anantharaman K (2022) METABOLIC: high-throughput profiling of microbial genomes for functional traits, metabolism, biogeochemistry, and community-scale functional networks. Microbiome 10(1):33

Zhou X, Wu F (2012) Effects of amendments of ferulic acid on soil microbial communities in the rhizosphere of cucumber (Cucumis sativus L.). European journal of soil biology 50:191–197

Zhuang ZY, Li SY (2013) Rubisco-based engineered Escherichia coli for in situ carbon dioxide recycling. Bioresour Technol 150:79–88

Zhuang X, Zhang Y, Xiao AF, Zhang A, Fang B (2022) Applications of synthetic biotechnology on carbon neutrality research: a review on electrically driven microbial and enzyme engineering. Front Bioeng Biotechnol 10:826008

Glacial Cryosphere, its Microbial Ecology and the Impact of Climate Change

13

Subir Kumar Shill, Manasi Debnath, Samikcha Rai, Santosh Kumar, Palash Hazra, Mingma Thundu Sherpa, and Sayak Das

Abstract

Water is essential for life on Earth, profoundly influencing human and animal existence throughout history. Glacial ecosystems, home to psychrophilic microbes, could be bio-indicators of climate change. Studies on microorganisms in Antarctic and Greenland ice sheets reveal that ancient prokaryotes are preserved in these icy environments, which resemble cold habitats on Mars and Europa. These microorganisms survival in extreme cold is attributed to the unique properties of their evolved proteins. Earth's Polar Regions, separated by climatic barriers, serve as excellent models for understanding similar ecological systems. Recognized as a significant habitable zone within Earth's biosphere, the cryosphere is integral to the global climate system and offers a model for extraterrestrial life exploration. Microorganisms in glacial environments endure intense light, UV radiation, and

S. K. Shill · S. Das (✉)
Department of Life Science and Bioinformatics, Har Gobind Khurana School of Life Sciences, Assam University, Silchar, Assam, India
e-mail: sayak.das@aus.ac.in

M. Debnath · S. Rai
Department of Geography, Nagaland University, Lumami, Nagaland, India

S. Kumar
Department of Microbiology, School of Life Sciences, Sikkim University, Gairigaon, Sikkim, India

P. Hazra
Institute of Bioresources and Sustainable Development, Regional Centre, Gangtok, Sikkim, India

M. T. Sherpa
Department of Microbiology, Nar Bahadur Bhandari Government College, Tadong, Sikkim, India

P. Pandey et al. (eds.), *Microorganisms Resilience to Climate Change*, Microorganisms for Sustainability 56, https://doi.org/10.1007/978-981-96-3748-5_13

seasonal temperature changes. Glacier ice uniquely preserves microbial life and historical climate records spanning hundreds of thousands of years. Since the end of the Little Ice Age around 1850 AD, Himalayan glaciers have been retreating amid rising temperatures. These glaciers cover 33,050 square kilometers, making up 4.8% of the world's glaciers and 28.8% of those in Central Asia. The current termini of Himalayan glaciers are between 3500 and 4500 meters above sea level. Smaller glaciers, less than 0.2 square kilometers, are the most threatened ones, and others are retreating by 10–60 meters annually. The retreat of glaciers in the Hindu-Kush Himalayas suggests a potential decrease in psychrophilic organisms as the ice melts. This glacier loss endangers a significant reservoir of beneficial microorganisms, highlighting the profound environmental consequences of glacial retreat. This chapter explores the complexities of the glacial cryosphere, looking at its distinct microbial communities and the long-term impacts of climate change on these ecosystems. How these communities adapt to severe cold and how they contribute to biological processes. Through examining these interrelated facets, our primary goal is to offer a thorough comprehension of the crucial connection between glaciers, the bacteria that live within them, and the wider consequences of a shifting climate.

Keywords

Glaciers · Climate change · Psychrophiles · Hindu-Kush · Himalayas

13.1 Introduction

The glaciers and icesheets are one of Earth's largest biomes. They serve as the main habitat for cold-adapted bacteria called psychrophiles and are essential to many important biogeochemical processes (Anesio and Laybourn-Parry 2012). Glaciers are the antediluvian repositories of ice and snow, covering the Earth's Surface and all the microbes within it. Worldwide the researchers are monitoring to track the changes in climate over time. Large volumes of ice from glaciers are melting due to quick and continuous changes in climate, which is slowing down ocean currents and rising sea levels steadily (Hancock 2024). Glaciers as well as sheets made up of ice are melting at a pace of around 780 Gt yr^{-1} (gigatons per year), releasing microorganisms, nutrients, and debris into aquatic, terrestrial, and marine habitats. Organic carbon (OC) is critical because it is transported to downstream ecosystems during glacial melting, enriching the environment with bioavailable OC. The OC obtained from glacier has an active microbial ecosystem that is found on melting glacier surfaces and it transmits meltwater with dissolved organic carbon (DOC) and particulate organic carbon (POC) respectively and microbial assemblages (Chifflard et al. 2019; Stevens et al. 2022).

Microbial populations trapped in glaciers may provide important insights into Earth's past and contribute to paleoclimatic data, much as historic trace gas records are used to predict past climates. Moreover, glacial microbes (true psychrophiles)

acts as models for understanding and deciphering the mechanisms behind survival and adaptation in cold environment. Microbiological studies of glaciers have recently attracted interest because of all these possibilities (Edwards et al. 2013). The surface of glaciers, sub glacial terrestrial, liquid water environments and in deeper ice (reaching up to 3–4 km at the poles) have all been found to contain microbial cells. Their number ranges from less than 100 to 10^6–10^7 cells mL^{-1}, depending on the depths, glacier altitudes and individual glaciers (Xiang et al. 2005; Zhang et al. 2016; Miteva et al. 2009; Loveland-Curtze et al. 2010). Both local and global environmental factors shape the composition of their populations and types within the ice (Edwards et al. 2013). A wide variety of microorganisms, such as bacteria, viruses, archaea, and unicellular eukaryotes, have colonized glaciers. Microorganisms in these regions, which are adapted to harsh environments, may be extremely vulnerable to climate change, as the consequences of climate change have been seen in glaciers (Garcia-Lopez and Cid 2017). Glacial habitats contain small number of cells compared to other environment but much vulnerable in any little alteration on or within glacier (Boetius et al. 2015; Garcia-Lopez and Cid 2017). Global comparisons of geographically varied glaciers have revealed significant differences in the microbial biomass and community composition (Xiang et al. 2009; García-Descalzo et al. 2014; Zhou et al. 2016; Garcia-Lopez and Cid 2017; Hassan 2019). Nevertheless, thawing permafrost may result in increased microbial activity and an increase in the thickness of the active layer, depending on the predicted global warming scenarios (Lopatina et al. 2013; Garcia-Lopez et al. 2016).

Glaciers are sensitive and 'high-confident' indicators of climate change. They respond to changes in temperature and snowfall: glacial length change indicates an indirect and delayed response to climate change, while glacier mass balance indicates a direct and un-delayed response to the climate. There are metabolically active microorganisms in snowpack's all around the planet, even in far-off places like the South Pole, but these are not as well investigated (Carpenter et al. 2000). In Alpine snowpack's, bacterial abundances range between 10^3 and 10^5 cells mL^{-1} (Hell et al. 2013). It is possible for precipitation and dry deposition to introduce bacteria into snowpack's (Xiang et al. 2009); however, it is unclear how much of a role depositional and post-depositional processes play in organizing bacterial populations in various settings. Microbial communities are the main colonizers on the newly exposed soils when a glacier retreats (Wagner et al. 2017; Bradley et al. 2015). These communities mediate fore field soil pH (Lopatina et al. 2013; Schulz et al. 2013; Bradley et al. 2015), build up initial carbon and nitrogen pools, and enhance soil stability through the release of exopolymeric substances. It has been demonstrated that the age of the soil and the distance from the glacier terminal reduce the taxonomic variety and abundance of the *nifH* gene (Brankatschk et al. 2011).

Changes in the biodiversity of these habitats are driven by the widespread retreat and shrinkage of ice sheets and glaciers. Cauvy-Fraunié and Dangles (2019), discovered a direct association between deglaciation and global biodiversity in their recent meta-analysis of the response of biodiversity worldwide. They came to the conclusion that as glaciers recede, local biodiversity rises because they observed a decrease in "specialist" species and an increase in "generalist" taxa in ecosystems near to glaciers. A study on biodiversity changes from glacier retreat which excludes glacial ecosystems is a red flag as it shows that glacial habitats are still mostly unstudied and most are also unnoticed by ecologists. This is particularly concerning considering the extent of habitat loss expected in glaciers this century. The supraglacial ecosystem on the surface of glaciers as well as ice sheets consists of many important habitats. Algae predominate in the topmost layer of ice. Cryoconite holes, which are cylinder-shaped depressions created by the preferred defrosting of dark material into the surface, are another biological hotspot on glacier surfaces. Within the otherwise extremely turbulent supraglacial environment, cryoconite holes provide a stable haven for complex groups of microorganisms that include bacteria, fungus, algae, and archaea. Subglacial lakes and aquatic brine pools are just two of the unusual, remote aquatic environments found in the subglacial world. Thousands of microbial taxa call these systems home, and considering their seclusion, it's likely that they harbor rare, undiscovered species as well. The Himalayas are home to thousands of distinct microbial species. The full extent of prokaryotic diversity in subglacial ecosystems remains largely unexplored. Temperate and tropical glacier ecosystems are especially at risk, as climate change drives their rapid retreat and potential disappearance. This is a serious ecological concern, as these ecosystems harbor many endemic species, yet their biodiversity is poorly understood.

Methane (CH_4), one of the most significant greenhouse gases, is becoming more prevalent as the industrial revolution progresses (Guo and Zhang 2023; Mar et al. 2022). To ascertain the CH_4 concentration, the source and sink of CH_4 must continue to be in equilibrium (Guo and Zhang 2023; Liu et al. 2021; Mar et al. 2022). The glacier and sub glacial ecosystems are potential reservoirs of CH_4, as they include carbon-rich sediments that aid in the creation and retention of CH_4. With the loss of glaciers and global warming, the concentration of CH_4 rises significantly. In addition, higher temperatures stimulate the decomposition of organic matter and elevate the metabolic activity of microbes, both of which contribute to the atmospheric release of greenhouse gases and, eventually, operate as a positive feedback process for climate change (Guo and Zhang 2023). Various researches have documented microbial activity in glacial ice leading to the production of excess greenhouse gases. Sowers (2001) studied N_2O levels in sections of the Vostok ice core and observed a 30% excess at depths linked to the penultimate glacial maximum, where higher bacterial counts and dust concentrations were also recorded. He suggested that nitrifying bacteria might have produced the N_2O *in situ*. However, no significant increases in CH_4 levels were detected at these depths. Fluckiger (2003) subsequently identified microbial activity in glacial ice as one of four possible causes for the N_2O anomalies. Campen et al. (2003) found extremely high levels of CO_2, CH_4, and N_2O *viz.* 32, 8, and up to 240 times higher than tropospheric levels in air samples from two depths of the Sajama Glacier in Bolivia. Tung et al. (2005) observed large amounts of CO_2 and CH_4 in silt-rich basal ice at the Greenland Ice Sheet Project (GRIP) site and found high microbial concentrations in similar ice at the nearby GISP2 site. Price and Sowers (2004) measured CH_4 and microbial levels in GISP2 basal ice and concluded that in situ metabolism at −9 °C explained the excess CH_4. While methanogenic archaea are strict anaerobes, many species show varying tolerance to oxygen. Viable methanogens have been cultured from environments considered aerobic but containing anaerobic microenvironments. Enrichment techniques have also detected methanogens in meltwater beneath an Arctic glacier (Skidmore et al. 2000). Methane trapped in the 3053-m-deep GIPS2 core provided an important insight of millennial-scale climate change over the last 110,000 yr. However, at several depths in the lowest 90 m of the ice core, the methane concentration is up to an order of magnitude higher than at other depths. At those depths methanogenic archaea were discovered with the in-situ metabolism of which accounts for the excess methane.

The Himalayas are considered biodiversity hotspots regions of the world, consists of Hindu Kush Mountain, which includes Nepal and Bhutan in their entire parts and parts of other six countries Afghanistan, Pakistan, India, Nepal, Bhutan, and China (Chettri et al. 2008) with a total surface area of more than 4.3 million sq. km. Apart from the Polar Regions, the Himalayan cryosphere have the largest snowfall deposits and are house of many prompt glaciers. The range of glaciers in the Himalayas is very large covering more than 33,000 sq. km (Williams 2013); for which, this region is sometimes referred as the "Third Pole of the world". The monsoons of the Indian subcontinent play a vital role in sustaining life for billions across Asia. They directly impact the annual snow accumulation in the Himalayas, which serves as a critical source of freshwater for the entire continent. Ten main rivers originating from the Himalayas create a crucial ecological buffer: the Yellow and Yangtze Rivers, originating from the Qinghai Plateau, China; Mekong, Salween, and Irrawaddy rivers, which run southward from Tibet into South-East Asia, feed water to South-East Asia. The Ganges, Brahmaputra, and Meghna rivers which account for a lifeline to India, Pakistan, and Bangladesh ensure the year-long water supply, and these rivers also provide electricity to about 1.5 billion people in the Himalayas (National Research Council 2012). The presented article is constructed to articulate about the impact of climate change on the microbes i.e. psychrophiles of the Himalayan glaciers provided an unprecedented rise in temperature causes meltdown of these ancient water reservoirs.

13.2 Climate Change and Sustainable Development Goals (SDGs)

Climate change is the most imperative and significant concern as it's the rate of change has multiplied by factors, and human activity associated with wars, possible nukes, and extreme solar flares experienced recently are some of the major contributing factors. The term "climate change" describes any variation in the region's average weather over time, whether brought on by human action or natural variability (National Geographic Society 2023; Santos and Bakhshoodeh 2021). Over the past few decades, human activity and industrialization have contributed to a progressive acceleration of climate change, which has resulted in an annual increase in the average surface temperature (Santos and Bakhshoodeh 2021). The term "climate change" refers to a long-lasting alteration in the state of the climate, usually lasting decades or more, that may be recognized (as an instance, by statistical assessment) by changes in the variability and/or mean of its characteristics. There are two possible causes of climate change: natural internal processes and external forcing like variations in the sun's cycle, volcanic eruptions, and ongoing human-caused changes to the land use or atmosphere's composition (IPCC 2019).

Global warming due to human activity is estimated to have increased by 1 °C, over pre-industrial levels. If the current pace of growth in global warming is maintained, it is expected to reach 1.5 °C between 2030 and 2052. Over the next two to three decades, anthropogenic emissions—which include greenhouse gas (GHG) emissions, precursors to GHGs and aerosols, fossil fuel combustion, deforestation, land use and land-use changes (LULUC), livestock production, waste management, and industrial processes—are not expected to contribute to additional warming of more than 0.5 °C (IPCC 2019). Since the middle of the twentieth century, the planet's average surface temperature has increased due to pollution from humans, particularly the combustion of fossil fuels, which has increased the amount of heat-trapping greenhouse gases that exist in the atmosphere. While human activity takes center stage, natural processes are equally responsible for climate change. These consist of internal variability (which can include cyclical ocean patterns like El Niño, La Niña, and the Pacific Decadal Oscillation) and external forcing (including variables such as volcanic activity, fluctuations in Earth's orbit, and changes in the Sun's energy output) (NASA 2024).

The Sustainable Development Goals (SDGs) of the United Nations (UN) are an international policy framework that focuses on incorporating collective action to solve social, economic, and environmental problems that impede the world's progress toward sustainability, with the goal of achieving sustainability by 2030 (Yamaguchi et al. 2023). The 5Ps make up the SDGs: Partnership, Prosperity, Peace, People, and Planet (Yamaguchi et al. 2023). Climate change is included into UN's SDGs as SDG13 (Sustainable Development Goal 13), or "Climate Action". The necessary actions to lessen and mitigate the consequences of climate change must be implemented in order to accomplish this goal. Furthermore, in order to address the issue on a wider scale, it need action at various organizational levels. The first volume of the sixth Intergovernmental Panel on Climate Change (IPCC), the evaluation Report (AR6) has shown how urgently the causes of climate change must be addressed, especially by reducing CO_2 emissions (Filho et al. 2023).

13.3 Glaciers, its Importance and Role

Earth's different spheres include the biosphere, atmosphere, cryosphere, hydrosphere, lithosphere, and geosphere. The cryosphere, which is located between the hydrosphere, lithosphere, and atmosphere, is where water is found in solid or super cooled condition and temperatures are at or below zero (Qin et al. 2018). The Greek word "krios," which means cold, is where the term "cryosphere" first appeared (NOAA 2023). The main elements of the cryosphere include snow, cover of snow, ice sheets, glaciers, ice caps, shelves of ice, icebergs, permafrost, ocean, river and lake ice, dense precipitation, etc. (Key et al. 2020; Qin et al. 2018). Around 10% of Earth's geographical area is composed of glaciers and sheets of ice, which retain 69% of freshwater naturally and act as a natural reservoir (Winkel et al. 2022). The majority of Asia's river systems originate from melt water discharged by the Himalayan glaciers, which supports a variety of ecosystems and ecosystem services in addition to supplying food and energy downstream (Lee et al. 2021).

The volume of glaciers is diminishing due to climate change, which also affects the size and seasonal water flow magnitude over time. These changes have direct effects on water supply, agriculture, tourism, and hydroelectric power generation. Precipitation (P), glacial melt water (old ice and snow; ISM), and shallow ground water (SGW) are the three potential sources of surface water. Throughout the rainy season, they come into the reservoir via a number of streams (i.e., ISM+ P+ SGW). Streams and reservoirs receive the flow of glacial melt water (old ice melt; IM) and shallow ground water (SGW) through arroyos during the dry season (Guido et al. 2016) as depicted in Fig. 13.1.

A glacier consists of six distinct layers, based on the areas: snow, firn, supraglacial ice, englacial ice, subglacial basal ice, and subglacial lakes.

(i) Snow – It plays a crucial role in maintaining the water balance in glacial ice. It collects excess water vapor from the surrounding atmosphere, which freezes and accumulates on the surface of glaciers (Kim et al. 2022).
(ii) Firn – Is the intermediate stage between snow and glacial ice, undergoing repeated cycles of freezing and thawing before transforming into glaciers under internal pressure (Kim et al. 2022; Hollmann 2014). Although firn is less dense than pure ice, the visible portion above the water is still only a small fraction of the total ice mass (Matsuoka 2020).

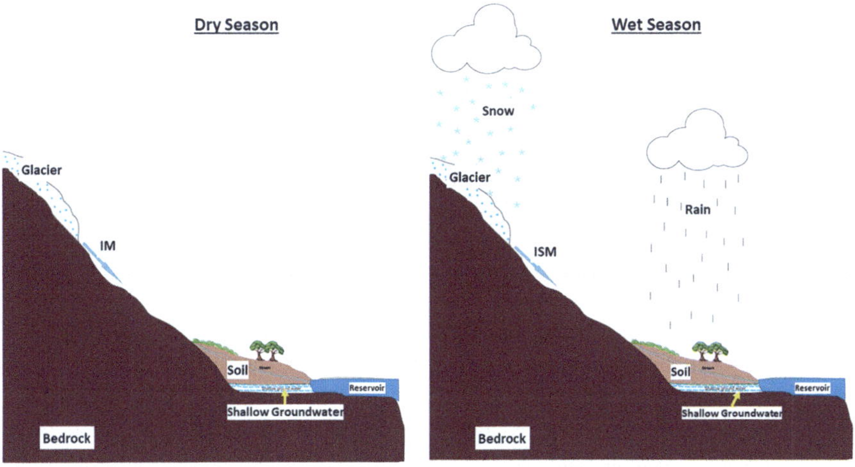

Fig. 13.1 Glacier and snowmelt water contribution to the freshwater reservoir during the dry and wet seasons

(iii) Supraglacial ice – The bottom layer that solar radiation can penetrate is supraglacial ice covered by firn, which permits a variety of dynamic microbial metabolism in the layer (Kim et al. 2022). Sea level fluctuations, ice dynamics, and subglacial processes are all connected to the ice surface conditions through supraglacial melting waters channels on ice sheets and glaciers. They are not yet incorporated in models of surface mass balance and runoff's influence on global sea level rise, despite their critical function in heat and water transmission (Pitcher and Smith 2019).
(iv) Englacial ice – The interior of a glacier, known as englacial ice, is defined by high pressure, minimal interstitial space, an almost complete absence of accessible water for biological activity, and a lack of available energy sources (Kim et al. 2022; Hodson et al. 2008; Irvine-Fynn et al. 2011).
(v) Subglacial ice – In contrast to other glacial formations, subglacial basal ice is a nutrient-rich layer that forms from accumulated nutrients from supraglacial and englacial ice (Kim et al. 2022; Hodson et al. 2008).
(vi) Subglacial lakes – Below basal ice, subglacial lakes regulate basal hydrology, ice movement, biogeochemical fluxes, and geomorphic activity. They are also a repository of historic climate conditions (Livingstone et al. 2022; Lepp et al. 2022).

According to Cauvy-Fraunié and Dangles (2019), the possible advantages include the management of climate change and freshwater for consumption at home, agriculture, industry, and hydropower. A "water tower" is the water delivery and storage that the mountain ranges provide to suit the environment's and people downstream water demands. A water tower (seasonally) generates more runoff from rain than its downstream region because to orographic rainfall and colder temperatures at high altitudes. By retaining water in glaciers, lakes, and snowpack, it also holds off the release of water. Water towers supply populations downstream with generally consistent water supplies due to their ability for absorption. Delivering glacial water that has melted during the hot, dry season is an aspect of the process (Immerzeel et al. 2020).

13.4 Current State and Climate Impact on Himalayan Hindu-Kush Glaciers

According to a global glacier database Randolph Glacier Inventory (RGI), there are around 198,000 glaciers worldwide which covers cover 726,000 km^2 which covers glacier outlines of the world but outside of ice sheets (GLIMS 2024). The latest version, RGI 7.0, includes 20 regions as shown in Fig. 13.2 worldwide with glaciers: Alaska, Western Canada and the US, Arctic Canada South, Arctic Canada North, Iceland, Greenland Periphery, Svalbard, Jan Mayen, Scandinavia, North Asia, Russian Arctic, Caucasus and the Middle East, South Asia East, South Asia West, Central Asia, Low Latitudes, New Zealand, Southern Andes, Sub Antarctic and Antarctic Mainland Antarctic Islands (GLIMS 2024).

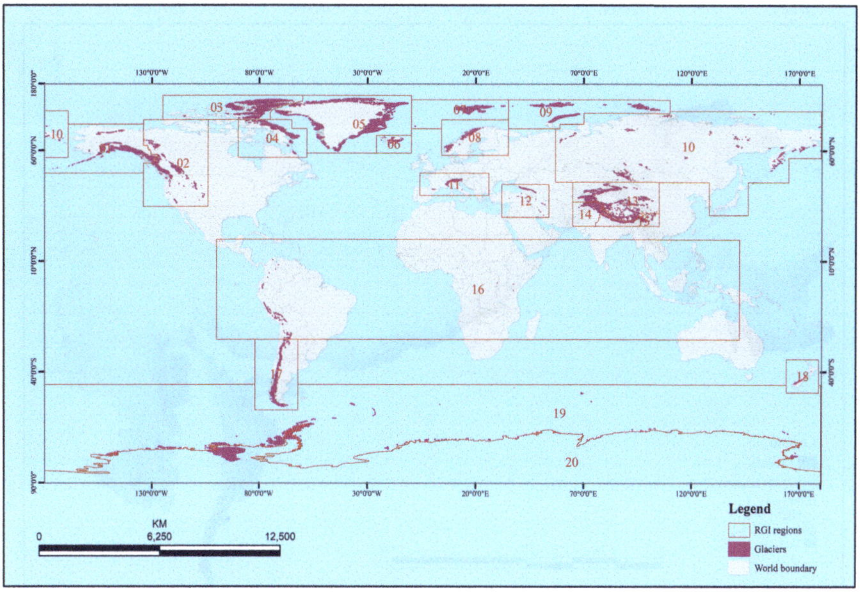

Fig. 13.2 Worldwide glacier distribution

From a hydrological perspective, glaciers are important water resources that greatly affect stream flow (Hock 2005). The amount of glacier ice that was formerly permanent has shrunk globally (Hugonnet et al. 2021), primarily leading to the creation of glacial lakes. Climate fluctuations and global climate change have had a profound influence on the environment of high alpine glaciers (Bajracharya et al. 2007). Between 2000 and 2019, the global glaciers lost an average of 267 gigatonnes (billion tons) of ice per year. The rate of glacial reduction in mass also significantly accelerated during this period. Glacial melt was responsible for up to 21% of the sea level rise that was seen throughout this period approximately 0.74 mm annually (Hugonnet et al. 2021).

There are about 18,600 glaciers in 50 World Heritage sites that have been identified. These glaciers encompass over 66,000 km^2, or about 10%, of the Earth's glaciated surface. Research investigations utilizing satellite data show that since 2000, these icebergs have been retreating more swiftly. Sea level rise is caused by more than 5% of the world's World Heritage glaciers. They decline in 58 billion tons of ice a year on average, which is equivalent to the whole amount of water that France and Spain drink together. Based on forecasts, nearly one third of World Heritage glaciered regions would see the disappearance of glaciers by the year 2050, irrespective of the climate scenario adopted. Approximately half of all sites might see the virtual disappearance of glaciation by 2100 in a scenario involving business-as-usual emissions (UNESCO 2022).

The Hindu Kush Himalaya (HKH), also known as High Mountain Asia (HMA) or the "Water Tower of Asia," boasts the highest density of melting glaciers as

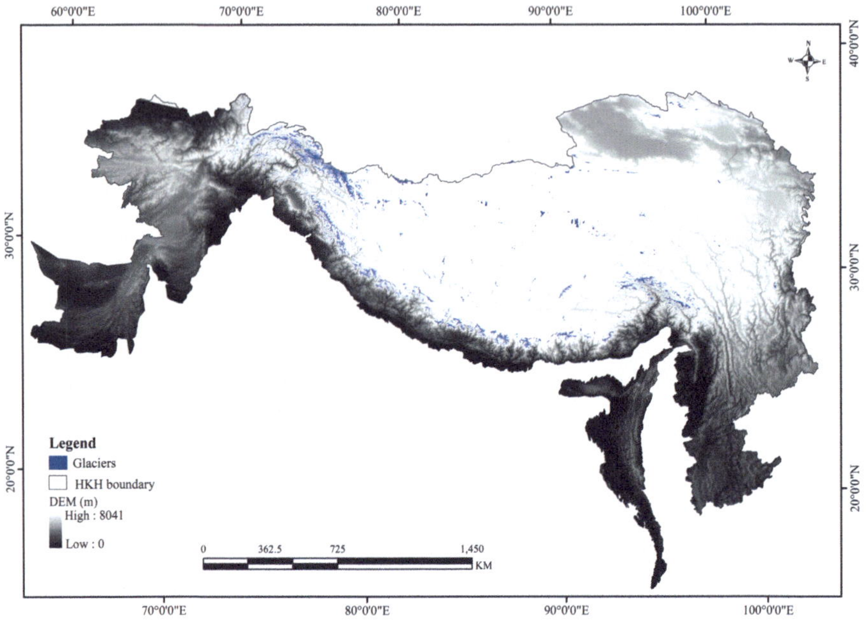

Fig. 13.3 Glacier distribution in the Third Pole region (Hindu-Kush-Himalaya and Tibetan plateau)

depicted in Fig. 13.3, over ten thousand glacier lakes, and abundant yearly snowfall outside of the polar region (Immerzeel et al. 2020; Kraaijenbrink et al. 2021; Lin et al. 2022; Romshoo et al. 2022; Banerjee et al. 2014; Banerjee 2017).

The HMA region has experienced climate change at a rate slightly higher than the global average in recent decades, significantly impacting its sensitive cryosphere (Yao et al. 2022; Banerjee et al. 2014). Increased glacier thinning caused by high-elevation surface warming has led to the rapid formation of glacier lakes, often dammed by ice or moraines (Adhikari et al. 2019; Banerjee et al. 2014), which poses hazards to downstream infrastructure, settlements, and industrial projects including hydropower, in the form of Glacier Lake Outburst Floods (GLOFs) (Zheng et al. 2021; Veh et al. 2022; Banerjee et al. 2014).

The Himalayas have one of the greatest amounts of glacier-stored water. Variations in extend of the glaciers (significant glacier retreat and thinning) in the Himalayas are understood to be one of the most noticeable effects of change in climate (Romshoo et al. 2022; Bahuguna et al. 2014). The observed thinning and retreat are characterized by variation throughout the Himalaya (Abdullah et al. 2020; Farinotti et al. 2020). Over the past 40–50 years, 13% of the nearly 11,000 km^2 of glaciated terrain in the Himalayas, mapped between 1962 and 2009 and selected based on major climatic zones, has disappeared. The mass loss of Himalayan glaciers is occurring at an alarming rate (Kulkarni and Karyakarte 2014).

Until 2010, the glaciers in the Karakoram Range remained stable, with some even growing. However, it has been observed that the majority of Karakoram glaciers are now also shrinking (Hugonnet et al. 2021). They highlight the concerning situation in the Himalayas, where glacier melt water has been a crucial water source during the dry season for major rivers like the Indus, Brahmaputra, and Ganges. While the accelerated melting currently benefits local populations, continued glacier loss could lead to significant food and water shortages in densely populated countries such as Bangladesh and India within a few decades.

Debnath et al. (2019) analyzed the glaciers in the Chhombo Chhu watershed of the Tista basin in the Sikkim Himalaya and found that their area decreased between 1975 and 2018 (Debnath et al. 2019). The glaciers in the Chhombo Chhu watershed are retreating faster than those in the western and central Himalayas. Large glaciers have lost more area than smaller ones since 1975, attributed to higher solar incidence, lower terminus elevation, gentle slopes, and warmer air temperatures. Borah et al. (2022) discovered that from 1988 to 2018, glaciers in the Mago Chu basin, Eastern Himalayas, decreased from covering 10% to 7% of the total catchment area, with about 20 glaciers disappearing and 61 new lakes forming. Kumar and Sharma (2023) studied 24 medium-sized glaciers in Sikkim between 1988 and 2018, finding an average retreat rate of 5.22 m/yr. Compared to similar-sized glaciers in Nepal, Himachal, and Uttarakhand, Sikkim's glaciers are retreating at a slower pace.

In numerous instances, retreating glaciers have led to the creation of separate glaciers detached from the main trunk. An examination of the Chenab, Parvati, and Baspa basins from 1962 to 2001 revealed a reduction in the overall glaciated area, but an increase in the number of glaciers due to fragmentation. Large glaciers in these basins lost 12% of their area, while small glaciers experienced a 38% reduction (Kulkarni et al. 2007). Chand et al. (2020) conducted a study on the Parvati glacier in the North-Western Himalaya and found significant retreat. Between 1973 and 1989, the glacier retreated at a rate of 70.9 ± 3.1 m per year, causing it to split into two. From 1965 to 2018, the average retreat was 37.5 ± 0.5 m per year. Additionally, between 1989 and 2018, the detached portion of the glacier retreated faster (28.4 ± 1.6 m per year) compared to the main trunk (18.7 ± 1.6 m per year). In the Chandra basin, the Gaglu glacier retreated from 1989 to 2015, vacating an area of 1576×10^3 m^2. This thinning is attributed to rising temperatures, as the glacier is highly sensitive to climate changes (Singh et al. 2020). Garg et al. (2019) evaluated multiple parameters of 23 glaciers in the Sikkim region of the Eastern Himalaya and found that they significantly retreated between 1991 and 2015. The study included glaciers of varying sizes, orientations, altitudinal ranges, spatial locations, and debris cover to provide a representative sample of the Sikkim Himalaya. Despite their differences, all glaciers experienced area changes at different rates, highlighting the complex, distinct, and multidimensional nature of individual glaciers' responses to climate change (Table 13.1).

Studies on various glaciers across different regions have shown that the retreat rates of individual glaciers are not uniform, even within the same basin or part of the Himalayas. This indicates that changes in each glacier are influenced not only by regional climatic conditions, such as rising temperatures, but also by geographic factors. While glaciers are currently either stagnant or retreating, the differing rates

Table 13.1 The current conditions of glaciers in the Hindu-Kush Himalayas

Sl No.	Name of the basin	No. of glaciers	Count study period	Area change (km^2)	Area change rate (km^2a^{-1})	References
1	Changme Khnagpu basin, Sikkim Himalaya	81	1975–2016	−20.72 ± 325	−0.453	(Debnath et al. 2019)
2	Chenab, Parvati and Baspa basins	466	1962–2001	From 1.4 to 0.32 (reduced)		(Kulkarni et al. 2007)
3	Sikkim Himalaya	24	1988–2018	3.95 ± 0.35	63.9 to 3.9 m yr^{-1}	(Kumar and Sharma 2023)
4	Parvati glacier, North-Western Himalaya	1	1965–2018	2.2 ± 0.01	37.5 ± 0.5 m a^{-1}	(Chand et al. 2020)
5	Sikkim Himalaya	23	1991–2015	231.64 ± 4.24 to 219.05 ± 4.26	17.78 ± 2.06 m a^{-1} (average)	(Garg et al. 2019)
6	Gaglu glacier, Chandra basin, Western Himalaya	1	1989–2015	37.65 ± 0.1(10^3m^2)	11.89 ± 0.12 m yr^{-1}	(Singh et al. 2022)

of area change and the underlying reasons for each glacier's behavior highlight the need for further research.

13.5 Applications of Psychrophiles Pertaining to Climate Change

Psychrophiles, also known as cold-loving bacteria, thrive at low temperatures. They are characterized by their ability to grow at 0 °C, with optimal growth occurring at 15 °C or below, a maximum growth temperature of 20 °C, and a minimum growth temperature at or below 0 °C (Anesio and Laybourn-Parry 2012; Moyer et al. 2017).

Channels of ice provide a liquid microenvironment within glaciers and ice sheets that could potentially host microbial life (Bradley et al. 2023; Garcia-Lopez and Cid 2017). Microorganisms contribute to the darkening of ice, which accelerates its melting and subsequently influences sea level rise (Bradley et al. 2023). However, microorganisms inhabiting ice veins have adapted to various physiochemical stressors such as low pH, reduced water activity, slower solute transport rates, and membrane damage caused by ice crystal formation. They have evolved intricate structural and functional strategies to adaptations to flourish in cold environments, as depicted in Fig. 13.4.

These adaptations include: (i) Producing highly efficient catalytic enzymes that operate effectively at low temperatures, (ii) Incorporating unsaturated fatty acids

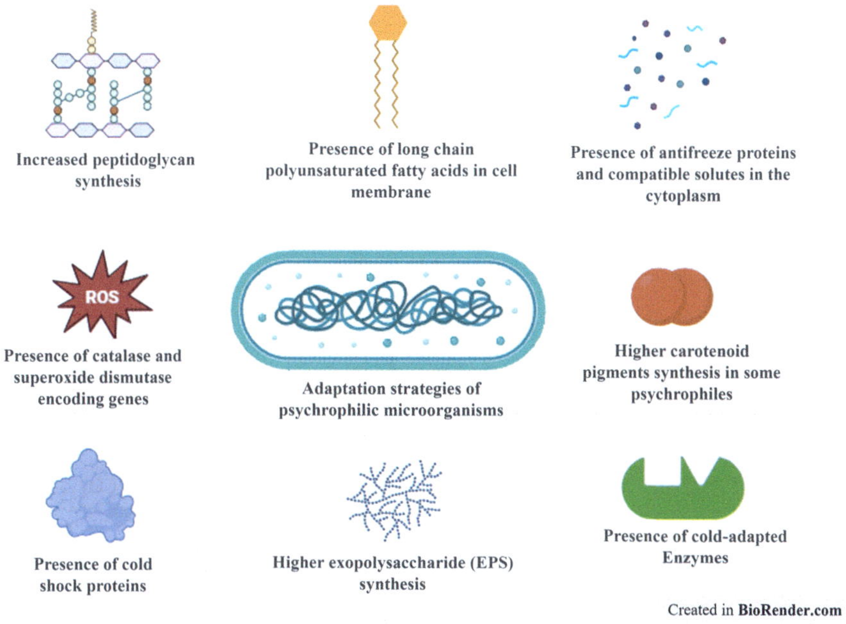

Fig. 13.4 Different adaptation strategies of psychrophilic microorganisms

into their cell membranes to enhance fluidity, (iii) Synthesizing specific proteins that facilitate enzymatic activity at low temperatures, (iv) Forming extracellular polymeric substances like sugars, polysaccharides, and antifreeze proteins that modify ice crystal structures and protect cells from freezing, (v) Reorganizing protein networks for improved functionality, and (vi) Utilizing pigments for energy production, stress resilience, and protection against UV radiation (Garcia-Lopez and Cid 2017). The number of bacteria in the wet regions of glaciers can vary significantly, ranging from 10^7 cells L-1 in clean snow and ice to 10^{11} cells L-1 in surface ice contaminated with external substances (Anesio et al. 2017). The dominant bacterial phyla reported from glacial habitats belong to Proteobacteria, Cyanobacteria, Actinobacteria, Bacteroidetes, and Firmicutes (Møller et al. 2013). Some of the predominant genera such as *Arthrobacter*, *Clavibacter*, *Mycobacterium*, *Methylobacterium*, *Sphingomonas*, *Acinetobacter*, *Pseudomonas*, *Chryseobacterium*, *Flavobacterium*, *Exiguobacterium*, *Paenibacillus*, and *Planococcus* have also been reported from various global glaciers (Loveland-Curtze et al. 2010). A compilation has been made of significant psychrophiles, known as "green energy microbes" utilized for diverse industrial and household purposes as detailed in Fig. 13.5.

These cold-adapted microorganisms have been isolated or documented in various glaciers (Table 13.2).

Here are the primary applications of these cold-loving microorganisms:

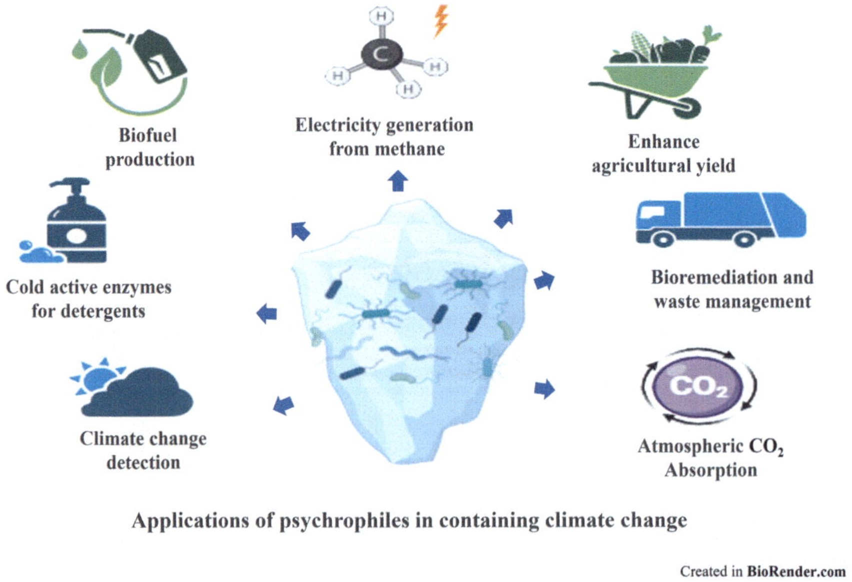

Fig. 13.5 Application of psychrophiles in mitigating climate change

Table 13.2 Summarizes significant psychrophiles investigated across diverse glaciers worldwide

Sl No.	Microbes	Temperature	Sources	Applications	References
1.	*Rhodococcus sp.*	<–5 to 30 °C	Collins glacier, Antarctica	Production of biodiesel	(Goordial et al. 2015; Sun et al. 2023; Wahlen et al. 2013)
2.	*Halocynthiibacter arcticus*	10–27 °C	Siberian permafrost of Kolyma glacier	Production of biodiesel	(Baek et al. 2015; Bakermans et al. 2006; Le et al. 2020; Novototskaya-Vlasova et al. 2012)
3.	*Pseudomonas fluorescens*	4–22 °C	Kanchengayao Glacier, North Sikkim, India	Production of biodiesel by converting animal fat or vegetable oil via using enzyme lipase	(De Pascale et al. 2009; Sherpa et al. 2019)
4.	*Clostridium sp.*	20–40 °C	Lower most region of Greenland Icesheet, Greenland glacier	Production of biodiesel from ethanol (from cellulosic waste) via using enzyme β-glucosidase.	(Barnard et al. 2010; Elshahed 2010; Visioli et al. 2014; Xiang et al. 2005)
5.	*Exiguobacterium antarcticum*	5–40 °C	Lake Fryxell, Antarctica	Production of ethanol by using enzyme β-galactosidases	(Crespim et al. 2016; Zhu et al. 2020)
6.	Bacterial strain GA024 (related to *Polaromonas jejuensis*)	4–22 °C	Glacier sediment and melted ice of Antarctica	Production of biohydrogen	(Alvarado-Cuevas et al. 2015; Ciok et al. 2018)
7.	Bacterial strain GA051 (related to *Janthinobacterium agaricidamnosum*)	4–30 °C	*Deschampsia antarctica* rhizosphere, sediments and ice melts of Antarctica	Production of biohydrogen	(Alvarado-Cuevas et al. 2015; Rajawat et al. 2019)
8.	*Klebsiella spp.*	4–37 °C	Antarctic seawater	Production of biohydrogen	(Mohammed et al. 2018 Rafiq et al. 2017)
9.	*Tetracladium sp.*	4–37 °C	Hailuogou Glacier, China and Mingyong Glacier, China	Production of biofuels	(Carrasco et al. 2017, 2019; M. Wang et al. 2015)

(continued)

Table 13.2 (continued)

Sl No.	Microbes	Temperature	Sources	Applications	References
10.	*Rhodotorula glacialis*	3–20 °C	Gulkana Glacier, Alaska	Potential precursor for biodiesel production	(Amaretti et al. 2010; Bibra et al. 2015; Uetake et al. 2012)
11.	*Arthrobacter sp.*	4–28 °C	Midui glacier, Tibet	Plant growth-promoting activity (IAA, siderophore, ACC deaminase, ammonia, and NF)	(Guo and Zhang 2023; Mukhia et al. 2024; Rizvi et al. 2021)
12.	*Bacillus firmus*	8–26 °C	Malan Glacier, Tibet, China & Sikkim Himalayan region.	Promotes the growth of plants, reduces the effects of abiotic salt stress, and helps plants resist against plant-parasitic nematodes like root-knot nematodes.	(Chiring Phukon et al. 2020; Wang et al. 2013)
13.	*Bacillus pumilus*	5–15 °C to 50 °C	Glacial soils of Thajwas Glacier, Himalayas, India	Solubilize the P and fix the N in soil and increase their transport to roots and thus acts as PGP (Plant Growth Promoters)	(Farooq et al. 2016; Farooq et al. 2021
14.	*Bacillus subtilis*	10–37 °C	Malan Glacier, Tibet, China	Biofertilizer effect on crop plant (plant growth hormones (IAA, GAs, cytokinin's and spermidines) which triggers plant growth are synthesized, enhances fruits and grains yield and Secretes ACC (1-aminocyclopropane-1-carboxylic acid) deaminase to inhibit plant senescence)	(Radhakrishnan and Hashem 2017; Xie et al. 2023; Zhang et al. 2001, 2024)
15.	*Paenibacillus sp.*	15–28 °C	Kafni glacier, Himalayas, India	Fixes nitrogen; solubilizes phosphorus; synthesizes siderophores; and plant hormone (indole, cytokinin's, and gibberellins) to improve biotic and abiotic stress responses of plants.	(de Lemos et al. 2023; Kishore et al. 2010; Xue et al. 2024)

16.	*Desemzia incerta*	20–30 °C	Soil and water samples of Leh Ladakh and Rohtang pass, India	IAA, GA, siderophores, NH_3, HCN, ACC deaminase	(Rizvi et al. 2021; Yadav et al. 2018, 2019)
.	*Bacillus licheniformis*	10–37 °C	Garhwal Himalayan region, India	Plant growth promoting enhancers such as- ACC deaminase, siderophores, IAA, GA, NH_3, HCN,	(Negi et al. 2005; Rizvi et al. 2021)
.	*Pseudomonas fragi*	7–25 °C	High altitude garlic rhizosphere	PGP enhancers—IAA and HCN	(Lebert et al. 1998; Rizvi et al. 2021; Selvakumar et al., 2009)
19.	*Rahnella sp.*	4–30 °C	Xinjiang Glacier, China	Plant growth promoters and effective biocontrol agents against numerous soil-borne pathogens.	(Núñez-Montero et al. 2021; Zouagui et al. 2024)
20.	*Sanguibacter antarcticus*	23–26 °C	Sea sand on King George Island, Antarctica.	Helps in plant growth and development.	(Hong et al. 2008; Ma et al., 2021; Rizvi et al. 2021)
21.	*Stenotrophomonas maltophilia*	5 –40 °C	Kanchengayao Glacier, North Sikkim, India	Potential to promote the wheat growth under biotic and abiotic (salt) stressors	(Sherpa et al. 2021; Singh et al. 2023)
22.	*Pseudomonas koreensis*	5–37 °C	Indian Himalayan cold desert	Phosphate biofertilizer to promote plant growth.	(Ibáñez et al. 2023; Wang et al. 2013)
23.	*Streptomyces sp.*	28–30 °C	Antarctic soils	PGP abilities and biocontrol agent.	(Gesheva 2010; Sivalingam et al. 2022)
24.	*Micromonospora sp*	20–37 °C	Antarctic soils	PGP abilities and biocontrol agent.	(Gesheva 2010)
25.	*Methanosarcina acetivorans*	35–40 °C	Nubra valley glacier, Ladakh	Converts CH_4 to acetate, produces enzyme Mcr.	(Aschenbach et al. 2013; Kondaveeti et al. 2019)

(continued)

Table 13.2 (continued)

Sl No.	Microbes	Temperature	Sources	Applications	References
26.	*Geobacter sulfurreducens*	30–35 °C	Damma Glacier, Switzerland	Convert acetate to Co2 to produce electricity. Produces electricity from hydrogen and acetate.	(Bond and Lovley 2003; Kondaveeti et al. 2019; Nash et al. 2018)
27.	*Shewanella putrefaciens*	4–40 °C	Antarctic coastal areas	Produces electricity from lactate and pyruvate.	(Bozal et al. 2002; Kim et al. 2002)
28.	*Rhodoferax ferrireducens*	2–35 °C	Kaldalónsjökull (Kal), a basalt-hosted glacier in Iceland	Produces electricity from glucose	(Jin et al. 2020; Yamasaki et al. 2018)
29.	*Xantobacter autotrophicus*	30–37 °C	Sediments of marine water	Produces zeaxanthin; used as food dye	(Meusel and Rehm 1993; Pander et al. 2020)
30.	*Rhodococcus opacus*	−5 to 30 °C	McMurdo Dry Valley permafrost, Antarctica & fuel contaminated soil	Used for the production of lipids for biofuel, bio-commodity chemicals, and nutrition.	(Goordial et al. 2015; Holder et al. 2020)

13.5.1 Renewable Alternative Energy/Biofuel

There is a growing global interest in developing sustainable and renewable alternative energy sources, driven by rapid industrial growth that relies heavily on fossil fuels. Concerns such as depletion of petroleum reserves, energy security, air pollution, and climate change further underscore the need for these alternatives (Kim et al. 2020).

Compared to fossil fuels, biofuels provide a more sustainable solution in the fight against global warming. They have the potential to significantly reduce emissions of greenhouse gases because they are produced from renewable resources such as plants and agricultural waste (Rial 2024). The term "Bioenergy" comprises various forms of solid, liquid, and/or gaseous substances that have been utilized as sources of energy, such as bio-ethanol, butanol, biodiesel, and biomethane (Zhu et al. 2020). Methane, ethanol, and hydrogen are examples of liquid or gas fuels that are often produced by microbes through the thermal, chemical, or fermentation decomposition of organic matter found in plants and animals. Biofuels include biodiesel, bioethanol, biobutanol and biomethane (Kim et al. 2020). According to Sun et al. (2023), *Rhodococcus* sp. were retrieved from the Collins Glacier in Antarctica and is utilized in the manufacturing of biodiesel (Wahlen et al. 2013). Certain species of *Rhodococcus* demonstrate remarkable capacity to produce substantial quantities of triacylglycerols (TAG) under conditions of carbon and nitrogen limitation. TAG can be converted into Single Cell-oil (SCO), which can then be used to produce biodiesel through trans esterification (Alvarez et al. 2021). *Halocynthiibacter arcticus* isolated from the marine sediment of the Arctic region (Baek et al. 2015) are used to prepare fatty-acid methyl ester (FAME) biodiesel (Le et al. 2020) and *Psychrobacter cryohalolentis* isolated from the Siberian permafrost of Kolyma glacier (Bakermans et al. 2006; Novototskaya-Vlasova et al. 2012) have potential uses in a variety of industrial domains, such as production of biodiesel (Le et al. 2020). In a low-energy process, animal fat or vegetable oil may be converted into biodiesel using enzyme lipase derived from *Pseudomonas fluorescens* in a cold environment (De Pascale et al. 2009). *Pseudomonas fluorescens* and other *Pseudomonas sp.* have been found to be reported in the Kanchengayao Glacier, North Sikkim, India (Sherpa et al. 2019). *Clostridium* sp. isolated from the lower most region of Greenland Ice sheet, Greenland glacier (Xiang et al. 2005), and the enzyme β-glycosidase derived from the *Clostridium sp.* possesses the ability to effectively produce biofuels from cellulosic waste at low temperatures by directly producing ethanol from cellulosic material (Elshahed 2010).

It has been reported that certain *Clostridium* sp. can ferment acetone–butanol–ethanol (ABE) (Visioli et al. 2014), with a yield of 3:6:1 (Barnard et al. 2010). Raw biomass, such as corn, sugar beet, and sorghum, undergoes partial conversion into hydrogen and butyric acid, which is then further transformed into butanol. Since biobutanol has 25% more energy than bioethanol, it has significant promise in the transportation sector (Barnard et al. 2010; Visioli et al. 2014).

Hydrogen, ethanol, and 2,3- butanediol can be produced using three agro-industrial residues: cheese whey powder (CWP), wheat straw hydro lysate (WSH), and sugarcane molasses (SCM). This process involves the Antarctic psychrophilic bacterium GA0F, isolated from sediments of an Antarctic glacier, which is closely genetically related to *Pseudomonas antarctica* (Alvarez-Guzmán et al. 2020). One potential sustainable energy source that shows promise is bio hydrogen, which has the potential to replace fossil fuels. Using dark- and photo-fermentation, water gas shift reactions, and microbial electrolysis cells to produce hydrogen from wastewater and organic solid wastes presents a promising method in energy production. This integrated approach not only produces bio hydrogen but also effectively treats waste materials, offering a pathway to carbon-neutral fuel production (Mogili et al. 2022).

In a study by Crespim et al. (2016), it was suggested that ethanol production could be enhanced by utilizing the β-galactosidase enzyme derived from *Exiguobacterium antarcticum* B7, isolated from Lake Fryxell in Antarctica (Crespim et al. 2016; Zhu et al. 2020b). Two psychrophilic bacterial strains were isolated from a variety of Antarctica samples, including glacier melted ice, glacier sediment, coastal mud, and the rhizosphere of *Deschampsia antarctica.* GA024 is linked to *Polaromonas jejuensis*, whereas GA051 is associated with *Janthinobacterium agaricidamnosum.* These strains have been identified as producers of bio hydrogen, achieving a maximum production of 253.3 mL and a production rate of 16.64 mL/L/h (Alvarado-Cuevas et al. 2015). The psychrotolerant *Klebsiella* sp. strain ABZ11, which was obtained from Antarctic seawater, is capable of producing biohydrogen (Mohammed et al. 2018). *Tetracladium* sp. reported to have been isolated from the Hailuogou Glacier, Sichuan and Mingyong Glacier, Yunnan, China (Wang et al. 2015). *Tetracladium* sp. produces glucoamylase, an enzyme that operates best at pH 6.0 and a temperature of 30 °C. This enzyme finds frequent use in various industrial applications, notably in the production of biofuel (Carrasco et al. 2017).

Psychrophilic yeast *Rhodotorula glacialis* has been isolated from a Gulkana Glacier in Alaska (Uetake et al. 2012) and is capable of surviving in temperatures at or below zero degrees, however it exhibits maximum development at 20 °C. The yeast cell is capable of producing significant quantities of lipids, reaching up to 68% of the total biomass. The lipid concentration of this substance is comparable to that of soybean oil and rapeseed oil, indicating its significant potential as a precursor for biodiesel production (Bibra et al. 2015).

13.5.2 Agriculture

Low temperatures exert a considerable adverse effect on food crop yields and plant development in agriculture. It is the primary environmental factor that negatively affects microbial biomass, community structure, composition, diversity, and leads to depletion of soil nutrient pools (Rizvi et al. 2021). Psychotropic microbes demonstrate several plant-growth-promoting (PGP) traits, including ACC deaminase activity, solubilization of potassium, zinc, and phosphorus, biological nitrogen fixation, and the synthesis of bioactive compounds such as gibberellic acids, ammonia, cytokinins, Fe-chelating agents, hydrogen cyanide, and indole-3-acetic acid. These microbes contribute to plant nutrient uptake, fostering plant growth, enhancing soil productivity, and promoting environmental health (Yadav et al. 2018). To achieve sustainable agriculture, it's crucial to adopt environmentally friendly approaches that either maintain or enhance current levels of crop and food production. Nitrogen-fixing bacteria and PGP microbes can directly influence plant development in cold climates (Yadav et al. 2019). One of the psychotropic microbes exhibiting multi-functional PGP attributes is *Arthrobacter* sp., isolated from the Midui glacier in Tibet (Liu et al. 2018), *Bacillus amyloliquefaciens* and *Bacillus firmus* (found in the Malan Glacier in Tibet, China) (Zhang et al. 2001), *Bacillus licheniformis* isolated from Garhwal Himalayan region (Negi et al. 2005), *Bacillus pumilus* (isolated from glacial soils of Thajwas Glacier, Himalayas) (Farooq et al. 2016, 2021), *Bacillus subtilis* (isolated from Malan Glacier, Tibet, China) (Zhang et al. 2001), *Desemzia incerta* (isolated from soil and water of Leh Ladakh and Rohtang pass) (Yadav et al. 2018), *Paenibacillus* sp. (isolated from Kafni glacier, Himalayas) (Kishore et al. 2010), *Pantoea dispersa, Pseudomonas fluorescens, Pseudomonas fragi* (isolated from high altitude garlic rhizosphere), *Rahnella sp*. (isolated from Xinjiang Glacier, China), *Sanguibacter antarcticus* (isolated from sediment on King George Island, Antarctica), and *Stenotrophomonas maltophilia* (isolated from Kanchengayao glacier's, North Sikkim) (Sherpa et al. 2021; Yadav et al. 2019).

Psychotropic endophytic bacteria reside within the internal tissues of their host plants and engage in various types of relationships, including symbiotic, mutualistic, commensalistic, and trophobiotic interactions. They produce metabolites that can be beneficial in medicine, agriculture, or industry, thereby enhancing plant growth, yield, and acting as bio control agents. Psychrophilic endophytes, adapted to cold environments, exhibit fungicidal and bactericidal properties, making them promising bio agents for protecting agricultural products during low-temperature storage (Nadeeva et al. 2023). The soil sample taken from Nevado del Ruiz (Caldas, Colombia) contains psychrophilic bacteria capable of fixing biological nitrogen and exhibiting phosphate-solubilizing activity. This trait holds promise for agricultural and biotechnological uses (Bolanos et al. 2023).

In the cold desert of the Indian Himalayas, researchers have identified *Pseudomonas koreensis*, a bacterium well-regarded for its ability to serve as a phosphate bio fertilizer. This strain has shown proficiency in phosphate solubilization and the production of beneficial PGP compounds like indole acetic acid and siderophores, particularly under low-temperature conditions. When applied to pea plants experiencing cold stress, this bio fertilizer significantly enhances plant growth and productivity, suggesting its potential as a valuable resource for agriculture in cold (Ibáñez et al. 2023).

Two types of actinomycetes, *Streptomyces* sp. and *Micromonospora* sp., were isolated from Antarctic soils. These actinomycetes demonstrate PGP abilities and act as bio control agents. They have shown the capability to inhibit the growth of *Cladosporium cladosporioides*, *Fusarium graminearum*, and *Botrytis cinerea* in laboratory conditions. These fungi are known pathogens responsible for causing plant diseases (Gesheva 2010). Two recent studies have demonstrated that *Streptomyces anulatus* 39 LBG09 and several *Streptomyces* species isolated from Antarctic soils exhibit inhibitory effects against plant-pathogenic bacteria such as

Erwinia amylovora, *Xanthomonas euvesicatoria*, *Burkholderia gladioli*, and *Pseudomonas syringae* pathovar syringae. Among these pathogens, *X. vesicatoria*, *X. euvesicatoria*, and *X. gardneri* are known to cause bacterial spot disease in pepper plants (Dimitrova et al. 2013; Encheva-Malinova et al. 2014, 2015).

Pseudomonas spp. isolated from soils on Greenwich Island (South Shetland Islands, Antarctic Peninsula) were found to possess PGP characteristics such as producing hydrogen cyanide, siderophores, indole-3-acetic acid (IAA), and solubilizing mineral phosphates. These isolates significantly enhanced the growth of *Triticum aestivum* seedlings at 14 °C, resulting in notable increases in both root and shoot length. Moreover, they exhibited inhibitory effects on the growth of three plant pathogens in laboratory tests (*Fusarium oxysporum*, *Phytophthora infestans* and *Pythium ultimum*) (Yarzábal 2020; Yarzábal et al. 2018).

13.5.3 Biomarkers of Climate Change

The significance of global warming on glacier microorganisms is notable because these organisms form the base of food webs and play critical roles in chemical weathering and carbon cycling processes (Garcia-Lopez and Cid 2017). Psychrophilic bacteria are organisms adapted to cold temperatures, capable of withstanding specific temperature ranges before reacting to sudden temperature shifts (S. Kumar et al. 2023; Kumar 2023). Researchers have documented diverse responses to increasing temperatures, including changes in cellular and community dynamics. Key adaptations at the individual level involve increased pigment production, Heat Shock Proteins (HSPs), and modifications in molecular interactions to cope with higher temperatures (Cid et al. 2017; García-Descalzo et al. 2014). Cold shock proteins (CSPs) are a type of protein that regulates transcription and translation in low temperatures, helping to preserve the single-stranded structure of RNA (Siddiqui et al. 2013; S. Kumar 2023; A. Kumar et al. 2009). Psychrophilic bacteria utilize antifreeze proteins (AFPs) to preserve cell membranes and control ice crystal formation, which helps them survive freezing temperatures effectively (Bagwell and Ricker 2019).

HSP, which has a molecular weight of 70 kilodalton (kDa), can act as bioindicators of stressful environments in both temperate and tropical climates (Yu et al. 2015). HSP70 proteins are highly sensitive components of cellular networks involving molecular chaperones and catalysts for protein folding processes. Acting as cellular thermometers, they play a crucial role in responding to various abiotic stresses like heat, salt, drought, low nutrient availability, and chemical exposure. In bacteria, the HSP70 homolog DnaK is instrumental. When cells experience stress, *DnaK* expression increases to facilitate the refolding of heat-damaged and misfolded proteins, which can be toxic to cells (Yu et al. 2015).

Using HSP70s as biomarkers offers a potential advantage in understanding biotic and ecological responses to climate change (Yusof et al. 2022). When organisms face prolonged periods of high ambient temperatures, the overexpression of HSP70 assists in mitigating the adverse effects of climate change (Yusof et al. 2022).

13.5.4 Microbial Carbon Sequestration Agents

Carbon sequestration involves capturing and storing CO_2 over an extended period to reduce greenhouse gas (GHG) emissions. Two main methods have been developed for this purpose, emphasizing long-term storage in geological formations: carbon capture and storage (CCS), and carbon capture and utilization (CCU) (Gayathri et al. 2021). The dominant microbial taxa found in Chinese glaciers—Actinobacteria, Cyanobacteria, and Proteobacteria—play a significant role in carbon sequestration (Cameron et al. 2012; Franzetti et al. 2016; Kong et al. 2019). Key microbial species have the capacity to convert atmospheric CO_2 through diverse pathways and processes. The advantages of carbon sequestration by microbes are manifold, including rapid growth and continuous cultivation in bioreactors, minimal competition for resources, high rates of CO_2 fixation, effective bioremediation of atmospheric CO_2, versatile production capabilities for various additives, ease of genetic modification, and potential applications in industrial bioprocessing (Gayathri et al. 2021).

Clostridium sp., an anaerobic bacterium, plays a crucial role in the carbon cycle, acid production, and breakdown of organic carbon. It utilizes the Wood-Ljungdahl pathway, also known as the Reductive acetyl CoA pathway, as the most direct and efficient thermodynamic method to fix CO_2 by incorporating it into acetyl-CoA. This process derives the necessary energy for CO_2 fixation from hydrogen molecules. Key enzymes involved in this pathway include acetyl CoA synthetase and carbon monoxide dehydrogenase, which catalyze the conversion of carbon monoxide to carbon dioxide, essential steps in acetyl-CoA production (Fast and Papoutsakis 2012; Liew et al. 2016). On the surface of the ice, the dominant members of the heterotrophic community include Bacteroidetes, Actinobacteria, and Proteobacteria (both α and β proteobacteria) (Anesio et al. 2017). Proteobacteria, a diverse phylum, have the capability to sequester carbon dioxide from the atmosphere through various biological pathways such as the Krebs cycle, TCA cycle, Calvin cycle, and reductive pentose phosphate cycle. They can also directly uptake carbon dioxide into their cytoplasm (Ding and Yokota 2004; Gayathri et al. 2021).

Most microbes rely on the breakdown of carbon nutrients for their growth. However, photoautotrophs, including chemoautotrophs, cyanobacteria, and some α-proteobacteria, can produce carbon compounds independently using inorganic carbon sources such as bicarbonate and CO_2 (Cavicchioli 2015; Wietrzyk-Pełka et al. 2020). *Cyanobium* (*Synechococcales*), *Trichormus*, and *Nostoc* (*Nostocales*) utilize light energy in the Calvin-Benson-Bassham (CBB) cycle to fix CO_2 into organic carbon compounds, producing triose phosphates from one molecule of CO_2 and ribulose 1,5-bisphosphate (Li et al. 2019). Certain cyanobacteria can enhance their carbon fixation capacity by upregulating the concentration of CO_2 (Price 2011). In Antarctic soils, chemoautotrophs including Chlorobiales and sulfate-reducing bacteria have been identified (Cavicchioli 2015). These microorganisms can use organic carbon by means of the reductive tricarboxylic acid cycle (rTCA cycle). Through this process, one molecule of oxalacetate is produced from four molecules of CO_2 (Cavicchioli 2015; Ma et al. 2022). Rawat et al. (2020) identified *Chloroflexi* sp. from the Gangotri glacier, *Armatimonadetes* sp. from the Roopkund glacier, and *Nitrospirae* sp. from the Pindari glacier (Rawat et al. 2020). These bacteria have also been found in Antarctica, suggesting the utilization of the reductive acetyl-CoA pathway (WL cycle) and the hydroxypropionate bicycle (3HP cycle) for converting two molecules of CO_2 into acetic acid or glyoxylate (Ma et al. 2022).

13.5.5 Methanogens

The main driver of climate change is CO_2, with methane (CH_4) and nitrous oxide (N_2O) contributing approximately 20% and 7% of the total net radiative forcing in the atmosphere, respectively (Yin et al. 2024). Microbial fuel cell (MFC) technology offers an effective approach for converting various substrates, such as biogas (CH_4), into energy. This technology utilizes gaseous substrates alongside a specialized and enriched microbial community (Kondaveeti et al. 2019). In 2011, Girguis and Reimers published a patent in the United States Patent Application Publication discussing methane-powered microbial fuel cells (Girguis and Reimers 2011). This has spurred researchers to explore the use of MFCs for converting methane (CH_4) into energy via a bio-electrochemical process.

An example of this is reverse methanogenesis, where anaerobic methane-oxidizing bacteria (ANME) can directly oxidize methane into electricity (Kondaveeti et al. 2019). Wood and colleagues investigated the impact of a microbial inoculum extracted from a wastewater treatment facility. This inoculum demonstrated the ability to generate electricity, representing a significant and sustainable energy source. *Methanosarcina acetivorans* plays a crucial role in converting CH_4 to acetate, facilitated by the enzyme Mcr (methylcoenzyme M reductase). Concurrently, *Geobacter sulfurreducens* and other microbial consortia can convert acetate to CO_2 while producing electricity. In experiments using a double-chamber MFC, electroactive *G. sulfurreducens* species, engineered *M. acetivorans*, and sludge resulted in an eight-fold increase in power density (160 mW/m^2 compared to control at 20 mW/m^2). Notably, the MFC fed with CH_4 achieved a 90% columbic efficiency (CE), indicating efficient retrieval of electrons produced during CH_4 conversion to current (Kondaveeti et al. 2019; McAnulty et al. 2017).

Rhodoferax spp., *Shewanella*, and *Geobacter* are examples of microorganisms capable of generating electricity from various organic substances. For instance, *G. sulfurreducens* utilizes hydrogen and acetate for electricity production (Bond and Lovley 2003), *S. putrefaciens* employs lactate and pyruvate (Kim et al. 2002), and *R. ferrireducens* generates electricity from glucose. Electrons transfer through mechanisms such as pili (nanowires), multi-heme complexes in cell membranes, or microbial-produced molecules (McAnulty et al. 2017; Yamasaki et al. 2018).

The two-stage MFC-AFMBR (Anaerobic Fluidized Bed Membrane Bioreactor) technology proves more efficient in treating domestic wastewater. This system typically comprises four MFCs and one AFBR operating at room temperature. MFCs directly generate electricity, whereas in AFBR, electricity production involves separate generation from biogas. This biogas may require purification to remove water and hydrogen sulfide, enhancing overall efficiency (Ren et al. 2014).

13.6 Conclusion

Glaciers serve as a significant habitat for microbial communities, comprising species that indicate biological conditions and make up 70% of the hydrosphere. The melting of glaciers due to global climate change and greenhouse gas effects is an issue of extensive debate, particularly in relation to local weather changes. Psychrophiles and psychrotolerants found in glaciers can provide valuable insights into the link between glacier melt and climate change. Climate change and global warming pose critical challenges to the environment today, evident in the Himalayan region where glaciers are already retreating rapidly. This retreat has led to the disappearance of numerous smaller glaciers. Harsh conditions such as low temperatures, UV radiation, katabatic winds, and limited water availability impact biotic systems within glacier habitats, which are primarily driven by microbial communities at the trophic level. Given that glaciers are crucial freshwater sources for Himalayan communities, their health is of utmost importance. Furthermore, psychrophilic microorganisms inhabiting glaciers and cryo-habitats have the potential to produce cryoenzymes with valuable ecological and biotechnological properties.

References

Abdullah T, Romshoo SA, Rashid I (2020) The satellite observed glacier mass changes over the Upper Indus Basin during 2000–2012. Sci Rep 10(1):14285. https://doi.org/10.1038/s41598-020-71281-7

Adhikari NP, Adhikari S, Liu X, Shen L, Gu Z (2019) Bacterial diversity in alpine lakes: a review from the third pole region. J Earth Sci 30(2):387–396. https://doi.org/10.1007/s12583-018-1206-5

Alvarado-Cuevas ZD, López-Hidalgo AM, Ordoñez LG, Oceguera-Contreras E, Ornelas-Salas JT, De León-Rodríguez A (2015) Biohydrogen production using psychrophilic bacteria isolated from Antarctica. Int J Hydrog Energy 40(24):7586–7592. https://doi.org/10.1016/j.ijhydene.2014.10.063

Alvarez HM, Hernández MA, Lanfranconi MP, Silva RA, Villalba MS (2021) Rhodococcus as biofactories for microbial oil production. Molecules 26(16):4871. https://doi.org/10.3390/molecules26164871

Alvarez-Guzmán CL, Balderas-Hernández VE, De Leon-Rodriguez A (2020) Coproduction of hydrogen, ethanol and 2,3-butanediol from agro-industrial residues by the Antarctic psychrophilic GA0F bacterium. Int J Hydrog Energy 45(49):26179–26187. https://doi.org/10.1016/j.ijhydene.2020.02.105

Amaretti A, Raimondi S, Sala M, Roncaglia L, De Lucia M, Leonardi A, Rossi M (2010) Single cell oils of the cold-adapted oleaginous yeast Rhodotorula glacialis DBVPG 4785. Microb Cell Factories 9:73. https://doi.org/10.1186/1475-2859-9-73

Anesio AM, Laybourn-Parry J (2012) Glaciers and ice sheets as a biome. Trends Ecol Evol 27(4):219–225. https://doi.org/10.1016/j.tree.2011.09.012

Anesio AM, Lutz S, Chrismas NAM, Benning LG (2017) The microbiome of glaciers and ice sheets. NPJ Biofilms Microbiomes 3(1):10. https://doi.org/10.1038/s41522-017-0019-0

Aschenbach K, Conrad R, Řeháková K, Doležal J, Janatková K, Angel R (2013) Methanogens at the top of the world: occurrence and potential activity of methanogens in newly deglaciated soils in high-altitude cold deserts in the Western Himalayas. Front Microbiol 4. https://doi.org/10.3389/fmicb.2013.00359

Baek K, Lee YM, Shin SC, Hwang K, Hwang CY, Hong SG, Lee HK (2015) Halocynthiibacter arcticus sp. nov., isolated from Arctic marine sediment. Int J Syst Evol Microbiol 65(Pt_11):3861–3865. https://doi.org/10.1099/ijsem.0.000507

Bagwell SN, Ricker JV (2019) Antifreeze proteins. Biol Forum 90(3):158–170. https://doi.org/10.2307/26867856

Bahuguna IM, Rathore BP, Brahmbhatt R, Sharma M, Dhar S, Randhawa SS, Kumar K, Romshoo S, Shah RD, Ganjoo RK (2014) Are the Himalayan glaciers retreating. Curr Sci 106(7):1008–1013

Bajracharya SR, Mool PK, Shrestha BR (2007) Impact of climate change on Himalayan Glaciers and Glacial Lakes: case studies on GLOF and associated hazards in Nepal and Bhutan. ICIMOD, Kathmandu

Bakermans C, Ayala-del-Río HL, Ponder MA, Vishnivetskaya T, Gilichinsky D, Thomashow MF, Tiedje JM (2006) Psychrobacter cryohalolentis sp. nov. and Psychrobacter arcticus sp. nov., isolated from Siberian permafrost. Int J Syst Evol Microbiol 56(6):1285–1291. https://doi.org/10.1099/ijs.0.64043-0

Banerjee A (2017) Brief communication: thinning of debris-covered and debris-free glaciers in a warming climate. Cryosphere 11(1):133–138. https://doi.org/10.5194/tc-11-133-2017

Banerjee S, Black R, Kniveton D, Kollmair M (2014) The changing Hindu Kush Himalayas: environmental change and migration. In: Global migration issues, pp 205–227. https://doi.org/10.1007/978-94-007-6985-4_9

Barnard D, Casanueva A, Tuffin M, Cowan D (2010) Extremophiles in biofuel synthesis. Environ Technol 31(8–9):871–888. https://doi.org/10.1080/09593331003710236

Bibra M, Wang J, Squillace P, Pinkelman, Rebecca J, Papendick S, Schneiderman S, Braband VW, Amar, Vinod S, Kumar S, Salem D, Sani R (2015) Biofuels and value-added products from extremophiles. In: Advances in biotechnology. I.K. International Publishing House, New Delhi

Boetiu A, Anesio AM, Deming JW, Mikucki JA, Rapp JZ (2015) Microbial ecology of the cryosphere: sea ice and glacial habitats. Nature reviews. Microbiology 13(11):677–690. https://doi.org/10.1038/nrmicro3522

Bolanos J, Buitrago J, Leal M, Tovar D, Ruiz E, Sanchez J (2023) Characterization of cultivable psychrophilic bacteria with phosphate-soluble activity and nitrogen fixation capacity, present in sediments of the Nevado Del Ruiz (Caldas, Colombia). In: III Congreso Latinoamericano de Astrobiolog´ Ia (2021), p 6567

Bond DR, Lovley DR (2003) Electricity production by *Geobacter sulfurreducens* attached to electrodes. Appl Environ Microbiol 69(3):1548–1555. https://doi.org/10.1128/AEM.69.3.1548-1555.2003

Borah SB, Das AK, Hazarika N, Basumatary H (2022) Monitoring and assessment of glaciers and glacial lakes: climate change impact on the Mago Chu Basin, Eastern Himalayas. Reg Environ Chang 22(4). https://doi.org/10.1007/s10113-022-01984-2

Bozal N, Montes MJ, Tudela E, Jiménez F, Guinea J (2002) Shewanella frigidimarina and Shewanella livingstonensis sp. nov. isolated from Antarctic coastal areas. Int J Syst Evol Microbiol 52(1):195–205. https://doi.org/10.1099/00207713-52-1-195

Bradley JA, Anesio AM, Singarayer JS et al (2015) SHIMMER (1.0): a novel mathematical model for microbial and biogeochemical dynamics in glacier forefield ecosystems. Geosci Model Dev Discuss 8:6143–6216

Bradley JA, Trivedi CB, Winkel M, Mourot R, Lutz S, Larose C, Keuschnig C, Doting E, Halbach L, Zervas A, Anesio AM, Benning LG (2023) Active and dormant microorganisms on glacier surfaces. Geobiology 21(2):244–261. https://doi.org/10.1111/gbi.12535

Brankatschk R, Töwe S, Kleineidam K et al (2011) Abundances and potential activities of nitrogen cycling microbial communities along a chronosequence of a glacier forefield. ISME J 5:1025–1037. https://doi.org/10.1038/ismej.2010.184

Cameron KA, Hodson AJ, Osborn AM (2012) Carbon and nitrogen biogeochemical cycling potentials of supraglacial cryoconite communities. Polar Biol 35(9):1375–1393. https://doi.org/10.1007/s00300-012-1178-3

Campen RK, Sowers T, Alley RB (2003) Evidence of microbial consortia metabolizing within a low-latitude mountain glacier. Geology 31:231–234
Carpenter EJ, Lin S, Capone DG (2000) Bacterial activity in South Pole snow. Appl Environ Microbiol 66(10):4514–4517. https://doi.org/10.1128/AEM.66.10.4514-4517.2000
Carrasco M, Alcaíno J, Cifuentes V, Baeza M (2017) Purification and characterization of a novel cold adapted fungal glucoamylase. Microb Cell Factories 16(1):75. https://doi.org/10.1186/s12934-017-0693-x
Carrasco M, Rozas JM, Alcaíno J, Cifuentes V, Baeza M (2019) Pectinase secreted by psychrotolerant fungi: identification, molecular characterization and heterologous expression of a cold-active polygalacturonase from Tetracladium sp. Microb Cell Factories 18(1):45. https://doi.org/10.1186/s12934-019-1092-2
Cauvy-Fraunié S, Dangles O (2019) A global synthesis of biodiversity responses to glacier retreat. Nat Ecol Evol 3(12):1675–1685. https://doi.org/10.1038/s41559-019-1042-8
Cavicchioli R (2015) Microbial ecology of Antarctic aquatic systems. Nat Rev Microbiol 13(11):691–706. https://doi.org/10.1038/nrmicro3549
Chand P, Jain SK, Thakur HP, Kumar S, Sharma MC (2020) Recessional pattern and surface elevation change of the Parvati Glacier, North-Western Himalaya (1965–2018) using remote sensing. Int J Remote Sens 41(24):9360–9392. https://doi.org/10.1080/01431161.2020.1798552
Chettri N, Shakya B, Thapa R, Sharma E (2008) Status of a protected area system in the Hindu Kush-Himalayas: an analysis of PA coverage. Int J Biodivers Sci Manag 4(3):164–178. https://doi.org/10.3843/Biodiv.4.3:4
Chifflard P, Fasching C, Reiss M, Ditzel L, Boodoo KS (2019) Dissolved and particulate organic carbon in Icelandic proglacial streams: a first estimate. Water 11(4):748. https://doi.org/10.3390/w11040748
Ciok A, Budzik K, Zdanowski MK, Gawor J, Grzesiak J, Decewicz P, Gromadka R, Bartosik D, Dziewit L (2018) Plasmids of Psychrotolerant Polaromonas spp. Isolated From Arctic and Antarctic Glaciers - Diversity and Role in Adaptation to Polar Environments. Frontiers in microbiology 9:1285. https://doi.org/10.3389/fmicb.2018.01285
Crespim E, Zanphorlin LM, de Souza FHM, Diogo JA, Gazolla AC, Machado CB, Figueiredo F, Sousa AS, Nóbrega F, Pellizari VH, Murakami MT, Ruller R (2016) A novel cold-adapted and glucose-tolerant GH1 β-glucosidase from Exiguobacterium antarcticum B7. Int J Biol Macromol 82:375–380. https://doi.org/10.1016/j.ijbiomac.2015.09.018
Debnath M, Sharma MC, Syiemlieh HJ (2019) Glacier dynamics in Changme Khangpu Basin, Sikkim Himalaya, India, between 1975 and 2016. Geosciences 9(6):259. https://doi.org/10.3390/geosciences9060259
de Lemos EA, Procópio L, da Mota FF, Jurelevicius D, Rosado AS, Seldin L (2023) Molecular characterization of Paenibacillus antarcticus IPAC21, a bioemulsifier producer isolated from Antarctic soil. Front Microbiol 14:1142582. https://doi.org/10.3389/fmicb.2023.1142582
De Pascale D, Di Prisco G, Marino G, Luisa Tutino M (2009) Cold-adapted Esterases and lipases: from fundamentals to application. Protein Pept Lett 16(10):1172–1180. https://doi.org/10.2174/092986609789071270
Dimitrova D, Dorkov P, Gocheva B (2013) Antibiotic complex, produced by an antarctic actinomycete strain S. anulatus 39 LBG09. Bulgarian J Agr Sci 19(2):72–76
Ding L, Yokota A (2004) Proposals of Curvibacter gracilis gen. nov., sp. nov. and Herbaspirillum putei sp. nov. for bacterial strains isolated from well water and reclassification of [Pseudomonas] huttiensis, [Pseudomonas] lanceolata, [Aquaspirillum] delicatum and [Aquaspirillum] autotrophicum as Herbaspirillum huttiense comb. nov., Curvibacter lanceolatus comb. nov., Curvibacter delicatus comb. nov. and Herbaspirillum autotrophicum comb. nov. International journal of systematic and evolutionary microbiology 54(6):2223–2230. https://doi.org/10.1099/ijs.0.02975-0
Edwards A, Pachebat JA, Swain M, Hegarty M, Hodson AJ, Irvine-Fynn TD, Rassner SM, Sattler B (2013) A metagenomic snapshot of taxonomic and functional diversity in an alpine glacier cryoconite ecosystem. Environ Res Lett 8:035003. https://doi.org/10.1088/1748-9326/8/3/035003

Elshahed MS (2010) Microbiological aspects of biofuel production: current status and future directions. J Adv Res 1(2):103–111. https://doi.org/10.1016/j.jare.2010.03.001

Encheva-Malinova M, Stoyanova M, Avramova H, Pavlova Y, Gocheva B, Ivanova I, Moncheva P (2014) Antibacterial potential of streptomycete strains from Antarctic soils. Biotechnol Biotechnol Equip 28(4):721–727. https://doi.org/10.1080/13102818.2014.947066

Encheva-Malinova M, Vancheva T, Badzhinerov N, Koleva V, Tishkov S, Bogatzevska N, Moncheva P (2015) Antimicrobial activity of antarctic streptomycetes against pepper bacterial spot causing agents. Annuaire De L'université De Sofia "st Kliment Ohridski" Faculte De Biologie 100:216–222

Farinotti D, Immerzeel WW, de Kok RJ, Quincey DJ, Dehecq A (2020) Manifestations and mechanisms of the Karakoram glacier Anomaly. Nat Geosci 13:8–16. https://doi.org/10.1038/s41561-019-0513-5

Fast AG, Papoutsakis ET (2012) Stoichiometric and energetic analyses of non-photosynthetic CO2-fixation pathways to support synthetic biology strategies for production of fuels and chemicals. Curr. Opin. Chem. Eng. 1:380–395

Farooq S, Nazir R, Uqab B (2016) Psychrophiles: their habitat and applications. J Himalayan Ecol Sustain Dev:27–36

Farooq S, Nazir R, Ganai SA, Ganai BA (2021) Isolation and characterization of a new cold-active protease from psychrotrophic bacteria of Western Himalayan glacial soil. Sci Rep 11(1):12768. https://doi.org/10.1038/s41598-021-92197-w

Filho WL, Wall T, Salvia AL, Dinis MAP, Mifsud M (2023) The central role of climate action in achieving the United Nations' Sustainable Development Goals. Sci Rep 13(1):20582. https://doi.org/10.1038/s41598-023-47746-w

Fluckiger J (2003) Ph.D. thesis (Univ. of Bern, Bern, Switzerland)

Franzetti A, Tagliaferri I, Gandolfi I, Bestetti G, Minora U, Mayer C, Azzoni RS, Diolaiuti G, Smiraglia C, Ambrosini R (2016) Light-dependent microbial metabolisms drive carbon fluxes on glacier surfaces. ISME J 10(12):2984–2988. https://doi.org/10.1038/ismej.2016.72

García-Descalzo L, García-López E, Alcázar A, Baquero F, Cid C (2014) Proteomic analysis of the adaptation to warming in the Antarctic bacteria Shewanella frigidimarina. Biochim Biophys Acta Proteins Proteomics 1844(12):2229–2240. https://doi.org/10.1016/j.bbapap.2014.08.006

Garcia-Lopez E, Alcazar P, Postigo M, Cid C (2016) "The effect of climate change on microbial communities from glaciers," in Glaciers: Formation, Climate Change and Their Effect, ed. Nova Science Publishers, Inc. (New York, NY: Nova Science Publishers, Inc 71–88

Garcia-Lopez E, Cid C (2017) Glaciers and ice sheets as analog environments of potentially habitable icy worlds. Front Microbiol 8,1407. https://doi.org/10.3389/fmicb.2017.01407

Garg PK, Shukla A, Jasrotia AS (2019) On the strongly imbalanced state of glaciers in the Sikkim, eastern Himalaya, India. Sci Total Environ 691:16–35. https://doi.org/10.1016/j.scitotenv.2019.07.086

Gayathri R, Mahboob S, Govindarajan M, Al-Ghanim KA, Ahmed Z, Al-Mulhm N, Vodovnik M, Vijayalakshmi S (2021) A review on biological carbon sequestration: a sustainable solution for a cleaner air environment, less pollution and lower health risks. J King Saud Univ Sci 33(2):101282. https://doi.org/10.1016/j.jksus.2020.101282

Gesheva V (2010) Production of antibiotics and enzymes by soil microorganisms from the windmill islands region, Wilkes Land, East Antarctica. Polar Biol 33(10):1351–1357. https://doi.org/10.1007/s00300-010-0824-x

Girguis P, Reimers CE (2011) Patent MFC. United States Patent Application Publication. https://patents.google.com/patent/US20110123835

GLIMS (2024) Randolph glacier inventory 7.0. Global Land Ice Measurements from Space Monitoring the World's Changing Glaciers

Goordial J, Raymond-Bouchard I, Ronholm J, Shapiro N, Woyke T, Whyte L, Bakermans C (2015) Improved-high-quality draft genome sequence of Rhodococcus sp. JG-3, a eurypsychrophilic Actinobacteria from Antarctic Dry Valley permafrost. Stand Genomic Sci 10(1):61. https://doi.org/10.1186/s40793-015-0043-8

Guido Z, McIntosh JC, Papuga SA, Meixner T (2016) Seasonal glacial meltwater contributions to surface water in the Bolivian Andes: a case study using environmental tracers. J Hydrol Reg Stud 8:260–273. https://doi.org/10.1016/j.ejrh.2016.10.002

Guo Y, Zhang S (2023) Potential effects of methane metabolic microbial communities on the glacial methane budget in the Northwestern Tibetan Plateau. Sustain For 15(9):7352. https://doi.org/10.3390/su15097352

Hancock L (2024) Why are glaciers and sea ice melting? World Wildlife Fund

Hassan N (2019) Analysis of the cell membrane fatty acids and characterization of the psychrotolerant serratia marcescens hi6 isolated from Hopar (Bualtar) glacier, Pakistan. Appl Ecol Environ Res 17(5):10.15666/aeer/1705_1191111924

Hell K, Edwards A, Zarsky J, Podmirseg SM, Girdwood S, Pachebat JA, Insam H, Sattler B (2013) The dynamic bacterial communities of a melting High Arctic glacier snowpack. ISME J 7:1814–1826. https://doi.org/10.1038/ismej.2013.51

Hock R (2005) Glacier melt: a review of processes and their modelling. Prog Phys Geogr 29(3):362–339

Hodson A, Anesio AM, Tranter M, Fountain A, Osborn M, Priscu J, Laybourn-Parry J, Sattler B (2008) Glacial ecosystems. Ecol Monogr 78(1):41–67. https://doi.org/10.1890/07-0187.1

Holder L, Duffy M, Opdyke B, Leventer A, Post A, O'brien P, Armand LK (2020) Controls since the mid-Pleistocene transition on sedimentation and primary productivity downslope of Totten Glacier, East Antarctica. Paleoceanography and Paleoclimatology 35(12):e2020PA003981

Hollmann H (2014) Seismic investigations of firn in Antarctica. PhD diss., University of Tasmania

Hong SG, Lee YK, Yim JH, Chun J, Lee HK (2008) Sanguibacter antarcticus sp. nov., isolated from Antarctic Sea sand. Int J Syst Evol Microbiol 58(Pt 1):50–52. https://doi.org/10.1099/ijs.0.65031-0

Hugonnet R, McNabb R, Berthier E, Menounos B, Nuth C, Girod L, Farinotti D, Huss M, Dussaillant I, Brun F, Kääb A (2021) Accelerated global glacier mass loss in the early twenty-first century. Nature 592(7856):726–731. https://doi.org/10.1038/s41586-021-03436-z

Ibáñez A, Garrido-Chamorro S, Vasco-Cárdenas M, Barreiro C (2023) From lab to field: biofertilizers in the 21st century. Horticulturae 9(12):1306. https://doi.org/10.3390/horticulturae9121306

Immerzeel WW, Lutz AF, Andrade M, Bahl A, Biemans H, Bolch T, Hyde S, Brumby S, Davies BJ, Elmore AC, Emmer A, Feng M, Fernández A, Haritashya U, Kargel JS, Koppes M, Kraaijenbrink PDA, Kulkarni AV, Mayewski PA, Baillie JEM (2020) Importance and vulnerability of the world's water towers. Nature 577(7790):364–369. https://doi.org/10.1038/s41586-019-1822-y

IPCC (2019) Global warming of 1.5°C: An IPCC Special Report on the impacts of global warming of 1.5°C above pre-industrial levels and related global greenhouse gas emission pathways, in the context of strengthening the global response to the threat of climate change, sustainable development, and efforts to eradicate poverty. In: Delmotte VM, Pörtner H-O, Skea J, Pirani A, Roz P, Yang C, Lonnoy E, Zhai P, Roberts D, Shukla PR, Péan C, Matthews JBR, Gomis MI, Tim W, Okia WM, Connors S, Zhou X, Maycock T, Tignor M (eds) Global warming of 1.5°C. In Press. https://www.ipcc.ch/sr15/

Irvine-Fynn TD, Hodson AJ, Moorman BJ, Vatne G, Hubbard AL (2011) Polythermal glacier hydrology: a review. Rev Geophys 49(4). https://doi.org/10.1029/2010RG000350

Jin Z, Yu J, Zhang F, Qiang X (2020) Glacial-interglacial variation in catchment weathering and erosion paces the Indian summer monsoon during the Pleistocene. Quaternary Science Reviews 248:106619

Key JR, Liu Y, Wang X, Letterly A, Painter TH (2020) Snow and ice products from ABI on the GOES-R series. In: The GOES-R series. Elsevier, pp 165–177. https://doi.org/10.1016/B978-0-12-814327-8.00014-7

Kim HJ, Park HS, Hyun MS, Chang IS, Kim M, Kim BH (2002) A mediator-less microbial fuel cell using a metal reducing bacterium, Shewanella putrefaciens. Enzyme Microb Technol 30(2):145–152. https://doi.org/10.1016/S0141-0229(01)00478-1

Kim EJ, Kim S, Choi H-G, Han SJ (2020) Co-production of biodiesel and bioethanol using psychrophilic microalga Chlamydomonas sp. KNM0029C isolated from Arctic sea ice. Biotechnol Biofuels 13(1):20. https://doi.org/10.1186/s13068-020-1660-z

Kim S, Lee H, Hur S-D, Sul WJ, Kim O-S (2022) Glaciers as microbial habitats: current knowledge and implication. J Microbiol 60(8):767–779. https://doi.org/10.1007/s12275-022-2275-9

Kishore KH, Begum Z, Pathan AAK, Shivaji S (2010) Paenibacillus glacialis sp. nov., isolated from the Kafni glacier of the Himalayas, India. Int J Syst Evol Microbiol 60(8):1909–1913. https://doi.org/10.1099/ijs.0.015271-0

Kondaveeti S, Mohanakrishna G, Lee J-K, Kalia VC (2019) Methane as a substrate for energy generation using microbial fuel cells. Indian J Microbiol 59(1):121–124. https://doi.org/10.1007/s12088-018-0765-6

Kong W, Liu J, Ji M, Yue L, Kang S, Morgan-Kiss RM (2019) Autotrophic microbial community succession from glacier terminus to downstream waters on the Tibetan Plateau. FEMS Microbiol Ecol 95(6). https://doi.org/10.1093/femsec/fiz074

Kraaijenbrink PDA, Stigter EE, Yao T, Immerzeel WW (2021) Climate change decisive for Asia's snow meltwater supply. Nat Clim Chang 11(7):591–597. https://doi.org/10.1038/s41558-021-01074-x

Kulkarni AV, Karyakarte Y (2014) Observed changes in Himalaya glaciers. Curr Sci 106(2):237244. https://www.jstor.org/stable/24099804

Kulkarni AV, Bahuguna IM, Rathore BP, Singh SK, Randhawa SS, Sood RK, Dhar S (2007) Glacial retreat in Himalaya using Indian remote sensing satellite data. Curr Sci 92(1):69–74. https://www.jstor.org/stable/24096824

Kumar S (2023) Antibiotic susceptibility profile in bacteria isolated from different natural thermal conditions of Sikkim (Doctoral dissertation). http://dspace.cus.ac.in/jspui/handle/1/7896

Kumar P, Sharma MC (2023) Frontal changes in medium-sized glaciers in Sikkim, India during 1988–2018: insights for glacier-climate synthesis over the Himalaya. iScience 26(10):107789. https://doi.org/10.1016/j.isci.2023.107789

Kumar A, Singh S, Poddar P, Prabhune A, Pundle A (2009) Effect of cultural conditions and media constituents on production of penicillin V Acylase and CTAB treatment to enhance whole-cell enzyme activity of Rhodotorula aurantiaca (NCIM 3425). Appl Biochem Biotechnol 157(3):463–472. https://doi.org/10.1007/s12010-008-8294-0

Kumar S, Najar IN, Sharma P, Tamang S, Mondal K, Das S, Sherpa MT, Thakur N (2023) Temperature–A critical abiotic paradigm that governs bacterial heterogeneity in natural ecological system. Environ Res 234:116547. https://doi.org/10.1016/j.envres.2023.116547

Le LTHL, Yoo W, Jeon S, Lee C, Kim KK, Lee JH, Kim TD (2020) Biodiesel and flavor compound production using a novel promiscuous cold-adapted SGNH-type lipase (HaSGNH1) from the psychrophilic bacterium *Halocynthiibacter arcticus*. Biotechnol Biofuels 13(1):55. https://doi.org/10.1186/s13068-020-01696-x

Lebert I, Begot C, Lebert A (1998) Growth of Pseudomonas fluorescens and Pseudomonas fragi in a meat medium as affected by pH (5.8-7.0), water activity (0.97-1.00) and temperature (7-25 degrees C). Int J Food Microbiol 39(1–2):53–60. https://doi.org/10.1016/s0168-1605(97)00116-5. PMID: 9562876

Lee E, Carrivick JL, Quincey DJ, Cook SJ, James WHM, Brown LE (2021) Accelerated mass loss of Himalayan glaciers since the Little Ice Age. Sci Rep 11(1):24284. https://doi.org/10.1038/s41598-021-03805-8

Lepp AP, Simkins LM, Anderson JB, Clark RW, Wellner JS, Hillenbrand CD, Smith JA, Lehrmann AA, Totten R, Larter RD, Hogan KA (2022) Sedimentary signatures of persistent subglacial meltwater drainage from Thwaites Glacier, Antarctica. Front Earth Sci 10:863200. https://doi.org/10.3389/feart.2022.863200

Li YJ, Ding YJ, Shangguan DH, Wang RJ (2019) Regional differences in global glacier retreat from 1980 to 2015. Advances in Climate Change Research 10(4):203–213

Liew F, Henstra AM, Winzer K, Köpke M, Simpson SD, Minton NP (2016) Insights into CO_2 fixation pathway of *Clostridium autoethanogenum* by targeted mutagenesis. MBio 7(3). https://doi.org/10.1128/mBio.00427-16

Lin L, Huang H, Zhang X, Dong L, Chen Y (2022) Hydrogen-oxidizing bacteria and their applications in resource recovery and pollutant removal. Sci Total Environ 835:155559. https://doi.org/10.1016/j.scitotenv.2022.155559
Liu Q, Xin Y-H, Chen X-L, Liu H-C, Zhou Y-G, Chen W-X (2018) Arthrobacter ruber sp. nov., isolated from glacier ice. Int J Syst Evol Microbiol 68(5):1616–1621. https://doi.org/10.1099/ijsem.0.002719
Liu S, Proudman J, Mitloehner FM (2021) Rethinking methane from animal agriculture. CABI Agric Biosci 2(1):22. https://doi.org/10.1186/s43170-021-00041-y
Livingstone SJ, Li Y, Rutishauser A, Sanderson RJ, Winter K, Mikucki JA, Björnsson H, Bowling JS, Chu W, Dow CF, Fricker HA (2022) Subglacial lakes and their changing role in a warming climate. Nat Rev Earth Environ 3(2):106–124
Lopatina A, Krylenkov V, Severinov K (2013) Activity and bacterial diversity of snow around Russian Antarctic stations. Res Microbiol 164:949–958. https://doi.org/10.1016/j.resmic.2013.08.005
Loveland-Curtze J, Miteva V, Brenchley J (2010) Novel ultramicrobacterial isolates from a deep Greenland ice core represent a proposed new species, *Chryseobacterium greenlandense* sp. Nov. Extremophiles 14:61–69. https://doi.org/10.1007/s00792-009-0287-6
Ma A, Zhang J, Liu G, Zhuang X, Zhuang G (2022) Cryosphere microbiome biobanks for mountain glaciers in China. Sustain For 14(5):2903. https://doi.org/10.3390/su14052903
Mar KA, Unger C, Walderdorff L, Butler T (2022) Beyond CO2 equivalence: the impacts of methane on climate, ecosystems, and health. Environ Sci Pol 134:127–136. https://doi.org/10.1016/j.envsci.2022.03.027
Matsuoka K (2020) Life of the Antarctic ice: from deep inland to the coast. Norwegian Polar Institute
McAnulty MJ, Poosarla VG, Kim K-Y, Jasso-Chávez R, Logan BE, Wood TK (2017) Electricity from methane by reversing methanogenesis. Nat Commun 8(1):15419. https://doi.org/10.1038/ncomms15419
Meusel M, Rehm HJ (1993) Biodegradation of dichloroacetic acid by freely suspended and adsorptive immobilized Xanthobacter autotrophicus GJ10 in soil. Appl Microbiol Biotechnol 40:165–171. https://doi.org/10.1007/BF00170446
Miteva V, Teacher C, Sowers T, Brenchley J (2009) Comparison of microbial diversity at different depth of the GISP2 Greenland ice core in relation to deposition climate. Environ Microbiol 11:640–656. https://doi.org/10.1111/j.1462-2920.2008.01835.x
Mogili NV, Murugesan N, Ayothiraman S, Gautam R, Deshavath NN, Reddy Erva R (2022) Biohydrogen production from wastewater and organic solid wastes. In: Waste-to-energy approaches towards zero waste. Elsevier, pp 165–195. https://doi.org/10.1016/B978-0-323-85387-3.00009-4
Mohammed A, Abdul-Wahab MF, Hashim M, Omar AH, Md Reba MN, Muhamad Said MF, Soeed K, Alias SA, Smykla J, Abba M, Ibrahim Z (2018) Biohydrogen production by Antarctic Psychrotolerant *Klebsiella* sp. ABZ11. Pol J Microbiol 67(3):283–290. https://doi.org/10.21307/pjm-2018-033
Møller AK, Søborg DA, Al-Soud WA, Sørensen SJ, Kroer N (2013) Bacterial community structure in High-Arctic snow and freshwater as revealed by pyrosequencing of 16S rRNA genes and cultivation. Polar Res 32:17390. https://doi.org/10.3402/polar.v32i0.17390
Moyer CL, Collins RE, Morita RY (2017) Psychrophiles and psychrotrophs. Elsevier. https://doi.org/10.1016/b978-0-12-809633-8.02282-2
Mukhia S, Kumar A, Kumar R (2024) Bacterial community distribution and functional potentials provide key insights into their role in the ecosystem functioning of a retreating Eastern Himalayan glacier. FEMS microbiology ecology 100(3):012. https://doi.org/10.1093/femsec/fiae012
Nadeeva G, Ionova N, Kravtsova O, Bagaeva T (2023) Psychotropic endophytes of cereals as potential bioagents. BIO Web Conf 67:02031. https://doi.org/10.1051/bioconf/20236702031
NASA (2024) What is climate change? The National Aeronautics and Space Administration
Nash MV, Anesio AM, Barker G, Tranter M, Varliero G, Eloe-Fadrosh EA, Nielsen T, Turpin-Jelfs T, Benning LG, Sánchez-Baracaldo P (2018) Metagenomic insights into diazotrophic com-

munities across Arctic glacier forefields. FEMS microbiology ecology 94(9):114. https://doi.org/10.1093/femsec/fiy114
National Geographic Society (2023) Climate change. National Geography
National Research Council (2012) Himalayan glaciers: climate change, water resources, and water security. The National Academies Press, Washington, DC. https://doi.org/10.17226/13449
Negi YK, Garg SK, Kumar J (2005) Cold-tolerant fluorescent Pseudomonas isolates from Garhwal Himalayas as potential plant growth promoting and biocontrol agents in pea. Current Science Association, pp 2151–2155
NOAA (2023) What is the cryosphere? National Oceanic and Atmospheric Administration (NOAA)
Novototskaya-Vlasova K, Petrovskaya L, Yakimov S, Gilichinsky D (2012) Cloning, purification, and characterization of a cold-adapted esterase produced by *Psychrobacter cryohalolentis* K5 ^{T} from Siberian cryopeg. FEMS Microbiol Ecol 82(2):367–375. https://doi.org/10.1111/j.1574-6941.2012.01385.x
Núñez-Montero K, Salazar R, Santos A, Gómez-Espinoza O, Farah S, Troncoso C, Hoffmann C, Melivilu D, Scott F, Barrientos Díaz L (2021) Antarctic Rahnella inusitata: a producer of cold-stable β-galactosidase enzymes. Int J Mol Sci 22(8):4144. https://doi.org/10.3390/ijms22084144
Pander B, Mortimer Z, Woods C, McGregor C, Dempster A, Thomas L, Maliepaard J, Mansfield R, Rowe P, Krabben P (2020) Hydrogen oxidising bacteria for production of single cell protein and other food and feed ingredients. Eng Biol 4(2):21–24. https://doi.org/10.1049/enb.2020.0005
Phukon LC, Chourasia R, Kumari M, Godan TK, Sahoo D, Parameswaran B, Rai AK (2020) Production and characterisation of lipase for application in detergent industry from a novel Pseudomonas helmanticensis HS6. Bioresource technology 309:123352. https://doi.org/10.1016/j.biortech.2020.123352
Pitcher LH, Smith LC (2019) Supraglacial streams and rivers. Annu Rev Earth Planet Sci 47:421–452
Price GD (2011) Inorganic carbon transporters of the cyanobacterial CO_2 concentrating mechanism. Photosynth Res 109(1–3):47–57. https://doi.org/10.1007/s11120-010-9608-y
Price PB, Sowers T (2004) Temperature dependence of metabolic rates for microbial growth, maintenance, and survival. Proc Natl Acad Sci USA 101:4631–4636
Qin D, Ding Y, Xiao C, Kang S, Ren J, Yang J, Zhang S (2018) Cryospheric science: research framework and disciplinary system. Natl Sci Rev 5(2):255–268. https://doi.org/10.1093/nsr/nwx108
Radhakrishnan R, Park JM, Lee IJ (2016) Enterobacter sp. I-3, a bio-herbicide inhibits gibberellins biosynthetic pathway and regulates abscisic acid and amino acids synthesis to control plant growth. Microbiological research 193:132–139. https://doi.org/10.1016/j.micres.2016.10.004
Radhakrishnan R, Hashem A (2017) Bacillus: a biological tool for crop improvement through bio-molecular changes in adverse environments. Front Physiol 8. https://doi.org/10.3389/fphys.2017.00667
Rajawat MVS, Singh R, Singh D, Saxena AK (2019) Psychrotrophs of the genus Janthinobacterium with potential to weather potassium aluminosilicate mineral. 3 Biotech 9(4):142. https://doi.org/10.1007/s13205-019-1672-1
Rawat J, Yadav N, Pande V (2020) Role of rhizospheric microbial diversity in plant growth promotion in maintaining the sustainable agrosystem at high altitude regions. In: Recent advancements in microbial diversity. Elsevier, pp 147–196. https://doi.org/10.1016/B978-0-12-821265-3.00007-4
Ren L, Ahn Y, Logan BE (2014) A two-stage microbial fuel cell and anaerobic fluidized bed membrane bioreactor (MFC-AFMBR) system for effective domestic wastewater treatment. Environ Sci Technol 48(7):4199–4206. https://doi.org/10.1021/es500737m
Rial RC (2024) Biofuels versus climate change: exploring potentials and challenges in the energy transition. Renew Sust Energ Rev 196:114369. https://doi.org/10.1016/j.rser.2024.114369
Rizvi A, Ahmed B, Khan MS, Umar S, Lee J (2021) Psychrophilic bacterial phosphate-biofertilizers: a novel extremophile for sustainable crop production under cold environment. Microorganisms 9(12):2451. https://doi.org/10.3390/microorganisms9122451

Romshoo SA, Murtaza KO, Abdullah T (2022) Towards understanding various influences on mass balance of the Hoksar Glacier in the Upper Indus Basin using observations. Sci Rep 12(1):15669. https://doi.org/10.1038/s41598-022-20033-w

Santos RM, Bakhshoodeh R (2021) Climate change/global warming/climate emergency versus general climate research: comparative bibliometric trends of publications. Heliyon 7(11):e08219. https://doi.org/10.1016/j.heliyon.2021.e08219

Schulz S, Brankatschk R, Dümig A, Kögel-Knabner I, Schloter M, Zeyer J (2013) The role of microorganisms at different stages of ecosystem development for soil formation. Biogeosciences 10:3983–3996

Sherpa MT, Najar IN, Das S, Thakur N (2019) The diversity of pseudomonas species from the accumulation zone of Kanchengayao Glacier, North Sikkim, India. J Pure Appl Microbiol 13(1):339–348. https://doi.org/10.22207/JPAM.13.1.37

Sherpa MT, Das S, Najar IN, Thakur N (2021) Draft genome sequence of Stenotrophomonas maltophilia strain P13 gives insight into its protease production and assessment of sulfur and nitrogen metabolism. Curr Res Microb Sci 2:100012. https://doi.org/10.1016/j.crmicr.2020.100012

Siddiqui KS, Williams TJ, Wilkins D, Yau S, Allen MA, Brown MV, Lauro FM, Cavicchioli R (2013) Psychrophiles. Annu Rev Earth Planet Sci 41(1):87–115. https://doi.org/10.1146/annurev-earth-040610-133514

Singh G, Nela BR, Bandyopadhyay D, Mohanty S, Kulkarni AV (2020) Discovering anomalous dynamics and disintegrating behaviour in glaciers of Chandra-Bhaga sub-basins, part of Western Himalaya using DInSAR. Remote Sens Environ 246:111885. https://doi.org/10.1016/j.rse.2020.111885

Singh DS, Singh AK, Dubey CA, Kumar D, Sangode SJ, Trivedi A, Singh J (2022) Multi-Proxy analysis in the Gangotri Glacier region, Garhwal Himalaya: Glacial Stratigraphy and the overview of snout retreat, geomorphic evolution, and palaeoclimate signatures. Journal of the Palaeontological Society of India 67(1):158–182

Singh M, Jayant K, Singh D, Suyal DC, Mitra A, Bhutani S (2023) Microbial adaptations at higher altitude for sustainable development: A review. Journal of Applied Pharmaceutical Science 13(6): 001–009

Sivalingam S, Murugesan GP, Dhulipala K, Kulkarni AV, Devaraj S (2022) Temporal fluctuations of siachen glacier velocity: a repeat pass sar interferometry based approach. Geocarto International 37(17):4888–4910

Skidmore ML, Foght JM, Sharp MJ (2000) Microbial life beneath a high arctic glacier. Appl Environ Microbiol 66(8):3214–3220. https://doi.org/10.1128/AEM.66.8.3214-3220.2000

Sowers T (2001) The N2O record spanning the penultimate deglaciation from the Vostok ice core. Journal of Geophysical Research 106:31903–31914

Stevens IT, Irvine-Fynn TDL, Edwards A, Mitchell AC, Cook JM, Porter PR, Holt TO, Huss M, Fettweis X, Moorman BJ, Sattler B, Hodson AJ (2022) Spatially consistent microbial biomass and future cellular carbon release from melting Northern Hemisphere glacier surfaces. Commun Earth Environ 3(1):275. https://doi.org/10.1038/s43247-022-00609-0

Sun M, Zhu B, Liu J, Qin K, Peng F (2023) Rhodococcus antarcticus sp. nov., isolated from a till sample of Collins glacier front, Antarctica. Int J Syst Evol Microbiol 73(7). https://doi.org/10.1099/ijsem.0.005940

Tung HC, Bramall NE, Price BP (2005) Microbial origin of excess methane in glacial ice and implications for life on Mars. Proc Natl Acad Sci 102:18292–18296. https://doi.org/10.1073/pnas.0507601102

Uetake J, Yoshimura Y, Nagatsuka N, Kanda H (2012) Isolation of oligotrophic yeasts from supraglacial environments of different altitude on the Gulkana Glacier (Alaska). FEMS Microbiol Ecol 82(2):279–286. https://doi.org/10.1111/j.1574-6941.2012.01323.x

UNESCO (2022) World Heritage Glaciers Sentinels of climate change. United Nations Educational, Scientific and Cultural Organization (UNESCO), Paris, France and the IUCN, International Union for Conservation of Nature and Natural Resources, Gland, Switzerland. https://doi.org/10.3929/ethz-b-000578916

Veh G, Lützow N, Kharlamova V, Petrakov D, Hugonnet R, Korup O (2022) Trends, breaks, and biases in the frequency of reported glacier lake outburst floods. Earths Future 10(3):e2021EF002426. https://doi.org/10.1029/2021EF002426
Visioli LJ, Enzweiler H, Kuhn RC, Schwaab M, Mazutti MA (2014) Recent advances on biobutanol production. Sustain Chem Process 2(1):15. https://doi.org/10.1186/2043-7129-2-15
Wagner KM, McReynolds CB, Schmidt WK, Hammock BD (2017) Soluble epoxide hydrolase as a therapeutic target for pain, inflammatory and neurodegenerative diseases. Pharmacol Ther 180:62–76. https://doi.org/10.1016/j.pharmthera.2017.06.006
Wahlen BD, Morgan MR, McCurdy AT, Willis RM, Morgan MD, Dye DJ, Bugbee B, Wood BD, Seefeldt LC (2013) Biodiesel from microalgae, yeast, and bacteria: engine performance and exhaust emissions. Energy Fuel 27(1):220–228. https://doi.org/10.1021/ef3012382
Wang X, Siegert F, Zhou AG, Franke J (2013) Glacier and glacial lake changes and their relationship in the context of climate change, Central Tibetan Plateau 1972–2010. Global and Planetary Change 111:246–257
Wang M, Jiang X, Wu W, Hao Y, Su Y, Cai L, Xiang M, Liu X (2015) Psychrophilic fungi from the world's roof. Persoonia 34:100–112. https://doi.org/10.3767/003158515X685878
Wietrzyk-Pełka P, Rola K, Szymański W, Węgrzyn MH (2020) Organic carbon accumulation in the glacier forelands with regard to variability of environmental conditions in different ecogenesis stages of High Arctic ecosystems. Sci Total Environ 717:135151. https://doi.org/10.1016/j.scitotenv.2019.135151
Williams MW (2013) The status of glaciers in the Hindu Kush-Himalayan region. Mt Res Dev 33(1):114–115. https://doi.org/10.1659/mrd.mm113
Winkel M, Trivedi CB, Mourot R, Bradley JA, Vieth-Hillebrand A, Benning LG (2022) Seasonality of glacial snow and ice microbial communities. Front Microbiol 13. https://doi.org/10.3389/fmicb.2022.876848
Xiang S, Yao T, An L, Xu B, Wang J (2005) 16S rRNA sequences and differences in bacteria isolated from the Muztag Ata Glacier at increasing depths. Appl Environ Microbiol 71(8):4619–4627. https://doi.org/10.1128/AEM.71.8.4619-4627.2005
Xiang SR, Shang TC, Chen Y, Jing ZF, Yao T (2009) Dominant bacteria and biomass in the Kuytun 51 Glacier. Appl Environ Microbiol 75(22):7287–7290. https://doi.org/10.1128/AEM.00915-09
Xie J, Yang T, Zhou S, Xu L, Ou X, Hu G (2023) Glacial fluctuations during the Last Glacial Maximum and Lateglacial in the Zhuxi and Songlong valleys, eastern Nyainqêntanglha Range, southeastern Tibet. Journal of Quaternary Science 38(4):580–596
Xue Y, Hong X, He X, Kang S, Wang S, Ding Y (2024) An evaluation of TLS-based glacier change assessment in the central Tibetan plateau. International Journal of Digital Earth 17(1):2375527
Yadav AN, Verma P, Sachan SG, Kaushik R, Saxena AK (2018) Psychrotrophic microbiomes: molecular diversity and beneficial role in plant growth promotion and soil health, pp 197–240. https://doi.org/10.1007/978-981-10-7146-1_11
Yadav AN, Yadav N, Sachan SG, Saxena AK (2019) Biodiversity of psychrotrophic microbes and their biotechnological applications. J Appl Biol Biotechnol 7(4):99–108. https://doi.org/10.7324/JABB.2019.70415
Yamaguchi NU, Bernardino EG, Ferreira MEC, de Lima BP, Pascotini MR, Yamaguchi MU (2023) Sustainable development goals: a bibliometric analysis of literature reviews. Environ Sci Pollut Res 30(3):5502–5515. https://doi.org/10.1007/s11356-022-24379-6
Yamasaki R, Maeda T, Wood TK (2018) Electron carriers increase electricity production in methane microbial fuel cells that reverse methanogenesis. Biotechnol Biofuels 11(1):211. https://doi.org/10.1186/s13068-018-1208-7
Yao T, Bolch T, Chen D, Gao J, Immerzeel W, Piao S, Su F, Thompson L, Wada Y, Wang L, Wang T, Wu G, Xu B, Yang W, Zhang G, Zhao P (2022) The imbalance of the Asian water tower. Nat Rev Earth Environ 3:618–632. https://doi.org/10.1038/s43017-022-00299-4
Yarzábal LA (2020) Perspectives for using glacial and periglacial microorganisms for plant growth promotion at low temperatures. In: Applied microbiology and biotechnology, vol 104(8). Springer, pp 3267–3278. https://doi.org/10.1007/s00253-020-10468-4

Yarzábal LA, Monserrate L, Buela L, Chica E (2018) Antarctic Pseudomonas spp. promote wheat germination and growth at low temperatures. Polar Biol 41(11):2343–2354. https://doi.org/10.1007/s00300-018-2374-6

Yin Y, Kara-Murdoch F, Murdoch RW, Yan J, Chen G, Xie Y, Sun Y, Löffler FE (2024) Nitrous oxide inhibition of methanogenesis represents an underappreciated greenhouse gas emission feedback. ISME J 18(1). https://doi.org/10.1093/ismejo/wrae027

Yu A, Li P, Tang T, Wang J, Chen Y, Liu L (2015) Roles of Hsp70s in stress responses of microorganisms, plants, and animals. Biomed Res Int 2015:1–8. https://doi.org/10.1155/2015/510319

Yusof NA, Masnoddin M, Charles J, Thien YQ, Nasib FN, Wong CMVL, Abdul Murad AM, Mahadi NM, Bharudin I (2022) Can heat shock protein 70 (HSP70) serve as biomarkers in Antarctica for future ocean acidification, warming and salinity stress? Polar Biol 45(3):371–394. https://doi.org/10.1007/s00300-022-03006-7

Zhang X, Yao T, Ma X, Wang N (2001) Analysis of the characteristics of microorganisms packed in the ice core of Malan Glacier, Tibet, China. Sci China Ser D Earth Sci 44(S1):369–374. https://doi.org/10.1007/BF02912008

Zhang B, Wu X, Zhang G, Li S, Zhang W, Chen X, Sun L, Zhang B, Liu G, Chen T (2016) The diversity and biogeography of the communities of Actinobacteria in the forelands of glaciers at a continental scale. Environ Res Lett 11(5). https://doi.org/10.1088/1748-9326/11/5/054012

Zhang B, Xu Y, Yan X, Pu T, Wang S, Yang X, Yang H, Zhang G, Zhang W, Chen T, Liu G (2024) The diversity and risk of potential pathogenic bacteria on the surface of glaciers in the southeastern Tibetan Plateau. The Science of the total environment 945:173937. https://doi.org/10.1016/j.scitotenv.2024.173937

Zheng L, Shi P, Song M, Zhou T, Zong N, Zhang X (2021) Climate sensitivity of high-altitude tree growth across the Hindu Kush Himalaya. For Ecol Manag 486:118963. https://doi.org/10.1016/j.foreco.2021.118963

Zhou J, He L, Gao Y, Han N, Zhang R, Wu Q, Li J, Tang X, Xu B, Ding J, Huang Z (2016) Characterization of a novel low-temperature-active, alkaline and sucrose-tolerant invertase. Sci Rep 6(1):32081. https://doi.org/10.1038/srep32081

Zhu D, Adebisi WA, Ahmad F, Sethupathy S, Danso B, Sun J (2020) Recent development of extremophilic bacteria and their application in biorefinery. Front Bioeng Biotechnol 8. https://doi.org/10.3389/fbioe.2020.00483

Zouagui R, Zouagui H, Aurag J, Ibrahimi A, Sbabou L (2024) Functional analysis and comparative genomics of *Rahnella perminowiae* S11P1 and Variovorax sp. S12S4, two plant growth-promoting rhizobacteria isolated from Crocus sativus L. (saffron) rhizosphere. BMC Genomics 25(1):289. https://doi.org/10.1186/s12864-024-10088-6

Climate Change and Antimicrobial Resistance

14

Kamal Singh, Ganesh Chandra Sahoo, and Swati Sharma

Abstract

Climate change and antimicrobial resistance (AMR) represent two of the most pressing global challenges of the twenty-first century, both of which pose serious risks to environmental stability, human health, and sustainable development. Climate change, driven by anthropogenic activities, is causing profound shifts in weather patterns, ecosystems, and biodiversity, leading to increased frequency and severity of extreme weather events, rising sea levels, and altered disease dynamics. These environmental disruptions exacerbate the emergence and spread of infectious diseases, including those caused by antimicrobial-resistant pathogens, by creating favorable conditions for their transmission and evolution.

Conversely, antimicrobial resistance, fueled by the overuse and misuse of antibiotics in veterinary medicine, agriculture, and human health, endangering the control of common illnesses and lowering the quality of standard medical operations. Moreover, the interconnectedness between climate change and AMR is increasingly evident, as climate-related environmental disruptions influence the emergence, transmission, and evolution of infectious diseases, including antimicrobial-resistant strains.

Recognizing the synergistic effects and underlying drivers of AMR and climate change need interdisciplinary and integrative approaches to addressing their intersecting concerns. Efforts to mitigate climate change must prioritize

K. Singh
Regional Viral Research and Diagnostic Laboratory, All India Institute of Medical Sciences, Patna, Bihar, India

G. C. Sahoo · S. Sharma (✉)
Department of Virology, ICMR-Rajendra Memorial Research Institute of Medical Sciences, Patna, Bihar, India
e-mail: swati.sharma@bhu.ac.in

P. Pandey et al. (eds.), *Microorganisms Resilience to Climate Change*,
Microorganisms for Sustainability 56, https://doi.org/10.1007/978-981-96-3748-5_14

adaptation strategies that enhance the resilience of ecosystems and communities against health threats posed by infectious diseases and antimicrobial resistance. Similar to this, One Health approaches should be used in AMR interventions, recognizing the interdependence of environmental, animal, and human health and integrating climate resilience as a crucial element of antimicrobial stewardship initiatives. By addressing these challenges holistically, we can build a more resilient and sustainable future for generations to come.

Keywords

Infectious diseases · Interconnections · Rising temperature · Mitigation · Adaptation

14.1 Introduction

Climate change and antimicrobial resistance (AMR) are recognized as urgent global challenges, each posing significant threats to public health, ecosystems, and economies (Ahmad et al. 2021). In an era characterized by rapid globalization and unprecedented environmental transformations, humanity is confronted with multifaceted challenges that transcend geographical borders and disciplinary boundaries (Held and McGrew 2003). Among these, two pressing issues stand out for their profound implications on human health, ecological balance, and socio-economic stability: climate change and AMR (Hensher 2020). While traditionally viewed as distinct phenomena, recent scientific evidence suggests an intricate interplay between these two crises, amplifying their collective impact on ecosystems, public health, and sustainable development (Salam et al. 2023).

Anthropogenic activities including burning fossil fuels, deforestation, and industrial emissions are the main causes of climate change, which has caused a series of environmental disturbances with far-reaching effects (Upadhyay 2020). Extreme weather events, changing precipitation patterns, and rising temperatures are changing ecosystems, jeopardizing biodiversity, undermining the resilience of communities worldwide (Glicksman 2008; Upadhyay 2020). The repercussions of these changes extend beyond environmental degradation, encompassing diverse sectors ranging from agriculture and water resource management to human health and national security (Barnett et al. 2010).

Concurrently, AMR has emerged as a silent but potent threat to global health systems and medical interventions (Salam et al. 2023). Resulting from the excessive utilization and misuse of antibiotics in agriculture, animal husbandry, and human health, AMR renders once-effective antibiotics and antimicrobial agents increasingly ineffective against bacterial infections (Kavya et al. 2023). This phenomenon not only makes treating common infections more challenging but also diminishes the effectiveness of routine medical procedures like surgeries, chemotherapy, and organ transplantation, potentially pushing modern medicine back to a pre-antibiotic era (Prestinaci et al. 2015).

While climate change and AMR are often discussed in isolation, their interconnectedness becomes apparent upon closer examination (Graham et al. 2019; Pepi and Focardi 2021). Climate-related environmental disruptions exert a significant impact on the emergence, transmission, evolution of infectious diseases, including antimicrobial-resistant pathogens (El-Sayed and Kamel 2020). For instance, changes in humidity and temperature might expand the geographical coverage of disease vectors, allowing vector-borne illnesses to spread to previously unaffected areas (Mojahed et al. 2022). Similar to this, severe weather conditions like hurricanes and floods can cause damage to sanitary systems, contaminating water supplies and raising the risk of waterborne illnesses, some of which may be resistant to antibiotics (Levy et al. 2018).

Furthermore, the complex interplay between climate change and AMR extends beyond direct environmental impacts to encompass broader socio-economic factors. Vulnerable populations disproportionately bear the brunt of crises, exacerbating existing health disparities and socio-economic inequalities (Magnano San Lio et al. 2023). The health effects of climate change and AMR are most likely to affect impoverished populations that lack access to clean water, proper sanitation, and healthcare facilities, which further entrenches cycles of poverty and illness (Nadimpalli et al. 2020).

A comprehensive and coordinated strategy that acknowledges the underlying causes and synergistic effects of both AMR and climate change is necessary to address their overlapping difficulties (Magnano San Lio et al. 2023). Efforts to mitigate climate change must prioritize adaptation strategies that enhance the resilience of ecosystems and communities against the health threats posed by infectious diseases and AMR (Bardosh et al. 2017). Similarly, interventions to combat AMR should adopt a One Health approach that recognizes the connection between environmental, animal, and human health, incorporating climate resilience as a key component of antimicrobial stewardship efforts (Mudenda et al. 2023).

As our planet grapples with unprecedented challenges, two distinct yet profoundly interlinked crises have emerged at the forefront of global concern—climate change and AMR. Therefore, this chapter seeks to unravel the intricate relationship between these phenomena, illuminating the mutual influences that demand a comprehensive understanding and collective action. By delving into the nexus of climate change and AMR, we will seek to comprehend the complexities that lie at the intersection of environmental transformation and global health.

14.2 Climate Change and its Impacts on Antimicrobial Resistance

14.2.1 Changing Climates, Evolving Microbes

Climate change is reshaping the very landscapes that microorganisms inhabit. Extreme weather occurrences, changed precipitation patterns, and temperatures all contribute to the evolution and adaptation of microbes (Ibáñez et al. 2023). The

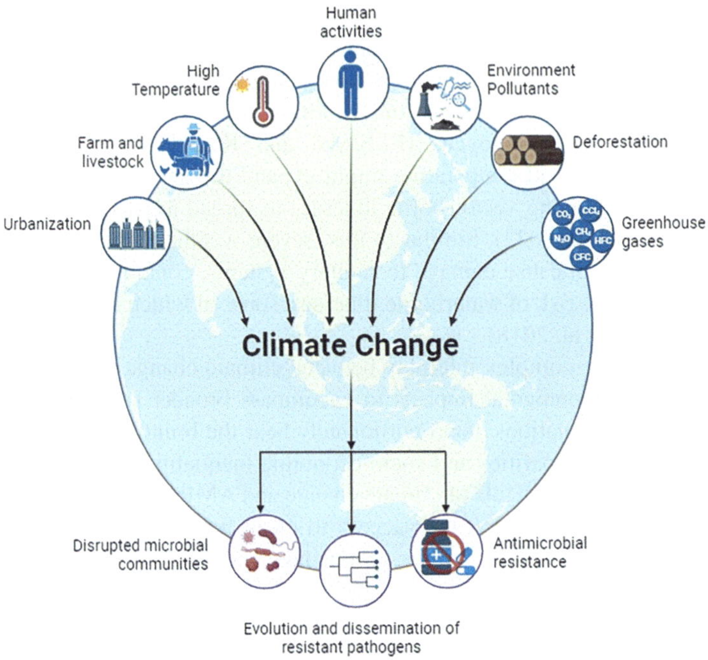

Fig. 14.1 The factors affecting the climate change and its consequences on microbial community

factors affecting the climate change and its consequences on microbial community have been shown in Fig. 14.1.

This section explores the direct and indirect pathways through which climate change influences the emergence and spread of antimicrobial resistant microbes.

As our planet's climate continues to undergo rapid transformations, the intricate bound between environmental shifts and microbial evolution unfolds with profound implications for human health and ecological balance. The interconnectedness of these phenomena has given rise to a dynamic interplay where changing climates act as catalysts for the evolution and dissemination of microbial pathogens (He et al. 2021).

One of the most visible manifestations of this interaction is the proliferation of infectious diseases in new geographic regions previously unaffected (Piscitelli and Miani 2024). Extreme weather conditions, changing rainfall patterns, and rising temperatures all contribute to the spread of vectors like ticks and mosquitoes, which makes it easier for diseases like Lyme disease, dengue fever, and malaria to infect once temperate areas (Fouque and Reeder 2019). As estimated, the risk of vector-borne diseases might increase to 500 million by 2050 (Ryan et al. 2019). Similarly, changes in temperature and humidity can influence the survival and persistence of microbial pathogens in water, soil, and air, amplifying the risk of waterborne and airborne diseases. For instance, studies have reported adaptation to the warmer temperatures has led to re-emergence of the disease caused by bacteria such as

Salmonella, Campylobacter, and *Vibrio cholera*. Elevated antibiotic resistance rate has been recorded for these bacteria resulting in restricted alternatives to treat antimicrobial resistant infections (Constantin de Magny 2009; Jobling 2019; Lipp et al. 2002; Omazic et al. 2019; Semenza and Menne 2009).

Moreover, rise in temperature is inextricably linked to AMR as it leads to increased microbial growth rates and increased horizontal gene transfer. In line with this, a higher prevalence of infections caused by gram-negative bacteria has been reported during the warmer months. Since these bacteria thrive around 32–37 °C, the elevated summer season due to the climate change crisis acts as a catalyst for these bacteria. This also provides a better environment for their adaptation and persistence. Hence, the cases of bloodstream infections (BSIs), health care associated infections (HAIs), intra-abdominal and surgical site infections caused by gram-negative pathogens have become much more common these days. Research indicates that a warming climate can influence the concentrations of heavy metals and biocides in soil and water, as well as their uptake by bacteria. These changes may drive antimicrobial resistance (AMR) through co-resistance mechanisms, where bacteria develop simultaneous resistance to multiple stressors (Huang et al. 2021; Magnano San Lio et al. 2023).

In addition, climate-induced disruptions to ecosystems and biodiversity can drive changes in host-pathogen dynamics, promoting the emergence and spread of antimicrobial-resistant microbes (Schmeller et al. 2020). Increased temperatures and habitat fragmentation can lead to shifts in species distributions and interactions, facilitating the transmission of resistant pathogens between humans, animals, and the environment (Simon et al. 2014). Furthermore, alterations to agricultural practices and land use patterns in response to climate change may exacerbate the overuse and misuse of antimicrobial agents, accelerating the development and dissemination of antimicrobial resistance.

14.2.2 Disrupted Ecosystems and Microbial Dynamics

Climate change disrupts the complex balance of ecosystems, affecting microbial communities and their interactions. It has been observed how ecological changes increase the challenges posed by AMR, leading to a complex web of consequences for both human health and the environment (Bissett et al. 2013; Ferri et al. 2017).

The integrity of ecosystems is intricately linked to the delicate balance of microbial communities that inhabit them. However, the rapid pace of environmental change, driven primarily by human activities such as deforestation, urbanization, and pollution, has disrupted ecosystems worldwide, triggering cascading effects on microbial dynamics with far-reaching consequences for ecosystem function and human well-being (Chakraborty and Chakraborty 2021; Harwood and Buckley 2008).

One of the most evident impacts of disrupted ecosystems on microbial dynamics is the loss of biodiversity, which serves as a critical buffer against the proliferation of pathogens and the emergence of antimicrobial resistance. Deforestation and habitat destruction diminish the natural habitats of diverse microbial species, reducing microbial diversity and resilience to environmental perturbations. Consequently, the loss of microbial diversity weakens ecosystem services such as nutrient cycling, soil fertility, and disease regulation, amplifying the risk of infectious diseases and AMR (Palit et al. 2022).

Furthermore, anthropogenic disturbances alter the physical and chemical properties of ecosystems, creating novel ecological niches and selective pressures that shape microbial communities. Pollution from agricultural runoff, industrial waste, and antibiotic residues disrupts aquatic and terrestrial ecosystems, fostering the proliferation of pollutant-tolerant and antibiotic-resistant microbes. In aquatic environments, nutrient pollution fuels algal blooms and hypoxic zones, creating conditions conducive to the growth and dissemination of waterborne pathogens and antimicrobial-resistant bacteria (Philippot et al. 2021; World Health Organization 2022).

The consequences of disrupted ecosystems on microbial dynamics extend beyond environmental health to encompass human health and socio-economic well-being. Exposure to contaminated water, soil, and air increases the risk of infectious diseases, while antimicrobial-resistant pathogens threaten the efficacy of medical treatments and agricultural practices. Moreover, the loss of ecosystem services undermines food security, livelihoods, and resilience to climate change, exacerbating inequalities and vulnerabilities in vulnerable communities (Galvani et al. 2016; Kusi et al. 2022).

14.3 Antimicrobial Resistance as a Driver of Climate Change

14.3.1 Agriculture and Antibiotics: A Silent Climate Culprit

Antimicrobial use, particularly in agriculture, contributes to the rise of resistant strains and concurrently impacts climate change. This section delves into the agricultural practices, livestock management, and the overuse of antibiotics, shedding light on their role in greenhouse gas emissions and environmental degradation.

The extensive use of antibiotics in agriculture presents a subtle but critical challenge to climate stability. In livestock farming, antibiotics are frequently administered to enhance growth and prevent disease outbreaks in overcrowded and unhygienic conditions. However, this practice accelerates the development of antimicrobial-resistant bacteria, which can spread to humans through food chains or environmental pathways, posing significant public health risks (Zalewska et al. 2021; Vidovic and Vidovic 2020).

Moreover, the production of antibiotics and their use in agriculture generate greenhouse gas emissions, primarily through energy-intensive manufacturing processes and the release of methane from livestock digestion. Additionally, the disposal of antibiotic-laden manure can contaminate waterways, further exacerbating environmental degradation (Stavi and Lal 2013).

14.3.2 Pharmaceuticals, Pollution, and Planetary Health

The manufacturing and disposal of pharmaceuticals, including antimicrobials, contribute to pollution and environmental degradation. The implications of pharmaceutical pollution on climate change and broader ecological consequences have been frequently studied (Felis et al. 2020).

The pharmaceutical industry plays a crucial role in healthcare, but its impact on planetary health extends beyond the confines of human medicine. Pharmaceuticals, including antibiotics, hormones, and painkillers, find their way into the environment through various pathways, such as manufacturing waste, improper disposal, and excretion by humans and animals. Once released, these compounds can persist in soil, water, and air, exerting profound effects on ecosystems and biodiversity (Sharma et al. 2022).

Pharmaceutical pollution poses a significant threat to aquatic organisms, disrupting endocrine systems, impairing reproduction, and altering behavior. Moreover, the presence of antibiotics in the environment contributes to the emergence and spread of antimicrobial resistance, undermining the effectiveness of medical treatments and exacerbating public health concerns (Kusi et al. 2022; Porretti et al. 2022).

Addressing pharmaceutical pollution requires coordinated efforts from multiple stakeholders, including policymakers, regulators, healthcare providers, and pharmaceutical manufacturers. Strategies to mitigate pharmaceutical pollution may include improving wastewater treatment technologies, implementing drug take-back programs, and promoting the development of eco-friendly pharmaceuticals with reduced environmental impact (Nijsingh et al. 2019).

By prioritizing planetary health in pharmaceutical production and consumption, the ecological footprint of health care practices can be reduced and the health of both humans and the environment can be protected for future generations.

14.4 A Holistic Approach—Policies and Strategies

14.4.1 Global Initiatives for Climate-Resilient Health

This section provides an overview of existing global initiatives addressing climate change and AMR. It examines frameworks, policies, and collaborations aimed at fostering a holistic approach to tackle these interconnected challenges.

Global initiatives for climate-resilient health are increasingly recognized as imperative in addressing the complex interplay between climate change and public health. One prominent example is the World Health Organization's (WHO) Global Climate and Health Alliance, which aims to raise awareness, build capacity, and mobilize action to protect health from climate change impacts. Through advocacy, research, and collaboration, this alliance seeks to integrate climate considerations into healthcare policies, strengthen health systems' resilience, and promote sustainable development strategies (Mendoza-Vasconez et al. 2022; World Health Organization 2004).

Additionally, the Paris Agreement, adopted by nearly every country in the world, emphasizes the importance of safeguarding human health in the context of climate change mitigation and adaptation efforts. Under the Paris Agreement, countries commit to reducing greenhouse gas emissions, enhancing adaptive capacity, and fostering international cooperation to address climate-related health risks (Mousavi et al. 2020).

Furthermore, initiatives such as the Climate and Clean Air Coalition and the Global Alliance for Clean Cookstoves prioritize interventions that simultaneously mitigate climate change and improve public health outcomes, particularly in vulnerable communities disproportionately affected by indoor and outdoor air pollution (Anenberg et al. 2013).

By fostering synergies between climate action and health promotion, these global initiatives play a crucial role in building resilience, reducing vulnerabilities, and advancing sustainable development pathways that prioritize human health in a changing climate.

14.4.2 The One Health Paradigm

The One Health approach recognizes the inextricable link between animal health, human health, and the environment. Consideration is given to how a multidisciplinary perspective can inform policies and interventions to simultaneously address climate change and AMR.

The One Health paradigm promotes the inherent links between environmental, animal, and human health, embodying an integrated approach. It highlights how various fields are interdependent and promotes interdisciplinary cooperation and systems-level approaches to successfully address complicated health issues (Conrad et al. 2013).

Central to the One Health approach is the recognition that many infectious diseases are zoonotic in nature, meaning they can be transmitted between humans and animals. Examples include diseases like Zika, Ebola, and COVID-19, all of which originated in animal populations before spilling over to humans. One Health seeks to enhance the prevention and control of zoonotic diseases by examining and addressing the ecological, social, and behavioral factors that drive disease emergence and transmission at the interface of human, animal, and environmental health (He et al. 2022; Dharmarajan et al. 2022).

Moreover, the One Health paradigm extends beyond infectious diseases to encompass broader health issues such as antimicrobial resistance, food safety, and environmental pollution. By integrating insights from diverse disciplines including medicine, veterinary science, ecology, epidemiology, and social sciences, One Health approaches can inform more holistic and sustainable strategies for promoting health and well-being across species boundaries (Mumford et al. 2023).

In an era characterized by rapid globalization, urbanization, and environmental degradation, embracing the One Health paradigm is essential for addressing emerging health threats, promoting resilience, and advancing planetary health goals for the benefit of current and future generations (Lueddeke 2015).

14.4.3 Sustainable Solutions for a Resilient Future

Drawing from successful case studies and innovative strategies, there have some sustainable solutions been discussed that integrate efforts to mitigate climate change and combat antimicrobial resistance. From sustainable agriculture practices to responsible antibiotic stewardship, here are few avenues suggested for building a resilient and healthier future (Kavya et al. 2023).

Sustainable solutions for a resilient future are essential for navigating the complex challenges posed by climate change, environmental degradation, and global health crises. At the heart of these solutions lies the imperative to balance human needs with the capacity of ecosystems to regenerate and support life in the long term (Nguyen et al. 2023).

One key aspect of sustainable solutions is the adoption of renewable energy sources such as wind, solar, and hydroelectric power, which mitigate greenhouse gas emissions and reduce dependence on finite fossil fuels. Transitioning towards a low-carbon economy not only mitigates climate change but also creates opportunities for green job creation and economic growth (Strielkowski et al. 2021).

Furthermore, sustainable land and water management practices, such as agroecology and integrated watershed management, promote soil health, conserve biodiversity, and enhance resilience to climate variability. By prioritizing regenerative agriculture, reforestation, and ecosystem restoration efforts, communities can mitigate the impacts of climate change while ensuring food security and safeguarding natural resources for future generations (Srivastav et al. 2021).

In addition, sustainable urban planning and infrastructure development play a crucial role in building climate-resilient cities and communities. By investing in public transportation, green spaces, and energy-efficient buildings, cities can reduce emissions, improve public health and enhance air quality outcomes while fostering social cohesion and economic prosperity (Lomba-Fernández et al. 2019). Ultimately, achieving a resilient future requires collective action and transformative change at local, national, and global levels. By embracing sustainability as a guiding principle, a world can be created where people, planet, and prosperity thrive in harmony with nature.

14.5 Conclusion

Concluding the exploration into the convergence of climate change and antimicrobial resistance, it becomes clear that a united and concerted effort is required to successfully address these interconnected challenges. In navigating the complex terrain of a changing climate and evolving pathogens, interdisciplinary approaches that integrate insights from climate science, microbiology, epidemiology, and public health are indispensable. By understanding and addressing the interconnected drivers of infectious diseases and AMR in the context of climate change, more effective strategies can be developed to mitigate their impacts and protect human health in a rapidly changing world. Furthermore, concerted efforts to restore and conserve ecosystems are essential for mitigating the impacts of disrupted microbial dynamics on human and environmental health. By preserving biodiversity, reducing pollution, and promoting sustainable land and water management practices, we can foster resilient ecosystems that support diverse microbial communities and contribute to the prevention of infectious diseases and antimicrobial resistance.

Finally, addressing the intersection of agriculture and antibiotics requires a multifaceted approach that promotes sustainable farming practices, reduces antibiotic use in livestock production, and enhances monitoring and regulation of antibiotic residues in the environment. By mitigating the environmental footprint of agriculture's antibiotic use, we can contribute to both climate resilience and public health. Only through collaborative and sustainable actions can we hope to mitigate the impacts of climate change and preserve the efficacy of antimicrobials, safeguarding the well-being of current and future generations.

References

Ahmad I, Malak HA, Abulreesh HH (2021) Environmental antimicrobial resistance and its drivers: a potential threat to public health. J Glob Antimicrob Resist 27:101–111. https://doi.org/10.1016/j.jgar.2021.08.001

Anenberg SC, Balakrishnan K, Jetter J, Masera O, Mehta S, Moss J, Ramanathan V (2013) Cleaner cooking solutions to achieve health, climate, and economic cobenefits. Environ Sci Technol 47:3944–3952. https://doi.org/10.1021/es304942e

Bardosh KL, Ryan SJ, Ebi K, Welburn S, Singer B (2017) Addressing vulnerability, building resilience: community-based adaptation to vector-borne diseases in the context of global change. Infect Dis Poverty 6:1–21. https://doi.org/10.1186/s40249-017-0375-2

Barnett J, Matthew RA, O'Brien KL (2010) Global environmental change and human security: an introduction. In: Global environmental change and human security. The MIT Press, pp 3–2

Bissett A, Brown MV, Siciliano SD, Thrall PH (2013) Microbial community responses to anthropogenically induced environmental change: towards a systems approach. Ecol Lett 16:128–139. https://doi.org/10.1111/ele.12109

Chakraborty SK, Chakraborty SK (2021) River pollution and perturbation: perspectives and processes. In: Riverine ecology volume 2: biodiversity conservation, conflicts and resolution, pp 443–530. https://doi.org/10.1007/978-3-030-53941-2_5

Conrad PA, Meek LA, Dumit J (2013) Operationalizing a One Health approach to global health challenges. Comp Immunol Microbiol Infect Dis 36(3):211–216. https://doi.org/10.1016/j.cimid.2013.03.006

Constantin de Magny G, Colwell RR (2009) Cholera and climate: a demonstrated relationship. Trans Am Clin Climatol Assoc 120:119–28

Dharmarajan G, Li R, Chanda E, Dean KR, Dirzo R, Jakobsen KS, Khan I, Leirs H, Shi Z, Wolfe ND, Yang R, Stenseth NC (2022) The animal origin of major human infectious diseases: what can past epidemics teach us about preventing the next pandemic? Zoonoses 2(1):989. https://doi.org/10.15212/ZOONOSES-2021-0028

El-Sayed A, Kamel M (2020) Climatic changes and their role in emergence and re-emergence of diseases. Environ Sci Pollut Res 27(18):22336–22352. https://doi.org/10.1007/s11356-020-08896-w

Felis E, Kalka J, Sochacki A, Kowalska K, Bajkacz S, Harnisz M, Korzeniewska E (2020) Antimicrobial pharmaceuticals in the aquatic environment-occurrence and environmental implications. Eur J Pharmacol 866:172813. https://doi.org/10.1016/j.ejphar.2019.172813

Ferri M, Ranucci E, Romagnoli P, Giaccone V (2017) Antimicrobial resistance: a global emerging threat to public health systems. Crit Rev Food Sci Nutr 57(13):2857–2876. https://doi.org/10.1080/10408398.2015.1077192

Fouque F, Reeder JC (2019) Impact of past and on-going changes on climate and weather on vector-borne diseases transmission: a look at the evidence. Infect Dis Poverty 8(03):1–9. https://doi.org/10.1186/s40249-019-0565-1

Galvani AP, Bauch CT, Anand M, Singer BH, Levin SA (2016) Human–environment interactions in population and ecosystem health. Proc Natl Acad Sci 113(51):14502–14506. https://doi.org/10.1073/pnas.1618138113

Glicksman RL (2008) Ecosystem resilience to disruptions linked to global climate change: an adaptive approach to federal land management. Neb Law Rev 87:833

Graham DW, Bergeron G, Bourassa MW, Dickson J, Gomes F, Howe A, Khan LH, Morley PS, Scott HM, Simjee S, Singer RS, Smith TC, Storrs C, Wittum TE (2019) Complexities in understanding antimicrobial resistance across domesticated animal, human, and environmental systems. Ann N Y Acad Sci 1441(1):17–30. https://doi.org/10.1111/nyas.14036

Harwood C, Buckley M (2008) The uncharted microbial world: microbes and their activities in the environment: this report is based on a colloquium, sponsored by the American Academy of Microbiology, convened February 9–11, 2007, in Seattle, Washington. American Society for Microbiology.; Available from: https://www.ncbi.nlm.nih.gov/books/NBK562611/, Washington, DC. https://doi.org/10.1128/AAMCol.9Feb.2007

He DC, He MH, Amalin DM, Liu W, Alvindia DG, Zhan J (2021) Biological control of plant diseases: an evolutionary and eco-economic consideration. Pathogens 10(10):1311. https://doi.org/10.3390/pathogens10101311

He J, Guo Z, Yang P, Cao C, Xu J, Zhou X, Li S (2022) Social insights on the implementation of One Health in zoonosis prevention and control: a scoping review. Infect Dis Poverty 11(03):1–11. https://doi.org/10.1186/s40249-022-00976-y

Held D, McGrew A (2003) The great globalization debate: an introduction. In: The global transformations reader an introduction to the globalization debate, 2nd edn. Polity Press, Cambridge, pp 1–42

Hensher M (2020) Human health and ecological economics. In: Sustainable wellbeing futures. Edward Elgar Publishing, pp 188–208

Huang L, Ahmed S, Gu Y, Huang J, An B, Wu C, Zhou Y, Cheng G (2021) The effects of natural products and environmental conditions on antimicrobial resistance. Molecules 26(14):4277. https://doi.org/10.3390/molecules26144277

Ibáñez A, Garrido-Chamorro S, Barreiro C (2023) Microorganisms and climate change: a not so invisible effect. Microbiol Res 14(3):918–947. https://doi.org/10.3390/microbiolres14030064

Jobling MG (2019) Trust but verify: uncorroborated assemblies of plasmid genomes from next-generation sequencing data are likely spurious comment on "Diverse Plasmids Harboring bla CTX-M-15 in Klebsiella pneumoniae ST11 Isolates from Several Asian Countries," by So Yeon Kim and Kwan Soo Ko (Microb. Drug Resist. 2019; 25 (2): 227–232; DOI: 10.1089/mdr.2018.0020). Microb Drug Resist 25(10):1521–1524. https://doi.org/10.1089/mdr.2018.0426

Kavya IK, Kochhar N, Ghosh A, Shrivastava S, Rawat VS, Ghorai SM, Sodhi KK, James A, Kumar M (2023) Perspectives on systematic generation of antibiotic resistance with special emphasis on modern antibiotics. Total Environ Res Themes 8:100068. https://doi.org/10.1016/j.totert.2023.100068

Kusi J, Ojewole CO, Ojewole AE, Nwi-Mozu I (2022) Antimicrobial resistance development pathways in surface waters and public health implications. Antibiotics 11(6):821. https://doi.org/10.3390/antibiotics11060821
Levy K, Smith SM, Carlton EJ (2018) Climate change impacts on waterborne diseases: moving toward designing interventions. Curr Environ Health Rep 5:272–282. https://doi.org/10.1007/s40572-018-0199-7
Lipp EK, Huq A, Colwell RR (2002) Effects of global climate on infectious disease: the cholera model. Clin Microbiol Rev 15(4):757–770. https://doi.org/10.1128/CMR.15.4.757-770.2002
Lomba-Fernández C, Hernantes J, Labaka L (2019) Guide for climate-resilient cities: an urban critical infrastructures approach. Sustain For 11(17):4727. https://doi.org/10.3390/su11174727
Lueddeke G (2015) Global population health and well-being in the 21st century: toward new paradigms, policy, and practice. Springer
Magnano San Lio R, Favara G, Maugeri A, Barchitta M, Agodi A (2023) How antimicrobial resistance is linked to climate change: an overview of two intertwined global challenges. Int J Environ Res Public Health 20(3):1681. https://doi.org/10.3390/ijerph20031681
Mendoza-Vasconez AS, McLaughlin E, Sallis JF, Maibach E, Epel E, Bennett G, Nogueira L, Thayer J, Dietz WH (2022) Advocacy to support climate and health policies: recommended actions for the Society of Behavioral Medicine. Transl Behav Med 12(4):535–542. https://doi.org/10.1093/tbm/ibac028
Mojahed N, Mohammadkhani MA, Mohamadkhani A (2022) Climate crises and developing vector-borne diseases: a narrative review. Iran J Public Health 51(12):2664. 10.18502/ijph.v51i12.11457
Mousavi A, Ardalan A, Takian A, Ostadtaghizadeh A, Naddafi K, Bavani AM (2020) Health system plan for implementation of Paris agreement on climate change (COP 21): a qualitative study in Iran. BMC Public Health 20:1–13. https://doi.org/10.1186/s12889-020-09503-w
Mudenda S, Chabalenge B, Daka V, Mfune RL, Salachi KI, Mohamed S, Mufwambi W, Kasanga M, Matafwali SK (2023) Global strategies to combat antimicrobial resistance: a one health perspective. Pharmacol Pharm 14(8):271–328. https://doi.org/10.4236/pp.2023.148020
Mumford EL, Martinez DJ, Tyance-Hassell K, Cook A, Hansen GR, Labonté R, Mazet JAK, Mumford EC, Rizzo DM, Togami E, Vreedzaam A, Parrish-Sprowl J (2023) Evolution and expansion of the One Health approach to promote sustainable and resilient health and well-being: a call to action. Front Public Health 10:1056459. https://doi.org/10.3389/fpubh.2022.1056459
Nadimpalli ML, Marks SJ, Montealegre MC, Gilman RH, Pajuelo MJ, Saito M, Tsukayama P, Njenga SM, Kiiru J, Swarthout J, Islam MA, Julian TR, Pickering AJ (2020) Urban informal settlements as hotspots of antimicrobial resistance and the need to curb environmental transmission. Nat Microbiol 5(6):787–795. https://doi.org/10.1038/s41564-020-0722-0
Nguyen TT, Grote U, Neubacher F, Do MH, Paudel GP (2023) Security risks from climate change and environmental degradation: implications for sustainable land use transformation in the Global South. Curr Opin Environ Sustain 63:101322. https://doi.org/10.1016/j.cosust.2023.101322
Nijsingh N, Munthe C, Larsson DJ (2019) Managing pollution from antibiotics manufacturing: charting actors, incentives and disincentives. Environ Health 18:1–17. https://doi.org/10.1186/s12940-019-0531-1
Omazic A, Bylund H, Boqvist S, Högberg A, Björkman C, Tryland M, Evengård B, Koch A, Berggren C, Malogolovkin A, Kolbasov D, Pavelko N, Thierfelder T, Albihn A (2019) Identifying climate-sensitive infectious diseases in animals and humans in Northern regions. Acta Vet Scand 61:1–12. https://doi.org/10.1186/s13028-019-0490-0
Palit K, Rath S, Chatterjee S, Das S (2022) Microbial diversity and ecological interactions of microorganisms in the mangrove ecosystem: threats, vulnerability, and adaptations. Environ Sci Pollut Res 29(22):32467–32512. https://doi.org/10.1007/s11356-022-19048-7
Pepi M, Focardi S (2021) Antibiotic-resistant bacteria in aquaculture and climate change: a challenge for health in the Mediterranean area. Int J Environ Res Public Health 18(11):5723. https://doi.org/10.3390/ijerph18115723

Philippot L, Griffiths BS, Langenheder S (2021) Microbial community resilience across ecosystems and multiple disturbances. Microbiol Mol Biol Rev 85(2):10–1128. https://doi.org/10.1128/MMBR.00026-20

Piscitelli P, Miani A (2024) Climate change and infectious diseases: navigating the intersection through innovation and interdisciplinary approaches. Int J Environ Res Public Health 21(3):314. https://doi.org/10.3390/ijerph21030314

Porretti M, Arrigo F, Di Bella G, Faggio C (2022) Impact of pharmaceutical products on zebrafish: an effective tool to assess aquatic pollution. Comp Biochem Physiol C Toxicol Pharmacol 261:109439. https://doi.org/10.1016/j.cbpc.2022.109439

Prestinaci F, Pezzotti P, Pantosti A (2015) Antimicrobial resistance: a global multifaceted phenomenon. Pathog Global Health 109(7):309–318. https://doi.org/10.1179/2047773215Y.0000000030

Ryan SJ, Carlson CJ, Mordecai EA, Johnson LR (2019) Global expansion and redistribution of Aedes-borne virus transmission risk with climate change. PLoS Negl Trop Dis 13(3):e0007213. https://doi.org/10.1371/journal.pntd.0007213

Salam MA, Al-Amin MY, Salam MT, Pawar JS, Akhter N, Rabaan AA, Alqumber MA (2023) Antimicrobial resistance: a growing serious threat for global public health. In: Healthcare, vol 11(13). MDPI, p 1946. https://doi.org/10.3390/healthcare11131946

Schmeller DS, Courchamp F, Killeen G (2020) Biodiversity loss, emerging pathogens and human health risks. Biodivers Conserv 29:3095–3102. https://doi.org/10.1007/s10531-020-02021-6

Semenza JC, Menne B (2009) Climate change and infectious diseases in Europe. Lancet Infect Dis 9(6):365–375. https://doi.org/10.1016/S1473-3099(09)70104-5

Sharma P, Rani L, Grewal AS, Srivastav AL (2022) Impact of pharmaceuticals and antibiotics waste on the river ecosystem: a growing threat. In: Ecological significance of river ecosystems. Elsevier, pp 15–36. https://doi.org/10.1016/B978-0-323-85045-2.00015-7

Simon JA, Marrotte RR, Desrosiers N, Fiset J, Gaitan J, Gonzalez A, Koffi JK, Lapointe F, Leighton PA, Lindsay LR, Logan T, Milord F, Ogden NH, Rogic A, Roy-Dufresne E, Suter D, Tessier N, Millien V (2014) Climate change and habitat fragmentation drive the occurrence of B orrelia burgdorferi, the agent of Lyme disease, at the northeastern limit of its distribution. Evol Appl 7(7):750–764. https://doi.org/10.1111/eva.12165

Srivastav AL, Dhyani R, Ranjan M, Madhav S, Sillanpää M (2021) Climate-resilient strategies for sustainable management of water resources and agriculture. Environ Sci Pollut Res 28(31):41576–41595. https://doi.org/10.1007/s11356-021-14332-4

Stavi I, Lal R (2013) Agriculture and greenhouse gases, a common tragedy. A review. Agron Sustain Dev 33:275–289. https://doi.org/10.1007/s13593-012-0110-0

Strielkowski W, Civín L, Tarkhanova E, Tvaronavičienė M, Petrenko Y (2021) Renewable energy in the sustainable development of electrical power sector: a review. Energies 14(24):8240. https://doi.org/10.3390/en14248240

Upadhyay RK (2020) Markers for global climate change and its impact on social, biological and ecological systems: a review. Am J Clim Chang 9(03):159–203. https://doi.org/10.4236/ajcc.2020.93012

Vidovic N, Vidovic S (2020) Antimicrobial resistance and food animals: influence of livestock environment on the emergence and dissemination of antimicrobial resistance. Antibiotics 9(2):52. https://doi.org/10.3390/antibiotics9020052

World Health Organization (2004) World report on knowledge for better health: strengthening health systems. World Health Organization. https://iris.who.int/handle/10665/43058

World Health Organization, Regional Office for Europe (2022) A health perspective on the role of the environment in One Health. World Health Organization. Regional Office for Europe. https://iris.who.int/handle/10665/354574. License: CC BY-NC-SA 3.0 IGO

Zalewska M, Błażejewska A, Czapko A, Popowska M (2021) Antibiotics and antibiotic resistance genes in animal manure–consequences of its application in agriculture. Front Microbiol 12:610656. https://doi.org/10.3389/fmicb.2021.610656

Part V

Conclusion

Resilience of Microorganisms in the Face of Climate Change: Key Conclusions

15

Nandita Das, Rajarshi Chowdhury, and Piyush Pandey

Abstract

Rising global temperatures and increased greenhouse gas emissions, primarily due to anthropogenic activities like the burning of fossil fuels, have led to significant changes in microbial ecosystems. Microorganisms play a vital role in mitigating climate change, influencing soil health, and promoting sustainable agriculture. Soil microorganisms are integral to carbon and nitrogen cycles, impacting the global carbon balance. Soils act as significant carbon sinks, containing more carbon than the atmosphere and biosphere combined. Microbes, particularly bacteria and fungi, regulate the decomposition of organic matter, which governs the amount of carbon stored in the soil. The balance between carbon inputs, such as plant residues and root exudates, and carbon losses due to microbial decomposition determines the soil's carbon storage capacity. Microbes also facilitate soil aggregation, which physically protects soil organic carbon from microbial degradation, thus prolonging its residence time. Carbon sequestration can be enhanced by promoting microbial activities that favour the formation of stable organic matter and by increasing plant-microbe interactions through the application of biofertilizers and organic amendments. The ability of soil microbes to adapt to changing environmental conditions, including temperature fluctuations and drought, is crucial for maintaining ecosystem functions. Moreover, microorganisms, such as rhizobacteria and mycorrhizal fungi, enhance plant resilience to climate-induced stresses, promoting crop growth and agricultural productivity. Bioremediation techniques utilizing microbes can also help in mitigating environmental contamination, particularly in soils polluted with hydrocarbons and heavy metals. Additionally, microbial-driven processes such

N. Das · R. Chowdhury · P. Pandey (✉)
Department of Microbiology, Soil and Environmental Microbiology Laboratory, Assam University, Silchar, India

P. Pandey et al. (eds.), *Microorganisms Resilience to Climate Change*,
Microorganisms for Sustainability 56, https://doi.org/10.1007/978-981-96-3748-5_15

as methanogenesis and methane oxidation are key to regulating greenhouse gases. Some microbes, including acetogenic and oxalotrophic bacteria, fix carbon by converting CO_2 into organic compounds, contributing to long-term carbon storage. Innovative strategies, such as introducing beneficial microbes into agricultural soils, enhancing plant-microbe interactions, and applying biochar, are being explored to increase carbon sequestration potential, thus reducing the impacts of global warming. This study highlights the essential functions of microorganisms in addressing the challenges posed by climate change and underscores their potential in fostering sustainable ecosystems and agriculture.

15.1 Consequences of Climate Change

The words "climate change" and "global warming" had been introduced by geochemist Wallace Broecker in 1975 to draw attention to the potential effects of rising carbon dioxide (CO_2) levels on temperature increases worldwide (Broecker 1975). Climate change is now universally acknowledged as the most significant threat to mankind (Steffen et al. 2007). A recent report from the Intergovernmental Panel on Climate Change (IPCC) indicates that globally 3.3 billion individuals are highly susceptible to climate change, while prevailing unsustainable development practices exacerbate the vulnerability of ecosystems and populations to climate-related hazards (IPCC 2022). However, anthropogenic activities have predominantly accelerated the climate change, mainly due to the burning of fossil fuels. Fossil fuels, including coal, oil, and gas, represent the primary contributor to global climate change, accounting for approximately 75% of worldwide greenhouse gas (GHG) emissions and nearly 90% of total CO_2 emissions (Yoro and Daramola 2020). According to IPCC, the oil and gas sector is responsible for 37% of the total global GHG (CO_2, methane, and nitrous oxide) emissions produced by the combustion of fossil fuels (IPCC 2022). The increase in global atmospheric temperature correlates with a significant rise in the frequency of natural disasters, largely attributed to hydrological and meteorological irregularities (Wang et al. 2016). From 1990 to 2019, the cumulative warming effect of anthropogenic greenhouse gases in the Earth's atmosphere increased by 45%, with CO_2 causing the most significant increase at 36% (Lamb et al. 2021). Increased CO_2 levels, and precipitation has potentially impact soil carbon and nitrogen dynamics along with N_2O emissions (Lugato et al. 2018). Further, the ongoing reduction in soil moisture levels may necessitate increased irrigation practices in agricultural systems, potentially resulting in diminished crop yields and contributing to desertification, which could have substantial impacts for food production (Kopittke et al. 2019). Therefore, predictions regarding the decline in global yields for wheat, maize, and rice suggest that yield losses may escalate by 10–25% for each degree Celsius increase in temperature (Deutsch et al. 2018). Brás et al. (2021) demonstrated that the frequency of heatwaves and droughts has led to a fourfold increase in agricultural losses over the past five decades, rising from 2.2% to 7.3%. The reduction in agricultural yield resulting from climate-related abiotic factors could exceed US$

170 billion each year, representing a considerable threat to worldwide food security (Razzaq et al. 2021). Changes in climate influence the various soil processes particularly those related to soil fertility. The effects of climate change on soils are expected mainly through alteration in soil moisture conditions and increase in soil temperature and CO_2 levels as a result of climate change (VijayaVenkataRaman et al. 2012). The increase in worldwide CO_2 emissions, measured at 8.4 Pg carbon (C) annually in 2010, has escalated from a rate of 1% per year between 1990 and 1999 to 2.5% per year during the decade from 2000 to 2009, acting as the main driver of global warming (Feng and Fang 2022). The effects of climate change, presently evident in agricultural systems, ecological networks, and forest biomes, are expected to demonstrate a variety of intricacies and multifaceted interactions (Kumar et al. 2024). Climate change has additional soil effects, including erosion which intensified by extreme weather events like heavy rainfall, droughts, heat waves, and storms (Bolan et al. 2023). Additionally, rising sea levels can alter coastal soil composition and introduce pollutants from marine environments, further affecting landmass reduction. The EEA's (The European Environment Agency) report on climate change adaptation and agriculture suggests that climate change could significantly impact the European agricultural sector, potentially causing a loss of up to 16% in EU agricultural income by 2050, with significant regional disparities (Paleari 2024).

Rising ozone concentrations and heightened temperatures may reduce CO_2 fertilisation, negatively impacting plant growth in elevated CO_2 conditions and potentially negating its effect (Allen et al. 2024). The Food and Agriculture Organisation of the United Nations (FAO) has recently published a map indicating that the upper 30 cm of global soil holds approximately twice the amount of carbon found in the entire atmosphere (Komal et al. 2024). Following the oceans, soil represents the second most significant natural carbon reservoir, exceeding the ability of forests and other forms of vegetation to sequester CO_2 from the atmosphere (Anisha 2024). Although soil microbes play essential roles in influencing Earth's climate by managing soil organic matter, which is the primary organic carbon reservoir in the terrestrial biosphere, our understanding of how climate change affects these microbes and their regulatory mechanisms is still considerably lacking (Wu et al. 2023). Microbes play a crucial role in influencing the amount of carbon stored within the soil system, as highlighted by a recent study that offers insights into strategies for addressing climate change and enhancing soil vitality for agricultural practices and the cultivation of food (Gougoulias et al. 2014). The soil microbiome regulates the biogeochemical cycling of macronutrients, micronutrients, and other essential components for the development of plants and animals (Sokol et al. 2022). Comprehending and forecasting the effects of climate change on soil microbiomes and the ecosystem services they provide is a significant challenge and a substantial opportunity as we focus our research endeavours on one of the most urgent issues confronting our planet (Jansson and Hofmockel 2020). This book offers significant insights into the utilisation of biological processes to achieve more sustainable agricultural practices, enhance carbon sequestration, and facilitate the restoration of the ecosystems. In light of the extraordinary environmental challenges currently confronting the globe, this extensive resource functions as a reference for academics, policy making, and conservations committed to formulating innovative approaches for achieving sustainable ecosystem in the future.

15.1.1 Microbes in Climate Change and Carbon Sequestration

Soils play a crucial role in the global carbon cycle. Estimates indicate that soils have contributed between 55 and 878 billion tonnes (GT) of carbon to the total atmospheric CO_2 concentrations (Kimble et al. 2016). Terrestrial ecosystems are integral to climate feedback processes via the emission and absorption of greenhouse gases (Singh et al. 2010). Soil also sequesters substantial amounts of carbon in live plants and soils, functioning as a crucial global carbon sink (Xue et al. 2024a, b). Research demonstrates that alterations in climate, including rising temperatures, heightened atmospheric CO_2 levels, increased nitrogen deposition, extreme storm occurrences, and conditions of flooding or drought, profoundly impact the composition and structure of microbial communities (Barbieri et al. 2023). Microorganisms engage in the consumption and production of significant greenhouse gases such as CO_2, CH_4, and N_2O, which play a role in the phenomenon of global warming (Filonchyk et al. 2024). For example, the thawing of permafrost alters soil conditions, which may significantly impact carbon reserves. As permafrost undergoes thawing, it liberates sequestered carbon that microorganisms could utilise for their metabolic processes or for the generation of CH_4 through methanogenesis (Romanowicz et al. 2023). Plants absorb carbon through photosynthesis and emit some back into the atmosphere through respiration. The carbon sequestered in plant tissue is ingested by animals or incorporated into the soil as organic matter (Feng et al. 2023). Soil carbon levels are primarily governed by the balance between organic matter inputs mainly derived from plant residues, roots, and root exudates and organic matter losses caused by decomposition, erosion, and leaching (Lei et al. 2023). Bacteria and fungi generally constitute more than 90% of the total soil microbial biomass and play a major role in the decomposition of soil organic matter (SOM) (Xue et al. 2023). SOM is a complex matrix of carbon-based compounds that includes decomposed plant and animal materials, microorganisms such as protozoa, fungi, and bacteria, and humus, which is characterized by carbon bound to soil minerals (Raza et al. 2023). Marine phytoplankton, including *Vicicitus* spp. and *Emiliania huxleyi*, contribute to half of global CO_2 photosynthetic incorporation, despite only representing 1% of the total photosynthetic biomass in the biosphere (Ibáñez et al. 2023). Methane, a potent greenhouse gas, is primarily produced by bacteria, accounting for 70–90% of its production. Recent increases in methane levels are partly due to human activities, including agricultural operations, landfill methanogenic microbes, coal mines, and natural gas extraction from various sources (Shindell et al. 2024). Methane-oxidizing microbes like *Methylococcus* and *Methylobacter* are vital for global equilibrium and have been identified as potential biofertilizers. They consume 40–60% of methane in wetland ecosystems, making them suitable for landfills and agricultural soils, contributing to environmental sustainability (Masood et al.

2021). Nitrogen comprises around 78% of the atmosphere and exists in several forms, including NH_3, NO, and N_2O, with N_2 being the major type (Rizwan et al. 2024). Nonetheless, it is not easily available to flora and fauna, entering the biosphere via biological fixation mediated by different bacteria such as *Azotobacter, Beijerinckia, Clostridium*, or *Rhizobium* or cyanobacteria (Streletskii et al. 2024). As a result, marine nitrogen fixation surpasses terrestrial nitrogen fixation by 30%. Numerous soil bacteria facilitate carbon sequestration via various processes; however, certain microbes exhibit accelerated metabolic rates, resulting in more rapid carbon sequestration (Espe et al. 2024). Soils may be enriched with these beneficial species by introducing environmentally respectful microbial formulations and by improving carbon sequestration in agricultural soil (Behl et al. 2024). CO_2 sequestration is linked to atmospheric pollution carbon credits, and the biomass generated by certain algal genera, like *Botryococcus sp., Scenedesmus*, and *Neochloris*, presents potential for energy fuel production and value-added products, integrates two principles of the circular bioeconomy (Raj et al. 2024). This cycle is crucial due to rising concerns about fossil fuel depletion and holds significance from both ecological and economic perspectives (Raj et al. 2024).

15.1.2 Microbes in Agriculture and Plant Growth

Climate fluctuations, including rising temperatures, erratic precipitation, glacier retreat, and severe weather events, are expected to affect various businesses (Rawat et al. 2024). These impacts may lead to technical issues, such as power supply disruptions, and pose more severe challenges to critical sectors like agriculture, forestry, livestock, and commercial fishing (Tchonkouang et al. 2024). These sectors, which are economically significant, experienced a 73% global value increase from 2000 to 2019, reaching USD 3.5 trillion in 2018 (Tick et al. 2024). Currently, farmers and scientists are focusing on the use of biofertilizers and sustainable tillage methods to safeguard soil health, diminish greenhouse gas emissions, and enhance carbon sequestration (Saha and Bauddh 2020). Nonetheless, the rise in disease outbreaks and the geographical proliferation of various pathogens is jeopardising the equilibrium between environmentally sustainable practices and the strategies to address these emerging climate change-related threats (Shafique et al. 2024). The relationship between microorganisms, atmospheric gases, and agricultural plants is complex, with differences between short-term laboratory and long-term field studies on climate change's effects on soil communities and nutrient dynamics (Andrade-Linares et al. 2021). Microbes have developed mechanisms to endure stress conditions, however, changes in plant species that exhibit greater resistance to fire or drought will consequently modify the microbial communities associated with their root systems (Berard et al. 2015). The modification of microbial respiration or denitrification rates in soil microorganisms due to climate change can influence microbial metabolism and nutrient feedback mechanisms, resulting in inaccurate measurements of these significant greenhouse gases, which may impede mitigation efforts (Adingo et al. 2021). Consequently, elevated temperatures enhance microbial decomposition activities, resulting in a rise in CO_2 emissions. Observations indicate that elevated atmospheric CO_2 levels promote the proliferation of rhizosphere-colonizing bacteria, including *Burkholderia* and *Pseudomonas*, along with fungi that enhance plant growth (Singh et al. 2021). In contrast, non-rhizospheric species such as *Bacillus* do not exhibit the same degree of stimulation. The rise in CO_2 levels promotes the proliferation of rhizobial populations and boosts the activity of nitrogen-fixing microorganisms in controlled environments; nonetheless, various resource constraints hinder the functioning of rhizobia in natural ecosystems (Singh et al. 2021). Conversely, elevated temperatures result in drought conditions, which in turn diminish the colonisation by arbuscular mycorrhizal fungi (Liu et al. 2023). Delgado-Baquerizo et al. (2020), showed that rising temperatures lead to a higher prevalence of soil-borne fungal pathogens. Additionally, Bebber (2015) indicated that many pathogens and pests have been migrating poleward at a rate of approximately 3 km per year since 1960. Climate change's effects on plant pests, crop yields, and biodiversity are challenging to measure, but bacterial and fungal diseases are expanding geographically, requiring a fundamental approach (Singh et al. 2023). Change on agricultural production involves the application of biostimulants (Clark et al. 2021). These substances improve plant tolerance and resilience against both biotic and abiotic stressors, demonstrating low toxicity and inhibiting the development of novel resistant strains of pests and diseases (Bhuyan et al. 2023). As a result, the biofertilizer market has exhibited steady expansion in recent years, with Europe establishing itself as the leading region in the global industry (Backer et al. 2018). A variety of microorganisms function as biostimulants by providing vital nutrients for crops: (i) via nitrogen fixation (*Azospirillum, Azotobacter*, or *Rhizobium*), (ii) through the solubilisation of minerals such as phosphorus, potassium, or zinc (*Pseudomonas, Bacillus*, or arbuscular mycorrhiza), or (iii) by producing siderophores (*Pseudomonas* and *Acinetobacter*) (Kumar et al. 2022). Certain microbial species have demonstrated the ability to protect crops from elevated temperatures. For instance, *Paraburkholderia phytofirmans* has been observed in potato crops, while *Bacillus* and *Azospirillum* species have been noted in wheat and soy in high temperature (Sarker et al. 2021). Microorganisms can fix significant amounts of nitrogen in the rhizosphere, achieving efficiencies of 52 mg N_2 g^{-1} malate (Kuan et al. 2016).

15.1.3 Microbes in Ecosystem Restoration and Remediation

Ecosystem restoration involves the processes aimed at revitalising environments that have been damaged, or destroyed, primarily through deliberate actions by humans (Macdonald and King 2018). Bioremediation involves the utilisation of various organisms, including prokaryotes, fungi, and plants, to mitigate pollution in ecosystems by sequestering toxic substances within their biological systems (Pandey et al. 2012). Microbial organisms remarkable influence impact the global environment, contributing to 50% of total oxygen generation, 75% of additional nitrogen, and 92% of nitrogen removal, with bacterial activity playing a crucial role in these processes (Kuypers et al. 2018). Microbes are essential in facilitating the transfer of nutrients from the soil to plants in forms that can be utilised effectively (Pandey and Dheeman 2022). The bacterium *Alcanivorax borkumensis* exhibits a remarkable ability to thrive in environments rich in hydrocarbons and plays an essential part in the bioremediation of oil spills (Singha et al. 2018a, b). Understanding the vital role of soil microorganisms for the growth of higher plants and the overall health of ecosystems is crucial for effective restoration efforts (Singha and Pandey 2021). The utilisation of plants that engage in nitrogen-fixing symbioses has been a common practice in restoration efforts (Singha and Pandey 2023). Mycorrhizal associations have been extensively investigated in restoration contexts, where their advantages have been consistently evidenced, although the requirement for active mycorrhizal introduction remains essential (Alaux et al. 2021). Soil microorganisms facilitate the mineralisation of organic substrates derived from litter, subsequently releasing nutrients through their heterotrophic processes (Singha et al. 2018a, b). Consequently, microbial biomass plays a crucial role in the preservation of soil quality, the cycling of nutrients, and the process of soil reclamation (Saccá et al. 2017). *Pseudomonas* species are known for their ability to degrade hydrocarbons in oil spills in both marine and terrestrial environments (Devi et al. 2017).

This bioremediation treatment technology can be applied using two distinct methodologies: *in situ* and *ex situ*. *In situ* techniques improve the biodegradation rate of organic pollutants in affected soil, sediment, surface water, or groundwater environments (Das et al. 2024). Ex situ techniques require substantial extraction of contaminated materials, usually soils or sediments, which are subsequently treated under controlled conditions in bioreactors, biopiles, compost heaps, ponds, or lagoons (Baker and Herson 1994). Further, the contamination of heavy metals is a significant issue due to their toxic properties and the potential risks they pose to animal and human health, as well as the environment (Das et al. 2021). Microorganisms like *Alcaligenes faecalis, Pseudomonas fluorescens*, and *Enterobacter cloacae* are known for their metabolic activities associated with the absorption of heavy metals, particularly arsenic and chromium (Kotoky and Pandey 2021). A recent method utilising microbial surfactants has been developed, providing an enhanced approach for addressing contamination in soil and water (Das et al. 2023). *Acinetobacter calcoaceticus* synthesizes biosurfactants that help in binding heavy metals like uranium. The rhamnolipid class of biosurfactants is effective in mitigating metal toxicity and removing heavy metals like cadmium, zinc, and lead from soil. This occurs through a complex interaction between Cd and rhamnolipid, modifying Cd uptake. Surfactins are used in dewatering peats at concentrations as low as 13 g/ton (Shakeri et al. 2021). In addition, rhizoremediation may significantly increase carbon sequestration and successfully remove petroleum hydrocarbons from the environment by completely mineralising the area (Ruley et al. 2022). According to Kotoky et al. (2018), soil-dwelling obligate hydrocarbonoblastic bacteria, including *Nesiotobacter, Nitratireductor, Acinetobacter, Marinobacter, Pseudoalteromonas,* and *Sphingomonas,* have the ability to break down polyaromatic hydrocarbons (PAH).

15.1.4 Microbes in Soil Functioning and Nutrient Cycling

Microorganisms are fundamental agents in the cycling of essential elements, including carbon, nitrogen, and phosphorus, across the soil (Debnath et al. 2023). They engage in the consumption and production of gases that contribute to the phenomenon of global warming (Dincă et al. 2022). Variations in temperature, humidity, or elemental flux resulting from climate change can significantly influence microbial community structure and metabolic activity (Xue et al. 2016). Soil aggregation is a crucial mechanism for microbial activity, facilitating the sequestration of carbon in soil (Bhattacharyya et al. 2022). It enhances soil structure, facilitates gas exchange, and improves water infiltration and retention. Soil aggregates protect soil organic carbon from degradation, enhancing its residence time by converting carbon compounds from labile to recalcitrant pools (Blanco-Canqui and Lal 2004). Microorganisms, particularly fungi and bacteria, play a crucial role in the development of micro and macro aggregates, facilitating the binding of soil particles through secretions and biomass (Costa et al. 2018). Bhattacharyya et al. (2022), investigated the interactions between soil microbial communities and organic matter, revealing various mechanisms affecting soil carbon storage. Bacteria and fungi constitute the predominant fraction of soil microbial biomass, exceeding 90%. Consequently, these groups, along with the ratio between them (F:B ratio), can significantly influence the flux of carbon within the soil ecosystem (Xu et al. 2020). A soil system dominated by fungi is proposed to indicate a higher carbon ratio, as fungal groups are believed to allocate more carbon to biomass per unit of substrate utilized compared to other primary decomposers, such as bacteria (Rousk and Frey 2015). Chemoautotrophic microbes, including acetogenic bacteria, are utilized to transform CO_2 into various organic compounds, illustrating one of the most ancient pathways for carbon fixation (Correa et al. 2023). Oxalotrophic bacteria represent a specific microbial group that has the capability to convert organic carbon molecules into recalcitrant pools (Syed et al. 2020). Oxalotrophic bacteria utilize oxalate as their exclusive source of carbon and energy, resulting in the production of carbonates through the oxidation and degradation of oxalate, a process known as the oxalate-carbonate pathway (Syed et al. 2020). Clay minerals, such as kaolinite and montmorillonite, in conjunction with metal oxides like ferric oxide and borotitanite, possess the ability to promote polymerization, leading to the generation of inert carbon oxide (Kumari and Mohan 2021). Soil pH level significantly impacts chemical and biochemical processes, including the microbial carbon cycle. Its buffering capacity helps maintain a stable pH level despite external disturbances (Ramesh et al. 2019). However, minor pH changes can cause significant electric charge variations, affecting carbon emissions (Cheng et al. 2008). Soil pH also affects the turnover and accumulation of soil organic carbon by altering the structure and metabolic features of bacteria in the soil community (Xie et al. 2022). Carbonate precipitation is a significant contributor to soil inorganic carbon (SIC), with cyanobacteria being particularly notable (Lamérand et al. 2022). Photosynthesis, a process that exchanges

HCO^{3-}/OH^- across the cellular membrane, increases the pH surrounding cells, making it favorable for carbonate precipitation. Extracellular polymeric substances (EPS) can sequester divalent cations like Ca^{2+} and Mg^{2+}, facilitating carbonate precipitation (Zhu and Dittrich 2016). Bacterial species like *Bacillus subtilis* absorb metal ions, enhancing carbonate precipitation. Recent studies highlight the significant impact of environmental conditions on the distribution of chemoautotrophic microorganisms, particularly in their roles in global climate change, methane production, methane oxidation, and the nitrogen cycle (Robles-Fernández et al. 2022) (Fig. 15.1).

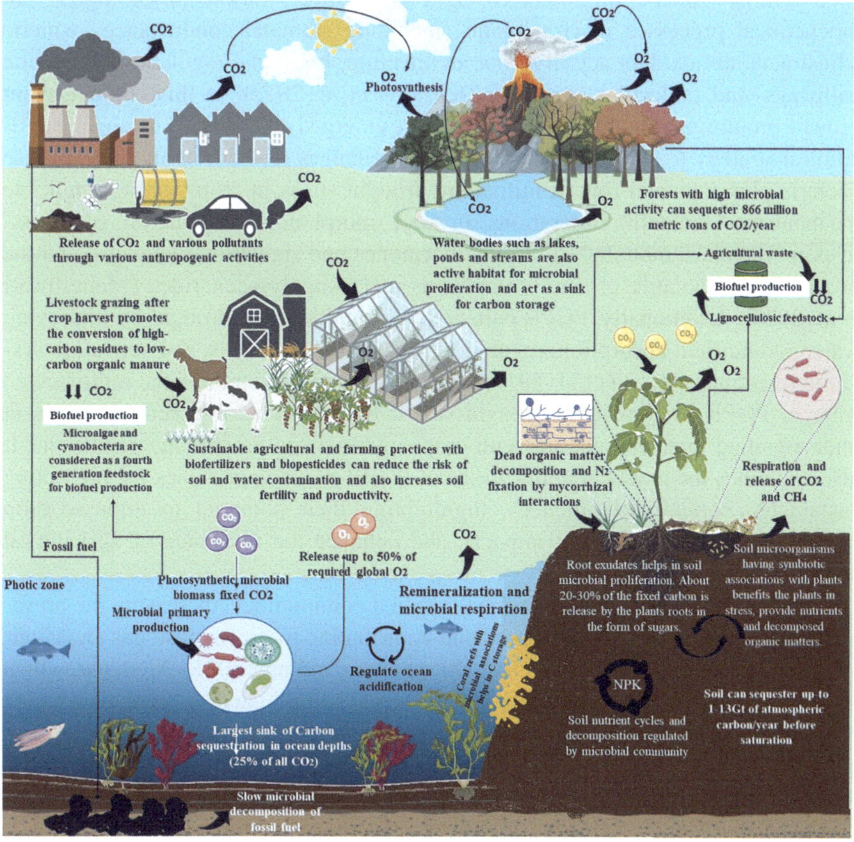

Fig. 15.1 Microbes play a crucial role in terrestrial and aquatic environments, regulating geochemical cycles, promoting plant growth, and reducing carbon sequestration. In aquatic environments, microbes are the largest carbon sink and contribute to ocean acidification. They also enhance crop yields and production, improving soil health. Soil and marine systems play a vital role as carbon sinks in carbon sequestration and mitigating climate change

15.2 Contributions and Considerations

A detailed discussion regarding climate change highlights its potential to alter microbial community composition, affect the production of greenhouse gases, and increase the occurrence of pathogens in both plant and animal populations (Hutchins et al. 2019). These observed variations may subsequently influence agricultural output and the well-being of human populations. On the other hand, application of microorganisms can to mitigate the harmful impacts associated with climate change has been discussed throughout the all-book chapters. The escalation of drought conditions and shifts in climate patterns pose a significant threat to agricultural productivity, emphasizing the urgent need for sustainable solutions to safeguard global food security (Raza et al. 2024). Drought stress affects various physiological and biochemical processes in crop plants, including stomatal conductance, osmotic adjustment, antioxidant activity, root architecture, hormonal regulation, metabolic pathways, and molecular responses (Khatoon et al. 2022). In this context, plant growth-promoting rhizobacteria (PGPR) have emerged as a critical sustainable agricultural strategy to address climate change challenges (Kshetri et al. 2019). These bacteria play a pivotal role in mitigating drought stress in crops through multiple mechanisms, including alterations in root morphology, regulation of stress-responsive genes, modulation of plant hormones and signalling pathways, and the production of volatile organic compounds and exopolysaccharides (Maheshwari et al. 2023). Additionally, PGPR enhance plant resilience to drought by improving water and nutrient absorption, regulating stress hormone levels, and activating systemic resistance (Nevita et al. 2018). Chapters 1 and 14, explores the multifaceted roles of PGPR in alleviating drought stress, with a focus on recent advances in understanding their interactions with crops under changing climatic conditions. It also examines the practical applications of PGPR-based approaches for promoting sustainable agricultural practices, highlighting their potential to improve crop yields, reduce environmental impacts, and enhance the resilience of agricultural systems vulnerable to drought.

Climate change is rapidly altering microbial communities, increasing the prevalence of plant and animal pathogens and contributing to greenhouse gas emissions such as carbon dioxide, methane, and nitrous oxide (Kang 2021). Methane, which traps heat 25 times more efficiently than CO_2, can be mitigated by regulating rumen microbes to produce less of it, thereby slowing global warming (Faizan 2024). For example, *Vibrio* spp. have expanded in warmer waters, while ocean acidification has led to the loss of microbial taxa and the emergence of new species (Brumfield et al. 2021). Despite these challenges, microbes also present solutions to combat climate change due to their rapid growth, large populations, and ability to share genetic information (Jansson and Hofmockel 2020). Phages can disrupt CO_2 fixation during infection, and methanotrophs can help reduce methane emissions, offering valuable tools for addressing global warming (Hutchins et al. 2019) was discussed in Chap. 2.

In-depth investigation of bacterial carbon fixation pathways, encompassing both natural and engineered processes, is crucial for comprehending their contribution to climate change mitigation (Mustafa et al. 2024). Utilizing these pathways allows for the improvement of carbon capture and storage, contributing to the regulation of atmospheric CO_2 concentrations and mitigating the effects of global warming (Nanda et al. 2016). Natural pathways include the Calvin-Benson-Bassham cycle, Reductive TCA cycle, Reductive acetyl-CoA cycle, and others, while engineered pathways such as the Crotonyl-CoA cycle, Synthetic formolase pathway, and Synthetic acetyl-CoA pathway are also examined. Chapter 3 discussed the significance of microbial carbon fixation and storage for maintaining atmospheric CO_2 balance and ecosystem health (Correa et al. 2023). The chapter emphasizes the importance of leveraging microbial carbon fixation for climate regulation, while addressing challenges and future directions for its application in mitigation strategies.

Climate change poses a significant challenge to humanity, directly affecting microorganisms that play critical roles in ecosystems and agriculture (Pandey et al. 2012). Soil microbes, particularly, are impacted by biotic and abiotic stresses, leading to disruptions in plant physiology, crop yields, and global food security (Debnath et al. 2023). These stresses disturb key biogeochemical cycles, including the carbon, nitrogen, and phosphorus cycles, and hinder plant-microbe interactions (Pandey and Dheeman 2022). However, beneficial microbes, such as mycorrhizal fungi, can improve plant resilience by enhancing nutrient acquisition, water absorption, and stress tolerance (Wahab et al. 2023). They regulate plant hormones, boost antioxidant systems, and induce immunities like Acquired Systemic Resistance (ASR) and Induced Systemic Resistance (ISR) (Riseh and Vazvani 2024). By strengthening plant defenses and improving soil health, microbes act as ecological engineers, playing a vital role in mitigating climate-related stress and supporting sustainable agriculture (Kotoky and Pandey 2020). Chapters 4 and 5 provide detailed discussions on the mechanisms through which microbes help plants cope with abiotic and biotic stresses, highlighting their importance for agricultural sustainability.

Iron is an essential micronutrient crucial for the growth and development of both plants and humans (Rout and Sahoo 2015). Its deficiency is a significant cause of death and disease, particularly in developing countries, with approximately one-fourth of the global population suffering from anemia, as reported by the World Health Organization (WHO) (Delil et al. 2018). Biofortification, which enhances iron concentration in edible food crops through agronomic intervention, plant breeding, or genetic selection, effectively combats this deficiency and hidden hunger (Kumari et al. 2022). Siderophore-producing plant growth-promoting rhizobacteria (PGPR), especially *Pseudomonas fluorescens*, play a vital role in improving plants' ability to absorb iron from the soil (Scavino and Pedraza 2013). Furthermore, plant roots are surrounded by a dynamic microbiome that contributes to plant health by aiding nutrient cycling, mineralization, and pathogen defense (Kotoky et al. 2018). Plants actively manipulate these interactions through host selection, recruiting specific microbes to optimize performance under various stressors (Kotoky et al. 2017). Advances in breeding techniques, such as gene editing, have led to improved crops that maximize the benefits of plant-microbe relationships, enhancing productivity and disease resistance (Shelake et al. 2019). Variations in root development, rhizodeposition rates, and nutrient uptake among different genotypes create a customized rhizomicrobiome shaped by selective pressures (Singh et al. 2020). To sustain a beneficial microbiota, plants employ various strategies for microbial recruitment at the root surface. Thus, Chaps. 6, 8 and 10 explore the complexities of these interactions and their implications for agricultural sustainability, highlighting their potential to enhance crop resilience and productivity.

Conventional CO_2 capture technologies are often inadequate and can contribute to secondary environmental pollution (Sifat and Haseli 2019). In contrast, bio-based remediation and environmentally friendly mitigation strategies have also been implemented to protect and restore environments contaminated by crude oil, with rhizoremediation gaining prominence for its effective in-situ applications (Bhuyan et al. 2023). Chapter 7, explores rhizoremediation approaches for mitigating crude oil contamination and strategies for CO_2 conversion aimed at reducing greenhouse gas emissions (Kotoky and Pandey 2020). Additionally, recent advancements in engineering bacteria to efficiently utilize CO_2 and other carbon compounds are highlighted (Kotoky et al. 2019). Modern sequencing technologies, such as metagenomics, transcriptomics, and proteomics, play a crucial role in identifying and characterizing microbial communities and metabolic pathways involved in these processes (Kotoky et al. 2018). These sequencing approaches provide deep insights into microbial ecology and function, facilitating the development of more efficient bio-based CO_2 conversion and bioremediation strategies (Das et al. 2024). Moreover, extremophiles present promising solutions to the escalating challenges of climate change, demonstrating resilience in extreme conditions and enhancing carbon sequestration and pollutant detoxification. Their unique abilities contribute to the bioeconomy and facilitate the development of innovative strategies for ecosystem resilience (Rajkumari et al. 2019). Integrating extremophiles into bio-based technologies exemplifies a strategic approach to advancing global sustainability efforts, highlighting their potential in mitigating climate change impacts while restoring degraded ecosystems is discussed in Chap. 9.

Chapter 11 explores the role of psychrophilic microbes in glacial ecosystems as potential bio-indicators of climate change. The survival of these microorganisms in extreme cold is due to their evolved proteins. The cryosphere, a significant habitable zone within Earth's biosphere, plays a crucial role in the global climate system and offers insights into extraterrestrial life exploration (Feller 2017). Glacier ice preserves microbial life and historical climate records spanning hundreds of thousands of years. The loss of glaciers in the Hindu-Kush Himalayas may lead to a decrease in psychrophilic organisms, threatening a significant reservoir of beneficial microorganisms and underscoring the environmental consequences of glacial retreat and the broader implications of a changing climate (Dhakar and Pandey 2020). On the other hand, the applications of thermophiles isolated from cow dung constitute a distinct and advantageous category of microorganisms that contribute substantially to sustainable agricultural practices, particularly in tropical regions like India and Pakistan, where summer temperatures can reach 42–45 °C (Moldagulova et al. 2020). Their capacity to thrive in elevated temperatures, along with their enzymatic functions, makes them significant assets for improving soil fertility, supporting organic agriculture, and fostering climate resilience (Meena and Jha 2018). Utilizing the capabilities of these thermophilic microbes has the potential to enhance sustainable agricultural methodologies and bolster food security in the context of a changing climate is discussed in Chap. 12.

Moreover, the health of individuals is significantly threatened by the global challenges of climate change and antimicrobial resistance (AMR) (Das et al. 2021). The environmental disruptions exacerbate the emergence and spread of infectious diseases, including those caused by antimicrobial-resistant pathogens (Baker et al. 2018). Antimicrobial resistance, fuelled by the overuse and misuse of antibiotics in human health, agriculture, and veterinary medicine, undermines the effectiveness of medical treatments, threatening the management of common infections and compromising the outcomes of routine medical procedures (Muteeb et al. 2023). Addressing these intersecting challenges requires integrated and interdisciplinary approaches that recognize their synergistic impacts and underlying drivers. Thus, Chap. 13 discussed the comprehensive details of tackling these issues holistically, to build a more resilient and sustainable future for generations to come.

This summary emphasizes the role of microorganisms in the global greenhouse gas balance and their various functions beyond atmospheric gas regulation. Microorganisms play a crucial role in maintaining soil health, promoting plant growth, and mitigating climate change impacts through natural processes like carbon sequestration and nutrient cycling (Kumari et al. 2023). The systematic association between climate change and microorganisms across various human sectors, including industrial and health, allows for the formulation of advanced strategies that contribute to environmental protection and promote sustainability for future generations.

References

Adingo S, Yu JR, Xuelu L, Li X, Jing S, Xiaong Z (2021) Variation of soil microbial carbon use efficiency (CUE) and its influence mechanism in the context of global environmental change: a review. PeerJ 9:e12131. https://doi.org/10.7717/peerj.12131

Alaux PL, Zhang Y, Gilbert L, Johnson D (2021) Can common mycorrhizal fungal networks be managed to enhance ecosystem functionality? Plants People Planet 3(5):433–444. https://doi.org/10.1002/ppp3.10178

Allen RJ, Gomez J, Horowitz LW, Shevliakova E (2024) Enhanced future vegetation growth with elevated carbon dioxide concentrations could increase fire activity. Commun Earth Environ 5(1):54. https://doi.org/10.1038/s43247-024-01228-7

Andrade-Linares DR, Zistl-Schlingmann M, Foesel B, Dannenmann M, Schulz S, Schloter M (2021) Short term effects of climate change and intensification of management on the abundance of microbes driving nitrogen turnover in montane grassland soils. Sci Total Environ 780:146672. https://doi.org/10.1016/j.scitotenv.2021.146672

Anisha GS (2024) Carbon storage in terrestrial ecosystems. In: Advances and technology development in greenhouse gases: emission, capture and conversion. Elsevier, pp 91–111. https://doi.org/10.1016/B978-0-443-19067-4.00008-5

Backer R, Rokem JS, Ilangumaran G, Lamont J, Praslickova D, Ricci E, Subramanian S, Smith DL (2018) Plant growth-promoting rhizobacteria: context, mechanisms of action, and roadmap to commercialization of biostimulants for sustainable agriculture. Front Plant Sci 9:1473. https://doi.org/10.3389/fpls.2018.01473

Baker KH, Herson DS (1994) Bioremediation (pp xvi+−375)
Baker S, Thomson N, Weill FX, Holt KE (2018) Genomic insights into the emergence and spread of antimicrobial-resistant bacterial pathogens. Science 360(6390):733–738. https://doi.org/10.1126/science.aar3777
Barbieri M, Barberio MD, Banzato F, Billi A, Boschetti T, Franchini S, Gori F, Petitta M (2023) Climate change and its effect on groundwater quality. Environ Geochem Health 45(4):1133–1144. https://doi.org/10.1007/s10653-021-01140-5
Bebber DP (2015) Range-expanding pests and pathogens in a warming world. Annu Rev Phytopathol 53(1):335–356. https://doi.org/10.1146/annurev-phyto-080614-120207
Behl K, Jaiswal P, Pabbi S (2024) Recent advances in microbial and nano-formulations for effective delivery and agriculture sustainability. Biocatal Agric Biotechnol 103180:103180. https://doi.org/10.1016/j.bcab.2024.103180
Berard A, Sassi MB, Kaisermann A, Renault P (2015) Soil microbial community responses to heat wave components: drought and high temperature. Clim Res 66(3):243–264. https://doi.org/10.3354/cr01343
Bhattacharyya SS, Ros GH, Furtak K, Iqbal HM, Parra-Saldívar R (2022) Soil carbon sequestration–an interplay between soil microbial community and soil organic matter dynamics. Sci Total Environ 815:152928. https://doi.org/10.1016/j.scitotenv.2022.152928
Bhuyan B, Kotoky R, Pandey P (2023) Impacts of rhizoremediation and biostimulation on soil microbial community, for enhanced degradation of petroleum hydrocarbons in crude oil-contaminated agricultural soils. Environ Sci Pollut Res 30(41):94649–94668. https://doi.org/10.1007/s11356-023-29033-3
Blanco-Canqui H, Lal R (2004) Mechanisms of carbon sequestration in soil aggregates. Crit Rev Plant Sci 23(6):481–504. https://doi.org/10.1080/07352680490886842
Bolan S, Padhye LP, Jasemizad T, Govarthanan M, Karmegam N, Wijesekara H, Amarasiri D, Hou D, Zhou P, Biswal BK, Balasubramanian R (2023) Impacts of climate change on the fate of contaminants through extreme weather events. Sci Total Environ 909:168388. https://doi.org/10.1016/j.scitotenv.2023.168388
Brás TA, Seixas J, Carvalhais N, Jägermeyr J (2021) Severity of drought and heatwave crop losses tripled over the last five decades in Europe. Environ Res Lett 16(6):065012. https://doi.org/10.1088/1748-9326/abf004
Broecker WS (1975) Climatic change: are we on the brink of a pronounced global warming? Science 189(4201):460–463. https://doi.org/10.1126/science.189.4201.460
Brumfield KD, Usmani M, Chen KM, Gangwar M, Jutla AS, Huq A, Colwell RR (2021) Environmental parameters associated with incidence and transmission of pathogenic vibrio spp. Environ Microbiol 23(12):7314–7340. https://doi.org/10.1111/1462-2920.15716
Cheng CH, Lehmann J, Engelhard MH (2008) Natural oxidation of black carbon in soils: changes in molecular form and surface charge along a climosequence. Geochim Cosmochim Acta 72(6):1598–1610. https://doi.org/10.1016/j.gca.2008.01.010
Clark IM, Hughes DJ, Fu Q, Abadie M, Hirsch PR (2021) Metagenomic approaches reveal differences in genetic diversity and relative abundance of nitrifying bacteria and archaea in contrasting soils. Sci Rep 11(1):15905. https://doi.org/10.1038/s41598-021-95100-9
Correa SS, Schultz J, Lauersen KJ, Rosado AS (2023) Natural carbon fixation and advances in synthetic engineering for redesigning and creating new fixation pathways. J Adv Res 47:75–92
Costa OY, Raaijmakers JM, Kuramae EE (2018) Microbial extracellular polymeric substances: ecological function and impact on soil aggregation. Front Microbiol 9:1636. https://doi.org/10.3389/fmicb.2018.01636
Das N, Kotoky R, Maurya AP, Bhuyan B, Pandey P (2021) Paradigm shift in antibiotic-resistome of petroleum hydrocarbon contaminated soil. Sci Total Environ 757:143777. https://doi.org/10.1016/j.scitotenv.2020.143777
Das N, Das A, Das S, Bhatawadekar V, Pandey P, Choure K, Damare S, Pandey P (2023) Petroleum hydrocarbon catabolic pathways as targets for metabolic engineering strategies for enhanced bioremediation of crude-oil-contaminated environments. Fermentation 9(2):1–31. https://doi.org/10.3390/fermentation9020196

Das N, Maheshwari DK, Pandey P (2024) Energy crop-based rhizoremediation and lignocellulosic biomass production as sustainable bioeconomy-driven solution for biofuel generation and waste mitigation. Prog Energy Combust Sci 103(July 2023):101161. https://doi.org/10.1016/j.pecs.2024.101161

Debnath S, Chakraborty S, Langthasa M, Choure K, Agnihotri V, Srivastava A, Rai PK, Tilwari A, Maheshwari DK, Pandey P (2023) Non-rhizobial nodule endophytes improve nodulation, change root exudation pattern and promote the growth of lentil, for prospective application in fallow soil. Front Plant Sci 14(April):1–19. https://doi.org/10.3389/fpls.2023.1152875

Delgado-Baquerizo M, Guerra CA, Cano-Díaz C, Egidi E, Wang JT, Eisenhauer N, Singh BK, Maestre FT (2020) The proportion of soil-borne pathogens increases with warming at the global scale. Nat Clim Chang 10(6):550–554. https://doi.org/10.1038/s41558-020-0759-3

Delil R, Tamiru D, Zinab B (2018) Dietary diversity and its association with anemia among pregnant women attending public health facilities in South Ethiopia. Ethiop J Health Sci 28(5):625–634. https://doi.org/10.4314/ejhs.v28i5.14

Deutsch CA, Tewksbury JJ, Tigchelaar M, Battisti DS, Merrill SC, Huey RB, Naylor RL (2018) Increase in crop losses to insect pests in a warming climate. Science 361(6405):916–919. https://doi.org/10.1126/science.aat3466

Devi KA, Pandey G, Rawat AKS, Sharma GD, Pandey P (2017) The endophytic symbiont-Pseudomonas aeruginosa stimulates the antioxidant activity and growth of Achyranthes aspera L. Front Microbiol 8(Sep):1–14. https://doi.org/10.3389/fmicb.2017.01897

Dhakar K, Pandey A (2020) Microbial ecology from the Himalayan cryosphere perspective. Microorganisms 8(2):257. https://doi.org/10.3390/microorganisms8020257

Dincă LC, Grenni P, Onet C, Onet A (2022) Fertilization and soil microbial community: a review. Appl Sci 12(3):1198. https://doi.org/10.3390/app12031198

Espe EL, Bouyoucef LS, Dozois JA, Yergeau E (2024) Tipping the plant-microbe competition for nitrogen in agricultural soils. iScience. https://doi.org/10.1016/j.isci.2024.110973

Faizan M (2024) Enteric methane production in ruminants: its effect on global warming and mitigation strategies-a review. Pak J Sci *76*(01):16–38. https://doi.org/10.57041/pjs.v76i01.1084

Feller G (2017) Cryosphere and psychrophiles: insights into a cold origin of life? Life 7(2):25. https://doi.org/10.3390/life7020025

Feng R, Fang X (2022) China's pathways to synchronize the emission reductions of air pollutants and greenhouse gases: pros and cons. Resour Conserv Recycl 184:106392. https://doi.org/10.1016/j.resconrec.2022.106392

Feng JC, Sun L, Yan J (2023) Carbon sequestration via shellfish farming: a potential negative emissions technology. Renew Sust Energ Rev 171:113018. https://doi.org/10.1016/j.rser.2022.113018

Filonchyk M, Peterson MP, Zhang L, Hurynovich V, He Y (2024) Greenhouse gases emissions and global climate change: examining the influence of CO_2, CH_4, and N_2O. Sci Total Environ 173359. https://doi.org/10.1016/j.scitotenv.2024.173359

Gougoulias C, Clark JM, Shaw LJ (2014) The role of soil microbes in the global carbon cycle: tracking the below-ground microbial processing of plant-derived carbon for manipulating carbon dynamics in agricultural systems. J Sci Food Agric 94(12):2362–2371. https://doi.org/10.1002/jsfa.6577

Hutchins DA, Jansson JK, Remais JV, Rich VI, Singh BK, Trivedi P (2019) Climate change microbiology—problems and perspectives. Nat Rev Microbiol 17(6):391–396. https://doi.org/10.1038/s41579-019-0178-5

Ibáñez A, Garrido-Chamorro S, Barreiro C (2023) Microorganisms and climate change: a not so invisible effect. Microbiol Res 14(3):918–947. https://doi.org/10.3390/microbiolres14030064

IPCC (2022) Fact sheets | Climate change 2022: impacts, adaptation and vulnerability. In Fact sheets | climate change 2022: impacts, adaptation and vulnerability. https://www.ipcc.ch/report/ar6/wg2/about/factsheets

Jansson JK, Hofmockel KS (2020) Soil microbiomes and climate change. Nat Rev Microbiol 18(1):35–46. https://doi.org/10.1038/s41579-019-0265-7

Kang S (2021) Microbes' many roles in climate change: contribution, consequence, mitigation, and model system. In: Microbes: the foundation stone of the biosphere, pp 187–194. https://doi.org/10.1007/978-3-030-63512-1_11

Khatoon Z, del Carmen Orozco-Mosqueda M, Huang S, Nascimento FX, Santoyo G (2022) Peptide antibiotics produced by bacillus species: first line of attack in the biocontrol of plant diseases. In: Islam MT, Rahman M, Pandey P (eds) Bacilli in Agrobiotechnology. Bacilli in climate resilient agriculture and bioprospecting. Springer, Cham. https://doi.org/10.1007/978-3-030-85465-2_2

Kimble JM, Lal R, Follett RR (2016) Agricultural practices and policy options for carbon sequestration: what we know and where we need to go. In: Agricultural practices and policies for carbon sequestration in soil. CRC Press, pp 519–526

Komal N, Zaman QU, Pantera A, Yasin G, Ashraf K, Nazir S, Baig MB (2024) Climate-smart agriculture: potential role in carbon sequestration and to address climate change under irrigated agro-ecosystems. In: Climate-smart and resilient food systems and security. Springer Nature Switzerland, Cham, pp 73–110. https://doi.org/10.1007/978-3-031-65968-3_4

Kopittke PM, Menzies NW, Wang P, McKenna BA, Lombi E (2019) Soil and the intensification of agriculture for global food security. Environ Int 132:105078. https://doi.org/10.1016/j.envint.2019.105078

Kotoky R, Pandey P (2020) Difference in the rhizosphere microbiome of Melia azedarach during removal of benzo(a)pyrene from cadmium co-contaminated soil. Chemosphere 258:127175. https://doi.org/10.1016/j.chemosphere.2020.127175

Kotoky R, Pandey P (2021) The genomic attributes of Cd-resistant, hydrocarbonoclastic Bacillus subtilis SR1 for rhizodegradation of benzo(a)pyrene under co-contaminated conditions. Genomics 113(1P2):613–623. https://doi.org/10.1016/j.ygeno.2020.09.057

Kotoky R, Singha LP, Pandey P (2017) Crossm resistant soil bacterium Serratia marcescens S2I7, which has the ability to degrade polyaromatic hydrocarbons 5(48):1–2

Kotoky R, Rajkumari J, Pandey P (2018) The rhizosphere microbiome: significance in rhizoremediation of polyaromatic hydrocarbon contaminated soil. J Environ Manag 217:858–870. https://doi.org/10.1016/j.jenvman.2018.04.022

Kotoky R, Nath S, Kumar Maheshwari D, Pandey P (2019) Cadmium resistant plant growth promoting rhizobacteria Serratia marcescens S2I7 associated with the growth promotion of rice plant. Environ Sustain 2(2):135–144. https://doi.org/10.1007/s42398-019-00055-3

Kshetri L, Naseem F, Pandey P (2019) Role of Serratia sp. As biocontrol agent and plant growth stimulator, with prospects of biotic stress management in plant. In: Sayyed R (ed) Plant growth promoting Rhizobacteria for sustainable stress management, Microorganisms for sustainability, vol 13. Springer, Singapore. https://doi.org/10.1007/978-981-13-6986-5_6

Kuan KB, Othman R, Abdul Rahim K, Shamsuddin ZH (2016) Plant growth-promoting rhizobacteria inoculation to enhance vegetative growth, nitrogen fixation and nitrogen remobilisation of maize under greenhouse conditions. PLoS One 11(3):e0152478. https://doi.org/10.1371/journal.pone.0152478

Kumar M, Ahmad S, Singh RP (2022) Plant growth promoting microbes: diverse roles for sustainable and ecofriendly agriculture. Energy Nexus 7:100133. https://doi.org/10.1016/j.nexus.2022.100133

Kumar D, Pandey V, Dixit S (2024) Agronomic strategies for enhancing Forest resilience to climate change. In: Forests and climate change: biological perspectives on impact, adaptation, and mitigation strategies. Springer Nature Singapore, Singapore, pp 385–420. https://doi.org/10.1007/978-981-97-3905-9_20

Kumari N, Mohan C (2021) Basics of clay minerals and their characteristic properties. In: Clay and clay minerals. IntechOpen, London

Kumari M, Sharma D, Sandeep S (2022) Biofortification of vegetable crops: an option for mitigating hidden hunger. Int J Econ Plants 9(3):184–193. https://doi.org/10.23910/2/2022.460a

Kumari A, Dash M, Singh SK, Jagadesh M, Mathpal B, Mishra PK, Pandey SK, Verma KK (2023) Soil microbes: a natural solution for mitigating the impact of climate change. Environ Monit Assess 195(12):1436. https://doi.org/10.1007/s10661-023-11988-y

Kuypers MM, Marchant HK, Kartal B (2018) The microbial nitrogen-cycling network. Nat Rev Microbiol 16(5):263–276

Lamb WF, Wiedmann T, Pongratz J, Andrew R, Crippa M, Olivier JG, Wiedenhofer D, Mattioli G, Al Khourdajie A, House J, Pachauri S (2021) A review of trends and drivers of greenhouse gas emissions by sector from 1990 to 2018. Environ Res Lett 16(7):073005. https://doi.org/10.1088/1748-9326/abee4e

Lamérand C, Shirokova LS, Bénézeth P, Rols JL, Pokrovsky OS (2022) Carbon sequestration potential of mg carbonate and silicate biomineralization in the presence of cyanobacterium Synechococcus. Chem Geol 599:120854. https://doi.org/10.1016/j.chemgeo.2022.120854

Lei X, Shen Y, Zhao J, Huang J, Wang H, Yu Y, Xiao C (2023) Root exudates mediate the processes of soil organic carbon input and efflux. Plan Theory 12(3):630. https://doi.org/10.3390/plants12030630

Liu Z, Yu Z, Song B, Li Y, Fang J, Guo Y, Jin J, Adams JM (2023) Elevated CO_2 and temperature increase arbuscular mycorrhizal fungal diversity, but decrease root colonization, in maize and wheat. Sci Total Environ 873:162321. https://doi.org/10.1016/j.scitotenv.2023.162321

Lugato E, Leip A, Jones A (2018) Mitigation potential of soil carbon management overestimated by neglecting N_2O emissions. Nat Clim Chang 8(3):219–223. https://doi.org/10.1038/s41558-018-0087-z

Macdonald E, King EG (2018) Novel ecosystems: a bridging concept for the consilience of cultural landscape conservation and ecological restoration. Landsc Urban Plan 177:148–159. https://doi.org/10.1016/j.landurbplan.2018.04.015

Maheshwari DK, Das A, Dheeman S, Pandey P (2023) An overall insight into the attributes, interactions, and future applications of "microbial consortium" for plant growth promotion with contemporary approaches. In: Maheshwari DK, Dheeman S (eds) Sustainable agrobiology, Microorganisms for sustainability, vol 43. Springer, Singapore. https://doi.org/10.1007/978-981-19-9570-5_1

Masood F, Ahmad S, Malik A (2021) Role of Methanotrophs in mitigating global warming. In: Lone SA, Malik A (eds) Microbiomes and the global climate change. Springer, Singapore. https://doi.org/10.1007/978-981-33-4508-9_4

Meena RP, Jha A (2018) Conservation agriculture for climate change resilience: a microbiological perspective. In: Microbes for climate resilient agriculture. Wiley, pp 165–190. https://doi.org/10.1002/9781119276050.ch8

Moldagulova N, Ayupova A, Sembayeva D, Duambekov M, Khassenova E, Nagyzbekkyzy E, Danlybayeva G, Sarsenova A (2020) Data on the isolation and identification of thermotolerant microorganisms from cow manure promising for organic waste processing. Data Brief 31:105761. https://doi.org/10.1016/j.dib.2020.105761

Mustafa G, Hussain S, Liu Y, Ali I, Liu J, Bano H (2024) Microbiology of wetlands and the carbon cycle in coastal wetland mediated by microorganisms. Sci Total Environ 175734:175734. https://doi.org/10.1016/j.scitotenv.2024.175734

Muteeb G, Rehman MT, Shahwan M, Aatif M (2023) Origin of antibiotics and antibiotic resistance, and their impacts on drug development: a narrative review. Pharmaceuticals 16(11):1615. https://doi.org/10.3390/ph16111615

Nanda S, Reddy SN, Mitra SK, Kozinski JA (2016) The progressive routes for carbon capture and sequestration. Energy Sci Eng 4(2):99–122. https://doi.org/10.1002/ese3.117

Nevita T, Sharma GD, Pandey P (2018) Differences in rice rhizosphere bacterial community structure by application of lignocellulolytic plant-probiotic bacteria with rapid composting traits. Ecol Eng 120(December 2017):209–221. https://doi.org/10.1016/j.ecoleng.2018.06.007

Paleari S (2024) The EU policy on climate change, biodiversity and circular economy: moving towards a nexus approach. Environ Sci Pol 151:103603

Pandey P, Dheeman S (2022) Conclusions: the Rhizobial Eminence for biological nitrogen fixation—revisited and refined. In: Maheshwari DK, Dobhal R, Dheeman S (eds) Nitrogen fixing bacteria: sustainable growth of non-legumes. Microorganisms for sustainability, vol 36. Springer, Singapore. https://doi.org/10.1007/978-981-19-4906-7_17

Pandey P, Bisht S, Sood A, Aeron A, Sharma GD, Maheshwari DK (2012) Consortium of plant-growth-promoting bacteria: future perspective in agriculture. In: Maheshwari D (ed) Bacteria in agrobiology: plant probiotics. Springer, Berlin/Heidelberg. https://doi.org/10.1007/978-3-642-27515-9_10

Raj T, Verma S, Kumar N, Agrawal R (2024) Role of woody biomass in carbon capture, circular bioeconomy, and biomanufacturing. In: Sustainable biorefining of Woody biomass to biofuels and biochemicals. Woodhead Publishing, pp 291–318. https://doi.org/10.1016/B978-0-323-91187-0.00002-3

Rajkumari J, Bhuyan B, Das N, Pandey P (2019) Environmental applications of microbial extremophiles in the degradation of petroleum hydrocarbons in extreme environments. Environ Sustain 2:311–328. https://doi.org/10.1007/s42398-019-00065-1

Ramesh T, Bolan NS, Kirkham MB, Wijesekara H, Kanchikerimath M, Rao CS, Sandeep S, Rinklebe J, Ok YS, Choudhury BU, Wang H (2019) Soil organic carbon dynamics: impact of land use changes and management practices: a review. In: Advances in agronomy, vol 156. Elsevier, pp 1–107. https://doi.org/10.1016/bs.agron.2019.02.001

Rawat A, Kumar D, Khati BS (2024) A review on climate change impacts, models, and its consequences on different sectors: a systematic approach. J Water Clim Change 15(1):104–126. https://doi.org/10.2166/wcc.2023.536

Raza T, Qadir MF, Khan KS, Eash NS, Yousuf M, Chatterjee S, Manzoor R, ur Rehman S, Oetting JN (2023) Unrevealing the potential of microbes in decomposition of organic matter and release of carbon in the ecosystem. J Environ Manag 344:118529. https://doi.org/10.1016/j.jenvman.2023.118529

Raza A, Safdar M, Adnan Shahid M, Shabir G, Khil A, Hussain S, Khan M, Aziz SUR, Azam S, Sattar J, Muhammad NE (2024) Climate change impacts on crop productivity and food security: an overview. In: Transforming agricultural management for a sustainable future: climate change and machine learning perspectives, pp 163–186. https://doi.org/10.1007/978-3-031-63430-7_8

Razzaq A, Wani SH, Saleem F, Yu M, Zhou M, Shabala S (2021) Rewilding crops for climate resilience: economic analysis and de novo domestication strategies. J Exp Bot 72(18):6123–6139. https://doi.org/10.1093/jxb/erab276

Riseh RS, Vazvani MG (2024) Unveiling methods to stimulate plant resistance against pathogens. Front Biosci-Landmark 29(5):188. https://doi.org/10.31083/j.fbl2905188

Rizwan M, Tanveer H, Ali MH, Sanaullah M, Wakeel A (2024) Role of reactive nitrogen species in changing climate and future concerns of environmental sustainability. Environ Sci Pollut Res:51147–51163. https://doi.org/10.1007/s11356-024-34647-2

Robles-Fernández A, Areias C, Daffonchio D, Vahrenkamp VC, Sánchez-Román M (2022) The role of microorganisms in the nucleation of carbonates, environmental implications and applications. Fortschr Mineral 2022(12):1562. https://doi.org/10.3390/min12121562

Romanowicz KJ, Crump BC, Kling GW (2023) Genomic evidence that microbial carbon degradation is dominated by iron redox metabolism in thawing permafrost. ISME Commun 3(1):124. https://doi.org/10.1038/s43705-023-00326-5

Rousk J, Frey SD (2015) Revisiting the hypothesis that fungal-to-bacterial dominance characterizes turnover of soil organic matter and nutrients. Ecol Monogr 85(3):457–472. https://doi.org/10.1890/14-1796.1

Rout GR, Sahoo S (2015) Role of iron in plant growth and metabolism. Rev Agric Sci 3:1–24

Ruley JA, Amoding A, Tumuhairwe JB, Basamba TA (2022) Rhizoremediation of petroleum hydrocarbon–contaminated soils: a systematic review of mutualism between phytoremediation species and soil living microorganisms. In: Phytoremediation. Elsevier, pp 263–296. https://doi.org/10.1016/B978-0-323-89874-4.00008-X

Saccá ML, Barra Caracciolo A, Di Lenola M, Grenni P (2017) Ecosystem services provided by soil microorganisms. In: Soil biological communities and ecosystem resilience. Springer International Publishing, pp 9–24. https://doi.org/10.1007/978-3-319-63336-7_2

Saha L, Bauddh K (2020) Sustainable agricultural approaches for enhanced crop productivity, better soil health, and improved ecosystem services. In: Bauddh K, Kumar S, Singh R, Korstad

J (eds) Ecological and practical applications for sustainable agriculture. Springer, Singapore. https://doi.org/10.1007/978-981-15-3372-3_1

Sarker A, Ansary MWR, Hossain MN, Islam T (2021) Prospect and challenges for sustainable management of climate change-associated stresses to soil and plant health by beneficial rhizobacteria. Stress 1(4):200–222. https://doi.org/10.3390/stresses1040015

Scavino AF, Pedraza RO (2013) The role of siderophores in plant growth-promoting bacteria. In: Bacteria in agrobiology: crop productivity. Springer Berlin Heidelberg, Berlin/Heidelberg, pp 265–285. https://doi.org/10.1007/978-3-642-37241-4_11

Shafique M, Khurshid M, Muzammil S, Arshad MI, Malik IR, Rasool MH, Khalid A, Khalid R, Asghar R, Baloch Z, Aslam B (2024) Traversed dynamics of climate change and one health. Environ Sci Eur 36(1):135. https://doi.org/10.1186/s12302-024-00931-8

Shakeri F, Babavalian H, Amoozegar MA, Ahmadzadeh Z, Zuhuriyanizadi S, Afsharian MP (2021) Production and application of biosurfactants in biotechnology. Biointerface Res Appl Chem 11(3):10446–10460. https://doi.org/10.33263/briac113.1044610460

Shelake RM, Pramanik D, Kim JY (2019) Exploration of plant-microbe interactions for sustainable agriculture in CRISPR era. Microorganisms 7(8):269. https://doi.org/10.3390/microorganisms7080269

Shindell D, Sadavarte P, Aben I, Bredariol TDO, Dreyfus G, Höglund-Isaksson L, Poulter B, Saunois M, Schmidt GA, Szopa S, Rentz K (2024) The methane imperative. Front Sci 2:1349770. https://doi.org/10.3389/fsci.2024.1349770

Sifat NS, Haseli Y (2019) A critical review of CO_2 capture technologies and prospects for clean power generation. Energies 12(21):4143. https://doi.org/10.3390/en12214143

Singh BK, Bardgett RD, Smith P, Reay DS (2010) Microorganisms and climate change: terrestrial feedbacks and mitigation options. Nat Rev Microbiol 8(11):779–790

Singh P, Patel R, Koka RK (2020) Chapter 3: Rhizomicrobiome: an unexplored engine under sustainable agroecosystem. Chief Editor Dr RK Naresh, 15, 39

Singh S, Kumar V, Dhanjal DS, Sonali, Dhaka V, Thotapalli S, Singh J, Al-Ani LKT, Aguilar-Marcelino L (2021) Rhizosphere biology: a key to agricultural sustainability. In: Current trends in microbial biotechnology for sustainable agriculture. Springer Singapore, pp 161–182. https://doi.org/10.1007/978-981-15-6949-4_7

Singh BK, Delgado-Baquerizo M, Egidi E, Guirado E, Leach JE, Liu H, Trivedi P (2023) Climate change impacts on plant pathogens, food security and paths forward. Nat Rev Microbiol 21(10):640–656. https://doi.org/10.1038/s41579-023-00900-7

Singha K, M., Pandey, P. (2023) Endophytic bacteria with host-supportive genetic determinants in their genomes induce growth and antioxidant activity related gene functions in transcriptome of black rice (Oryza sativa L.). Environ Exp Bot 213:105396. https://doi.org/10.1016/j.envexpbot.2023.105396

Singha LP, Pandey P (2021) Rhizosphere assisted bioengineering approaches for the mitigation of petroleum hydrocarbons contamination in soil. Crit Rev Biotechnol 41(5):749–766. https://doi.org/10.1080/07388551.2021.1888066

Singha B, Mazumder PB, Pandey P (2018a) Characterization of plant growth promoting rhizobia from root nodule of two legume species cultivated in Assam, India. Proc Natl Acad Sci India Sect B: Biol Sci 88(3):1007–1016. https://doi.org/10.1007/s40011-016-0836-6

Singha LP, Sinha N, Pandey P (2018b) Rhizoremediation prospects of Polyaromatic hydrocarbon degrading rhizobacteria, that facilitate glutathione and glutathione-S-transferase mediated stress response, and enhance growth of rice plants in pyrene contaminated soil. Ecotoxicol Environ Saf 164(August):579–588. https://doi.org/10.1016/j.ecoenv.2018.08.069

Sokol NW, Slessarev E, Marschmann GL, Nicolas A, Blazewicz SJ, Brodie EL, Firestone MK, Foley MM, Hestrin R, Hungate BA, Koch BJ (2022) Life and death in the soil microbiome: how ecological processes influence biogeochemistry. Nat Rev Microbiol 20(7):415–430. https://doi.org/10.1038/s41579-022-00695-z

Steffen W, Crutzen PJ, McNeill JR (2007) The Anthropocene: are humans now overwhelming the great forces of nature. Ambio-J Hum Environ Res Manag 36(8):614–621. https://doi.org/10.1525/9780520964297-051

Streletskii RA, Astaykina AA, Belov AA, Cheptsov VS, Vetrova AA (2024) Beneficial soil microorganisms and their role in sustainable agriculture. In: Sustainable Agricultural Practices. Academic, pp 293–333. https://doi.org/10.1016/B978-0-443-19150-3.00013-8

Syed S, Buddolla V, Lian B (2020) Oxalate carbonate pathway—conversion and fixation of soil carbon—a potential scenario for sustainability. Front Plant Sci 11:591297. https://doi.org/10.3389/fpls.2020.591297

Tchonkouang RD, Onyeaka H, Nkoutchou H (2024) Assessing the vulnerability of food supply chains to climate change-induced disruptions. Sci Total Environ 171047:171047. https://doi.org/10.1016/j.scitotenv.2024.171047

Tick A, Akaev A, Devezas TC, Sarygulov A, Petryakov A, Evgenevich AI (2024) Assessing Decarbonization approaches across major economies. Energies 17(17):4381. https://doi.org/10.3390/en17174381

VijayaVenkataRaman S, Iniyan S, Goic R (2012) A review of climate change, mitigation and adaptation. Renew Sust Energ Rev 16(1):878–897. https://doi.org/10.1016/j.rser.2011.09.009

Wahab A, Muhammad M, Munir A, Abdi G, Zaman W, Ayaz A, Khizar C, Reddy SPP (2023) Role of arbuscular mycorrhizal fungi in regulating growth, enhancing productivity, and potentially influencing ecosystems under abiotic and biotic stresses. Plan Theory 12(17):3102. https://doi.org/10.3390/plants12173102

Wang L, Yuan X, Xie Z, Wu P, Li Y (2016) Increasing flash droughts over China during the recent global warming hiatus. Sci Rep 6(1):30571. https://doi.org/10.1038/srep30571

Wu H, Cui H, Fu C, Li R, Qi F, Liu Z, Yang G, Xiao K, Qiao M (2023) Unveiling the crucial role of soil microorganisms in carbon cycling: a review. Sci Total Environ 909:168627. https://doi.org/10.1016/j.scitotenv.2023.168627

Xie Z, Yu Z, Li Y, Wang G, Liu X, Tang C, Lian T, Adams J, Liu J, Liu J, Herbert SJ (2022) Soil microbial metabolism on carbon and nitrogen transformation links the crop-residue contribution to soil organic carbon. npj Biofilms Microb 8(1):14. https://doi.org/10.1038/s41522-022-00277-0

Xu Z, Zhang T, Wang S, Wang Z (2020) Soil pH and C/N ratio determines spatial variations in soil microbial communities and enzymatic activities of the agricultural ecosystems in Northeast China: Jilin Province case. Appl Soil Ecol 155:103629. https://doi.org/10.1016/j.apsoil.2020.103629

Xue K, Yuan MM, Shi JZ, Qin Y, Deng YE, Cheng L, Wu L, He Z, Van Nostrand JD, Bracho R, Natali S (2016) Tundra soil carbon is vulnerable to rapid microbial decomposition under climate warming. Nat Clim Chang 6(6):595–600

Xue T, Zhang L, Yang F, Cheng C, Wang J, Sang Q, Yang S, Zhao X, Zhang L (2024a) Carbon sink and soil organic carbon sequestration mechanisms in vineyards. J Clean Prod 143217:143217. https://doi.org/10.1016/j.jclepro.2024.143217

Xue Z, Qu T, Li X, Chen Q, Zhou Z, Wang B, Lv X (2024b) Different contributing processes in bacterial vs. fungal necromass affect soil carbon fractions during plant residue transformation. Plant Soil 494(1):301–319. https://doi.org/10.1007/s11104-023-06277-z

Yoro KO, Daramola MO (2020) CO_2 emission sources, greenhouse gases, and the global warming effect. In: Advances in carbon capture. Woodhead Publishing, pp 3–28. https://doi.org/10.1016/B978-0-12-819657-1.00001-3

Zhu T, Dittrich M (2016) Carbonate precipitation through microbial activities in natural environment, and their potential in biotechnology: a review. Front Bioeng Biotechnol 4:4. https://doi.org/10.3389/fbioe.2016.00004

The manufacturer's authorised representative in the EU is Springer Nature Customer Service Centre GmbH, Europaplatz 3, 69115 Heidelberg, Germany. If you have any concerns regarding our products, please contact ProductSafety@springernature.com

Printed and bound by CPI Group (UK) Ltd, Croydon, CR0 4YY
01/10/2026
02225928-0014